S0-CAS-611

Analytical Calorimetry

Volume 3

A Continuation Order Plan is available for this series. A continuation order will bring delivery of each new volume immediately upon publication. Volumes are billed only upon actual shipment. For further information please contact the publisher.

Analytical Calorimetry

Volume 3

Edited by

Roger S. Porter

Head, Polymer Science and Engineering
University of Massachusetts
Amherst, Massachusetts

and

Julian F. Johnson

Department of Chemistry and Institute of Materials Science
University of Connecticut
Storrs, Connecticut

PLENUM PRESS • NEW YORK AND LONDON

The Library of Congress cataloged the first volume of this title as follows:

**American Chemical Society Symposium on Analytical Calorime-
try, San Francisco, 1968.**
  Analytical calorimetry; proceedings. Edited by Roger S.
Porter and Julian F. Johnson.   New York, Plenum Press, 1968.
  ix, 322 p.   illus.   26 cm.
  Jointly sponsored by the Division of Analytical Chemistry and the Division
of Polymer Chemistry.
  Includes bibliographies.

  1. Calorimeters and calorimetry—Addresses, essays, lectures.   2. Thermochemistry—Addresses, essays, lectures.   I. Porter, Roger Stephen, 1928-
      ed. II. Johnson, Julian Frank, 1923-   ed. III. American Chemi-
cal Society.   Division of Analytical Chemistry.   IV. American Chemical Society.   Division of Polymer Chemistry.   V. Title.
QD511.A4   1968   547'.308'6   68-8862
                                                            MARC

  Library of Congress

CHEMISTRY

Proceedings of the American Chemical Society Symposium on
Analytical Calorimetry held in Los Angeles, California, April 1974

Library of Congress Catalog Card Number 68-8862

ISBN 0-306-35243-5

© 1974 Plenum Press, New York
A Division of Plenum Publishing Corporation
227 West 17th Street, New York, N.Y. 10011

United Kingdom edition published by Plenum Press, London
A Division of Plenum Publishing Company, Ltd.
4a Lower John Street, London, W1R 3PD, England

All rights reserved

No part of this book may be reproduced, stored in a retrieval system, or transmitted,
in any form or by any means, electronic, mechanical, photocopying, microfilming,
recording, or otherwise, without written permission from the Publisher

Printed in the United States of America

QD 79
T 38 A51
v. 3
CHEM

Preface

The research reported in the third volume of Analytical Calorimetry covers a wide variety of topics. The variety indicates the sophistication which thermal analysis is reaching and additionally the ever widening applications that are being developed.

Advances in instrumentation include: microcalorimeter design, development and refinement of titration calorimetry, definition of further theory of scanning calorimetry, studies of the temperature of resolution of thermistors, and a refinement of the effluent gas analysis technique and its application to agricultural chemicals as well as organic materials.

A wide variety of applications is reported. These cover the fields of polymeric materials, dental materials, inorganic proteins, biochemical materials, gels, mixed crystals, and other specialized areas.

Contributions also include applications of important related techniques such as thermomechanical and thermogravimetric analysis. The contributions to this Volume represent papers presented before the Division of Analytical Chemistry at the Third Symposium on Analytical Chemistry held at the 167th National Meeting of the American Chemical Society, March 30 - April 5, 1974.

April 15, 1974

Julian F. Johnson
Roger S. Porter

20199

Contents

The Construction of a Microcalorimeter and Measurement
of Heats of Solution of Stretched Glassy Polystyrene   1
    Y. Takashima, K. Miasa, S. Miyata, and K. Sakaoku

Recent Advances in Titration Calorimetry   7
    L.D. Hansen, R.M. Izatt, D.J. Eatough,
    T.E. Jensen, and J.J. Christensen

Theory of Differential Scanning Calorimetry - Coupling
of Electronic and Thermal Steps 17
    J.H. Flynn

On the Temperature Resolution of Thermistors 45
    P.W. Carr and L.D. Bowers

The Development and Application of an Ultra-Sensitive
Quantitative Effluent Gas Analysis Technique 57
    P.A. Barnes and E. Kirton

A Small, Mini-Computer-Automated Thermoanalytical
Laboratory 69
    E. Catalano and J.C. English

Steady State Technique for Low Temperature Heat Capacity
of Small Samples
    R. Viswanathan 81

Fusion as an Opportunity for Calorimetrically Probing
Polymer Conformations and Interactions in the Bulk State . . . 89
    A.E. Tonelli

Application of Differential Scanning Calorimetry for the
Study of Phase Transitions 103
    W.P. Brennan

Importance of Joint Knowledge of ΔH^O and ΔS^O Values for
Interpreting Coordination Reactions in Solution 119
    G. Berthon

The Determination of the Heat of Fusion by Differential
Scanning Calorimetry and by Measurements of the Heat
of Dissolution 127
    E. Marti

Novel Methods for Gas Generating Reaction 137
    A.A. Duswalt

DSC: New Developments in Clinical Analysis and
Physiochemical Research 147
    B. Cassel and T. Ohnishi

The Use of Thermal and Ultrasonic Data to Calculate
the Pressure Dependence of the Gruneisen Parameter 165
    Sister R. Urzendowski and A.H. Guenther

The Use of Thermal Evolution Analysis (TEA) for the
Determination of Vapor Pressure of Agricultural Chemicals . . 185
    R.L. Blaine and P.F. Levy

Thermal Evolution Analysis of Some Organic Materials 199
    E.W. Kifer and L.H. Leiner

High Sensitivity Enthalpimetric Determination of Olefins . . . 207
    L.A. Williams, B. Howard, and D.W. Rogers

Biochemical and Clinical Applications of Titration
Calorimetry and Enthalpimetric Analysis 217
    A.C. Censullo, J.A. Lynch, D.H. Waugh, and J. Jordan

Recent Analytical Applications of Titration Calorimetry . . . 237
    R.M. Izatt, L.D. Hansen, D.J. Eatough, T.E. Jensen,
    and J.J. Christensen

Calorimetric Studies of Pi-Molecular Complexes 249
    W.C. Herndon, J. Feuer, and R.E. Mitchell

Calorimetric Investigation of the Pyridine -
Chloroform Complex 283
    G.L. Bertrand and T.E. Burchfield

Application of Thermal Analysis as a Substitute for
Standard ASTM Polymer Characterization Test 293
    P.S. Gill and P.F. Levy

CONTENTS

Application of Scanning Calorimetry to Petroleum
Oil Oxidation Studies 305
     F. Noel and G.E. Cranton

The Detection of Impurities by Thermal Analysis 321
     H.J. Ferrari and N.J. Passarello

Thermomechanical Analysis of Dental Waxes in the
Penetration Mode 349
     J.M. Powers and R.G. Craig

Hot-Stage Electron Microscopy of Clay Minerals 363
     D.L. Jernigan and J.L. McAtee, Jr.

Heats of Immersion of Hydroxyapatites in Water 381
     H.M. Rootare and R.G. Craig

Heats of Solution of Apatites, Human Enamel and
Dicalcium-Phosphate in Dilute Hydrochloric Acid 397
     R.G. Craig and H.M. Rootare

Verification of the Ionic Constants of Proteins by
Calorimetry 407
     M.A. Marini, C.J. Martin, R.L. Berger, and L. Forlani

Calorimetric Investigation of Chymotrypsin Ionization
Reactions 425
     C.J. Martin, B.R. Sreenathan, and M.A. Marini

DSC Study of the Conformational Transition of
Poly-γ-Benzyl-L-Glutamate in the System:
1,3-Dichlorotetrafluoroacetone-Water 443
     J. Simon, G.E. Gajnos, and F.E. Karasz

A Thermometric Investigation of the Reaction between
Proteins and 12-Phosphotungstic Acid 457
     P.W. Carr, E.B. Smith, S.R. Betso, and R.H. Callicott

Differential Scanning Calorimetry Studies on DNA Gels 465
     H.W. Hoyer and S. Nevin

Applications of Group Enthalpies of Transfer 473
     R. Fuchs

Thermodynamics of Intermolecular Self-Association of
Hydrogen Bonding Solutes by Titration Calorimetry 479
     E.M. Woolley and N.S. Zaugg

Rapid Quantitative Method for Bound Water Determination
in Aqueous Systems Using Differential Scanning Calorimeter . . 489
    E. Karmas and C.C. Chen

Effect of Dehydration on the Specific Heat of Cheese Whey . . 497
    E. Berlin and P.G. Kliman

Solid State Reaction Kinetics IV: The Analysis of Chemical
Reactions by Means of the Weibull Function 505
    E.A. Dorko, W. Bryant, and T.L. Regulinski

Preliminary Results on the Nature of n in Equation
(dx/dt)=k(a-x)^n as Applied to Three Solid Thermal
Decomposition Reactions 511
    D.M. Speros and H.R. Werner

Kinetics of an Anhydride-Epoxy Polymerization as Determined
by Differential Scanning Calorimetry 537
    P. Peyser and W.D. Bascom

Characterization of Energetic Mixed Crystals by Means of
TGA and DSC Calorimetry 555
    S.I. Morrow

Heat of Fusion of Crystalline Polypropylene by Volume
Dilatometry and Differential Scanning Calorimetry 569
    J.A. Currie, E.M. Petruska, and R.W. Tung

Self-Seeded PE Crystals; Melting and Morphology 579
    I.R. Harrison and G.L. Stutzman

The Enthalpy of Fusion of Low Molecular Weight Polyethylene
Fractions Crystallized from Dilute Solution 593
    S. Go, F. Kloos, and L. Mandelkern

Crystallization of Polyethylene from Xylene Solutions
under High Pressure 603
    S. Miyata, T. Arikawa, and K. Sakaoku

Thermal Properties of the Polylactone of Dimethylketene,
A New Model Polymer 611
    E.M. Barrall II, D.E. Johnson, and B.L. Dawson

Melting Behavior of Some Oligomers of Heterocyclic
Polymers by Differential Scanning Calorimetry 621
    H. Kambe and R. Yokota

Thermogravimetric Analysis of Polymethylmethacrylate and
Polytetrafluoroethylene 629
    J.A. Currie and N. Pathmanand

Applications of Quantitative Thermal Analysis to
Molecular Sieve Zeolites 649
    W.H. Flank

Some Recent Research on Thermal Properties of Milk Fat
Systems 661
    J.W. Sherbon

Thermal Behavior of Chemical Fertilizers 671
    C. Giavarini

Positional Effects of the Phosphate Group on Thermal
Polymerization of Isomeric Uridine Phosphates 685
    A.M. Bryan and P.G. Olafsson

Thermal Decomposition of Cementitious Hydrates 697
    J.N. Maycock, J. Skalny, and R.S. Kalyoncu

Thermal Decomposition Studies of Sodium and Potassium
Tartrates 713
    A.C. Glatz and A. Pinella

The Enthalpy and Heat of Transition of Cs_2MoO_4 by Drop
Calorimetry 723
    D.R. Fredrickson and M.G. Chasanov

The Dissociation Energy of NiO and Vaporization and
Sublimation Enthalpies of Ni 731
    M. Farber and R.D. Srivastava

Thermodynamic Properties of REX_3, $AuCu_3$-Type, Intermetallic
Compounds 743
    A. Palenzona and S. Cirafici

Thermal and Microscopical Study of the Condensed Phase
Behavior of Nitrocellulose and Double Base Materials 757
    S.I. Morrow

Thermal Analysis of Dicyclopentadienyl Zirconium
Diborohydride and Bis(Triphenylphosphine) Borohydrido
Copper (I) 777
    B.D. James, B. Annuar, J.O. Hill, and R.J. Magee

Slow Reversion of Potassium Nitrate 787
    P.D. Garn

Thermoanalytic Measurements of Fire-Retardant ABS Resins . . . 797
    H.E. Bair

List of Contributors 807

Index 815

THE CONSTRUCTION OF A MICROCALORIMETER AND MEASUREMENT OF

HEATS OF SOLUTION OF STRETCHED GLASSY POLYSTYRENE

Y. Takashima, K. Miasa, S. Miyata and K. Sakaoku

Faculty of Technology, Tokyo University of Agriculture
and Technology, Koganei, Tokyo, Japan

SYNOPSIS
    A conduction-type twin microcalorimeter was constructed in
order to measure very small amounts of heat, and the microcalorimeter was applied to measure the heat of solution (ΔH) of
stretched polystyrene (glassy state). The area of the thermogram
was proportional to the supplied electric energy into the microcalorimeter in a range of 0 to 0.3 calory. As the calibration,
ΔH for KCl was measured at 30°C in molar fraction, 1 : 5,500. The
standard deviation was ± 0.4%. The ΔH of stretched glassy polystyrene was measured in toluene at 30°C. The specific volume of
the samples was measure by floating method of each stretched
sample and unstretched sample. The measured value of the specific
volume did not depend on the stretched temperature and the draw
ratio. However, the value of measured ΔH of stretched (at 110°C)
samples decreased with the increase of the draw ratio: it decreased about 50 cal/mole at the draw ratio 3 as compared with that
of unstretched sample.

INTRODUCTION
    Flory[1] and other researchers have developed the theory of
rubber elasticity and revealed that the elasticity depended mainly
on entropy of the randomly coiled molecules in the network and
that the internal energy has only small effect on the elasticity.

    A number of experiments of rubber elasticity have been
carried out, and the amount of the internal energy was estimated
based on those theories. The obtained values are scattered by researchers because of their indirect measurements. In this study,
the internal energy contribution on stretching was estimated

directly from the measurements of heat of solution for stretched
glassy polymer by using the following equation:

$$\Delta H^{\Delta\alpha} = \Delta E_p^{\Delta\alpha} + P\Delta V$$

$$= \Delta E_v^{\Delta\alpha} + \left(\frac{\partial E}{\partial V}\right)_\tau \Delta V + \tfrac{1}{2}\left(\frac{\partial^2 E}{\partial V^2}\right)_\tau \left(\Delta V\right)^2 + \cdots + P\Delta V \qquad\qquad (1)$$

$$\approx \Delta E_v^{\Delta\alpha} + \left(P + Pi\right)\Delta V$$

where,

$\Delta H^{\Delta\alpha}$: the enthalpy difference between the stretched sample
and the unstretched sample.

$\Delta E_v^{\Delta\alpha}$: the difference of internal energy at constant volume.

ΔV : the change of the specific volume.

Pi: the internal pressure., P : the external pressure.

According to equation (1) if the enthalpy difference of glass-
transition is the same for the stretched sample with various draw
ratios, the internal energy contribution caused by stretching,
$\Delta E_v^{\Delta\alpha}$, can be estimated from measurements of heats of solution, internal pressure, and specific volume.

Experimental

A microcalorimeter was constructed to measure very small
quantities of heat as shown in Figure 1. It was a conduction type

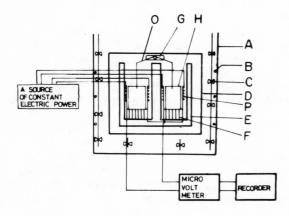

Fig. 1-Calorimeter:
A-water tank
B-heater
C-fan
D-steel box
E-aluminum block
F-thermomodule
G-small motor
H-sample cell
O-string

twin calorimeter based on the principle first developed by Tian
and Calvet.[2] On the construction of this microcalorimeter, the
following modifications had been made in this laboratory. (1) :
In order to suppress the heat flow in the microcalorimeter as small
as possible, the inside of microcalorimeter had not fans and shafts

for stirring. Therefore the stirring of solvents in a twin cells
was carried out by rocking the whole microcalorimeter at an angle
of $\pm$ 30° with a rate of five times per minute. (2) : In order to
keep temperature of environment of the microcalorimeter at constant,
a well thermostated large water tank (90 cm X90 cm), which was controlled $\pm$ 0.007°C at 30°C, was prepared. (3) : In order to detect
the very small temperature-difference between the twin cells as
precisely as possible, thermomodules, which consisted of 16 coupled
P and N elements of semiconductor, was equipped instead of thermocouples. The thermal electromotive force was 6.5 mV/deg and
the sensitivity of temperature was $4\text{X}10^{-5}$degrees. (4) : In order
to avoid the heat liberated by breaking an ampoule to dissolve the
sample, the sample sealed by mercury was designed. The detail of
the sample cell is shown in Figure 2.

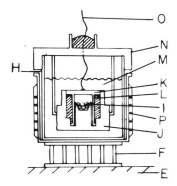

Fig. 2-Sample cell: I, sample;
J, sample holder; K, cap;
L, mercury; M, solvent;
N, lid; O, string;
P, base heater

The heat of solution was measured by the following procedures.
0.06 gram of stretched or unstretched film of polystyrene, as described later, was placed in the sample holder (J) and then covered with the cap which was sealed by mercury. The same operation
except placing the sample was carried out at the reference cell.
Forty grams of toluene were poured into the twin cells having
teflon lids (N). The microcalorimeter was then submerged in the
water tank, and rocking was started. After about 24 hours at 30°C
to attain complete temperature equilibrium of both cells, the caps
were opened in order to dissolve the sample by winding up the
strings through the lids with a small motor. The temperature difference between the twin cells was recorded against the time for 4
hours in order to dissolve the sample completely. The integrated
area of the thermogram compared with that of the thermogram due to
a supplied electric energy and then the heat of solution of the
sample was determined.

The films of polystyrene as the samples in this study were
made from the powder of atactic polystyrene, which was purified
by reprecipitation with methanol-toluene mixture. The purified
powder ($M[\eta]=1.82\times10^5$)[3] was molded (at 180°C, 100 Kg/cm^2) into films of a thickness of 0.6 mm with a laboratory hot press. The
obtained films were cut into a dimension of 20 mm X 50 mm. The
cut films were stretched with a constant drawing velocity of
15 cm/min and then the stretched films were quenched in ice water.

RESULTS AND DISCUSSION

The drift of the base line on the thermogram was not observed
for 24 hours. The blank test was carried out to detect the unexpected heats owing to the unwanted asymmetry of the two cells and
the influence of a small motor in the microcalorimeter. However
those heats were not observed. The relation between the recorded
area of thermogram against the supplied electric energy is shown
in Figure 3,

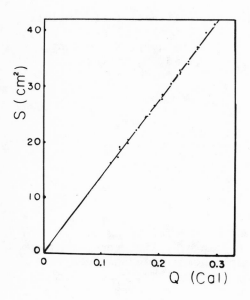

Fig. 3-Area of thermogram vs.
supplied heat.

in which the linearity was held between 0 and 0.3 cal. An example
of observed thermogram for the heat of solution of KCl and the obtained values is shown in Figure 4 and Table 1. The average value,
which was 3981 cal/mol in molar fraction of 1/5500[4], was in good
agreement with the results of V. P. Vasilev. The standard deviation was ± 0.4%.

The measured ΔH of the stretched polystyrene films is shown in
Figure 5, in which each spot is the average value in three times
measurements. All the values of measured heat for the stretched

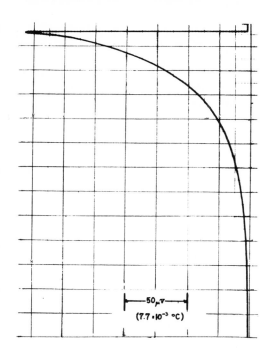

Fig. 4-Thermogram of ΔH for
    KCl.

Table I- ΔH of KCl at 30°C (molar frac. 1/5500)

KCl (mg)	36.5	33.3	39.8	39.5	Ave.
ΔH cal/mole	3952	4003	3955	4015	3981

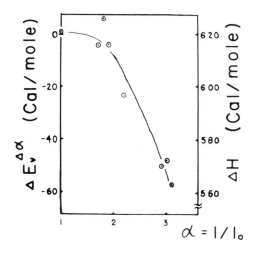

Fig. 5-ΔH of drawn poly-
    styrene

films were exothermic. The ΔH of the stretched film of draw ratio
3 was 50 cal/mol smaller than that of the unstretched film. The
measured specific volumes of the films are shown in Figure 6.
Those values were constant within an accuracy of 0.001 cc/g. $Pi\Delta V$
in equation (1) was calculated 8 cal/mol from using the average
value of the internal pressure in the glassy state ($Pi,\tilde{g}=58$ cal/cc)
and liquid state ($Pi,\ell=100$ cal/cc) of polystyrene. Thus,
$\Delta H^{\Delta 3}=-50$ cal/mol and $\Delta E_v^{\Delta 3}=-50\pm8$ cal/mol were obtained.

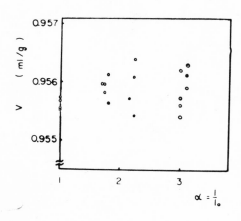

Fig. 6-Specific volume of
        drawn polystyrene.

The decrease of internal energy with increasing draw ratio
was also shown by U. Bianchi[5] and F. H. Müller based on the results
of heat of solution, although A. Ciferri et al.[6] suggested the
opposite conclusion from the thermoelasticity.

If affine deformation is assumed for the stretching, molecules
could be also elongated. It is easily considered that the number
of energetic stable trans conformations are increased. Then the
internal energy of the stretched sample is supposed lower than
that of unstretched one. This consideration is reasonable with the
results of this study.

REFERENCES

1. P. J. Flory, J. Chem. Phys. 18, 108 (1950).
2. E. Calvet and H. Prat, Recent Progress in Microcalorimeter,
    Pergamon Press, London, 1963.
3. F. Damsso and G. Moraglio, J. Polymer Sci., 24, 161 (1957).
4. V. P. Vasilev and G. A. Lobnov, Zh. Neorgan. Khim., 11 (4),
    703 (1966).
5. U. Bianchi, C. Cuniberti, E. Pedemonte, and C. Rossi, J.
    Polymer Sci., A-2, 7, 855 (1969).
6. A. Ciferri and T. A. Orofino, J. Phys. Chem., 68, 3136 (1964).

RECENT ADVANCES IN TITRATION CALORIMETRY

Lee D. Hansen, Reed M. Izatt, Delbert J. Eatough
Trescott E. Jensen, and James J. Christensen

Departments of Chemistry and Chemical Engineering
and Contribution 57 from the Center for Thermochemical
Studies, Brigham Young University, Provo, Utah 84602

INTRODUCTION

Calorimeters applicable for the measurement of thermodynamic
parameters by the continuous titration technique have existed for
about fifteen years. During this time, problems associated with
the components of the calorimeter (i.e. constant temperature bath,
constant rate buret, reaction vessel, temperature sensing circuit,
and data analysis procedure) have gradually been solved so that
the continuous titration method now gives results comparable in
accuracy to those obtained in conventional solution calorimeters.
Inexpensive research quality titration calorimeters have become
available commercially in the last few years.

The availability of equipment and a growing acceptance of
titration calorimetry as a convenient and reliable analytical
technique suggest that this method may provide unique solutions
to calorimetric and analytical problems in specific areas. The
justifications for this are that the method is rapid, sensitive
and general, and that the equipment is relatively inexpensive and
lends itself well to automation. Many applications, however, call
for the use of small volume reaction vessels (< 5 ml) because of
cost or availability limitations for one or more of the reacting
species. Reducing the volume of solution means that less material
is needed for the same accuracy of determination, or that more concentrated solutions can be studied with the same amount of material
with consequently larger temperature changes and hence increased
accuracy.

Although the design of the small vessel is simply a matter of scaling down the large vessel to the required size, other difficulties are encountered. One difficulty is that the modulus of heat loss (κ) increases exponentially as the vessel size decreases and more exact calculations are required to correct for heat exchange during the experiment. Another difficulty that must be overcome is that of delivering very small (~ 0.2 ml) quantities of titrant at a constant (± 0.1 percent) rate over relatively long time periods (10 to 30 minutes).

This paper describes the development of isoperibol and isothermal titration calorimeters which operate with 3 and 4 ml, respectively, of solution in the reaction vessel and of burets capable of delivering < 0.2 ml of titrant over a 20 minute period with a variation in delivery rate of < 0.1%.

EQUIPMENT AND DATA ANALYSIS

Calorimeter

The calorimeter used was a Tronac Model 1000[1] equipped with a small head as shown in Figure 1.

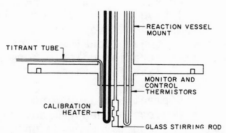

Figure 1. Titration Calorimeter Head for Small Reaction Vessels. The stirrer, thermistors, and calibration heater protrude into the solution in the reaction vessel when it is in place. (Reproduced by permission from reference 2).

Isothermal Reaction Vessel and Control Circuit[2]

A schematic diagram of the miniature isothermal reaction vessel is shown in Figure 2. The reaction vessel consists of a 5 ml platinum cup (4 ml effective solution volume) inside a stainless-steel container. Under the cup and inside the container are a 100 Ω wafer control heater and a Peltier thermoelectric

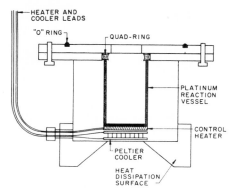

Figure 2.   Schematic of isothermal titration reaction vessel.
(Reproduced by permission from reference 2).

cooler located adjacent to each other.   This arrangement provides
better temperature control because heat flow from the heater to
the cooler through the solution in the reaction vessel is eliminated.   The bottom of the stainless-steel container is provided
with a metal heat dissipation surface to facilitate dissipation
of heat from the Peltier cooler.   The container, fitted with an
0-ring seal to prevent water leaks, is clamped in position on
the stainless steel head shown in Figure 1.

    A block diagram of the isothermal control circuit is
shown in Figure 3.

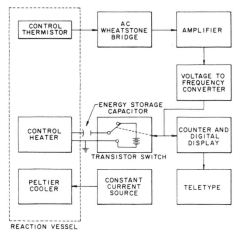

Figure 3.   Block diagram of isothermal control circuit.
(Reproduced by permission from reference 2).

The reaction vessel is held at a constant temperature
($\pm$ 2 X 10^{-5}°C) by maintaining the Peltier cooler at a constant
cooling rate while varying the heat input of the control heater.
A constant current source supplies current to the Peltier cooler.
The control heater is supplied with fixed energy pulses (selectable
from 0.02 to 20 µcal/pulse) through a transistor switch. The
rate of pulses supplied to the input heater is determined by the
control thermistor which is in one arm of an AC Wheatstone bridge
circuit. The control thermistor therefore directly regulates the
amount of heat supplied to the reaction vessel via the control
heater. The average frequency of pulses (which can vary from
0 to 20,000 sec^{-1}) supplied to the heater is monitored and displayed in digital form by a time averaging counter. Conversion
of the measured frequency data to energy is a simple matter of
multiplying the frequency by the time interval to get the total
number of pulses and then multiplying this result by the energy
per pulse obtained in calibration experiments.

Isoperibol Reaction Vessel

A scale drawing of the 3 ml isoperibol reaction vessel
is shown in Figure 4. This Dewar fits on the same head (Figure 1)
as does the isothermal reaction vessel.

The design of a Dewar for use as a reaction vessel in titration calorimetry requires that a satisfactory optimum be reached
between two opposing criteria. First, the thermal response time
must be kept short (< 1 sec) and second, the value of the modulus
of heat loss, κ, should be kept small, preferably less than 0.005
min^{-1}.

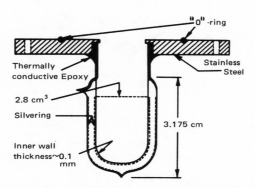

Figure 4.   Glass Dewar reaction vessel for isoperibol titration
calorimetry.

We have not been able to simultaneously meet both of these
criteria (i.e. time response < 1 sec and κ < 0.005 min^{-1}) in
Dewars smaller than about 25 ml. Since the first criteria must
be met it was necessary to develop an accurate method for calculating the heat loss from Dewars with large heat leak constants.

In previous derivations of equations describing the heat
exchange between the Dewar and its surroundings during a thermometric titration, it has been assumed that (1) κ, (2) the product
of κ and the thermal equivalent, ε, and (3) the power input from
all sources except chemical effects and heat exchange were constant.[3] These assumptions cause negligible error so long as κ
and consequently the ratio of heat exchanged to total heat
measured is small. As can be seen in Figure 5, the ratio of the
heat lost from the Dewar to the heat kept in the Dewar rises
rapidly as Dewar size decreases below 25 ml.

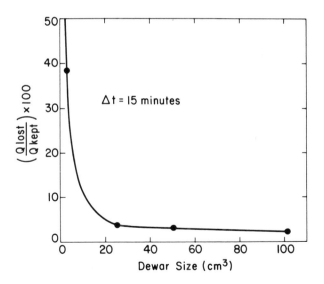

Figure 5. Plot of the ratio of the heat lost to the heat kept
for isoperibol calorimeter Dewars with volumes of 3 to 100 ml.

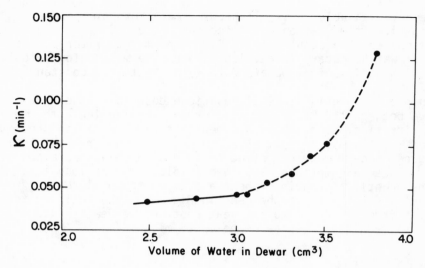

Figure 6: Plot of κ vs. the volume of water in the Dewar.

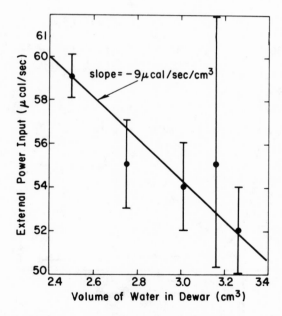

Figure 7: Plot of the external power (P) input vs. the volume of water in the Dewar.

A plot of κ vs. the volume of water in the 3 ml Dewar is shown in Figure 6. The plot becomes linear below 3.0 ml. For this reason we chose to titrate 2.7 ml of solution with 0.25 ml of titrant. This enables us to describe κ as a function of titrant added, V_p, as

$$\kappa_p = \kappa_\ell + \alpha V_p \tag{1}$$

where κ_ℓ is the heat leak constant with 2.7 ml in the Dewar and α is a calibration constant.[4]

The sum of the external power inputs, P_p, was also determined as a function of the volume of water in the Dewar. The results are shown in Figure 7. We conclude that P_p may be described as a linear function of the volume of titrant added, i.e.

$$P_p = P_\ell + \beta V_p \tag{2}$$

Combining equations (1) and (2) with equations previously derived[3,4] for heat exchange corrections and setting $T_{surroundings} = T_{bath}$ gives equation (3).

$$q_{HL,p} = q_{HL,\ell} - \varepsilon_p \alpha V_p \Delta T_p - \beta V_p - (q_{HL,\ell} - q_{HL,t} - \varepsilon_t \alpha V_t \Delta T_t - \beta V_t) \times$$

$$(\varepsilon_p \Delta T_p - \varepsilon_\ell \Delta T_\ell) \ / \ (\varepsilon_t \Delta T_t - \varepsilon_\ell \Delta T_\ell) \tag{3}$$

where $q_{HL,p}$ is the rate of heat loss at point p in the titration; $q_{HL,\ell}$ and $q_{HL,t}$ are the rates of heat loss at the midpoint times of the lead and trail slopes, respectively; ε_p is the thermal equivalent of the calorimeter system at point p, α and β are calibration constants; V_p is the titrant added to point p; V_t is the total titrant volume added; $\Delta T_p, \Delta T_\ell,$ and ΔT_t are, respectively, the temperature difference between the bath and the reaction vessel at point p, at the midpoint time of the lead slope, and at the midpoint time of the trail slope; and ε_ℓ and ε_t are the thermal equivalents during the lead and trail periods.[4] Integration of equation (3) over time from the beginning of the titration to point p allows correction for the heat exchanged with the surroundings to be made.[3,4]

Buret

The buret used was constructed from a 1 ml Gilmont micrometer syringe and is driven by a stepping motor controlled by a frequency divider circuit (Tronac, Model 822) which is controlled in turn by the AC line frequency. The delivery rate of this buret is very precise ($\pm$ 0.04 percent, standard deviation of the mean of six determinations of the delivery rate using 0.25 ml samples). However, it is difficult to eliminate miniscule but significant leakage around the buret piston unless the buret is assembled with sufficient care. A 0.25 ml total volume adaptation of the Gilmont buret was successfully tried, but proved to be too fragile for routine use.

RESULTS ON TEST SYSTEMS

To check the accuracy and precision of the instruments, several determinations of the heat of ionization of water were made using the reaction of $HClO_4$ with NaOH. In all cases the acid (titrant) was titrated into the base (titrate). The results obtained with the isothermal and isoperibol calorimeters are given in Tables 1 and 2, respectively. The uncertainty given in each case is the standard deviation of the mean of each series of runs.

Table 1.   Heat of Ionization of H_2O at 25°C as Determined in a 4 ml Isothermal Calorimeter[a]

$$-\Delta H/kcal\ mol^{-1}$$

μ = 0.0097[b]	μ = 0.0050[c]	μ = 0.00097[d]
13.47	13.37	13.30
13.29	13.32	13.18
13.47	13.42	13.43
13.35	13.32	13.36
13.37		13.47
13.42		
Av. = 13.39±0.03	Av. = 13.36±0.02	Av. = 13.35±0.05

---

[a]Neutralization of NaOH with $HClO_4$.   [b]In each run 0.5 ml of 0.2059 M $HClO_4$ was titrated into 4.0 ml of 0.01030 M NaOH.   [c]In each run 0.5 ml of 0.2059 M $HClO_4$ was titrated into 4.0 ml of 0.005150 M NaOH.   [d]In each run 0.5 ml of 0.02026 M $HClO_4$ was titrated into 4.0 ml of 0.001031 M NaOH.

The results from the isothermal calorimeter (Table 1), ΔH_w
= -13.39±0.03, -13.36±0.02, and -13.35±0.05 kcal/mole at ionic
strengths of 0.0097, 0.0050, and 0.00097, respectively, are in
excellent agreement with previously determined values of -13.38,
-13.36 and -13.35 kcal/mole at these respective ionic strengths.[5]
The total heats measured in these runs were 0.55, 0.28, and 0.055
cal, respectively.

The results obtained using the isoperibol calorimeter (Table
2) ΔH_w= -13.44±0.04 at μ = 0.010 and -13.33±0.10 at μ = 0.0029,
are in good agreement with previously determined values of -13.38
and -13.36 at the respective ionic strengths.[5] Total heats
measured in these runs were 0.28 and 0.077 cal, respectively.

Not unexpectedly, the results from the isothermal calorimeter
are somewhat more precise than those from the isoperibol instrument. There do not appear to be any significant systematic errors
present in the operation of either instrument.

Table 2.   The Heat of Ionization of H_2O at 25° as Determined in
a 3 ml Isoperibol Titration Calorimeter[a]

$$-\Delta H/kcal \ mol^{-1}$$

μ = 0.010[b]	μ = 0.0029[c]
13.40	13.10
13.35	13.06
13.34	13.27
13.36	13.12
13.61	13.68
13.50	13.55
13.52	13.56
Ave. = 13.44±0.04	Ave. = 13.33±0.10

[a]Neutralization of NaOH with HCl.   [b]In each run 0.25 ml of 0.2581
M HCl was titrated into 2.7 ml of 0.01030 M NaOH.   [c]In each run
0.25 ml of 0.06727 M HCl was titrated into 2.7 ml of 0.00287
M NaOH.

SUMMARY

The major advantage of titration calorimetry over batch calorimetry is the increased amount of data which can be obtained with the same amount of effort and material using the titration technique. Considering that each data point in a calorimetric titration run contains nearly as much information as all of the data from a batch run, the titration method is 50 to 100 times more efficient in the use of materials and time. Continuously variable micro-flow calorimeters require about the same amount of time and effort per data point but generally require about twice as much material as the calorimeter described in this paper.

In comparison to the isothermal instrument the isoperibol calorimeter is simpler, less expensive, and easier to operate. The major advantage of the isoperibol instrument is its rapid response time ($\sim$ two seconds). The response time of the isothermal instrument is about two to three minutes which can obscure fine features in a thermogram for a complex chemical system with several end points. The sensitivity of the isothermal instrument, however, is about 10 times better (0.02 mcal/minute as compared to 0.2 mcal/minute). Also, the useful length of an isoperibol run is about 20 minutes while the isothermal method can be used for runs extending into days.

ACKNOWLEDGMENT

This work was supported by NSF Grant GF-33536X and by NIH Grants GM18811-02, GM18816-03, and AM15615-02 from the U.S. Public Health Service.

REFERENCES

1. Literature describing this instrument is available from Tronac, Inc., 1804 South Columbia Lane, Orem, Utah 84057
2. Christensen, J. J., Gardner, J. W., Eatough, D. J., Izatt, R. M., Watts, P. J., and Hart, R. M., Rev. Sci. Instrum., 44, 481 (1973)
3. Eatough, D. J., Christensen, J. J., and Izatt, R. M., Thermochimica Acta, 3, 219 (1972).
4. Hansen, L. D., Jensen, T. E., Mayne, S., Eatough, D. J., Izatt, R. M., and Christensen, J. J., J. Chem. Thermo., submitted for publication.
5. Parker, V. B., "Thermal Properties of Aqueous Uni-univalent Electrolytes", NSRDS-NBS 2, U.S. Government Printing Office, Washington, D. C., 1965

THEORY OF DIFFERENTIAL SCANNING CALORIMETRY -- COUPLING OF ELECTRONIC AND THERMAL STEPS

Joseph H. Flynn

Institute for Materials Research, National Bureau of Standards, Washington, D. C. 20234

ABSTRACT

A model system for differential scanning calorimetry (DSC) is developed in which the electronic response of the instrument is coupled with the heat flow across an interface. Equations are derived which relate the time constants for this two-step process with the thermal properties of the sample and the amplitudes, areas, slopes and dwell times of DSC traces. The cases discussed include first and second order transitions, partial and total "supercooling" and effects of a temperature dependence of the heat capacity and the rate of temperature change. The magnitude of the lag terms of these cases is determined from typical experimental data. The equations permit an independent determination of the interfacial time constant and an assessment of the limits for the validity of the theoretical model.

INTRODUCTION

An extremely simple model which assumes the linear dependence of heat flow upon a temperature difference (Newton's Law of Cooling) has been used as a basis for the theoretical discussion of differential scanning calorimetry (DSC) [1-6].

The flow of heat through the region between the calorimeter heater and the sample material often is the rate determining step. Therefore, despite the many paths and modes of heat dissipation and the geometric complexities of all real thermal analysis systems, equations derived from this simple model have been applied successfully to explain the shape of DSC traces and establish general criteria for their evaluation [1-6].

However, the power difference output of many DSC instruments
passes through electronic components whose time constants are
slower than those due to thermal effects [2, 4]. Therefore, this
electronic response must be included in any theory from which
quantitative relationships between the power difference output
and enthalpic events occurring in the sample are to be obtained.
This is especially true if these relationships are to be used for
instrumental calibration.

A two step model coupling the electronic and thermal effects
is developed in this paper and applied to first and second order
transitions, partial and total supercooling and cases for which
the heat capacity of the sample or the rate of temperature change
is temperature dependent.

MODEL SYSTEM

A typical DSC consists of two matching covered calorimeter
cups, each containing a heater and a temperature sensor. A
standard container of high thermal conductivity, holding the
sample material, is placed within the "sample" calorimeter cup in
good thermal contact with its surface. Usually an empty container
is placed similarly in the "reference" calorimeter cup. The two
calorimeters are situated in an environment designed to promote
stable and reproducible steady state heat flow conditions. The
temperature sensors in each cup are maintained at almost equal
temperatures by apportioning the electric power to their heaters.
This difference in power is amplified and read out. The temperature readout is the mechanically programmed temperature. The
total electrical energy to the heater bridge is increased or
decreased to keep the average temperature in the sample and
reference cups in correspondence with the programmed temperature.

It is assumed for this model that the base line is linear
and stable, i. e., the heat leaks of, and heat capacity differences
between the two empty calorimeters are linear and reproducible
functions of temperature. Also, since DSC measurement is usually
designed so that the region of interest is after "steady state"
conditions for temperature change have been closely approached,
transient pulses or lags affecting the measurement of base line
stability or temperature calibration generally will be neglected.

Therefore, the only instrumental time constant taken into
account is that for the power ordinate response. When rectangular
light pulses on the empty calorimeter surface were used as a step
forcing function, the power difference readout, X, was found to
build up and decay according to first order kinetics [2]. There-

fore $\dot{X}$ is driven by the simple linear equation

$$- \dot{X} = \tau_x^{-1} (X - \Delta\dot{q}) \tag{1}$$

where τ_x is the power response time constant, t, the time and the forcing function is the net heat flux, $\Delta\dot{q}$, is given by

$$\Delta\dot{q} = (dq/dt)_S - (dq/dt)_R . \tag{2}$$

$(dq/dt)_S$ and $(dq/dt)_R$ are the net heat fluxes between the calorimeter surface and the container for the sample and reference calorimeters, respectively.

The reference and sample calorimeters are assumed to be at the same isotropic temperature, T_C. The net heat fluxes from the calorimeters to the enclosed containers from all interfaces between them are assumed to be represented by linear (Newton's Law) relationships, viz.,

$$(dq/dt)_R = R_R^{-1} (T_C - T_R) \tag{3}$$

and

$$(dq/dt)_S = R_S^{-1} (T_C - T_S) . \tag{4}$$

R_R and R_S are the respective thermal resistances between the calorimeters and containers and T_R and T_S are isotropic temperatures of the containers and their contents.

The system is at isothermal equilibrium at zero time so that $T_C = T_R = T_S = T_o$. The rate of temperature change of the calorimeters, $\dot{T}$, is assumed to be constant for most cases, so that

$$T_C(t) = T_o + \dot{T} t \tag{5}$$

during the scan.

The temperature of the reference and sample containers are functions of the net heat flux into them, so that

$$T_R(t) = T_o + \int_o^t C_R^{-1} (dq/dt)_R \, dt \tag{6}$$

and

$$T_S(t) = T_o + \int_o^t C_S^{-1} (dq/dt)_S dt , \qquad (7)$$

where C_R and C_S are the heat capacities of the reference container (plus sample), respectively.

The assumptions of this model are embodied in eqs. 1 – 7. They include: first order kinetic response of the net power readout to the net heat fluxes (eq. 1); heat leak reproducibility of the two calorimetric vessels (eq. 2); Newton's Law of Cooling for the net heat exchange between each calorimeter and its container, and uniform temperatures in the calorimeters and the sample and reference containers (eqs. 3 – 4); a linear heating rate (eq. 5); and differential energy conservation for the sample and reference containers (eqs. 6 – 7).

The solutions to eqs. 1 – 7 for a number of cases are discussed in the subsequent sections.

. a. No Transitions: Solving eqs. 1 – 7 assuming C_S, C_R and T constant during a DSC scan, one obtains for the temperature, heat fluxes and power readout

$$T_R = T_C - \dot{T} \tau_R [1 - \exp (-t/\tau_R)] \qquad (8)$$

$$T_S = T_C - \dot{T} \tau_S [1 - \exp (-t/\tau_S)] \qquad (9)$$

$$\Delta q = \dot{T} C_S[1 - \exp (-t/\tau_S)] - \dot{T} C_R[1 - \exp (-t/\tau_R)] \qquad (10)$$

$$X = \dot{T} C_S \left\{ 1+[\tau_x/(\tau_S-\tau_x)]\exp(-t/\tau_x)-[\tau_S/(\tau_S-\tau_x)]\exp(-t/\tau_S) \right\}$$

$$- \dot{T} C_R \left\{ 1+[\tau_x/(\tau_R-\tau_x)]\exp(-t/\tau_x)-[\tau_R/(\tau_R-\tau_x)]\exp(-t/\tau_R) \right\} \qquad (11)$$

where T_C is given by eq. 5 and $\tau_S = C_S R_S$ and $\tau_R = C_R R_R$ are the interfacial time constants for the sample and reference containers.

If the scan is terminated at time

$$T_C = T_o + \dot{T} t_2 \qquad (12)$$

$$T_R = T_C - \dot{T} \tau_R[1-\exp(-t_2/\tau_R)]\exp[-(t-t_2)/\tau_R] \qquad (13)$$

$$T_S = T_C - \dot{T} \tau_S [1-\exp(-t_2/\tau_S)]\exp[-(t-t_2)/\tau_S] \tag{14}$$

$$\begin{aligned}\Delta q &= \dot{T} C_S [1-\exp(-t_2/\tau_S)]\exp[-(t-t_2)/\tau_S]\\ &\quad - \dot{T} C_R[1-\exp(-t_2/\tau_R)]\exp[-(t-t_2)/\tau_R]\end{aligned} \tag{15}$$

$$\begin{aligned}X &= \dot{T} C_S\Big\{[\tau_S/(\tau_S-\tau_x)][1-\exp(-t_2/\tau_S)]\exp[-(t-t_2)/\tau_S]-[\tau_x/(\tau_S-\tau_x)]\\ &\quad [1-\exp(-t_2/\tau_x)]\exp[-(t-t_2)/\tau_x]\Big\} - \dot{T} C_R\Big\{[\tau_R/(\tau_R-\tau_x)]\\ &\quad [1-\exp(-t_2/\tau_R)]\exp[-(t-t_2)/\tau_R]-[\tau_x/(\tau_R-\tau_x)][1-\exp(-t_2/\tau_x)]\\ &\quad \exp[-(t-t_2)/\tau_x]\Big\} \qquad \text{(all for } t \geq t_2) \tag{16}\end{aligned}$$

The heat flux, $\Delta \dot{q}$, and the power readout, x, are plotted as a function of time for this case in figure 1.

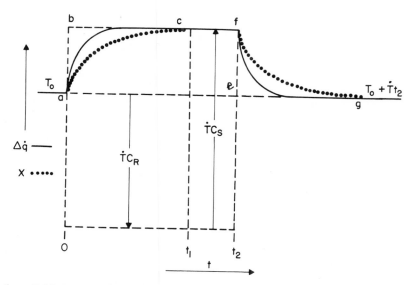

Fig. 1. DSC Trace (Differential Power vs. Time) No Transitions C_S and C_R constant; $\dot{T}$ = constant for $0 \leq t \leq t_2$.

Eqs. 8 - 11 approach steady state values as the scan progresses, i. e.,

$$T_R = T_C - \dot{T} \ \tau_R \tag{17}$$

$$T_S = T_C - \dot{T} \ \tau_S \tag{18}$$

and $\Delta \dot{q} = X = \dot{T} \ C_S - \dot{T} \ C_R$

(all for $t >> \tau_R$, τ_S, τ_x) $(0 \leq t \leq t_2)$ \hfill (19)

At the termination of the scan, $(t \geq t_2, \ \dot{T} = 0)$, T_R and T_S approach T_C while $\Delta \dot{q}$ and x decay to the base line (eqs. 12 - 16).

Two DSC scans may be performed that differ only in that for one of them a sample is sealed in the sample container. Then, from eqs. 17 - 19, the difference in steady state amplitudes or areas of the two scans yields, for this simple case, the heat capacity of the sample material.

The lag area abc, in figure 1, between the power readout x and its extrapolated steady state value at the beginning of the scan, and the "catch up" area efg at the termination may be obtained from integration of eq. 11 and 16. The areas are equal and simple functions of τ_S and τ_x. These areas might be used to evaluate τ_S, as τ_x is easily measurable [2]. However, in some instruments, the initial and terminal portions of the scan are complicated by the presence of other transient effects and, in some cases, by deliberately designed augmentation of heater voltages to compensate for the lag effects.

Because of the above complications, the subsequent cases begin at steady state scan conditions where the transient exponential terms of eqs. 8 - 11 and 13 - 16 may be neglected. Transient coupling terms resulting from enthalpic changes during the scan, of course, are retained in the development.

The appendices include two other transitionless cases of some interest. Appendix A treats a case in which the heat capacity changes linearly with temperature. In Appendix B, the rate of temperature change of the calorimeter cup is not constant. This latter case is important since, in some instruments, the calibration curve for programmed temperature vs calorimeter temperature has a large slope.

b. <u>Second Order Transition</u>: Glass transition temperatures, (T_G), and other phenomena involving sudden heat capacity changes are often determined from DSC traces. The physical model developed in this section is that of a sharp second order transition during which the heat capacity of the sample and its container change from C_S to C_S' at time, t_1, removed from initial transients. Hence, eq. 7 changes discontinuously to

$$T_S = T_S(t_1) + \int_{t_1}^{t} (C_S')^{-1}(dq/dt)_S \, dt$$

$$(t \geq t_1) \tag{21}$$

Solution of eqs. 1 - 6, 21 yields

$$T_S = T_o + \dot{T} t - \dot{T} \tau_S' \left\{1-\exp[-(t-t_1)/\tau_S']\right\} - \dot{T} \tau_S \exp[-(t-t_1)/\tau_S'] \tag{22}$$

$$\Delta q = \dot{T} C_S' \left\{1-\exp[-(t-t_1)/\tau_S']\right\} - \dot{T} C_S \exp[-(t-t_1)/\tau_S'] - \dot{T} C_R \tag{23}$$

$$X = \dot{T} C_S' \left\{1-[\tau_S'/(\tau_S'-\tau_x)]\exp[-(t-t_1)/\tau_S']+[\tau_x/(\tau_S'-\tau_x)]\right.$$
$$\left. \exp[-(t-t_1)/\tau_x]\right\} + \dot{T} C_S \left\{[\tau_S'/(\tau_S'-\tau_x)]\exp[-(t-t_1)/\tau_S']\right.$$
$$\left. -[\tau_x/(\tau_S'-\tau_x)]\exp[-(t-t_1)/\tau_x]\right\} - \dot{T} C_R. \tag{24}$$

$$(t \geq t_1)$$

where $\tau_S' = R_S C_S'$ and, by definition,

$$T_G = T_S(t_1) = T_o + \dot{T} t_1 - \dot{T} \tau_S \tag{25}$$

Eqs. 5 and 17 for T_C and T_R remain unchanged. Eqs. 22 - 24 approach new steady state values of

$$T_S = T_C - \dot{T} \tau_S' \tag{26}$$

$$X = \Delta q = \dot{T} C_S' - \dot{T} C_R \tag{27}$$

$$(t - t_1 \gg \tau_x, \tau_S, \tau_S', \tau_R)$$

The equations for the decay to isothermal values at the termination of the scan, $(t = t_2)$, differ from eqs. 14 - 16 of the previous case only in that τ_S is replaced by τ_S'.   Figure 2 represents a schematic DSC trace for this case.

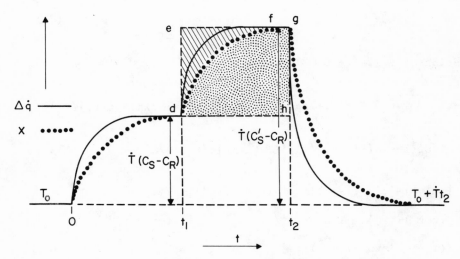

Fig. 2.   DSC Trace (Differential Power vs. Time) Second Order Transition $(C_S \longrightarrow C_S')$ at t_1. C_R constant; $\dot{T}$ constant for $0 \leq t \leq t_2$.

The area between the DSC trace and the extrapolation of its steady state value to time, t_1, (the hatched area, def), is equal to $\dot{T}$ $(C_S'-C_S)$ $(\tau_S' + \tau_x)$.   This lag area may be used to determine τ_S' in cases for which the second order transition kinetics is rapid enough for the thermal and electronic lags to be rate limiting.

However, glass transitions proceed by mechanisms which are complex and generally sluggish. Because of this, the change in amplitude of a DSC trace is an unreliable parameter upon which to base the determination of T_G. Guttman [7] has developed a method for T_G determination based on area (enthalpy) differences, which, therefore, depends only upon initial and final states of the substance. He thus avoids problems associated with the kinetics of transition. T_G is determined from a trace as in figure 2 from [8].

$$T_G = T_2 - \kappa \,[\text{AREA dfgh (shaded)}]/(C_S'-C_S) \qquad (28)$$

where T_2 is the temperature at time, t_2, κ is an enthalpic calibration factor, AREA dfgh is the area between the trace and its pretransitional extrapolation and $C_S'-C_S$ is the heat capacity change measured from the difference in steady state amplitudes before and after the transition.

Guttman's method does not compensate for electronic and thermal lags so it is instructive to determine these lags for the model developed here. (Lag areas are independent of transition kinetics since the lags follow first order kinetics.) The enthalpy difference between the final state and the extrapolated initial state over the time interval from t_1 to t_2 from eqs. 5, 18, 19, 26 is

$$(C_S'-C_S)(T_2-T_G)=(C_S'-C_S)[T_S(t_2)-T_S(t_1)]=\dot{T}(C_S'-C_S)(t_2-t_1+\tau_S-\tau_S') \quad (29)$$

The area calculated from the difference in integrals over t_1 to t_2 for eqs. 24 and 19 is

$$\int_{t_1}^{t_2} [X-\dot{T}(C_S-C_R)]dt=\dot{T}(C_S'-C_S)(t_2-t_1+\tau_S-\tau_S')-\dot{T}(C_S'-C_S)(\tau_x+\tau_S) \quad (30)$$

Thus, two additions to the measured area in Guttman's method are necessary from this model -- a $\dot{T}$ $(C_S'-C_S)$ τ_x term for the electronic lag and a $\dot{T}$ $(C_S'-C_S)$ τ_S $[=\dot{T}$ $(\tau_S'-\tau_S)\dot{C}_S]$ "catch up" term for the thermal time constant change. The second term of eq. 30 decreases T_G, calculated from eq. 28 by $\dot{T}$ $(\tau_x+\tau_S)$. Organic polymers have large effective τ_S values (see section e). This correction to T_G equals two degrees for $\tau_S+\tau_x = 12$ s and $\dot{T} = 0.1667$ K/s. It is tacitly assumed by the use of eq. 29 that T_G and T_2 are calibrated correctly, i. e., T_G is calibrated for a thermal lag, $\dot{T}$ τ_S, and T_2, for $\dot{T}$ τ_S' .

c. First Order Transition: No Heat Capacity Change: The thermodynamic quantities of interest for a first order transition are the temperature of transition, T_T, and the net enthalpy change in the region of T_T, the enthalpy of transition, ΔH_T. These need to be related to the parameters of a DSC trace. The case discussed here is a "sharp" transition, i. e., the increment of the sample temperature over which the transition occurs is much smaller than the temperature increment of the calorimeter during which heat must flow to or from the sample to match ΔH_T.

In this first case, C_S remains constant throughout the scan and it is assumed that a steady state heat flow as represented by eqs. 17 – 19 has been approached by the beginning of the transition. The sample reaches the transition temperature, T_T, at time, t_T, and it remains at this temperature until the transition is complete at time, t_C. Thus, during the transition, from eqs. 1 – 6, 17 – 19),

$$T_S = T_T = T_o + \dot{T} t_T - \dot{T} \tau_S \tag{31}$$

$$\Delta q = \dot{T} R_S^{-1} (t-t_T) + \dot{T} (C_S-C_R) \tag{32}$$

$$X = \dot{T} R_S^{-1} \left(t-t_T-\tau_x \left\{ 1 - \exp [-(t-t_T)/\tau_x] \right\} \right) + \dot{T} (C_S-C_R) \tag{33}$$

$$(t_T \leq t \leq t_C)$$

After t_C, $\dot{T}_S$ catches up, Δq decays and x peaks and then decays according to

$$T_S = T_C - \dot{T} \tau_S - \dot{T} (t_C-t_T) \exp [-(t-t_C)/\tau_S] \tag{34}$$

$$\Delta q = \dot{T} R_S^{-1}(t_C-t_T) \exp [-(t-t_C)/\tau_S] + \dot{T} (C_S-C_R) \tag{35}$$

$$X = \dot{T} R_S^{-1}(t_C-t_T) \left\{ [\tau_S/(\tau_S-\tau_x)]\exp[-(t-t_C)/\tau_S]-[\tau_x/(\tau_S-\tau_x)] \right.$$
$$\left. \exp[-(t-t_C)/\tau_x] \right\} - \dot{T} R_S^{-1}\tau_x \left\{ 1 - \exp[-(t_C-t_T)/\tau_x] \right\} \exp[-(t-t_C)/\tau_x]$$
$$+ \dot{T} (C_S-C_R)$$

$$(t \geq t_C) \tag{36}$$

As $t-t_C$ becomes much greater than τ_S or τ_x, eqs. 34 – 36 approach their pretransitional steady state values given by eqs. 18 – 19. Thus the trace has the same steady state base line before and after the transition.

A schematic trace for this case is illustrated in figure 3.

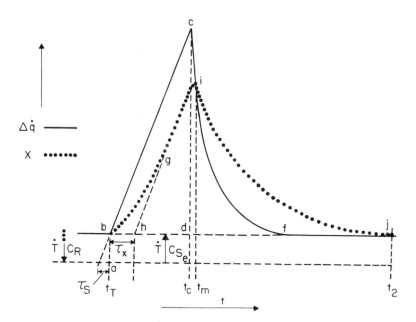

Fig. 3. DSC Trace (Differential Power vs. Time) First Order
Transition at t_T. C_S, C_R and $\dot{T}$ constant.

The heat of transition, ΔH_T, equals the total heat supplied
the sample at T_T between t_T and t_C, i. e.

$$\Delta H_T = \int_{t_T}^{t_C} (dq/dt)_S \, dt = 1/2 \ \dot{T} \ R_S^{-1}(t_C - t_T)^2 + \dot{T} \ C_S(t_C - t_T) \quad (37)$$

This is the area abcde in figure 3. The area between $\Delta\dot{q}$ and the
steady state base line, $\dot{T} \ (C_S - C_R)$, is obtained from integration
of eqs. 32 and 35,

$$\Delta H_T = \int_{t_T}^{t_C} [\Delta \dot{q} - \dot{T}(C_S - C_R)] \, dt + \int_{t_C}^{t_2} [\Delta \dot{q} - \dot{T}(C_S - C_R)] \, dt$$

$$= 1/2 \, \dot{T} \, R_S^{-1} \, (t_C - t_T)^2 + \dot{T} \, R_S^{-1} \, \tau_S \, (t_C - t_T) \qquad (38)$$

$$(t_2 - t_C \gg \tau_S)$$

Therefore the peak area [area bcf in figure 3] equals ΔH_T. The "catch up" enthalpy, [area dcf], the second term in eq. 38, is equal to the second term in eq. 37, [area abde]. The latter is the heat capacity integral over the dwell time, $(t_C - t_T)$, during which the sample remained at T_T. The area, bgij, between the power difference peak and the steady state base line is obtained by integration of eqs. 33 and 36 from whence it can be shown that

$$\int_{t_T}^{t_2} [X - \dot{T}(C_S - C_R)] \, dt = \Delta H_T, \qquad (39)$$

$$(t_2 - t_m \gg \tau_x, \; \tau_S).$$

Therefore ΔH_T is determined exactly from the measured DSC peak area for this case with a constant base line.

The slope of the leading edge of the x peak approaches a steady state value from which the thermal resistance between the calorimeter and the sample, R_S, can be determined, since, from eqs. 1, 32, 33,

$$\dot{X} = \dot{T} \, R_S^{-1}$$

$$(t - t_T \gg \tau_x) \; (t_T \leq t \leq t_C) \qquad (40)$$

The asymptote $[(t - t_T) \gg \tau_x]$ of the peak slope intersects the steady state base line at point h in figure 3 at time, $t_T + \tau_x$. The temperature associated with this time is called the "onset" temperature of the transition. The temperature of the calorimeter at the onset, $T_{C,0}$, is

$$T_{C,0} = T_0 + \dot{T} \, t_T + \dot{T} \, \tau_x. \qquad (41)$$

This may be related to the transition temperature, T_T, of the sample from eq. 31. Therefore, at the onset

$$T_{C,0} = T_T + \dot{T} (\tau_x + \tau_S). \tag{42}$$

This is a basic equation for temperature calibration from transition peaks.

Van't Hoff's Law is often applied to the ascending "tail" of a melting peak to determine sample purity. The area between this ascending "tail", the asymptotic slope and the base line, area bgh, is obtained from eq. 33 by integration between t_T and t_C.

$$\text{AREA bgh} = \dot{T} R_S^{-1} \tau_x \int_{t_T}^{t_C} \exp [-(t-t_T)/\tau_x] = \dot{T} R_S^{-1} \tau_x^2 \tag{43}$$

$$(t_C - t_T \gg \tau_x).$$

This area due to instrumental lag, demonstrates that there is a lower limit for methods which determine purity by this method. If R_S = 100 s· K/cal, τ_x = 5 s and $\dot{T}$ = 0.01 K/s, the area in eq. 43 is 2.5 mcal (1 cal = 4.1840 j). (Purity determinations from methods in which areas are obtained by isothermal interruption of the scan avoid this limitation [9].)

The power-difference readout, X, reaches a maximum at a time, t_m, greater than t_C, where $X = \Delta\dot{q}$, so from eqs. 35 – 36,

$$\ln [1+(\tau_S-\tau_x)/(t_C-t_T)]=[(\tau_S-\tau_x)/\tau_S\tau_x] (t_m-t_C) . \tag{44}$$

The time, t_d, from the onset to the X-peak maximum is

$$t_d = t_m - t_T - \tau_x . \tag{45}$$

Eqs. 44 and 45, eq. 40 for the steady state slope, and eq. 38 for the peak area may be combined to obtain an exact relationship between τ_S and τ_x. However this is not of practical utility as $t_m - t_C$ is very short. A precise determination of t_m is not possible as there is a rounding of the peak from the breakdown of thermal equilibrium in the sample near the completion of the transition, that is, the assumption of temperature isotropy no longer holds.

The peak area, ΔH_T, [from eq. 38], the steady state slope, $\dot{T} R_S^{-1}$ [from eq. 40] may be related to τ_x, τ_S and measurable parameters by approximating $t_C - t_T$ by $t_m - t_T$, thus

$$\text{peak area/slope} \simeq 1/2 \ (t_d + \tau_x)^2 + (t_d + \tau_x)\tau_S$$

$$\tag{46}$$

$$(t_c - t_m \ll t_C - t_T).$$

Literal application of eq. 46 to DSC melting peaks of tin and indium yielded values for τ_S that changed inversely with heating rate. Thus it appears that the approximations of eq. 46 and/or the assumptions of this model do not hold in the region of peak maximum.

However, an utilizable equation relating τ_S and τ_x to experimental parameters is of special interest. τ_x is obtained easily from experiment [2] and R_S from the peak slope. Thus an experimental value for $C_S (= R_S^{-1} \tau_S)$ may be determined. If the experimental C_S matches the sum of the heat capacities of the sample and its container, then the heat flow model assumed in this paper may be used with confidence as a basis for instrument calibration. A practical relationship between τ_x and τ_S is developed successfully in section e where "supercooling" peaks are discussed.

d. <u>First Order Transition with a Heat Capacity Change</u>: Many DSC traces exhibit a shift in the steady state base line from before to after a first order transition. This results from a change in heat capacity of the sample during the transition. Therefore, it is assumed for this case that at the transition temperature, T_T, at time, t_T, the heat capacity of the sample and its container changes from C_S to C_S'. Otherwise the conditions and symbols are the same as in the previous case c.

The sample remains at T_T until time, t_C, so eqs. 31 – 33 still apply over t_T to t_C. The total heat supplied to the sample during this time, ΔH_T, again equals eq. 37.

For $t \geq t_C$, however, the sample temperature changes as a function of C_S' [eq. 21]. The solution of eqs. 1 – 6, 21 , using eqs. 31 – 33 at t_C as boundary conditions, yields

$$T_S = T_C - \dot{T} \ \tau_S' \left\{ 1 - \exp[-(t-t_C)/\tau_S'] \right\} - \dot{T} \ (t_C - t_T + \tau_S)$$

$$\exp[-(t-t_C)/\tau_S' \ ] \tag{48}$$

$$\Delta \dot{q} = \dot{T} \, C_S' \left\{ 1 - \exp[-(t-t_C)/\tau_S'] \right\} + \dot{T} \, R_S^{-1} \, (t_C - t_T)$$

$$\exp[-(t-t_C)/\tau_S'] + \dot{T} \, C_S \, \exp\,[-(t-t_C)/\tau_S'] - \dot{T} \, C_R \tag{49}$$

$$X = \dot{T}(C_S' - C_R) + \dot{T}[C_S' - C_S - R_S^{-1}(t_C - t_T)] \left\{ [\tau_x/(\tau_S' - \tau_x)]\exp[-(t-t_C)/\tau_x] \right.$$

$$\left. - [\tau_S'/(\tau_S' - \tau_x)]\exp[-(t-t_C)/\tau_S'] \right\} - \dot{T} \, R_S^{-1} \tau_x \left\{ \exp\,[-(t-t_C)/\tau_x] \right.$$

$$\left. - \exp\,[-(t-t_T)/\tau_x] \right\} \tag{50}$$

$$(t \geq t_C) \quad .$$

Eqs. 48 and 49 decay and eq. 50 peaks and decays to steady state values given by eqs. 26 – 27 as $t - t_C$ becomes much greater than τ_x and τ_S'.

A schematic DSC trace for this case is illustrated in figure 4.

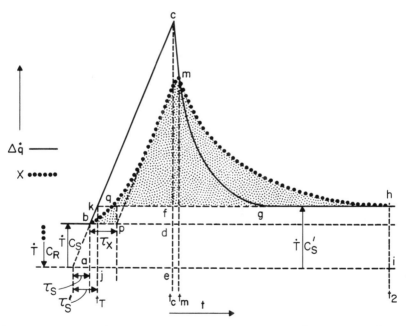

Fig. 4. DSC Trace (Differential Power vs. Time) First Order . Transition with Heat Capacity Change, $C_S \rightarrow C_S'$, at t_T. C_R and T constant.

This case includes a shift in the base line during the transition. It is of great interest to calculate how this shift can be taken into account in the determination of ΔH_T from the peak area.

As in case c, the heat of transition equals the total heat flux to the sample between t_T and t_C as given by eq. 37. This is the area abcde in figure 4. Eq. 51 gives the total heat flux into the sample between t_C and a time, t_2, for which $t_2-t_C >> \tau_S'$; i. e., t_2 is removed from the kinetics of transition. From integration of eq. 49,

$$\int_{t_C}^{t_2} (dq/dt)_S \, dt = \text{AREA edcghi} = \dot{T} \, C_S' (t_2-t_T)-\dot{T} \, C_S' \, (\tau_S'-\tau_S) \, .$$

(51)

The sample temperature changes during this time interval from $T_o + \dot{T} \, (t_T-\tau_S) = T_T$ at t_C to $T_o + \dot{T} \, (t_2-\tau_S')$ at t_2. Thus the second term in eq. 51 results from the increase in temperature lag.

Eq. 51, the "catch up" area equals the heat capacity-temperature integral, area jkhi $= \dot{T} \, C_S' \, (t_2-t_T-\tau_S' + \tau_S)$. Subtraction of this area from the total peak areas of eqs. 37 and 51 leaves a residual peak area equal to ΔH_T. Therefore ΔH_T is the area between the Δq peak and the extrapolated base line of the final state, (area kcg), plus the trapezoidal area, abkj $[= 1/2 \, \dot{T} \, (C_S'+C_S)(\tau_S'-\tau_S)]$.

However, the experimentally obtained area under the X peak is more relevant. It is obtained from integration of eqs. 33 and 50, minus their steady state values, between t_T and t_2, viz.,

$$\int_{t_T}^{t_C} [X - \dot{T}(C_S-C_R)] \, dt + \int_{t_C}^{t_2} [X - \dot{T} \, (C_S'-C_R)] dt = \text{AREA bmhfd}$$

$$= \dot{T} \, R_S^{-1} \, (t_C-t_T)^2 + \dot{T} \, \tau_S' \, (t_C-t_T) - \dot{T} \, (C_S'-C_S) \, (\tau_x-\tau_S')$$

$$[(t_2-t_C) >> \tau_x, \, \tau_S'] \, .$$

(52)

An extrapolation of the final base line from t_C back to the onset at point, q, removes the area pqfd $[= \dot{T} \, (C_S'-C_S)(t_C-t_T-\tau_x)]$ from eq. 52. Therefore,

$$\Delta H_T = \text{AREA bqp} + \text{AREA qmh} + \dot{T}\,(C_S'-C_S)\,\tau_S' . \qquad (53)$$

The enthalpy of transition equals the two shaded areas in figure
4 plus the last term in eq. 53 which corrects the shaded area for
the area lag during the heat capacity change. There is no power
lag term, $\dot{T}\,(C_S'-C_S)\,\tau_x$, as it was compensated for by extrapolation to the onset, rather than to t_T.

The lag term in eq. 53 is independent of the transition
kinetics and is inherent in any model which is based on heat flow
across an interface. The same term may be applied as a correction
to Guttman and Flynn's method [10] as t_T lies before the transition
kinetics and t_2 beyond them. Fortunately, τ_S' is often less than
one second so that the correction term in eq. 53 is small in
comparison to ΔH_T, even at rapid temperature change. Ignoring it
is equivalent to calculating ΔH_T at $T_T + \dot{T}\,\tau_S'$.

In general, any change in heat capacity changes the temperature lag and results in an enthalpy lag. This was also the
case for second order transitions and is observed again in
Appendix A.

    e. <u>First Order Transition with Partial "Supercooling"</u>:
Substances undergoing temperature change often remain in a metastable region of their initial state beyond the transition
temperature. This phenomenon occurs frequently in freezing
transitions so the term "supercooling" is used here to designate
it. If sufficient heat has passed to or from the material to
equal or exceed the enthalpy of transition, then,once begun, the
transition will proceed quite rapidly to completion. We name
this case "total supercooling" and discuss it in the next section.

For the case of partial supercooling, the heat exchanged
between the sample and the calorimeter before the commencement of
the transition is insufficient to bring the transition to completion.
The temperature of the sample returns to T_T and the transition
proceeds as further heat is exchanged with the sample.

The system is assumed again to be at steady state scan
conditions given by eqs. 17 - 19. At time, t_i, the transition
begins and proceeds at a rate which is very rapid compared with
the thermal and electronic responses, until the temperature of
the sample returns to T_T. The fraction of the sample undergoing
"instantaneous" transition is equal to μ for $\mu \leq 1$, where

$$\mu = \dot{T} C_S (t_i - t_T)/\Delta H_T \tag{54}$$

$$\mu < 1, \text{ partial supercooling}$$

$$\mu \geq 1, \text{ total supercooling}$$

where $t_i - t_T$ is the induction period.

The solution of eqs. 1 - 6 for the above boundary conditions again give eqs. 31 and 32 for T_S and Δq , and $t \geq t_i$. However, the power difference is

$$X = \dot{T} R_S^{-1}(t-t_T) - \dot{T} R_S^{-1} \tau_x \left\{ 1 - \exp[-(t-t_i)/\tau_x] \right\} - \dot{T} R_S^{-1}(t_i - t_T)$$

$$\exp [-(t-t_i)/\tau_x] + \dot{T} (C_S' - C_R) \tag{55}$$

$$(t_i \leq t \leq t_C')$$

as coupling of X with Δq does not begin until t_i.

Beyond t_C' , the transition is complete and T_S catches up, Δq decays to its steady state value and X peaks and decays according to

$$T_S = T_o + \dot{T} t - \dot{T} \tau_S - \dot{T} (t_C' - t_T) \exp [-(t-t_C')/\tau_S] \tag{56}$$

$$\Delta q = \dot{T} R_S^{-1}(t_C' - t_T) \exp [-(t-t_C')/\tau_S] + \dot{T} (C_S - C_R) \tag{57}$$

$$X = \dot{T} R_S^{-1} (t_C' - t_T) \left\{ [\tau_S/(\tau_S - \tau_x)] \exp[-(t-t_C')/\tau_S] - [\tau_x/(\tau_S - \tau_x)] \right.$$

$$\left. \exp [-(t-t_C')/\tau_x] \right\} - \dot{T} R_S^{-1} (t_i - t_T) \exp [-(t-t_T)/\tau_x]$$

$$- \dot{T} R_S^{-1} \tau_x \left\{ 1 - \exp [-(t_C' - t_i)/\tau_x] \right\} \exp [-(t-t_C')/\tau_x] + \dot{T} (C_S - C_R)$$

$$(t \geq t_C'). \tag{58}$$

This case is illustrated in figure 5.

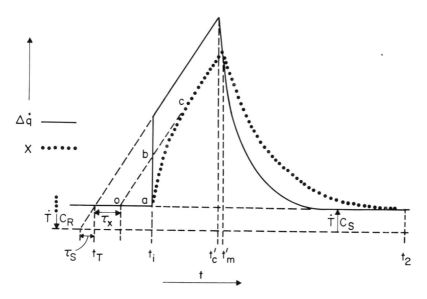

Fig. 5.   DSC Trace (Differential Power vs. Time) First Order
Transition with Partial Supercooling, C_S, C_R and $\dot{T}$ constant.

The asymptote for $t-t_i \gg \tau_x$ to eq. 55 yields the identical
slope and onset as in eqs. 40 and 41 for case c with no supercooling. This most important feature of this case is illustrated
by the extrapolation cbo, in figure 5. Thus slopes of partial
supercooling peaks have been used to obtain the onset for the
calibration of cooling scans for a DSC instrument [4].

The amplitude of the heat flux peak at t_i is $\dot{T} R_S^{-1}(t_i-t_T) =$
$\tau_S^{-1} \mu \Delta H_T$, and its maximum at t_c' is $\dot{T} R_S^{-1} (t_c'-t_T)$. The power
slope asymptote is displaced from these values by an amplitude
$\dot{T} R_S^{-1} \tau_x$. Integration of eq. 57 yields a catch up area for the
descending $\Delta\dot{q}$ peak equal to $\dot{T} C_S(t_c'-t_T)$. The experimental
induction time, t_{ind}, between the extrapolated onset and the start
of the X peak is related to t_T by

$$t_{ind} = t_i - t_T - \tau_x \quad . \qquad (59)$$

The lag area abc in figure 5 is obtained from integration of eq. 55.

$$\text{AREA abc} = \text{slope.} \ t_{ind} \cdot \tau_x \qquad (60)$$

This equation may be used to calculate τ_x if the induction period is of sufficient duration to obliterate any effect of impurities on the leading tail of the peak.

The area between the power curve and steady state base line again equals ΔH_T. Eq. 44 again holds for t_m' and t_C'. Other relationships between peak areas, slopes and maxima yield nothing significant which has not been noted in previous cases.

f. <u>First Order Transition with Complete "Supercooling"</u>: If the heat exchanged between the sample and calorimeter during the induction period is equal to or greater than the enthalpy of transition [$\mu \geq 1$ in eq. 54], then the transition will proceed rapidly to completion. It is assumed that the transition takes place at an explosive rate, i. e., much faster than the thermal and electronic steps of this model system. (Several of the equations developed in this section are applied to another explosive transformation -- chemical detonation.)

The transition takes place at time, t_C'', and the sample temperature changes from its steady state value, $T_o + \dot{T} t_C'' - \dot{T} \tau_S$, to $T_o + \dot{T} t_C'' - \dot{T}\tau_S - \Delta H_T/C_S$. The solution of eqs. 1 - 6 for this boundary condition gives

$$T_S = T_o + \dot{T} t - \dot{T} \tau_S - (\Delta H_T/C_S)\exp[-(t-t_C'')/\tau_S] \qquad (61)$$

$$\Delta q = (\Delta H_T/\tau_S) \ \exp \ [-(t-t_C'')/\tau_S] + \dot{T} \ (C_S - C_R) \qquad (62)$$

$$X = [\Delta H_T/(\tau_S - \tau_x)] \left\{ \exp \ [-(t-t'')/\tau_S] - \exp \ [-(t-t'')/\tau_x] \right\}$$

$$+ \dot{T} \ (C_S - C_R) \qquad (63)$$

$$(t \geq t_C'').$$

These equations decay to their steady state values in eqs. 18 - 19 for $t-t_C'' \gg \tau_S$, τ_x. This case illustrated in figure 6.

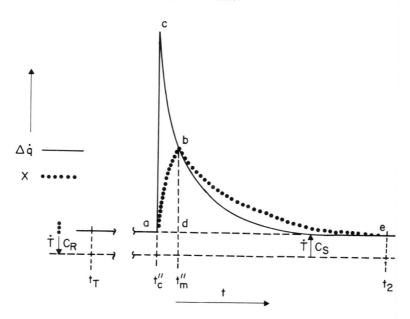

Fig. 6. DSC Trace (Differential Power vs. Time) First Order Transition with Complete Supercooling, C_S, C_R and $\dot{T}$ constant; $x_m = bd$, $\tau_m = ad$.

The area between the X peak and its steady state value, Area abe in figure 6, from integration of eq. 63, is

$$\text{AREA abe} = \int_{t_C''}^{t_2} [X - \dot{T} (C_S - C_R) \, dt = \Delta H_T \qquad (64)$$

$$(t_2 - t_C'' \gg \tau_x, \tau_S).$$

The maximum amplitude of the Δq peak at t_C'' is $\Delta H/\tau_S$ but this quantity is not measurable unless $\tau_x \ll \tau_S$. However, x_m, the power peak amplitude at t_m'', equal to bd in figure 6, is given by eq. 63 for $t = t_m''$. The time to reach the maximum is $\tau_m = ad = t_m'' - t_C''$.

The three experimental parameters, area abe, x_m and τ_m are related at the maximum to τ_x and τ_S by

$$\tau_m = [\tau_x \tau_S / (\tau_x - \tau_S)] \ln (\tau_x / \tau_S) \qquad (65)$$

and

$$\ln (\text{AREA abe}/x_m) = (\tau_x \ln \tau_x - \tau_S \ln \tau_S)/(\tau_x - \tau_S). \qquad (66)$$

These equations permit the most facile determination of τ_S and τ_x from a DSC trace. Fortunately, this is the case most likely to be closely adhered to by a DSC system, as the large driving forces resulting from severe supercooling insure that transition kinetics will not be rate limiting. Also the applicability of these equations to an experimental case may be verified by testing the T independence of the parameters.

The instrumental time constant, τ_x, is easily determined by a separate experiment [2], so the time constant for thermal flux, τ_S, may be calculated from eqs. 65 or 66. As the thermal resistance, R_S, is obtained from the slopes of melting peaks, an effective or fictive heat capacity (= $R_S^{-1} \tau_S$) for the thermal flow may be calculated. Comparison of this effective heat capacity with the actual heat capacity values of the sample and container tests the quantitative validity of the theoretical model.

Freezing scans for a 0.0064 g sample of tin encapsulated in a 0.0224 g aluminum pan and cover yielded peak maxima and times independent of cooling rates from 0.01 to 1.33 K/s . τ_S was determined to be 1.05 s from the area to amplitude ratios [eq. 66] and from τ_x = 1.7 s in reference [2]. The thermal resistance from the heating peak slopes was 99.5 Ks/cal so the effective heat capacity was 0.0106 cal/K. The heat capacity of the tin and its aluminum container were calculated to be 0.00575 cal/K, about a factor of two smaller. The DSC peaks from the exothermic detonation of heads of kitchen matches at 630 K also yielded an effective to calculated heat capacity ratio of approximately two to one.

This theoretical model combines all thermal steps into a single interface. The larger effective heat capacities of these instantaneous heat pulses indicate that the surface region of the calorimeter is sufficiently perturbed by them so that the effective thermal interface appears to be within the calorimeter.

On the other hand, time constants for the supercooling of several samples of indium, at $\sim$ 425 K, which were sandwiched between two sheets of polystyrene, were 2.5 − 6 seconds at −1.33 K/s and even larger at slower rates of cooling. Heating peak slopes yielded R_s values about ten times those for indium alone. The calculated effective heat capacities were factors of two or more smaller than those calculated from the weights of indium, polystyrene and aluminum. Heat flow through the sample has become at least partially rate limiting for this case so it is not surprising that the effective thermal interface appears to have moved into the sample.

These considerations forebode many complicating factors when heat flow lags are to be taken into account in instrument calibration. All thermal lags are lumped into a single time constant, τ_s, in the physical model used here. It is assumed also that the sample and its container have no effect on the external heat leak of the calorimeter vessel. Thus a scan with the sample calorimeter empty should produce a base line such as line ajei in figure 4. If the asymptote of the slope of the X peak is extrapolated to this base line, the projection of the extrapolation on the time axis yields a value for τ_s. This method of extrapolation is used in the calibration of a DSC instrument [11] to correct for the thermal lag. However, comparisons of the equations for this physical model with experimental results, such as in the preceding calculations indicate that the position of the effective thermal interface depends upon the thermal properties of the sample and these equations are not quantitatively applicable. Accurate instrumental calibration requires substances whose thermal diffusional properties are similar to those of the material of interest.

SUMMARY

The analysis of this model which couples thermal and electronic steps produces considerably more adequate means for interpreting and calibrating DSC traces than were available from considerations of heat flow alone.

Several features which received emphasis in this development were:

1. Independent determination of electronic and thermal time constants, especially from "supercooling" peaks [e. g., eq. 66].

2. Correct bounding of areas to equate them to thermodynamic properties, [e. g., eqs. 30 and 53].

    3. Lower limits imposed on determinations by the lag terms
[e. g., eq. 43].

    4. Critical relationships from which to assess the
applicability of the model from experimental parameters.

    5. Inevitability of lag terms in all DSC systems in which
heat capacity changes are taking place and necessity for
quantitative corrections to the enthalpy and temperature to
account for these lags [cases b, d, Appendix A].

In general, enthalpies calculated from areas under DSC traces
should be quite accurate if base line shifts are taken into
account properly [10], as lag corrections are small for these
cases, especially at slow rates of scan. Heat capacity and glass
transition temperature measurements should be compared at several
T to test for thermal lag effects. Errors in the calculation of
temperature inevitably result from differences between thermal
time constants of materials.

    Appendix A: Temperature Dependent Heat Capacity: The
simplifying assumption in the previous cases that the heat
capacity is constant is never valid. If the heat capacity of the
sample and its container, C_S, is assumed to be linear with
temperature, this substitution into eq. 7 followed by integration
over $C_S(T)$ results in an extremely complex expression. As the
initial and terminal boundary conditions of a scan are, in any
event, ambiguous, it is more convenient to assume that at constant
rate of temperature change,

$$C_S(t) = C_{S,0} + a \dot{T} t \tag{1A}$$

eqs. 1A, 4, 5 and 7 yield

$$T_S = T_C - [\dot{T} R_S C_S(t)/(1+a \dot{T} R_S)]\left\{ 1 - \left[\frac{C_{S,0}}{C_S(t)}\right]^{(1+1/a \dot{T} R_S)}\right\} \tag{2A}$$

$$(dq/dt)_S = [\dot{T} C_S(t)/(1+a \dot{T} R_S)]\left\{1 - \left[\frac{C_{S,0}}{C_S(t)}\right]^{(1+1/a \dot{T} R_S)}\right\} \tag{3A}$$

Eq. 2A reduces to eq. 9 as a approaches zero. For small values of a $T R_S$, these equations eventually approach steady state values, viz.,

$$T_S = T_C - \dot{T} R_S C_S(t)/(1+a \dot{T} R_S) \tag{4A}$$

and

$$(dq/dt)_S = \dot{T} C_S(t)/(1 + a R_S \dot{T}) . \tag{5A}$$

Coupling eq. 5A with the power readout, [eq. 1], for C_R = constant, one obtains, finally, at steady state temperature change,

$$X = \left\{ \dot{T} [C_S(t) - a \dot{T} \tau_x]/(1 + a R_S \dot{T}) \right\} - \dot{T} C_R. \tag{6A}$$

The two terms of interest in eq. 6A are the interfacial heat flux--heat capacity temperature coefficient coupling term, a $R_S \dot{T}$, and the instrumental – heat capacity coupling term, a $\dot{T} \tau_x$. The effect of these terms upon the amplitude of a DSC trace is estimated for the following, rather extreme, case: 10 mg sample of ice at 233 K, where C_S = 0.435 cal/gK, a = dC_S/dT = 0.0022 cal/g K^2, R_S = 100 s·K/cal, τ_x = 5 s and $\dot{T}$ = 1 K/s. Substitution of these values into eqs. 6A gives

$$\dot{T}^{-1} x + C_R = \frac{4.35 \times 10^{-3} - 0.11 \times 10^{-3}}{1 + 0.0022} = 4.23 \times 10^{-3} \text{ cal/K, so the}$$

amplitude differs from the true heat capacity by about three percent. The dominant source of error for this example is the electronic coupling term in the numerator. The lag term, a $R_S \dot{T}$, may become significant for poor thermal conductors. R_S is calculated to be $\sim$ 1200 s·K/cal from the slopes of melting peaks for indium sandwiched between 5 mil polystyrene sheets. Therefore, when ratios of amplitudes of DSC traces are used to determine the specific heat of a poor thermal conductor, the value may be several percent in error if it is obtained from comparison with a reference material of high thermal conductivity.

This case emphasizes a fact that follows from the law of energy conservation. There is not only a transient area lag, recoverable at the cessation of the scan, for cases involving instantaneous changes in heat capacity or rate of temperature change, but there exists also a constant, steady state lag term for scans in which these variables change continuously. Therefore, for all real systems whose heat capacities are temperature dependent, values for the heat capacity determined from the amplitude of the scan are modified by coupling terms similar to those discussed in this section.

Appendix B: Temperature Dependent Rate of Temperature Change: The programmed temperature, T_p, of some DSC instruments increases linearly, e. g.,

$$T_p = T_o + \dot{T}_p t \qquad (1B)$$

$$(\dot{T}_p = \text{constant}).$$

If T_p is coupled linearly with the output of a platinum resistance thermometer, the temperature at the surface of the calorimeter is related to T_p by a quadratic,

$$T_C = p + q T_p + r T_p^2 . \qquad (2B)$$

Hence the rate of temperature change at the calorimeter surface, $\dot{T}_C$, changes linearly with time, viz.,

$$\dot{T}_C = \alpha + \beta t$$

where

$$\alpha = \dot{T}_p (q + 2 r T_o) \qquad (3B)$$

and

$$\beta = 2 r \dot{T}_p^2$$

$$(\text{at } t = 0, \ T_p = T_C = T_S = T_R = T_o).$$

One obtains, ignoring transient terms, for the steady state solution, i. e., for $t \gg \tau_S$, τ_R, τ_x, from eqs. 1 – 4, 3B and 6 – 7

$$T_R = T_C - \dot{T}_C \tau_R + \ddot{T}_C \tau_R^2 \qquad (4B)$$

$$T_S = T_C - \dot{T}_C \tau_S + \ddot{T}_C \tau_S^2 \qquad (5B)$$

$$\Delta q = \dot{T}_C (C_S - C_R) - \ddot{T}_C (C_S \tau_S - C_R \tau_R) \qquad (6B)$$

and

$$X = \dot{T}_C(C_S - C_R) - \ddot{T}_C[C_S(\tau_x + \tau_S) - C_R(\tau_x + \tau_R)] \tag{7B}$$

where $\ddot{T}_C = d^2T_C/dt^2 = \beta$, the temperature acceleration at the calorimeter surface.

Eqs. 6B and 7B may be written in terms of measurable variables: the programmed temperature, T_p, its constant rate of change, $\dot{T}_p$, and the temperature calibration coefficients, p, q and r. Then,

$$\Delta q = \dot{T}_p \, C_S[q + 2 \, r \, (\dot{T}_p - \ddot{T}_p \tau_S)] - \dot{T}_p C_R[q + 2 \, r(\dot{T}_p - \ddot{T}_p \tau_R)] \tag{8B}$$

$$X = \dot{T}_p C_S[q + 2 \, r \, \dot{T}_p - 2 \, r \, \ddot{T}_p(\tau_S + \tau_x)] - \dot{T}_p C_R \tag{9B}$$

$$[q + 2 \, r \, \dot{T}_p - 2 \, r \, \ddot{T}_p \, (\tau_R + \tau_x)]$$

The first term in the brackets, $q + 2 \, r \, \dot{T}_p = \dot{T}_C / \dot{T}_p$, is the calibration correction for $\dot{T}_p$. The temperature calibration curve for a specific DSC instrument yielded q = .573 and r = 4.23 x 10^{-4} K^{-1}. Thus, $\dot{T}_C / \dot{T}_p$ varied from .83 at 300 K to 1.17 at 700 K. It is essential to make this calibration correction on such instruments when the absolute slope or amplitude is to be determined.

The second term in the brackets of eq. 9B couples the temperature acceleration with the thermal and electronic lags. This term for the above example, even at a quite rapid rate of temperature change of 1 K/s, is less than 1% of $q + 2 \, r \, T_p$.

REFERENCES

[1]   A. P. Gray, "Analytical Calorimetry", R. S. Porter and
J. F. Johnson, Eds., p 209, Plenum Press, New York (1968).

[2]   J. H. Flynn, "An Analytical Evaluation of Differential
Scanning Calorimetry (DSC)","Status of Thermal Analysis",
O. Menis, Ed., NBS Special Publ. #338, p 119 (1970).

[3]   W. P. Brennan, "Theory and Practice of Thermoanalytical
Calorimetry", Ph.D. Thesis, Princeton U. (1970).

[4]   J. H. Flynn, "Thermal Analysis (3 ICTA)" Vol. 1, p 127,
H. G. Wiedemann, Ed.Birkhauser Verlag, Basel und Stuttgart (1972).

[5]   G. Adam and F. H. Muller, Kolloid-Z. Z. Polym. 192, 29
(1963).

[6]   R. N. Goldberg and E. J. Prosen, Thermochimica Acta, 6, 1
(1973).

[7]   C. M. Guttman, Personal Communication. (For a description
of the method see ref. [8].

[8]   J. H. Flynn, "Thermodynamic Properties from Differential
Scanning Calorimetry by Calorimetric Methods", Thermochimica
Acta, 6, (1973), in press.

[9]   A. P. Gray, Thermal Analysis Application Study No. 3, The
Perkin-Elmer Corp., Norwalk, Conn., (1973) (Order No. TAAS-3).

[10]   C. M. Guttman and J. H. Flynn, Anal. Chem. 45, 408 (1973).

[11]   Anon.,Thermal Analysis News Letter No. 5, The Perkin-Elmer
Corp., Norwalk, Conn., (Order No. TAN-5).

ON THE TEMPERATURE RESOLUTION OF THERMISTORS

Peter W. Carr and Larry D. Bowers

Department of Chemistry

University of Georgia, Athens, Georgia 30602

INTRODUCTION

The past fifteen years have witnessed the development of reaction calorimetry as an analytical methodology. The applicability of such techniques as thermometric enthalpy titration (TET) (1) and direct injection enthalpimetry (DIE) (2) to clinical and biochemical problems has greatly enhanced this development. All of the above mentioned techniques most commonly utilize a thermistor as the temperature sensing element. It is well known that thermistors are among the most sensitive and simple temperature transducers available. The detection limit of these devices is controversial. Recently Lampugnani and Meites (3) have reported an 0.6 μ° C limit, while Smith, Barnes, and Carr (4) and Dohner, Wachter and Simon (5) have reported a value of 3.5 μ° C. Considerably higher values have also been reported (6). Thus an investigation was initiated into the effect of various physical factors which might influence thermistor sensitivity.

Although the precise measurement of noise is difficult, essentially noise free electronics with a well defined band width were used to qualitatively assess the effect of thermistor resistance, mode and rate of stirring, and the boundary between the thermistor and the solution as well as the importance of various noise sources. Since noise is a difficult quantity to measure, the procedures necessary for consistent reporting of limits of detection were investigated. Parameters evaluated were total sampling time, the sampling rate, and the sampling interval.

EXPERIMENTAL

The apparatus used for the noise studies was an electronically shielded and thermally insulated non-differential DC Wheatstone bridge. Since the stirring process and the self-heating of the thermistor add heat to the solution, the circuitry shown in Figure 1 was devised to allow the noise to be observed on top of a steadily changing baseline. This was accomplished by amplifying the signal from the Wheatstone bridge with a Keithley Model 140 Precision Nanovolt DC Amplifier, thus enhancing the noise and rendering the noise contribution from subsequent electronics negligible. Since it was found that the slope due to heat sources in the dewar biased the observed noise, a ramp generator was placed in the circuit to eliminate this effect. The single pole low pass filter (RC = 1 sec) was utilized to minimize chopper noise from the electronics and more importantly to define the system band width. The operational amplifiers were mounted in an MPI Console Model MP-1001. A Leeds and Northrup Speedomax XL (100 microvolt full scale) recorder was used to monitor the output of the circuitry. The bridge voltage was measured under load by a Hewlett Packard Model 3440A Digital Voltmeter.

Thermistors used were of the bead in glass probe type, Models 41A1 (10KΩ) and 51A1 (100 KΩ) (Victory Engineering Corporation, Springfield, New Jersey). The thermistors were rigidly mounted in a glass tube and connected to the bridge via miniature coaxial cable. Several thermistors were mounted in sealed glass tubes utilizing either air or a mercury pool as a buffer between the thermistor and the glass-solution interface.

The adiabatic calorimeter used in these studies was similar to that designed by Christensen et. al. (7). The experiments were conducted in a large water bath which was controlled to 0.001°C by a proportional controller.

THEORY

Noise will be defined here as any apparently random fluctuation of the voltage developed by the Wheatstone bridge shown in Figure 1. For a DC signal, this variation can be best represented as the root-mean-square (rms) value of the signal:

$$N_{rms} = \sqrt{\frac{S_i^2}{n}} \qquad (1)$$

where S_i is the deviation of the signal from the calculated least squares line. It can be readily seen that the rms noise is computed in a manner similar to the

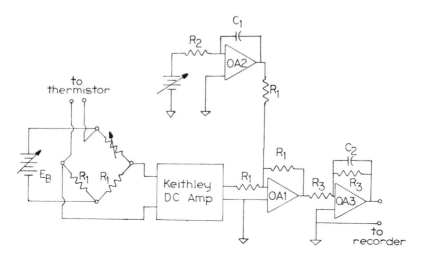

Figure 1. Schematic Diagram of Noise Measurement System
$R_1 = 100 \text{ K}\Omega$, $R_2 = 3.3 \text{ M}\Omega$, $R_3 = 1\text{M}\Omega$, $C_1 = 1 \text{ µf}$, $C_2 = 10 \text{ µf}$

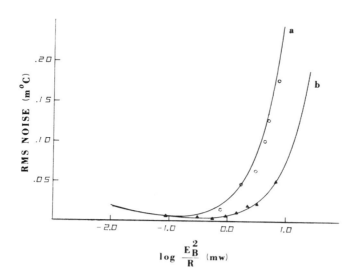

Figure 2. Plot of RMS Noise vs. Power Applied to the Thermistor.
Curve a, 10 K Ω thermistor, Curve b, 100 K Ω thermistor

standard deviation of a series of samples. If the variation in the signal is truly random, any instantaneous fluctuation will be no greater than 2.5 times the standard deviation at the 99% confidence interval. Hence, the rms noise can be estimated by taking one fifth of the peak-to-peak noise. A third method of calculating the noise is given in Equation 2:

$$N_{ave\ p-p} = \sqrt{\frac{\sum_{i=1}^{n} S_i}{n}} \tag{2}$$

This average peak-to-peak value determines the average excursion from the least square representation of the signal.

Effects such as circuit component warm up, drift in the applied bridge voltage (E_B), and variations in the rate of stirring will appear as very low frequency noise or drift. To some extent, slow or long term variation in the recorded signal may be negated by fitting the data to a least squares straight line. Very long term fluctuations (drift) of thermistors have been extensively investigated (8). The factors which follow may be identified as components of the short term fluctuations. The phenomena have been grouped in terms of their dependence upon E_B which has a very pronounced effect on the short term temperature resolution.

Voltage Independent Noise Sources

1. Pick up noise (e_{PN}) from 60 Hz lines, stray capacitance, inadequate shielding. In general the larger the source impedance is, the greater the pick up noise.

2. Johnson noise (e_{JN}). This is defined by the well known equation (9) given below, and some typical values for source impedances (R) and band widths (ΔF) encountered in titration calorimetry are summarized in Table I. This is the minimum voltage noise which can be observed; it can not be reduced by better shielding.

$$e_{JN} = \sqrt{4kTR\Delta F} \tag{3}$$

3. Flicker or 1/f noise. Semiconductors such as transistors and thermistors show a noise component whose magnitude increases inversely with frequency (9). The source of this noise is not well understood but is believed to be related to the presence of discontinuities in the conduction band and are therefore not seen in wire wound resistors.

Table I

Effect of Bond Width and Impedance on Johnson Noise

Johnson Noise Voltage (microvolts, rms)[a]	Bandwidth (Hz)	Impedance (Ω)
0.0129	1	10^4
0.0407	10	10^4
0.0407	1	10^5
0.129	10	10^5
0.129	1	10^6
0.407	10	10^6

[a]Computed from equation 1 with $T = 300°K$. $k = 1.38 \cdot 10^{-23}$ joule/$°K$.

Noise Sources Dependent Upon the First Power of the Applied Voltage

A thermistor placed in a thermally inhomogeneous fluid will show irregular temperature changes as the fluid moves by the sensor. In titration calorimetry there are a number of heat sources (and sinks) including sitrring, chemical reaction and poor adiabaticity which may act as essential, constant sources of heat. In the absence of instantaneous mixing these heat fluxes will create a thermally inhomogeneous solution and thereby act as a noise source. The temperature equivalent noise due to inhomogeneity is termed ΔT_{IN} and it will produce the apparently random voltage e_{IN} given below:

$$e_{IN} = \tfrac{1}{4} \alpha E_B \Delta T_{IN} \qquad (4)$$

where α is the temperature coefficient of the thermistor (4%/$°C$) and the numerical factor derives from an analysis of an equal arm bridge (10). The magnitude of ΔT_{IN} is set by the size of the heat fluxes, the mode of stirring, and by the heat capacity of the Dewar.

Noise Sources Dependent upon the Second Power of the Applied Voltage

1. The thermistor itself must be viewed as a heat source since most of the electrical power delivered to it is transferred to the ambient fluid whose temperature is being monitored. To a first approximation the contribution this source makes to ΔT_{IN} will be proportional to the applied power (and therefore to E_B^2).

2. The internal temperature of a thermistor T_T is invariably higher than that of the surrounding fluid. The steady state self-heating is given below (11)

$$T_{sh} = \frac{P}{\delta}$$

where P is the applied power and δ is the thermistor's dissipation constant. It is known that δ depends upon the rate and mode of stirring (12). Any fluctuation in δ will appear as noise at the bridge output.

Both of these noise sources generate an apparent temperature fluctuation (ΔT_τ) which is dependent upon E_B^2. Therefore the equivalent voltage fluctuation (e_τ) will be:

$$e_{\tau N} = \frac{1}{4} \alpha \, \Delta T_\tau \, E_B = \text{const} \cdot E_B^3 \tag{6}$$

There are certainly other conceivable noise sources, e.g. fluctuating thermal emf's, changes in the bridge components and in E_B. The above discussion was presented to justify the form of the function which is used to describe the dependence of temperature noise on the applied bridge potential. The temperature equivalent (ΔT_N) to the total voltage noise (e_N) may be calculated as:

$$\Delta T_N = \frac{4 \, e_N}{\alpha \, E_B} \tag{7}$$

The assumption that all of the noise sources are independent predicates that they must add as statistical variances (13) thus,

$$\Delta T_N = \sqrt{\frac{A}{E_B^2} + B + C \, E_B^4} \tag{8}$$

where A is a coefficient related to the voltage independent noise, B is related to noise sources which are proportional to the applied voltage and C is due to the power dependent noise sources. The mathematical form of equation 8 is shown in Figure 2; it is quite evident that an optimum voltage and minimum temperature resolution should exist. These are easily obtained from equation 8.

$$E_B^{opt} = \left(\frac{A}{2C} \right)^{\frac{1}{6}} \tag{9}$$

$$\Delta T_N^{min} = \sqrt{B + 3 \left(\frac{A}{2} \right)^{\frac{2}{3}} C^{\frac{1}{3}}} \tag{10}$$

RESULTS AND DISCUSSION

Before beginning a detailed study of the effect of bridge voltage it was essential to determine whether the data acquisition and data treatment procedures might give biased representations of the real system noise. In preliminary work we did indeed note that rather low estimates of noise were obtained when measured on a steeply inclined baseline. Many of the previous estimates were obtained under just these conditions (3,6) and may therefore be overly optimistic estimates of the temperature resolution. For this reason the data was obtained with the system shown in Figure 1.

Table II
Comparison of Various Noise Reporting Techniques[a]

Voltage	RMS (μ v)	Mean Peak-to-Peak (μ v)	Peak-to-Peak (μ v)
3.0	0.40	-----	1.9
4.5	0.90	0.72	3.4[b]
6.0	2.6	2.0	9.3[b]
7.5	3.6	2.9	9.1[b]
9.0	4.9	3.8	25

[a]For thermistor in sealed glass tube, mercury buffer.

[b]The peak-to-peak noise was measured over a 75 sec. interval for these
values, which may explain the lower P-P/RMS ratio (see 16).

In order to establish how to best report the noise, the data of Table II
were obtained. This table compares the rms noise (vide supra), the average
peak-to-peak noise and the peak-to-peak variations defined by drawing an
envelope closing ∽ 99% of the data over on interval of ∽75 sec. The data
of Table II indicates that the rms and average peak-to-peak noise are very
similar, but as expected the peak-to-peak variations are considerably
larger. Although it is computationally simpler to measure the peak-to-peak
noise it is much less reproducible then either the rms or mean peak-to-peak
noise. Since rms noise is statistically better defined all subsequent noise
estimates were obtained via equation 1.

The Nyquist (14) criteria essentially states that a signal should be sampled at a rate of at least twice the highest frequency component of the signal. We felt that sampling over long time intervals might produce some
averaging of the signal; therefore the effect of both the sample duration or
interval and sampling rate were studied.   The data are summarized in Table
III. Although it might appear that the noise decreases with a decrease in
sampling rate and varies with sample interval, F tests at the 90% confidence
level indicate that sampling rate has no effect and only one pair of the
sampling interval data were different. Despite these statistical considerations
we adopted a standard interval of 50 sec and rate of one data point per three
sec. This is admittedly a compromise between the dictates of the Nyquist
criteria and our ability to obtain and posses large volumes of data.

Table III
Effect of Sampling Interval and Rate on Measured Noise

ΔT rms ($\mu°C$)[a]	Sample Rate[b] (Hz)
15	2.1
15	1.1
14	0.54
13	0.27

Table III (cont.)

T rms (μ°C)[c]	Sample Interval (sec)
11	4.7
14	9.4
16	18.8
17	28.2
16	37.5
15	46.9

[a]Total time of sampling (interval)   50 sec.

[b]Defined as the inverse of the time between consecutive data points.

[c]Sampling rate 2.13 Hz.

The results of our measurement of the temperature resolution of both a 10 KΩ and 100 KΩ thermistor as calculated from equation 7 as shown in Figure 2. The continuous curve is an empirical one based upon only the measurement of the noise with no bridge voltage, i.e. $E_B = 0$ and a measurement of the noise at very high applied power (9 and 27 volts for the 10 KΩ and 100 KΩ thermistors respectively). We assumed that the B term of equa-

Table IV
Typical Noise Parameters[a]

Device	Impedance (kΩ)	$e_{n,o}$ (μv, rms)	A[c]	C[d]
wire wound resistor	10	0.047	22	0.008
thermistor	10[e]	0.059	35	7.7
	10[f]	0.058	34	0.084
	10[g]	0.054	35	0.12
	100[e]	0.19	380	0.0057

[a]All data was obtained at a sampling rate of 1.04 Hz over an interval of 75 sec.

[b]The measured noise with a dead short in place of the bridge driving voltage (E_B).

[c]See equation 8. Calculated from the noise at $E_B = 0$. Units are $\mu°C^2 \cdot volt^2$.

[d]See equation 8. Calculated for the noise at power greater than 8 milliwatts. Units are: $\mu°C^2 \cdot volt^{-4}$.

[e]Thermistor in direct contact with water.

[f]Thermistor encased in air.

[g]Thermistor encased in Hg.

tion 8 was zero. It is clear that a function of the form of equation 8 does indeed fit the data in the region of the minima and that at least under our experimental conditions thermal inhomogeneity is not a significant noise source. The abscissa of Figure 2 is proportional to the applied power it is clear that a 100 KΩ thermistor has a much smaller C term, i.e. noise dependent upon the power dissipated, than a 10 KΩ thermistor. This is also noted in Table IV.

It is interesting to note that even though a 100 KΩ thermistor has a very much reduced C term there is no very significant improvement in ΔT_{min}. This is a direct consequence of two factors. A 100 KΩ thermistor (in an equal arm bridge) picks up more noise than a 10 KΩ thermistor; this is reflected in both the third and fourth columns of Table III. In addition and most importantly ΔT_{min} depends upon the sixth root of the C term. If the B term is dropped from equation 10 it may be rearranged to give:

$$\Delta T_{min} = 2.2 \, A^{\frac{1}{3}} C^{\frac{1}{6}} \tag{11}$$

Equation 11 was used only with the data of Table III to generate estimates of the minimum temperature resolution and the optimum applied voltage. It is evident that due to the functional form of this equation the noise is most easily improved by decreasing the pick up noise to below the Johnson noise. Under our present conditions the pick up noise for both the 10 and 100 KΩ thermistor is 4-5 times as large as the Johnson noise (see Table I; 1 Hz bond width). The last two entries in Table V indicate the ultimate temperature resolution of a thermistor i.e. when the zero voltage noise is reduced to the Johnson noise level.

Table V
Minimum Temperature Resolution[a] for Various Thermistors

Device	T_{min} ($\mu°C$)	E_B^{opt} (volt)
10 K thermistor[b]	10	1.28
100 K thermistor[b]	6.6	6.30
10 K thermistor[c]	4.6	2.72
10 K thermistor[d]	4.9	2.57
10 K thermistor[e]	3.6	0.77
100 K thermistor[e]	2.3	3.8

[a] Calculated from the data of Table IV and equations 9 and 11.

[b] Immersed directly in water.

[c] Air enclosed.

[d] Mercury enclosed.

[e] Calculated using the Johnson noise of Table I with 1 Hz for band width.

The dependence of the C term on the dissipative characteristics of the thermistor was assessed by use of thermistors encased in air and in mercury. This increased the thermal resistance and heat capacity of the device respectively. The data of Figure 3 show that both these changes have a very marked effect on the C term but only slightly improve the temperature resolution. No significant effect on pick up noise was noted.

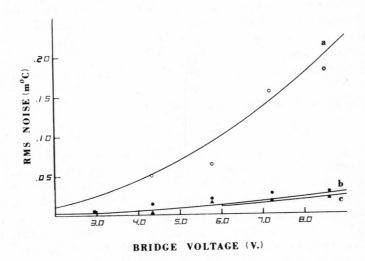

Figure 3. Plot of RMS Noise vs. Applied Bridge Voltage Curve a, 10 KΩ thermistor in water; Curve b, 10 KΩ thermistor in mercury; Curve c, 10 KΩ thermistor in air.

It would seem that estimates of the noise below 1 μ°C are in error since it would take a vast improvement in the C term or a decrease in the A term to below the Johnson limit. Furthermore other noise sources, such as thermal inhomogeneity, which are not evident at the temperature resolution obtained in this work may come into play. Recent work of Hepler (15) indicates that the use of positive temperature coefficient thermistors, with temperature coefficients as large as 40%/°C, may improve the temperature resolution since they should be subject to much the same noise sources as encountered in the present work.

In view of the difficulties in precisely measuring the variances of a signal we believe that equation 8 provides a sufficiently simple, realistic representation of the dependence of the noise on the applied voltage to provide an operational method for choosing bridge voltage. The optimum voltage may be calculated via equation 9 once an estimate of A and C are obtained from a measurement of the noise at zero and very high bridge power respectively.

ACKNOWLEDGMENTS

The studies described here were supported by the National Institutes of Health Grant GM 17913.

REFERENCES

1. H. J. V. Tyrrell and A. E. Beezer, Thermometric Titrimetry, Chapman and Hall, London (1968).
2. J. C. Wasilewski, P. T. S. Pei, and J. Jordan, Anal. Chem., 36, 2131 (1964).
3. L. Lampugnani and L. Meites, Thermochem. Acta, 5, 351 (1972).
4. R. E. Dohner, A. H. Wachter, and N. Simon, Helv. Chim. Acta, 50, 2193 (1967).
5. E. B. Smith, C. S. Barnes and P. W. Carr, Anal. Chem., 44, 1663 (1972).
6. T. Meites, L. Meites and J. N. Jaitley, J. Phys. Chem., 73, 3801 (1969).
7. J. J. Christensen, I. M. Izatt and L. D. Hansen, Rev. Sci. Instrum., 36, 486 (1968).
8. E. C. Robertson, R. Raspet, J. H. Swartz and M. E. Lillard, Geological Survey Bulletin 1203-B, U.S. Government Printing Office, Washington, D.C.
9. E. J. Bair, Introduction to Chemical Instrumentation, McGraw-Hill, New York, (1962) p. 255.
10. P. W. Carr, Crit. Rev. Anal. Chem., 5, 519 (1972).
11. F. J. Hyde, Thermistors, Illiffe Books, London (1971), pp. 115-141.
12. R. A. Rasmusson, Rev. Sci. Instrum., 32, 38 (1962).
13. Bair, op. cit., p. 251.
14. Ibid, pp. 254-6.
15. R. J. Reilley and L. G. Hepler, J. Chem. Ed., 49, 514 (1972).
16. V. D. Landon, Proc. IRE, 50 (February, 1941).

THE DEVELOPMENT AND APPLICATION OF AN ULTRA-SENSITIVE
QUANTITATIVE EFFLUENT GAS ANALYSIS TECHNIQUE

P. A. BARNES AND E. KIRTON

DEPARTMENT OF CHEMISTRY, LEEDS POLYTECHNIC

CALVERLEY STREET, LEEDS LS1 3HE, YORKSHIRE, ENGLAND

SUMMARY

A technique has been developed for both quantitative and qualitative thermal analysis using a katharometer to measure evolved gases quantitatively and a differential freezing technique to aid in the identification of the gases produced. The technique results in a great increase in sensitivity when compared with DTA and conventional EGA techniques using a hot wire detector, and offers the advantages of simplicity and low cost. To offset this there are certain limitations but nevertheless the method is applicable to many problems in quantitative differential thermal analysis. The procedure enables DTA equipment to be used conventionally or with simultaneous monitoring of either the total gas evolved, or a specific gas. In the latter case, with minor modifications to the technique, an increase in sensitivity of $> 1000X$ can be obtained, making the detection and measurement of trace quantities of impurity possible. The method has been applied successfully to the identification and determination of carbonate impurity in silver (II) oxide, at levels of well below 1% CO_2.

INTRODUCTION

Differential thermal analysis is an invaluable technique for both the quantitative and qualitative analysis of materials. However, the former application is subject to a number of difficulties due mainly to the irreproducibility of instrumental and sample parameters. To a large extent the former problems have been overcome by improved design [1] and by the advent of DSC.

Given good instrumentation, one of the limiting factors in DTA is the method of "information retrieval" from the reacting sample. The ΔT sensor can detect changes occurring in the sample only via a number of complex temperature gradients. The variations thus encountered can be reduced, but not eliminated, by careful sample preparation.

Further, best results are obtained at low heating rates when, unfortunately, the sensitivity of conventional DTA is reduced. Attempts to overcome this by increased amplification of the ΔT signal are often frustrated by an unacceptable rise in noise.

The effects of these problems can be minimised if, instead of relying on the enthalpy change of the process under study to provide the signal, with the accompanying difficulties in changing heat capacity, area of contact with the sample pan etc., the change in some other property is measured. If the property chosen is such that the information yielded can be 'stored' in some way until the reaction is complete, it can then be examined under optimum conditions, rather than those dictated by the requirements of DTA. Consequently there should be a gain in sensitivity and precision. The measurement of evolved gases provides such a method when used with an evolved-gas storage facility. An early method [2] of detecting evolved material, which also acts as a material store, relies on the absorption of gas by a suitable solution, the amount of gas evolved being found by titration. The method has obvious drawbacks, although the sensitivity could be improved using modern instrumental titrimetric methods.

The present work makes use of a differential freezing technique, the coolant being chosen to specifically remove the evolved component of interest from the purge or carrier gas. When the evolution of the gas is complete the coolant is removed, the frozen material rapidly vaporises and is then swept into the gas detector. This presents a greatly increased concentration of the evolved gas to the detector, with a consequent increase in sensitivity.

It is clear that for the approach to be successful the gas monitored must be evolved quantitatively, a fact which limits quantitative applications of the method. However, with suitable materials the inherent sensitivity of the technique and the ease with which results can be obtained are distinct advantages.

One of the problems of conventional EGA, which is largely used for qualitative work, is that diffusion effects often cannot be neglected. However, in the present case, this does not apply as the total amount of a specific component is required and this does not depend on the rate of evolution of the gas.

Of the many types of commonly used gas detector system a microkatharometer was selected as it offered high sensitivity together with ease of use and low cost. Such a hot-wire detector responds to all gases. However, an important advantage of the differential freezing technique is that it goes some way towards meeting the requirements for a specific detector. Nevertheless it is recognised that in a number of cases, where freezing or other forms of trapping, e.g. adsorption are not specific, then mass spectrometry or infrared spectroscopy would be invaluable as means of detection and analysis.

The development of the method outlined above was in response to a problem of thermal analysis which did not yield to conventional techniques. Silver (II) oxide is a material of both commercial importance [3] and academic interest [4]. It is difficult to prepare in a pure form due to the absorption of atmospheric carbon dioxide to give silver carbonate. Some interest in the analysis of this oxide by thermoanalytical techniques has been shown previously [5, 6] but the published results revealed difficulties due to the overlap of the silver (II) oxide and silver carbonate decompositions.

The reactions found are

$$2 \text{ Ag O} \longrightarrow \text{Ag}_2\text{O} + \tfrac{1}{2}\text{O}_2 \quad \dots \quad \dots \quad \dots \quad (1)$$
$$\text{Ag}_2\text{O} \longrightarrow 2 \text{ Ag} + \tfrac{1}{2}\text{O}_2 \quad \dots \quad \dots \quad \dots \quad (2)$$
$$\text{and } \text{Ag}_2\text{CO}_3 \longrightarrow \text{Ag}_2\text{O} + \text{CO}_2 \quad \dots \quad \dots \quad \dots \quad (3)$$

Previous work [7] suggested that analysis for silver carbonate impurity in silver (II) oxide by DTA could only be achieved at high concentrations ($> 5\%$) for the reason given above.

It has been shown theoretically that whilst the silver (II) oxide decomposition was unaffected by pressure [8] that of silver carbonate, being essentially a reversible process, was shifted to lower temperatures in vacuum and to higher temperatures in an elevated pressure of carbon dioxide [9]. However, in addition to decomposition endotherms, silver carbonate exhibits two pressure independent endothermic phase changes in the region of silver (II) oxide decomposition exotherm which make quantitative analysis in a CO_2 atmosphere by DTA difficult. Analysis by TG in a carbon dioxide atmosphere is a possibility as the phase change endotherms would not affect the weight loss but it would be complicated by the high pressure required. Further, the silver (II) oxide continues to decompose over a wide temperature range [5] and so accurate results would be difficult to achieve. A further complication is that the silver carbonate formed by adsorption by the oxide would most probably be of the "reactive" type [10]

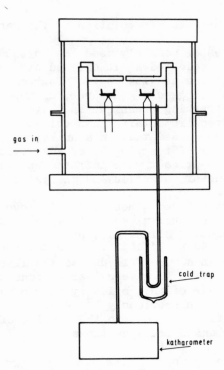

gas in

cold trap

katharometer

Fig.1.   Schematic diagram of the DTA cell of a Stanton Redcroft
          671B instrument, modified to give a small swept volume.
          Also shown are the cold trap and katharometer.

which decomposes as much as 100K below the normal material.

    In addition, as both silver carbonate and silver (II) oxide
produce the same decomposition product, silver (I) oxide, neither
can be analysed in the presence of the other by the breakdown of
that product.   Under normal conditions silver (II) oxide absorbs
relatively small amounts of carbon dioxide [6], and hence its
accurate analysis by DTA is very difficult.   Therefore, it was
felt justified to attempt to extend the range of the determination
of carbon dioxide in silver (II) oxide to much lower limits using
the evolved gas analysis technique outlined above.

                            EXPERIMENTAL

    Silver (II) oxide and silver carbonate were obtained from
B.D.H. Ltd. and were used without further purification.   A Stanton
Redcroft 671B DTA apparatus was used for the DTA and EGA work after

modification as shown in Fig. 1.   In order to keep the swept
volume between the sample chamber and the detector to a minimum,
the carrier gas entered the instrument via the vacuum port.   A
small capillary tube, normally used as a purge-gas entry point
into the sample chamber, was used to remove the gas.   The flow
meters were by-passed and the exit tube continued, via a needle
valve, and 1/8" o.d. copper tube, to the "U" shaped cold trap
and finally to the gas detector.   By reversing the flow-path in
this way the swept volume, including that of the sample chamber
was kept below 10 cm^3.

The 671B was used with open aluminium dishes to contain the
sample and reference materials and this, together with the small
sample size (1-20mg) and small dead space, reduced both diffusion
effects and the time lag in DTA-EGA experiments between the DTA
peak and katharometer response to negligible proportions.   The
DTA-EGD and DTA-EGA results were displayed on a dual channel
Philips 8010 recorder.   The microkatharometer, type DK223 and
its control unit, type GC197, were supplied by Taylor Servomex Ltd.

The freezing points of the gases to be separated by the trap
determine the choice of coolant.   It must provide a temperature
such that the vapour pressure of the component to be stored is
negligible, while that of other gases evolved and the carrier gas
should be sufficiently high for them to pass through unhindered.
In the present work it was found that at liquid nitrogen temperature
(77K), carbon dioxide (vapour pressure < 1 x 10^{-7} torr) was
quantitatively removed at the flow rate used, while oxygen (vapour
pressure ca. 100 torr) was not trapped.   The carrier gas, either
helium or nitrogen, was not affected by the trap.   The solid
carbon dioxide produced was found to vaporise rapidly as the trap
was warmed up.

The heating rate was standardised throughout at 20K per
minute and the carrier gas flow rate used was 35 cm^3 per minute.
This was found to be a suitable compromise between the opposing
requirements of a minimum time lag between the DTA and katharometer
response for which a high flow rate is needed and those of a stable
DTA trace and suitable katharometer sensitivity for which the
optimum flow rate is low.

For the calibration experiments, using nitrogen as the carrier
gas, the voltage applied to the katharometer filaments was 2.5V
and the attenuation set at 4X.

RESULTS

In order to establish the temperature range of the
decompositions under study, simultaneous DTA-EGD experiments were

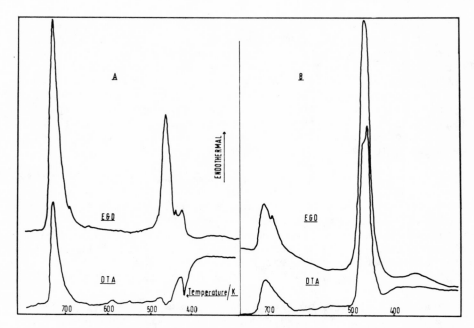

Fig.2.   DTA and EGD traces for (A) AgO and (B) Ag$_2$CO$_3$.
Heating rate, 20K min^{-1};   carrier gas, helium.

carried out on silver (II) oxide and silver carbonate (Fig.2).
These show that the main thermal events recorded on the DTA trace
are accompanied by evolution of gas.   In further experiments
using simultaneous DTA-EGA the gases evolved were identified by
applying liquid nitrogen to the cold trap to freeze out carbon
dioxide so that only oxygen was registered by the katharometer
(Fig.3).   As expected, the EGA curve, B, for silver carbonate
shows only one peak, due to the decomposition of silver (I) oxide
to silver, the carbon dioxide peak being suppressed.   The silver
(II) oxide curve, A, is largely unchanged as oxygen is evolved in
each of the two decomposition steps.   The difference between the

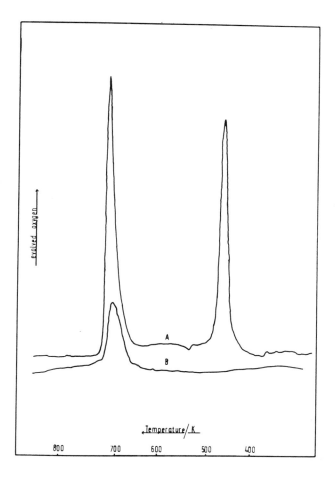

Fig. 3. EGA curves for (A) AgO and (B) Ag$_2$CO$_3$. Coolant, liquid nitrogen. The evolved gas detected is oxygen. Heating rate, 20K min^{-1}; carrier gas helium. The oxygen peak of curve (B) is apparently reduced in size due to greater attenuation of the signal.

EGA and the EGD traces for the silver (II) oxide should reveal the presence of carbon dioxide from silver carbonate impurity. However, as the amount of the impurity is low (< 3%) it is not easy to observe any significant difference and hence the level of silver carbonate present in this sample cannot be readily determined using conventional DTA-EGA.

The next stage was to investigate the use of the cold trap in storing evolved carbon dioxide. In Fig. 4a, the results of a typical experiment on silver (II) oxide are shown. During this

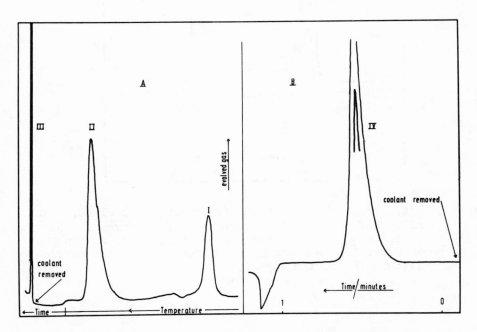

Fig. 4.   EGA curves for AgO containing Ag_2CO_3 impurity showing (A) response to O_2 (peaks I and II) during freezing, and enhanced response to CO_2 (peak III) on removal of the coolant. Carrier gas, helium.
(B) a typical curve obtained in the analysis of AgO for Ag_2CO_3 showing the CO_2 response (peak IV). Carrier gas, nitrogen. Weight of CO_2 giving rise to peak IV was $2x10^{-4}$g.

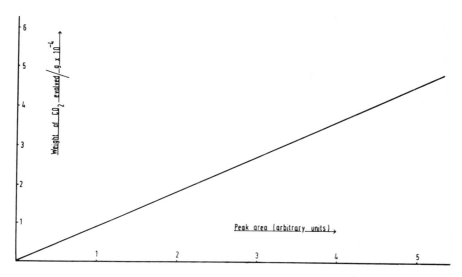

Fig. 5. Calibration curve of CO_2 peak area/weight of CO_2 evolved. Flow rate 35 cm^3 of nitrogen per minute.

DTA-EGA experiment the two oxide decompositions are clearly shown by the oxygen evolution peaks I and II, the evolved gases being passed through the cold trap to remove any carbon dioxide impurity. After complete decomposition of the material to metallic silver (ca. 775K), the coolant was removed and the carbon dioxide allowed to vaporise. The resulting microkatharometer response, peak III, went off scale and subsequent work showed that even full attenuation of the katharometer control unit (a factor of 2,500x) could not prevent this. The problem was solved by changing the carrier gas from helium to nitrogen which reduced the sensitivity of the detector sufficiently.

The results of a further experiment on silver (II) oxide in which the cold trap was used again to store carbon dioxide impurity are shown in Fig. 4b. In this experiment the carrier gas was nitrogen.

After complete decomposition of the solid the trap was warmed, the frozen carbon dioxide allowed to vapourise and the resulting katharometer response recorded as shown. (Peak IV). The excellent stability of the baseline is seen clearly. This particular curve was obtained from a sample containing $2x10^{-4}$ g of carbon dioxide. Greater sensitivity was obtained by decreasing the attenuation and increasing the voltage applied to the filaments. In this way usable peaks could be obtained from $2x10^{-6}$ g of CO_2.

Having established that carbon dioxide could be detected with

sufficient sensitivity, it remained to show the method to be
quantitative.   The total amount of carbon dioxide evolved should
be proportional to the area under the peak.   As the evolution was
rapid, a chart speed of 100 mm per minute was required to obtain
a reasonable peak width.   A calibration graph was constructed by
plotting the peak areas due to the evolution of carbon dioxide
against the weight of the gas evolved from accurately known
quantities of silver carbonate.   A straight line plot was
obtained (Fig.5).

As no value for the carbon dioxide content of silver (II)
oxide was available, samples of the "pure" material were subjected
to the EGA method outlined above and a value of 0.45% carbon
dioxide was obtained.   To check the method a mixture of silver
carbonate and silver (II) oxide was prepared and analysed using
the same technique.   The result, allowing for the minor correction
due to the carbon dioxide already present in the oxide, gave the
percentage of carbon dioxide as 8.3 (1), the expected value being
8.4 (0).

Thus the technique proved to be capable of extending the
range of the thermoanalytical determination of silver carbonate
impurity from the 5% level reported previously to below a level of
0.1% in a sample of 10 mg.   As this was more than adequate for the
analysis of the silver (II) oxide in use no detailed work was
undertaken using helium as the carrier gas.   However, preliminary
results indicate that a level of 1×10^{-8} g of carbon dioxide could
be readily measured by taking advantage of the higher sensitivity
obtainable with this purge gas.

The investigation was limited to the study of carbon dioxide
evolution.   The sensitivity of the method for other gases would
depend on their thermal conductivity.   Lower limits of detection
are found for materials of less thermal conductivity and vice versa.

In quantitative work the carrier gas flow rate must be
controlled as the katharometer is a flow-sensitive instrument.
The rise in temperature of the gas in the DTA sample chamber causes
a change in viscosity and hence flow rate and this is superimposed
on random fluctuations.   The problem can be overcome by the use
of suitable mass-flow control units provided the added swept volume
is sufficiently low.   A limiting factor in precise work is the
response time of the gas detector and the recording instrument.
Problems from this source can be reduced, by decreasing the rate of
evaporation of the frozen evolved gas, at the expense of signal
strength.

In conclusion, the differential freezing technique can be
applied in different ways to yield additional information.
Provided the evolved material is condensible the differential

freezing procedure will increase the sensitivity of conventional EGD techniques by at least two orders of magnitude. In addition, by varying the coolant and observing the changes on the EGD patterns qualitative information can be obtained in many cases, with the enhanced sensitivity mentioned above. Finally, if the gas is evolved quantitatively, measurements can be made of trace materials with an accuracy as good as that achieved by normal DTA on much larger samples.

REFERENCES

1. R.C. Mackenzie, Differential Thermal Analysis, Academic Press London and New York, 1970, pp. 117-121.

2. F. Paulik, J. Paulik, L. Erday, Talanta 13 (1966) 1405.

3. A.S. McKie and D. Clark, Proc. 3rd Int. Symp. Batteries, Bournemouth, 1962, p. 285.

4. J.A. McMillan, Chem. Rev. 62 (1962) 65.

5. R.E. Klausmeier, U.S. Dept. Comm. Office Tech. Serv. A.D. 255 (1961) p. 225. (C.A. 60 (1964) 7673L).

6. J.C. Jack and T. Kennedy, J. Thermal Analysis 3 (1971) 25.

7. P.A. Barnes and R.M. Tomlinson, accepted for publication in the Journal of Thermal Analysis.

8. P.A. Barnes and A. Barner, unpublished results.

9. P.A. Barnes and F.S. Stone, Thermochim. Acta 4 (1972) 105.

10. P.A. Barnes and F.S. Stone, Proc. 6th Int. Symp. Reactivity of Solids, Schenetady, New York, 1968, p. 261. Published by John Wiley and Sons Inc., 1969.

A SMALL, MINI-COMPUTER-AUTOMATED THERMOANALYTICAL LABORATORY[*]

E. Catalano and J. C. English[**]

Lawrence Livermore Laboratory
University of California

Livermore, California 94550

INTRODUCTION

Several years ago the need for a thermoanalytical laboratory that provided services over a broad area was recognized at LLL. The objectives were to provide thermophysical and thermochemical data for a variety of unique problems. The quality of the data required vary with the needs of the requestor and the problem under investigation. The nature of the type of measurements is known to change with time and the services are therefore required to change also. The services provided by such a laboratory would include:

- Differential thermal analyses (DTA) of organic and inorganic systems over ranges of 4 to 2400 K.

- Thermogravimetric analyses (TGA) in the range from 300 to 1300 K.

- Differential scanning calorimetry (DSC) over a range from 150 to 500 K.

---

[*]This work was performed under the auspices of the U.S. Atomic Energy Commission.

[**]Present address: Chemistry Department, Purdue University, West Lafayette, Indiana 47907.

- Simultaneous differential thermal and thermogravimetric analyses of systems that decompose and mass spectroscopy of the decomposition products over a range from 300 to 1300 K.

- Thermodynamic functions of solids from specific heat and magnetic measurements over a range from 3 to 300 K.

- Calorimetric determination of "free" water in materials.

- Bomb calorimetry.

- Detonation calorimetry.

- Crystal growth and phase boundary determinations.

Equipment immediately available for our use included a Du Pont* Model 900 DTA with a model 950 TGA, a Perkin-Elmer Model DSC-1, and a Mettler simultaneous DTA-TGA with a Balzers quadrupole mass spectrometer QMG-101. Several laboratory constructed items were also available: DTA's for the regions from 20 to 300 K and from 300 to 1870 K, crystal growth apparatus, calorimetric and magnetic susceptibility apparatus for the region from 3 to 300 K, and a calorimeter for "free" water determinations.

We decided a mini-computer-automated laboratory could meet our objectives within the manpower and fairly severe budget constraints imposed on us. Also entering into our decision was the fact that there were several mini-computer-automated facilities in the LLL Chemistry Department.[1-4] Both timeshared and experiment-dedicated computer systems were being evaluated. Because we were able to profit from the experience of the Chemistry Department, we chose neither a timeshared nor a dedicated system. We developed our own hybrid system.

PHILOSOPHY OF OPERATION
AND TECHNICAL ORGANIZATION

The formation and organization of a mini-computer-automated laboratory is more dependent on the philosophy of how

---

*Reference to a company or product name does not imply approval or recommendation of the product by the University of California or the U.S. Atomic Energy Commission to the exclusion of others that may be suitable.

it is to operate than on strictly technical considerations. However, there is a strong interplay between philosophy of operation and technical considerations; the organization reflects this interplay.

In a way, a thermoanalytical laboratory is an excellent candidate for mini-computer automation because of one overwhelming technical fact: the required data rates are slow, on the order of 10 data points per second. These slow data rates allow the computer to do a number of tasks such as data acquisition, operator interaction, and control.

We imposed the following operational requirements for the total system:

1. The system must be viable, i.e., blocks of components can be changed to satisfy any job requirement.

2. The system must allow total operator interaction for any experiment.

3. The computer language must be of a high level to require minimal operator training.

4. The master systems program must be flexible enough to handle all jobs by use of program options.

5. The system would,in a special sense, be dedicated to the experiment.

6. Any commercial apparatus could be run independent of the system.

7. A priority list of apparatus interfacing into the system must be based on both the anticipated sample loads and the expense of the analyses.

8. Data processing would be on a time-available basis (both computational time and calendar time availability).

9. All data storage would remain within the system.

The viability requirement for the operational system is simply an extension of the same viability requirement that must be met by any stable laboratory. It also implies constant system upgrading within fixed limits. The system should be able to in-

corporate changes with a minimum of down time and interaction
with other components. Severe system changes lead to long down
times and an overemphasis on systems development. This, in
turn, leads to degradation of the analytical services required.
Therefore, a middle of the road policy is essential; it is also
essential to consider the computational system only as a laboratory tool.

With the operational requirements, we can organize the
system.

An examination of the thermal measuring apparatus reveals
that besides slow data rates, they have in common very similar
physical observables: usually low level electromotive forces,
time, and slow pulse counting. For control of such apparatus,
the modes are usually confined to on-off switches or relays,
stepping motors, and programmed power supplies.

We chose to use low-level digital voltmeters, frequency
counters, time generators,decoders, devices such as quartz
thermometers, power supplies, digital-to-analog converters,
programmable power supplies, stepping motors, and relays as
parts of the data acquisition and control systems.

Because of operational requirement 6, much of the system
interfacing for control becomes exceedingly simple.

Technical decisions have to be keyed around the most difficult job the system will be required to perform: the measurement of heat capacity at low temperatures. Several technical
decisions are required in addition to the partial decision implicit
in requirement 5 above. The bases for making these decisions
are not unique; furthermore, the decisions we made may not be
optimal. It was our contention that a true systems analysis would
be extremely costly in manpower. We felt that it would be less
costly to make a small number of system decisions that were
poor and then improve our choices. Actually, this process fits in
with requirement 1 quite well.

The choice of the PDP 8/I central processing unit (CPU) was
based on our operational requirements and the fact that a number
of DEC 8's were operational at this Laboratory. We started with
one digital data acquisition and control system devoted to the
calorimetry problem. Once the hardware and software were
proven, a second system was acquired and immediately dedicated
to service work. Although the individual components are not
identical, functionally, the two systems are alike. Both are dedicated to thermoanalytical services and research.

COMPUTER SYSTEMS HARDWARE

The hardware interface serves four functions: to input data, to output commands, to interrupt processing, and to provide timing control. The system hardware block diagram is shown in Fig. 1.

The input data can be either analog signals or digital words. The analog-to-digital (A/D) conversion technique is determined by the signal level and the frequency spectrum of the analog input signals. Most of the signals involved in thermal properties analyses are low level (1 μV to 10 mV) and low frequency. For these signals, A/D conversions are best performed by an integrating digital voltmeter (IDVM). The low-level signals also require the use of a crossbar-type input multiplexer, which provides a floating, fully guarded system and eliminates ground loops. The dynamic range and flexibility of signal measurements are extended by adding autoranging and programmable integration time for the IDVM. Digital input words are generated by other A/D techniques or by inherently digital signal devices such as counters.

The digital multiplexer is needed to multiplex the digital words onto the computer input-output (I/O) bus. The word size is determined by the computer. Each device on the multiplexer input must have an address, or more than one address if the input word is longer than the computer word. When a particular device is addressed, its data is transferred to the computer. The multiplexer was designed so that additional devices could be incorporated.

The second function of the interface is to output commands, which can be either analog or digital signals. The analog outputs might be used to control programmable power supplies. The digital word is held in a storage register for the digital-to-analog conversion. Digital output words are also held in storage registers called command registers. The digital word can be encoded, represent an address, or each bit of the word can represent a particular function. A second type of digital output is in the form of pulses. Pulses are used for strobing, serial transmission, etc.

The third function of the interface is interrupt processing. An interrupting device must be able to signal the computer that an interrupt is desired, and provide identification so the device can be acknowledged. The interrupt identification is provided by the status interrupt register. Each bit in this register identifies an interrupting device. The register is read into the computer by an interrupt handling program and the interrupt is serviced.

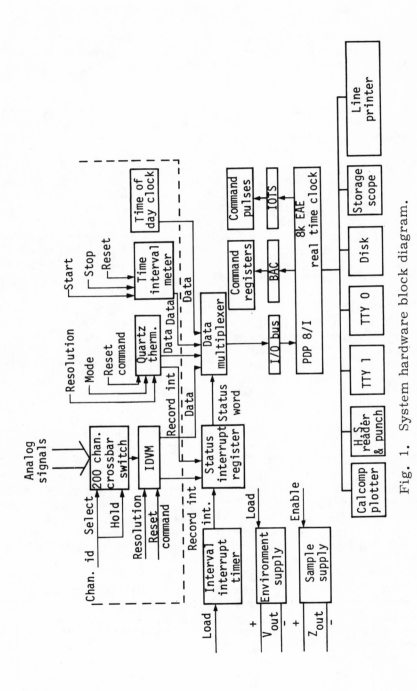

Fig. 1. System hardware block diagram.

The final function of the interface is to provide timing signals. Asynchronous timing can be provided by the computer real time clock. However, timing pulses or intervals that must be synchronized to an external event are provided by the interface. A temperature-stabilized crystal-controlled clock and the 1 pulse per second from the time-of-day clock provide the basis for the timing control.

The crystal clock drives a decade divider, the outputs of which can be selected under program control. The outputs can be used to encode A/D conversion or to signal the computer via the interrupt.

COMPUTER SYSTEM SOFTWARE

The software was designed in conjunction with the hardware to provide a flexible, easy to use system. Some of the same constraints and requirements considered in the hardware design also apply to the software design. A major consideration is the input data rate. The low data rates (10 pts/sec maximum) allow the use of a high-level language. The systems software, CALDAC (Calorimetry Data Acquisition and Control), was built around an interpretive language, FOCAL 69, developed by DEC. This language provides an easy to learn command set and a floating point package. Additional commands needed to handle the hardware and to acknowledge interrupts are provided by CALDAC.

The system provides two types of interrupts: status and non-status. Status interrupts are handled by the CALDAC text program. The interrupt specifies a line number to be executed. The digital voltmeter command (Table 1), for example, causes a status interrupt when a reading is complete. The line number to be executed is specified in the command. Nonstatus interrupts use only machine language handlers. Devices such as the teletype (TTY), disk, and high-speed reader provide nonstatus-type interrupts.

A mass storage device was needed to store the large amounts of generated data. A moving head disk with removable platter cartridges provided ample storage. Seven data arrays with 32K data points each are provided on each cartridge. Each of the seven arrays has a fast core image of 128 data points. A disk-core swap occurs only when a data point outside of the core image is accessed. This technique assures fast response and a minimum of disk swaps.

CALDAC contains all of the command structure to handle data acquisition and control for all of the experimental thermal apparatus. However, to fulfill the requirements posed by the

Table 1.   Summary of system commands.

Integrating digital voltmeter
[V (A, B, C)

A = Voltage channel to be read
    (0-199)
B = Resolution on DVM 0-3
    0 = Lowest resolution
    3 = Highest resolution
C = Group number to handle the
    DVM encode interrupt. In
    other words, FOCAL will exe-
    cute Group C when DVM read-
    ing is complete. Reading is
    stored as variable X0.

Time-of-day clock
[C

Read the time-of-day clock and
    store reading as variable X1.

Quartz thermometer
[Q(A, B, C)

A = Resolution
    0 = 0.1%
    1 = 0.01%
    2 = 0.001%
B = Mode
    0 = temperature probe 1
    1 = temperature probe 2
C = Group number to handle ther-
    mometer encode interrupt,
    similar to DVM command.
    Reading is stored as variable
    X2.

Time interval meter
[T

Read the time interval meter then
    reset the count. Reading is
    stored as variable X3.

Interval time out
[I(A, B, C)

A = Clock number 0, 1, 2
B = Number of seconds in interval
C = Group number to handle inter-
    rupt after time-out interval

Sample power supply
[S (A, B)

If $A \neq 0$, start interval timer and
    set power supply to a value
    proportional to B.
If A = 0, stop interval timer
$0 \leq B < 1024$

Environment power supply
[E(A)

Set power level on environment
    power supply to a value pro-
    portional to A.
$0 \leq A < 1024$

Table 1 (continued)

Rate of interrupt [R(A, B)	If A $\neq$ 0, enable clock interrupts and then select rate according to B. If A = 0, disable interrupts.
CalComp plotter commands [U(A)	If A = 0, raise pen. If A $\neq$ 0, lower pen.
[P(A, B)	Move pen to coordinates (A, B). 0 $\leq$ A $\leq$ 2047 0 $\leq$ B $\leq$ 1000
Disk data point swap SFN(I) = X SX = FN(I)	Store and retrieve value X in array N, address I.
Type command T#A	A must be a number 0, 1, or 2. If A = 0, output on console TTY If A = 1, output on data TTY. If A = 2, output on high-speed punch.

philosophy of operation, a very flexible formal software structure is required so that the operator can put together a given program which is appropriate to the problem at hand and yet maintaining operator interaction. This formal structure contains sets of utility programs for data handling capabilities. Examples of utility programs are:

- Automatic scaling and plotting of data

- Curve fitting

- Data deletion and augmentation.

- Baseline construction

- Integration

- Differentiation

- Display of data

The main function of the formal structure software is to handle data from its acquisition to its final output forms. In a sense, it is a data traffic flow director and a bookkeeper. Figure 2 is a block diagram of the structure. Three file sets are indicated: scratch, protected, and table files. The scratch files are temporary files; the table files act as bookkeeper for the protected files. The data identification, starting and ending locations, and the structural format of the data are kept in the table files. The table files are up-dated automatically. The data flow is as shown. The SAVE, DELETE, LISTING, and UTILITY programs

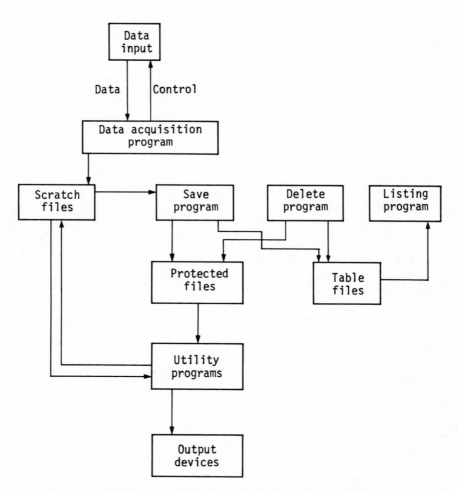

Fig. 2.   Formal software structure block diagram.

are written in the CALDAC language and are called in as subroutines.

SPECIAL FEATURES OF THE FACILITY

The analytical service most often requested is for DTA data in the temperature region from 300 to 1100 K. To service all requests would require operating all of the commercial instruments we have on a 24 hour basis. Therefore, we constructed a dual 8-sample DTA apparatus with two furnaces and sample chamber units. The samples are run sumultaneously and the computer system provides the time ramp furnace control. It is possible to run up to 32 samples in a normal work day.

The computer generated furnace control has been extended to the region of exceptional furnace temperature stability. This feature is of particular importance for crystal growth. For example, using Pt/90Pt, 10Rh thermocouples, stabilities of ±0.1 μV at about 1000°C over a period of several weeks were obtained.

The requirement for curve fitting in a least squares manner arises frequently. Since mini-computers are generally unable to do the necessary matrix inversions well because word size, storage capacity, or both are limited, we adapted Forsythe's method of curve fitting[5] in a least squares manner. Hamming[6] discusses the advantages of this method which include neither having to do matrix inversions nor having to have equally spaced data, etc. and develops the necessary recursive relationships. We have applied the method to the numerical handling of germanium resistance thermometer data for use in low-temperature calorimetry.[7]

COST AND PRODUCTIVITY

Each system has approximately $35,000 of computer and related peripherals. One system has approximately $5,000 and the other about $10,000 of hardware purchased for data acquisition.

The CALDAC language and associated interfaces required about 3 man years from start to the debugged productive stage. The laboratory has a steady state direct manpower of 6: 2 professional chemists, 2 chemist technicians, 0.5 electronic engineer, 0.5 electronic technician, and 1 mechanical technician. They handle all of the work from routine DTA's to research level projects such as low-temperature calorimetry and systems software development.

CONCLUSION

This facility meets the analytical service requirements and the philosophical operational requirements. It is exceptionally successful in optimizing individual productivity and reducing tedium associated with routine service work by cutting down the number of analyses repeated due to operator misjudgment and accidents. Furthermore, it delivers a generally higher quality of data than services supplied in a manual manner by the same operators. Especially valuable is the fact that it releases time for research activities and that there is enough flexibility to continually add to or change the particular variety of problems at hand.

REFERENCES

1. R. E. Anderson, Modern Chemical Automation, Lawrence Livermore Laboratory, Rept. UCRL-71335, 1969.

2. R. E. Anderson, Time-Share Versus Dedicated Systems and Hardware-Software Tradeoffs, Lawrence Livermore Laboratory, Rept. UCRL-72093, 1969.

3. J. W. Frazer, Management of Computer Automation in the Scientific Laboratory, Lawrence Livermore Laboratory, Rept. UCRL-72162, 1969.

4. J. W. Frazer, et al., On-line Interactive Data Processing, Lawrence Livermore Laboratory, Rept. UCRL-72921, 1971.

5. G. E. Forsythe, J. Soc. Ind., Appl. Math. 5, 74 (1957).

6. R. W. Hamming, Numerical Methods for Scientists and Engineers (McGraw-Hill, New York, 1962) Chapters 17, 18.

7. E. Catalano, B. L. Shroyer, and J. C. English, Rev. Sci. Inst. 41, 1663 (1970).

STEADY STATE TECHNIQUE FOR LOW TEMPERATURE HEAT CAPACITY OF SMALL SAMPLES

R. Viswanathan*

Department of Applied Physics and Information Science

University of California, San Diego, La Jolla, Calif.

ABSTRACT

Reported here is an elegant method based on steady state or ac calorimetry technique, using laser beam as heat source, to measure absolute heat capacity of tiny samples of mass 1-100mgm. The results on high purity copper, gold, nickel and isotopes of molybdenum are given to show that the absolute accuracy of this method is $\sim \pm 2\%$, at least comparable, if not better than conventional heat pulse techniques, which require bulky samples ~ 100 times larger. The possible variations of this method will also be discussed.

I. INTRODUCTION

The adiabatic[1] and the semi-adiabatic[2] techniques used in heat capacity measurements require large samples (1-100gms.) with very good thermal isolation from their surroundings. Many of the homogeneous intermetallic compounds and nearly all condensed films, amorphous or disordered materials are often available only in small quantities. For such samples of typical mass 1-100mgm. the steady state or ac calorimetry technique is most suitable. In this paper is described an apparatus, based on the ac technique, for measuring the absolute heat capacity at low temperatures (1.5-20°K).

---

*Supported by U. S. Atomic Energy Commission under Contract AEC-AT(0 4-3)34.

2. METHOD

Though not exploited fully till recently, this technique for heat capacity measurements has been known for quite some time.[3] Hence only the principle of the method is given here. More details are found elsewhere.[4,5,6,7]

An oscillatory heat input of frequency ω introduced onto a specimen of heat capacity C, coupled to a heat sink by a weak link characterised by a thermal conductivity K, will develop a temperature difference ΔT in the sample given by

$$\Delta T = \Delta T)_{dc} + \Delta T)_{ac}$$

The dc part $\Delta T)_{dc}$ is $\propto 1/K$ and the ac part $\Delta T)_{ac}$ is $\propto 1/\omega C \times [1 + (\omega\tau_1)^{-2} + (\omega\tau_2)^{+2}]^{-1/2}$ where τ_1 is the external time constant between the sample and heat sink and τ_2 is the internal time constant of the sample and thermometer together. If the frequency ω is so chosen that $\omega\tau_1 \gg 1 \gg \omega\tau_2$ then $\Delta T)_{ac} \propto 1/\omega C$. Typically τ_1 is $\sim$ 2-5 seconds, $\tau_2 \sim$ 1-5 milliseconds and hence ω can be between 5-100cps.

If the sample is heated resistively[4,5] the heat input is known exactly; but special care is to be taken to minimize the heat capacity of the heater in comparison to that of the sample. If it is heated optically[6,7] the exact heat input is hard to measure, due to the problems associated with the beam divergence and optical absorption of the sample. In such a situation, a comparison technique with two equal intensity laser beams shining on two samples, one of which is a known standard, will be most suitable.

3. APPARATUS

Fig. 1 gives the sketch of the low temperature probe. Two chopped laser beams of equal intensity are made to shine on two samples kept inside two chambers in an evaculated calorimeter immersed in liquid helium. Then the heat capacity of the sample is given by

$$C_1 \equiv C_2 \ \Delta T_2/\Delta T_1)_{ac}$$

where subscript 1 refers to the experimental sample and 2 to a standard sample such as copper or silver, whose heat capacity C_2 is accurately known. 5 mil. Au-.07%Fe vs. Cu thermo-couples attached to the samples monitor the $\Delta T)_{ac}$ and $\Delta T)_{dc}$ through a phase sensitive detector (HR-8 of Princeton Applied Research) and a microvoltmeter (K-150B of the Keithley Instruments) respectively. The

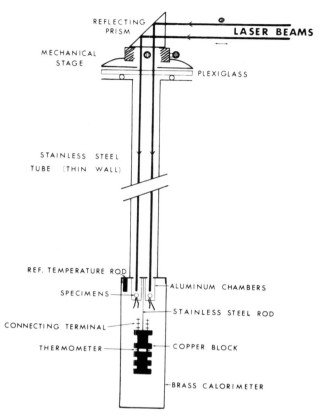

Fig. 1. Low temperature probe for ac calorimetry.

reference junction is embedded on a copper block which is connected to the helium bath by a 1/8" stainless steel rod and whose temperature is measured by a cryocal CR-1000 calibrated Ge-thermometer. The sample temperatures are increased by heating the copper block with a 300Ω non-inductively wound manganin heater. A third thermocouple between the copper block and the reference rod at the bath temperature is used to check the calibration of the thermocouples periodically, compared to the published tables.[8]

The alignment of the beams is achieved by manipulating the reflecting prism mounted on a rotating mechanical stage, with lateral motions. The two samples are mounted symmetrically with nearly

equal lengths of the thermocouple wires, which are the only weak
thermal links between the samples and the heat sink. The equal
intensity of the two aligned beams is ascertained by laterally moving the prism so that each of the two beams shines on the same
sample and by monitoring the $\Delta T)_{dc}$.

Fig. 2 shows the optical arrangement used to obtain two equal
intensity beams. The unpolarized 1mW beam from a helium-neon laser
is split into two beams of equal intensity, but of perpendicular
polarizations, by a Wollaston prism. The beam diameters (typically
1/16") and the separation between the beams are adjusted using proper lenses. The chopping is done either by a rotating linear polar-
izer or a mechanical chopper of variable frequency. The chopping
frequency in the present measurements is 10cps and the changes in
ac signal scale well with frequency in the range 5-40cps. Typically the $\Delta T)_{ac}$ is $\sim$ 1-10m°K and $\Delta T)_{dc}$ $\sim$ 0.1°K.

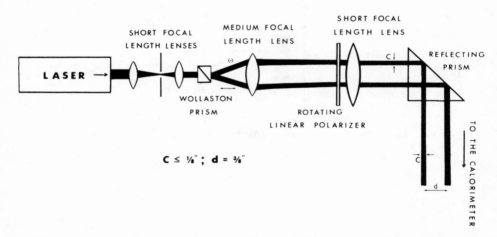

Fig. 2. Optical set-up for two equal intensity laser beams.

4.   SAMPLE PREPARATION

The standard sample is high purity (99.999%) copper from
Johnson-Matthey. The thermocouples are attached by copper plating
the wires onto the samples, using a current of 2mA for 10-15 minutes in a standard plating solution. The addenda is thus only

1-2mg. of copper, whose heat capacity is small and well-known and hence is corrected for easily. The correction in the present measurements is typically $\lesssim$ 5% of the total heat capacity of the samples. The electroplating solution can be either acidic or alkalline depending on which one does <u>not</u> react with the sample chemically. It must be necessary to have a good thermal contact between the thermocouple and the sample for the ac technique to work satisfactorily, and hence the surface of the sample is to be kept clean before plating. Sometimes it is necessary to anodise the sample surface to remove any oxide layer. For semiconductors and insulators it may be necessary to use a Nickel or Gold "Strike"-- an evaporated thin layer of Ni or Au--prior to plating.

Alternatively, the thermocouples can be attached by spot welding or by thermocompression bonding. For larger samples ($\sim$50-100 mgm) the use of 0.1-1 mgm. of Indium with ultrasonic soldering will be reasonably satisfactory. One can also use the arrangement of a sapphire platform for certain types of materials.

The samples are often coated with a thin layer of graphite spray to keep their surfaces similar for absorption of the laser beams. The differences in the heat input, if any, due to sample surfaces or due to unequal length of termocouple wires can be taken into account in the calculations by using the ratio $\Delta T_1/\Delta T_2)_{dc}$.

5. RESULTS AND DISCUSSION

Figures 3 and 4 give the data on high purity copper, gold and nickel in conventional C/T vs T^2 plots. The random scatter is $\sim \pm$ 2%. The least square fitted values of the electronic heat capacity coefficient γ and the Debye temperature θ_D have a precision of 1/2%, whereas the absolute accuracy is $\sim \pm$ 2%, comparable to that of the conventional techniques which use samples $\sim$ 100 times larger.

In the present set-up the maximum heat capacity that can be measured is $\sim$ 500μJ/$^\circ$K. This corresponds to the heat capacity of 1×10^{-3} mole. of copper at $\sim$ 20°K. By decreasing the sample mass, increasing the intensity of the laser beam and by choosing thinner thermocouple wires of 2 mil. diameter (to decrease the thermal conductivity of the weak link), the measurements can be extended to higher temperatures.

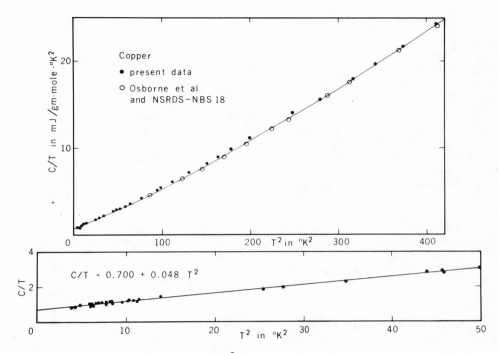

Fig. 3.   C/T vs T^2 for high purity copper.

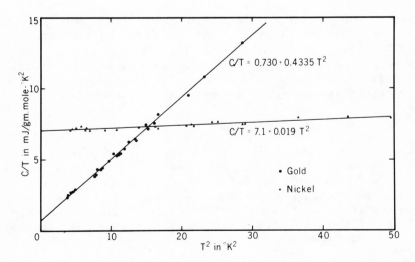

Fig. 4.   C/T vs T^2 for high purity gold and nickel.

In Table 1 are listed the γ and θ_D values of copper, gold, and nickel as compared with the critically evaluated literature data[9,10,11] and the agreement is found to be very good. Also listed in Table 1 are the γ and θ_D values of the enriched isotopes of molybdenum[12] which show, as would be expected, that the γ-values are not mass dependent and that the θ_D-values are $\propto M^{-1/2}$, following the well-known Debye law.[13]

Table 1. Electronic heat capacity coefficient, γ and Debye temperature, θ_D from ac technique as compared to literature data.[9,10,11]

Sample	γ in mJ/gm.mole.$°K^2$		θ_D in $°K$	
	Present	Literature	Present	Literature
Copper	0.700	0.695	343	344.5
Gold	0.730	0.729	165	162.5
Nickel	7.1	7.1	467	470
Mo^{92}	1.75	–	458	–
Mo^{96}	1.75	–	449	–
Mo^{100}	1.75	–	440	–

It must be pointed out that this apparatus can also be used for measuring the heat capacity of small samples and films, using the relaxation technique,[14] which has certain advantages over the ac technique. Details of this will be published elsewhere.[15]

ACKNOWLEDGEMENT

The author would like to thank Dr. R. R. Oder and Prof. T. H. Geballe for valuable suggestions at the beginning of this work, Dr. A. W. Lohmann for suggesting the optical set-up and Dr. Luo for his interest throughout this work.

REFERENCES

1. P. H. Keesom and N. Pearlman, Methods in Experimental Physics, Ed. K. Lark'-Horovitz and V. A. Johnson, Academic Press, N. Y. (1959), Vol. 6, Part A, Sec. 5.1.2.1.
2. C. A. Luengo, Ph.D. Thesis, Universidad Nacional de Cuyo, San Carlos de Bariloche Argentina (1972).
3. Y. A. Kraftmakher, Zh. Prikl. Mekhan. i. Tekhn. Fiz. 5, 176 (1962).
4. P. F. Sullivan and G. Seidel, Phys. Rev. 173, 679 (1968).

5. R. D. Hempstead, Ph.D. Thesis, University of Illinois, Urbana (1970).

6. P. Handler, D. E. Mapother and M. Rayl, Phys. Rev. Lett. 19, 356 (1967).

7. D. S. Simons, Ph.D. Thesis, University of Illinois, Urbana (1973).

8. L. L. Sparks and W. J. Hall, Cryogenic Temperature Tables III, NBS Report 9721 (1969), pp. 16.

9. G. T. Furukawa, W. G. Saba and M. L. Reilly, National Standard Ref. Data Series, NBS18, April 1968, pp. 20.

10. D. W. Osborne, H. E. Flowtow and F. Schreiner, Rev. Sci. Instrum. 38, 159 (1967).

11. F. Heiniger, E. Bucher and J. Müller, Phys. Kondens. Materie. 5, 243 (1966).

12. R. Viswanathan, H. L. Luo and J. J. Engelhardt, Proc. 13th International Conf. on Low Temp. Phys. (LT-13), 1972 (in press).

13. E. S. R. Gopal, Specific Heats at Low Temperatures, Plenum Press. N. Y. (1966), pp. 34.

14. R. Bachmann, F. J. DiSalvo, Jr., T. H. Geballe, R. L. Greene, R. E. Howard, C. N. King, H. C. Kirsch, K. N. Lee, R. E. Schwall, H. U. Thomas and R. B. Zubeck, Rev. Sci. Instrum. 43, 205 (1972).

15. R. Viswanathan and C. T. Wu, (to be published).

FUSION AS AN OPPORTUNITY FOR CALORIMETRICALLY PROBING POLYMER

CONFORMATIONS AND INTERACTIONS IN THE BULK STATE

Alan E. Tonelli

Bell Laboratories

Murray Hill, New Jersey 07974

INTRODUCTION

It is a well documented[1] fact that the fusion or melting of crystalline polymers is a first-order phase transition between two polymeric states in equilibrium: the crystalline and molten amorphous, or liquid states. Consequently, the melting temperature T_m is well defined and given by

$$T_m = \Delta H_u / \Delta S_u \qquad (1)$$

where ΔH_u and ΔS_u are the differences between the enthalpy and entropy, respectively, of the crystalline and molten polymer phases in equilibrium at T_m. Unlike calorimetric studies of the glass transition in polymers, for example, whose interpretations in terms of polymer chain conformations and interactions suffer from a lack of knowledge of the state of the polymer chains both above and below this transition, because the polymer chains may not be in equilibrium,[2] the fusion process affords an opportunity for studying the effect of molecular structure on the melting temperature and, more importantly, upon both components of the ratio $\Delta H_u / \Delta S_u$ which determine T_m.

Knowledge of the state or conformation of the polymer chains in either the crystalline or molten phases together with T_m, ΔH_u ΔS_u permits certain deductions to be made concerning the state of the polymer chains in that phase for which information is lacking. It may be worthwhile at this point to briefly describe the conformations of polymer chains in the crystalline and molten phases.

Without addressing questions of the size and regularity of the chain folds, one can still safely say that in both bulk and solution crystallized polymer samples chain folded lamellae contain the overwhelming fraction of crystalline polymer chains.[1] Each polymer chain in the interior of a lamella adopts a regular conformation, either planar zigzag or helical. Excluding crystal defects, the conformation of a polymer chain in the crystalline phase is fixed and in a state of low entropy.

In the molten phase, on the other hand, it is apparent[3] that each polymer chain is free to adopt a myriad of different conformations because of the absence of intra- and intermolecular excluded volume effects.[4] Each chain in the molten bulk may assume any of the conformations available to the same chain if it were present in dilute solution at its θ-temperature. In contrast to the crystalline phase, polymer chains in the molten phase are not conformationally restricted by interchain interactions and usually possess a significant conformational entropy.

In an attempt to obtain information concerning bulk polymer chain conformations and interactions from the fusion process, we separate ΔH_u and ΔS_u into two independent contributions.[1,5,6]

$$\Delta H_u = \left(\Delta H_u\right)_v + \Delta H_u \tag{2}$$

$$\Delta H_u = \Delta E_u = \left(\Delta E_u\right)_v + \Delta E_v \tag{3}$$

$$\Delta S_u = \left(\Delta S_u\right)_v + \Delta S_u \tag{4}$$

where $\left(\Delta E_u\right)_v$ and $\left(\Delta S_u\right)_v$ are the energy and entropy associated with melting at constant volume, and ΔE_v and ΔS_v are the contributions to the energy and entropy of fusion made by volume expansion.

Presumably the constant volume contribution to the fusion process is attributable to the conformational changes in the polymer chains (increased conformational freedom) which accompany their melting, and is an intramolecular process. Thus, we tentatively write

$$\left(\Delta E_u\right)_v = \Delta E_{conf.} = E_a - E_c \tag{5}$$

$$\left(\Delta S_u\right)_v = \Delta S_{conf.} = S_a - S_c \tag{6}$$

where the subscripts a and c denote the molten amorphous and crystalline states, respectively. On the other hand, the volume expansion contribution to the fusion process consists of the increase in energy and entropy attendant upon the increase in the average polymer chain separation and is an intermolecular process. ΔS_v can be obtained[7,8] from

$$\Delta S_v = \Delta V_u (\partial P/\partial T)_v = (\alpha/\beta)\Delta V_u \tag{7}$$

where α, β and ΔV_u are the thermal expansion coefficient at T_m, the compressibility at T_m, and the volume change on melting, respectively.

It is apparent (see Eqs. 1-7) that from knowledge of the polymer chain conformation in the crystal we can deduce information such as the conformational energy and entropy about the molten amorphous polymer chains if we know ΔH_u, T_m, α, β and ΔV_u. Furthermore, if we are able to calculate $E_{conf.}$ and $S_{conf.}$ (see below), then it becomes possible, by comparison with the experimental values, to learn something about the conformations of and the interactions between polymer chains in the melt and in the crystal.

DESCRIPTION AND EXPERIMENTAL JUSTIFICATION OF THE METHOD

If the crystalline polymer chain conformation is assigned a zero intramolecular energy, then the change in intramolecular or conformational energy $\Delta E_{conf.}$ and entropy $\Delta S_{conf.}$ are reduced* to

$$\Delta E_{conf.} = E_a - E_c^0 = E_a \tag{8}$$

$$\Delta S_{conf.} = S_a - S_c^0 = S_a \tag{9}$$

E_a and S_a, the intramolecular or conformational energy and entropy of a polymer chain in the molten amorphous state at T_m, can be calculated from the usual statistical thermodynamic relations[12]

$$E_a = \left(\frac{RT^2}{Z}\right) dZ/dT \tag{10}$$

---

* Only the polymer backbone conformations are considered in the fusion process. Side-chain motion is ignored, an assumption supported by dynamic mechanical and NMR measurements[9-11] which indicate the onset of side-chain mobility in crystalline polymers well below their melting temperatures.

$$S_a = R[\ell_n Z + (T/Z)dZ/dT] \tag{11}$$

providing the configurational partition function Z and its temperature coefficient are known.

The configurational partition function and its temperature
coefficient can be evaluated by adopting the rotational isomeric
state model[13] of polymer chains and utilizing the matrix methods
of Flory and Jernigan.[14,15] Each backbone bond in the chain is
allowed to adopt a small number (usually 3) of rotational states
whose probability of occurrence normally depends on the rotational
states of its nearest neighbor bonds. A statistical weight matrix
U is constructed[14,15] from the Boltzmann factors of the energies,
$u_{\alpha,\beta} = \exp(-E_{\alpha\beta}/RT)$, appropriate to the various pairwise dependent
rotational states. For bond i,

$$U_i = \begin{array}{c} \\ i-1 \\ \alpha \\ \\ \beta \\ \\ \gamma \end{array}
\begin{array}{ccc} i\quad \alpha & \beta & \gamma \end{array}
\left[\begin{array}{ccc} u_{\alpha\alpha} & u_{\beta\beta} & u_{\alpha} \\ u_{\beta\alpha} & u_{\beta\beta} & u_{\beta\gamma} \\ u_{\gamma\alpha} & u_{\gamma\gamma\beta} & u_{\gamma\gamma} \end{array}\right] \tag{12}$$

The sum of the Boltzmann factors, or u's, of all the possible conformations, which is the partition function Z, is obtained through
sequential matrix multiplication.

$$Z = J*(\prod_{i=2}^{n-1} U_i)J \tag{13}$$

where $J* = [100...0]$ $(1\times v)$; $J = \left[\begin{array}{c} 1 \\ 1 \\ . \\ . \\ 1 \end{array}\right]$ $(v\times 1)$ $\tag{14}$

and v is the number of rotational states permitted for each of the
n bonds. In a similar fashion,

$$dZ/dT = G*(\prod_{i=2}^{n-1} U_{T,i})G \tag{15}$$

$$\text{where } G^* = [J^*00...0] \ (1\times2v); \quad G = \begin{array}{c} 0 \\ 0 \\ \cdot \\ \cdot \\ \cdot \\ 0 \\ J \end{array} \quad (2v\times1) \tag{16}$$

$$U_{T,i} = \begin{bmatrix} U_i & U_{T,i} \\ 0 & U_i \end{bmatrix} \tag{17}$$

and

$$U_{T,i} = dU_i/dT \tag{18}$$

In the above treatment it has been assumed that the constant-volume contribution to the fusion process is attributable exclusively to the increase in the number and energy of conformations available to an isolated chain in the melt. Flory has noted, however, in his lattice treatment[16] of flexible chain molecules, that when isolated chains are constrained to occupy the same region of space, spatial overlap must be avoided. This packing requirement results in a decrease in the entropy which is of the order -R/N, where N is the number of backbone bonds in a polymer segment, or the number of bonds which gives a unit ratio of the length to the breadth of the segment.

On the other hand, Starkweather and Boyd[6] proposed an additional contribution to the constant-volume entropy of fusion suggested to originate in the appearance of long-range disorder in all liquids. Based on calculations performed on several metals by Oriani,[8] they propose a value on the order of R/2 to R e.u./mol of backbone bonds for the long range disorder contribution.

We believe that both of these nonconformational contributions to the constant-volume entropy of fusion largely cancel each other, and the constant-volume entropy of fusion is closely approximated by the gain in conformational entropy upon melting as calculated here for isolated polymer chains. This belief is supported by the following comparison of experimental constant-volume entropies of fusion with the calculated changes in isolated chain conformational entropies for several polymers.

The conformational entropies ΔS_{conf} of a variety of polymers in the molten amorphous state at T_m were calculated as outlined above and are compared with the experimental total ΔS_v, and constant-volume $(\Delta S_u)_v$ entropies of fusion in Table 1. Appropriate statistical weight matrices U were taken from Flory[14] after adjustment to their melting temperatures. Of the twelve polymers whose calorimetric and thermodynamic data permit evaluation of their entropy of fusion at constant volume, only cis-1,4-polyisoprene and polytetrafluoroethylene (PTFE) show discrepancies with the calculated conformational contributions to the entropy of fusion. In the case of PTFE, conformational polymorphism is known[26] to occur in the crystal at room temperature and slightly below T_m. This would result in a non-zero conformational entropy for the crystalline chains, and might explain why the calculated conformational entropy exceeds the measured value at constant-volume. In addition, the rotational isomeric state model for PTFE[14] has not been verified experimentally.

In general, the agreement between $(\Delta S_u)_v$ and ΔS_{conf} is quite satisfactory. This agreement[27] lends support to the identification of the constant-volume contribution to the fusion process with the increase in the number and energy of conformations available to the polymer chain upon melting. It also appears that neglect or cancellation of the Flory[16] and Starkweather and Boyd[6] corrections to $(\Delta S_u)_v$ is justified.

SPECIFIC APPLICATIONS OF THE METHOD

In this section we briefly describe several specific examples of the type of conformational information that can be obtained by applying the method of separating the fusion process into its inter- and intramolecular components.

Conformational Order in Amorphous Polyethylene

In an attempt to test the consistency of the "bundle" theory[28] of polymer liquids with the conformational entropy of a molten polymer chain, the effect of introducing all trans bond or planar zigzag stretches upon the conformational entropy of polyethylene was calculated.[29] A polymer "bundle" is envisioned as an aggregate of portions of different polymer chains in parallel alignment. Parallel alignment of polyethylene chains requires all trans bond stretches. By comparing the experimental constant-volume entropy of fusion with the conformational entropy calculated as a function of the length of the all trans bond stretches, it was concluded[29] that no more than 5% of the bonds in polyethylene can participate in such ordered regions without lowering the calculated conformational entropy significantly below that measured[6,17] during melting at constant volume. If polymer bundles do indeed exist they are few in number and small in size.

TABLE I

ENTROPIES OF FUSION

Polymer	T_m, °C	ΔS_u^+	ΔS_v	$(\Delta S_u)_v$	$\Delta S_{conf.}$
Polyethylene[6,17]	140	2.29-2.34	0.46-.52	1.77-1.84	1.76
Isotactic Polypropylene[18,19]	208	1.50	0.44-.65	0.85-1.09	0.96
cis-1,4-poly-isoprene[20]	28	0.87	0.45	0.43	1.34
trans-1,4-poly-isoprene[21]	74	2.19	0.91	1.28	1.37
Polyoxymethylene[22]	183	1.75	0.35	1.40	1.50
Polyoxyethylene[23]	66	1.78	0.37	1.41	1.70
Polyethylene-*terephthalate[24]	267	1.46	0.29	1.17	1.07
Polytetrafluoro-ethylene[6]	327	1.14	0.37	0.77	1.60
cis-1,4-poly-butadiene[24]	5	1.92	0.43	1.49	1.38
Polyethylene-adipate[25]	65	1.48	0.38	1.10	1.04
Polyethylene-suberate[25]	75	1.50	0.38	1.12	1.16
Polyethylene-sebacate[25]	83	1.54	0.38	1.16	1.24

+All entropies are given in e.u./mol of backbone bonds.

*Benzene ring is treated as a single bond.

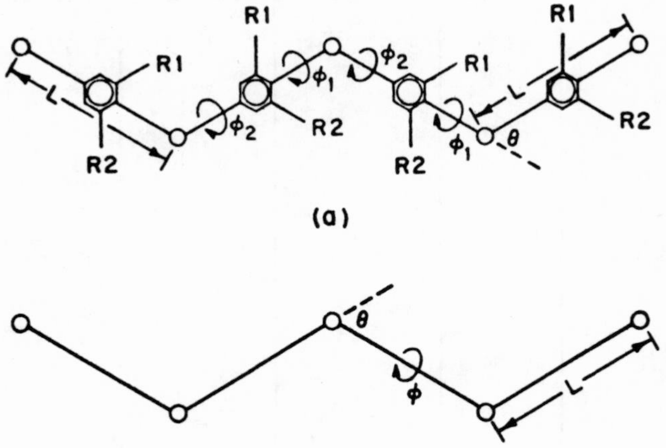

FIGURE 1

(a) A portion of a 2,6-disubstituted-1,4-phenylene
     oxide chain ($R_{1,2}$ = H or CH_3 or C_6H_5) in the
     planar zigzag conformation, where $\varphi_1 = \varphi_2 = 0°$.
     All phenyl rings are coplanar in this reference
     conformation, and φ_1, φ_2 assume positive values
     for right-handed rotations.[14]

(b) A portion of the same chain where the phenyl
     rings are replaced by virtual bonds L linking
     the ether oxygen atoms. The virtual bond
     rotation angle φ is taken as 0° in this con-
     formation and adopts positive values for right-
     handed rotations.[14]

Conformational Characteristics of Isotactic Polypropylene

The conformational entropy and energy at T_m = 208°C were calculated[30] for isotactic polypropylene as a function of the energy differences between the backbond bond rotational states, which enter into the Boltzmann factors of the statistical weight matrix U (see Eqs. 10-13). It was found that energy differences in agreement with those deduced previously by Abe, et al.[31-34] from dilute solution data were required to reproduce the entropy of fusion measured[18,19] at constant volume. Thus, the comparison of calculated conformational entropies with the measured constant-volume entropy of fusion served to confirm the validity of the rotational isomeric state model of isotactic polypropylene derived by Flory et al.[14,32,34] On the other hand, two more recent rotational isomeric state models of isotactic polypropylene put forward by Heatley[35] and Boyd and Breitling,[36] respectively, were found to lead to calculated conformational entropies and energies in excess of the total measured entropy and enthalpy of fusion. It was concluded that both models were in error, a conclusion subsequently verified and explained by Flory[37] with respect to Heatley's model.

Source of the Disparity Between the T_m's of Aliphatic Polyamides and Polyesters

Calculation[38] of the conformational entropies of several aliphatic polyamides (nylons) and polyesters, their comparison parison with the measured total and constant-volume entropies of fusion, and comparison of their experimental enthalpies of fusion[25,39-41] led to the observation that the nylons melt at significantly higher temperatures due to smaller entropies of fusion. In addition, it was also possible to conclude that polymorphism or conformational disorder in nylon crystals is the most likely source for their smaller entropies of fusion.

Phenyl Group Disorder in Poly(2,6-disubstituted-1,4-phenylene oxide) Crystals

Approximate potential energy estimates were used[42] to show that nearly all the rotational states about the virtual bonds spanning the phenyl rings and connecting the ether oxygen atoms (see sketch in Fig. 1) in the poly(2,6-disubstituted-1,4-phenylene oxides) (PPO's) were appreciably accessible. Conformational entropies calculated[42] on the basis of these energy estimates exceeded the total measured entropies[43-47] of fusion, as was also observed for the nylons. Since it is possible to achieve isoenergetic rotations of the phenyl groups around their virtual bonds without altering the relative orientation of the two virtual bonds flanking any given rotating phenyl group, it was suggested that such a disordering of phenyl rings in the crystal might account for the small observed entropy of fusion. An X-ray diffraction study[48] of drawn PPO fibers lends support to this interpretation.

SUMMARY AND CONCLUSIONS

The procedure of separating the total entropy of fusion into intra- and intermolecular contributions, and associating the gain in conformational entropy upon melting with the entropy of fusion measured at constant volume seems to be valid. Utilization of the rotational isomeric state model of polymer chains together with recently developed matrix multiplication techniques, enables facile calculation of polymer chain conformational entropies and energies. The rotational isomeric state model for a given polymer can be checked by this procedure, a check which is geometry independent, i.e., independent of bond lengths and angles, positions of the rotational states, etc., because only the number and energies of the rotational isomeric states are required to evaluate $\Delta S_{conf.}$ and $\Delta E_{conf.}$. In addition, comparison of the calculated conformational and measured constant-volume contributions to the entropy and enthalpy of fusion can sometimes lead to conclusions regarding the state of conformational order, or disorder, and the interactions occurring between polymer chains in the crystalline and the molten polymer phases.

REFERENCES

1. L. Mandelkern, "Crystallization of Polymers," McGraw-Hill Book Co., New York, 1964, Chaps. 1, 2, 5, 9.

2. M. C. Shen and A. Eisenberg, Prog. Solid State Chem., $\underline{3}$, 407 (1966).

3. P. J. Flory, International Symposium on Macromolecules, Helsinki, July, 1972.

4. P. J. Flory, "Principles of Polymer Chemistry," Cornell University Press, Ithaca, New York, 1953, Chaps. 10, 14.

5. L. Mandelkern, Chem. Rev., $\underline{56}$, 903 (1956).

6. H. W. Starkweather, Jr., and R. H. Boyd, Jr., J. Phys. Chem., $\underline{64}$, 410 (1960).

7. J. C. Slater, "Introduction to Chemical Physics," McGraw-Hill Book Co., New York, 1939.

8. R. A. Oriani, J. Chem. Phys., $\underline{19}$, 93 (1951).

9. A. E. Woodward, A. Odajima and J. A. Sauer, J. Phys. Chem., 65, 1384 (1961).

10. K. S. Chan, G. Ranby, H. Brumberger, and A. Odajima, J. Poly. Sci., 61, 529 (1962).

11. I. Kirshenbaum, R. B. Isaacson, and W. C. Feist, J. Poly. Sci., Part B, 2, 897 (1964).

12. T. L. Hill, "Introduction to Statistical Thermodynamics, Addison-Wesley Publishers, Inc., Reading, Mass., 1960, Chap. 1.

13. M. V. Volkenstein, "Configurational Statistics of Polymeric Chains," English Translation, Interscience Publishers, Inc., New York, 1963, Chap. 3.

14. P. J. Flory, "Statistical Mechanics of Chain Molecules," Interscience Publishers, Inc., New York, 1969, Chaps. I, III-VI.

15. P. J. Flory and R. L. Jernigan, J. Chem. Phys., 42, 3509 (1965).

16. P. J. Flory, Proc. Roy. Soc., Ser. A, 234, 60 (1956).

17. F. A. Quinn, Jr., and L. Mandelkern, J. Am. Chem. Soc., 80, 3187 (1958).

18. J. G. Fatou, Eur. Polym. J., 7, 1057 (1971).

19. G. C. Fortune and G. N. Malcolm, J. Phys. Chem., 71, 876 (1967).

20. P. E. Roberts and L. Mandelkern, J. Am. Chem. Soc., 77, 781 (1955).

21. L. Mandelkern, F. A. Quinn, Jr. and P. E. Roberts, ibid., 78, 926 (1956).

22. I. Kirshenbaum, J. Polym. Sci., Part A, 3, 1869 (1965).

23. G. N. Malcolm and G. L. P. Ritchie, J. Phys. Chem., 66, 852 (1962).

24. G. Allen, J. Appl. Chem., 14, 1 (1967).

25. S. Y. Hobbs and F. W. Billmeyer, Jr., J. Poly. Sci., Part A-2, 8, 1387 (1970).

26. E. S. Clark and L. T. Muus, Z. Krist., 117, 119 (1962).

27. A. E. Tonelli, J. Chem. Phys., 52, 4749 (1970).

28. Yu. A. Ovchinnikov, G. S. Markova and V. A. Kargin, Polymer
     Sci., (USSR), 11, 369 (1969).

29. A. E. Tonelli, J. Chem. Phys., 53, 4339 (1970).

30. A. E. Tonelli, Macromolecules, 5, 563 (1972).

31. A. Abe, R. L. Jernigan and P. J. Flory, J. Am. Chem. Soc.,
     88, 631 (1966).

32. P. J. Flory, J. E. Mark and A. Abe, ibid, 88, 639 (1966);
     J. Poly. Sci., Part C, No. 3 973 (1965).

33. A. Abe, Polymer J., 1, 232 (1970); J. Am. Chem. Soc., 90,
     2205 (1968).

34. P. J. Flory, Macromolecules, 3, 613 (1970).

35. F. Heatley, Polymer, 13, 218 (1972).

36. R. H. Boyd and S. M. Breitling, Macromolecules, 5, 279 (1972).

37. P. J. Flory, J. Poly. Sci., Part A-2, 11, 621 (1973).

38. A. E. Tonelli, J. Chem. Phys., 54, 4637 (1971).

39. F. Rybniker, Chem. Listz, 52, 1024 (1948).

40. R. P. Evans, H. R. Mighton and P. J. Flory, J. Am. Chem. Soc.,
     72, 2018 (1950).

41. G. B. Geschele and L. Crescentini, J. Appl. Polymer Sci., 7,
     1349 (1963).

42. A. E. Tonelli, Macromolecules, 5, 558 (1972); ibid, 6, 503
     (1973).

43. J. M. O'Reilly and F. E. Karasz, J. Poly. Sci., Part C, 14,
     49 (1966).

44. F. E. Karasz, J. M. O'Reilly, H. E. Bair and R. A. Kluge,
     "Analytical Colorimetry," R. S. Porter and J. F. Johnson,
     Eds., Plenum Press, New York, 1968, p. 59; J. Poly. Sci.,
     Part A-2, 6, 1141 (1968).

45. A. R. Shultz and C. R. McCullough, *ibid*, 7, 1977 (1969);
     ibid, 10, 307 (1972).

46. W. Wrasidlo, Macromolecules, 4, 642 (1971); J. Poly. Sci.,
     Part A-2, 10, 1719 (1972).

47. J. M. Barrales-Rienda and J. G. M. Fatou, Kolloid-Z.U.Z.
     Polymere, 244, 317 (1971).

48. J. Boon and E. P. Magre, Makromol. Chem., 126, 130 (1960);
     ibid, 136, 267 (1970).

APPLICATION OF DIFFERENTIAL SCANNING CALORIMETRY FOR THE STUDY OF PHASE TRANSITIONS

William P. Brennan

Perkin-Elmer Corporation

Main Avenue, Norwalk, Connecticut 06856

Introduction

The identification and characterization of phase transitions in organic, inorganic and polymeric materials is of obvious and general importance. For example, the technological importance of the temperature that marks a discontinuity in the chemical or mechanical properties of a material will frequently determine the usefulness of a material for a given application. The melting point is one such temperature; however, it frequently occurs that the temperature of a solid-solid phase transition is more important in this respect.

In this sense a phase is defined as any homogeneous and physically distinct part of a chemical system which is separated from other parts of the system by definite boundary surfaces [1]. Both simple and complex materials may exist under particular conditions in a variety of phases. The laws which govern the equilibrium coexistence of different phases of the same material and the equilibrium transformation of the material from one phase to another are deduced from the first principles of thermodynamics and are embodied in the well-known "Phase Rule," $F = C-P+2$, where C is the number of distinct chemical species present, P is the number of phases, and F is the number of degrees of freedom or variable factors such as temperature, pressure and concentration necessary to specify the physical state of the system [2].

The fact that each phase of a particular system is physically
distinct and therefore has its own set of physical properties de-
mands that the characterization of a chemical system include a
study of its phase behavior. The determination of the melting
point and the normal boiling point of a new chemical compound is,
of course, routine and the simplest example of phase characterization. However, other equally important phase transitions; for
example, those between polymorphic forms, being visually less
obvious, are often missed in the cursory examination of new
materials.

Aside from its scientific importance and the practical necessity of fully understanding the behavior of a chemical system under
influences such as temperature, pressure and composition, the
study of phase behavior often has very direct economic significance.
In our own experience we have encountered numerous cases where
literally carloads of chemical intermediates have been discarded
or rejected for failure to pass a simple melting point acceptance
test only to discover that the material existed in an unsuspected
metastable crystal form melting sharply a few degrees below the
normal melting point. In such cases the material is acceptable in
terms of purity, and the reduced melting point only indicated that
under the crystallization conditions used for that particular lot a
lower melting crystal form was preferentially produced.

Study of Phase Behavior

Phase behavior is amenable to study by a variety of techniques
including optical microscopy, X-ray diffraction, dilatometry, classical thermal analysis, and DTA. However, there is little doubt that
the relatively new technique of Differential Scanning Calorimetry
(DSC) is rapidly supplanting or supplementing these slower, more
cumbersome and less informative methods. With DSC and a few
milligrams of sample, phase behavior can be investigated under
precise temperature control; transitions can be observed either
heating, cooling or isothermally; temperatures can be measured
to an accuracy of $0.1°$; energies of transition can be determined to
within a few tenths of a percent; and a permanent recording is obtained representing the rate of transformation as a function of
temperature. In addition, by analysis of fusion peak shapes, the
absolute purity of single component samples may be determined

[3] and the displacement of the recording from the instrument baseline may be converted directly into specific heat data, again accurate to better than 1%. No other single technique yields such a wealth of information on so small a sample in such a short time.

A final point of importance in connection with phase behavior relates to the diversity of transition "types" which are observed in practice. The predictions of the "Phase Rule" which apply strictly to ideal systems in thermodynamic equilibrium are not always observed in real systems for a variety of reasons--crystal defects, surface free energy effects, nucleation phenomena and the like very often influence phase transitions to the point that their nature is more determined by such factors than by thermodynamic considerations. The melting behavior of semi-crystalline polymers is an extreme and well-known example.

In the following we have selected a few examples of phase transition studies by DSC to illustrate the power of the technique and the diverse nature of phase behavior in real systems. A further example, a very detailed investigation of KNO_3, has previously been published in Thermal Analysis Application Study No. 1 [4].

Experimental

A Perkin-Elmer Model DSC-2 Differential Scanning Calorimeter was used for all measurements. With an instrument of this type, both transition temperature and transition energy are obtained simultaneously. The DSC-2 was calibrated for temperature using materials with known temperatures of fusion. The improved temperature linearity of the Model DSC-2, as compared with the Model DSC-1B, necessitates the use of only two such materials. Calibration for energy was made by measurements on accurately weighed samples of high purity indium taken to have a heat of fusion of 6.79 cals/gram. Since the calorimetric response of the instrument has been shown to be independent of temperature, or other experimental variables, a single point calibration is sufficient [5]. Using standard procedures, ΔH values are expected to be accurate to within 0.5% and specific heat values to within 1.0% [6]. All samples were precisely weighed with a Perkin-Elmer Model AD-2 Autobalance.

FIGURE 1

DSC thermogram illustrating
the rotational and melting
transitions in dotriacontane.

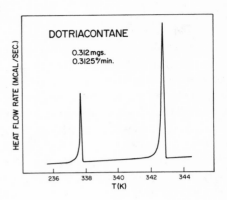

FIGURE 2

The specific heat of a metal
complex in the region of a solid-
solid transition.

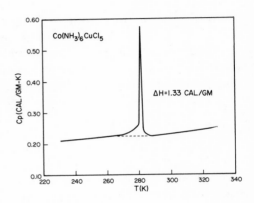

FIGURE 3

The specific heat of a metal
complex in the region of a
solid-solid transition (expanded
scale).

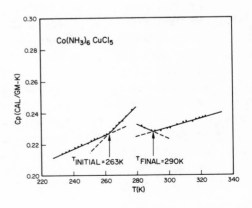

Results

1. Dotriacontane

Linear aliphatic hydrocarbons frequently exhibit one or more solid-solid transitions of the normal enantiotropic type; that is, the transitions are observed on heating and cooling. Dotriacontane in particular is frequently used in DTA and DSC as a standard material for assessing the resolution capabilities of instrumentation. In addition to the melting transition at approximately 343K (70°C), a solid-solid transition occurs near 338K (65°C). This transition represents the onset of a cooperative rotational or "crankshaft" mobility of the chains within the crystal. Analogous transitions are found in numerous molecules having long aliphatic structural units--fats, soaps and waxes are among the major classes of materials where one or more such transitions are commonly observed close to the terminal melting point.

The thermogram of dotriacontane (Figure 1) shows the solid-solid transition and the melting transition as relatively sharp peaks having a curved pre-transition region and a sharp termination. The pre-transition curvature is an expected effect of impurities for the melting peak. The similar appearance of the rotational transition is coincidentally similar in this case as peak broadening due to impurities is not usually manifested for solid-solid transitions as we shall see later. Less pure samples of dotriacontane show substantial broadening of the melting peak but have relatively little effect on the shape of the solid-solid transition.

The energy associated with the two transitions in dotriacontane (approximately 20 cals/gram and 40 cals/gram, respectively) is typical for heats of transition and fusion. Heats of fusion for the majority of compounds appear to lie in the range of 20 to 50 calories per gram. Heats of solid-solid transitions are, however, more variable and frequently much smaller, sometimes being only a few tenths of a calorie per gram. Structural changes may involve only slight rearrangement of the lattice with correspondingly small energy changes.

2. Hexa-aminecobalt (III)-pentachlorocuprate (II)

As indicated earlier, peak shapes for solid-solid transitions are frequently anomalous in the sense that they may exhibit either pre-transition or post-transition effects or both and not have the simple structure shown by typical fusion peaks such as that of dotriacontane. An example is the well-known quartz inversion transition near 570°C where the specific heat increases rapidly over an extended temperature range prior to the relatively sharp transition peak [7].

The metal complex compound [8] illustrated in Figure 2 shows both pre-, and post-transition effects where the specific heat function neither rises sharply to the transition peak nor drops immediately to a linear baseline. Such effects are common. An excellent reference for those interested in further detail may be found in the review of Westrum and McCullough [9].

Anomalous peak shapes raise the question of the proper procedure for measuring the peak area or heat of transition. How much of the pre-, and post-transition effects should be counted as heat of transition and how much as ordinary specific heat is not always clear. In such cases it is important to carefully define and illustrate the procedure used in determining the transition heat. In this example the data were plotted on an expanded specific heat scale and the points where the specific heat clearly departed from a linear function were determined (Figure 3). Returning to Figure 2, a straight line was drawn between these points to define the peak area. A ΔH of 1.33 cals/grams was found, a relatively small energy compared to typical heats of fusion.

3. Nickel Nitrate Hexahydrate

Since information on the phase behavior of nickel nitrate hexahydrate is virtually nonexistent, we undertook a brief investigation. Utilizing supplemental information from the Perkin-Elmer TGS-1 thermogravimetric system, we believe the following results fairly represent the phase behavior of this system over the temperature range studied.

The hexahydrate sample was placed in a hermetically sealed sample pan for these experiments. Gold sample pans were used instead of the usual aluminum sample pans because the material was found to react with aluminum.

Two representative runs are shown in Figure 4. Run A at the top of the figure represents the behavior of the sample on cooling, while Run B on the bottom of the figure represents the behavior on heating.

A description of the type of transition occurring at each of the represented peaks was arrived at only after a series of experiments. For example, the sample was heated to various temperatures in the range studied until one or more of the transitions occurred and immediately cooled at a controlled rate to observe the number, temperatures and energies of any appearing peaks. For this type of experiment to be successful, it is necessary for the instrument to revert to the cooling mode with no overshoot. Conversely, the sample was cooled to various levels, the cycle reversed, and the sample heated to various levels. Again, to be successful, the sample must always be under strict temperature control.

After completion of these experiments and an examination of the temperatures and energies involved, the following conclusions were arrived at. On cooling from 380K, the sample exhibits only a single exothermic thermal event labeled I, due to the precipitation of the tetrahydrate from a saturated solution. On heating from 260K, the sample clearly exhibits three thermal events represented by II, III, and IV. The exothermic event, labeled II, is due to the recombination of the tetrahydrate with two moles of water to reform the hexahydrate. The first-order transition, labeled III, is an invariant transformation (i. e., occurring isothermally) and is due to the hexahydrate forming the tetrahydrate and two moles of water. This is immediately followed by endotherm IV due to the formation of a saturated solution.

4. Nickel-Iron Alloy (76% Ni - 24% Fe)

Other types of phase change have been discovered which show quite a different character from those previously discussed. In these there seems to be no difference of volume between the two forms of the substance and also little or no difference in enthalpy, i. e., zero or almost zero latent heat. The transition is evident simply by a sharp change in the heat capacity and compressibility, and these properties vary rather rapidly as the transition point is approached.

FIGURE 4

DSC thermogram illustrating
the phase behavior of nickel
nitrate hexahydrate in the
heating and cooling mode.

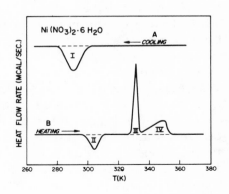

FIGURE 5

The specific heat of a nickel-
iron alloy in the region of its
Curie Point.

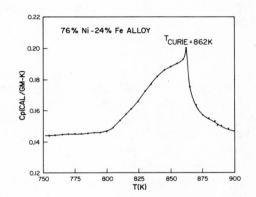

FIGURE 6

DSC thermogram illustrating
the effect of an impurity on the
phase behavior of cyclopentane.

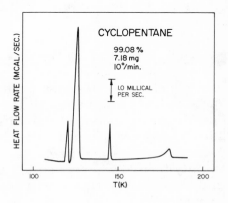

An example of this type is the transition which occurs in ferromagnetic materials. The magnetic domains in these materials are known to become disoriented over characteristic temperature ranges where materials more or less sharply transform to the paramagnetic state. These characteristic temperatures are generally known as Curie points. Since energy must be supplied in order to disrupt polarization, the specific heat would be expected to be anomalous at the Curie point.

To illustrate this anomalous behavior, the specific heat of a nickel-iron alloy (76% Ni-24% Fe) was measured in the region of its Curie point (Figure 5). With no prior indication, the specific heat begins a rapid and anomalous rise at about 800K. This increase in specific heat continues to a maximum value at its Curie point (T=862K), and falls off rapidly at the completion of the transition. Thermal studies on such materials may provide insight into the electrical contributions to specific heat and also into the nature of phase transitions in general.

5. Cyclopentane

Cyclopentane is a classical example of the group of materials known as plastic crystals. These materials are characterized by one or more solid-solid transitions, unusually low entropies of fusion and higher melting points than expected. Cyclopentane has two solid-solid transitions at 122K and 139K and melts at 179K. Due to these well-defined transition temperatures, the material is well suited for the calibration of the DSC in this low temperature region. It is particularly convenient since solid-solid transitions do not supercool as much as liquid-solid transitions.

However, in the course of evaluating cyclopentane as a temperature standard, it was found that the sample must be very pure since even small levels of impurity produced gross changes in the thermogram (Figure 6). An additional peak is observed at 118K due to a metastable crystal form apparently induced by the impurity, even though the sample is better than 99% pure. Moreover, the melting range is very broad. Since the heat of fusion of cyclopentane is only 2 calories per gram, this change in the purity of 1% will actually depress its melting point by 4°. Since impurities in the sample account for these changes, the sample was subsequently purified to a purity of better than 99.99% by a gas chromatographic technique. Figure 7 shows the DSC scan of the purified

FIGURE 7

DSC thermogram illustrating
the phase behavior of ultra-
high purity cyclopentane.

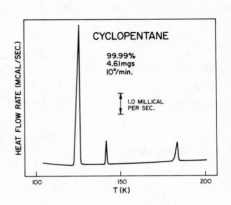

FIGURE 8

DSC thermogram illustrating
the phase behavior of cholesteryl myristate in the heating
and cooling mode.

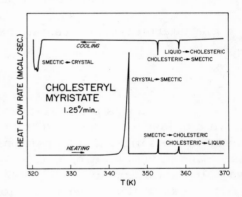

FIGURE 9

DSC thermogram of cholesteryl
palmitate illustrating the
detection of a "monotropic"
transition.

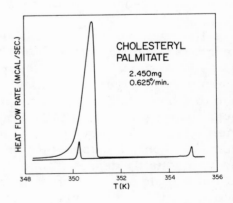

cyclopentane. Now, only the two transitions are observed and the melting point is sharp as expected.

6. Cholesteryl Myristate

A liquid crystal substance can exhibit more than one type of intermediate or mesomorphic structure as the temperature is raised. The phase transitions between the various mesomorphic forms occur at a thermodynamically defined temperature as the liquid crystal undergoes a change in internal order at the point of the phase transition. Transitions from solid to mesophase, from mesophase to isotropic liquid and between multiple mesophase structures are first-order transitions.

Figure 9 illustrates the behavior of a cholesteryl myristate sample in both the heating and cooling mode. While there is nothing particularly novel about a thermogram of cholesteryl myristate [10]; it does illustrate another class of materials where the study of phase behavior is important.

7. Cholesterol Palmitate

Phase behavior under conditions of programmed cooling is just as important in the characterization of materials as that observed with heating runs. Very often different crystal forms are produced depending on cooling rate. In addition, some transitions are observable only in the cooling mode. For example, cholesteryl palmitate on heating exhibits a large crystal to liquid crystal transition followed by a relatively weak liquid crystal to liquid phase change. On cooling, however, the fact that the major transition supercools, allows the observation of a second liquid crystal transition which occurs below the major transition temperature. Having detected this lower temperature transition, it is now possible to observe it by heating as well (Figure 9). This is accomplished by cooling the material and then reversing the program before the sample has a chance to crystallize. Figure 9 shows two scans of the same sample made under high resolution conditions. The upper thermogram shows the normal crystal-cholesteric transition and the lower shows the monotropic smectic-cholesteric change which occurs only a few tenths of a degree lower [11]. Both, of course, show the cholesteric-liquid transition near 355K.

FIGURE 10

DSC thermogram illustrating
phase transitions in a lipid-
water system.

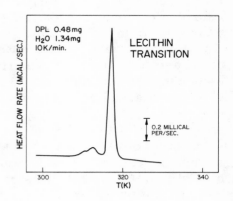

FIGURE 11

DSC thermogram illustrating
the effect of thermal history
on solid-solid transitions in
PTFE.

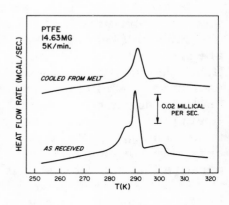

8. Dipalmitol Lecithin

The class of biological materials known as phospholipids are of particular interest to biochemists for the remarkable thermal transitions which occur in these systems often near body temperature [12]. Lipids are surfactants consisting of a hydrophilic polar head and one or more hydrophobic aliphatic tails. In solution these lipids form a two-dimensional array in which all the lipids are arranged in parallel. As a result of thermal treatment, lipids undergo cooperative processes similar to the transitions occurring in liquid crystals.

The system chosen to demonstrate this phenomena was that of a commercially available lecithin, 1, 2 dipalmitol-L-phosphatidycholine, in water (Figure 10). The main thermal transition consists in the "melting" of the two-dimensional array to form an array where the long range structure remains intact but the tails have additional liquid-like degrees of freedom. The lower temperature pre-transition peak has been attributed to various other phenomena [12].

9. Polytetrafluoroethylene

Few polymers exhibit the type of solid-solid phase transitions that are common in many organic materials. Probably the best known polymer that exhibits this phenomena is polytetrafluoroethylene [13]. The importance of these transitions in PTFE is increased because these transitions occur over a relatively short interval near room temperature, a normal use temperature for these materials.

To illustrate these transitions, ordinary PTFE "tape" was used (Figure 11). The first run on this sample produced the "as received" thermogram at the bottom of the figure. This sample was subsequently melted, cooled and rescanned over the same temperature range and produced the "cooled from melt" thermogram at the top of the figure. The difference between these two runs, both in the number of peaks present and their intensity, is no doubt due to thermal history. Most probably the sample was initially somewhat more crystalline than after the first heating, accounting for lower intensity in the "cooled from melt" thermogram which is consistent with the interpretation of these transitions. The presence of a third peak in the lower thermogram is probably due to an annealing effect which occurred during the processing of the material.

Conclusions

Differential Scanning Calorimetry is particularly well suited for the investigation of phase transitions occurring in organic, inorganic, and polymeric materials. DSC not only allows the determination of the temperature and energy of any phase transition but also absolute purity for single component systems. Particularly advantageous is the ability of the DSC to examine a sample under precisely controlled cooling conditions or isothermally. Because of the fundamental and applied importance of phase transitions, DSC will continue to prove its unique value.

Acknowledgements

The author would like to express his appreciation to Mr. Robert Fox of the Hall Chemical Co., Arab, Alabama, for supplying the nickel nitrate hexahydrate sample, and to Dr. Ivan Bernal of Brookhaven National Laboratory for the metal complex sample.

References

1. Glasstone, S., "Textbook of Physical Chemistry," D. Van Nostrand Co., New York (1950).

2. Denbigh, K., "The Principles of Chemical Equilibrium," Cambridge University Press, London (1964).

3. Marti. E. F., Thermochim. Acta, 4, 173 (1973).

4. Gray, A. P., Thermal Analysis Application Study No. 1 Perkin-Elmer Corp., Norwalk, Conn.

5. Schwenker, Jr., R. F., and Whitwell, in "Analytical Calorimetry," Vol. 1, Plenum Press, New York (1968) p. 249.

6. Brennan, W.P., and Gray, A.P., Thermal Analysis Application Study No. 9, Perkin-Elmer Corp., Norwalk, Conn.

7. O'Neill, M.J., and Fyans, R. L., Report MA-9, Perkin-Elmer Corp., Norwalk, Conn.

8. Long II., T.V., et al., Inorg. Chem., 9, 459 (1970).

9. Westrum, Jr., E.F., and McCullough, J.P., in "Physics and Chemistry of the Organic Solid State," Vol. I., Interscience Publishers, New York (1963), p. 1.

10. Davis, G.J., Porter, R.S., and Barrall II, E.M., Mol. Cryst. Liq. Cryst., 11, 319 (1970).

11. Vogel, M.J., Barrall II, E.M. and Mignosa, C.P., Mol. Cryst. Liq. Cryst., 15, 49 (1971).

12. Ladbrook, B.D. and Chapman, D., Chem. Phys. Lipids, 3, 304 (1969).

13. Araki, Y., J. Appl. Poly. Sci., 9, 421 (1965).

IMPORTANCE OF JOINT KNOWLEDGE OF $\Delta H°$ AND $\Delta S°$ VALUES FOR INTERPRETING COORDINATION REACTIONS IN SOLUTION

Guy BERTHON

Laboratoire de Thermodynamique chimique et Electrochimie

40, Avenue du Recteur Pineau - 86022 POITIERS (France)

INTRODUCTION

In the study of proton ionization and complexing of amines with transition metal ions, the nature and properties of the bonds, and the influence of substituents are mostly discussed in terms only of the free energies (or stability constants) for the corresponding reactions (1)(2).

However, during the past ten years, advances in experimental calorimetry have permitted the individual determination of enthalpies and entropies, allowing a better interpretation of the effects observed. This field of research has thus been considerably advanced (3)(4)(5)(6).

The present paper aims to emphasize the importance of knowing both enthalpies and entropies when interpreting the effects exerted upon the bond formation.

In point of fact, it often occurs that the variations of each of these thermodynamic functions are compensated for in the free energy expression : when free energy alone is determined, this compensation may prevent the discovery of interesting phenomena.

This affirmation is here evidenced by means of exemplary cases related to H^+-piperidine (and its methyl-derivatives), Ag^+-pyridazine and Ag^+-thiourea systems.

EXPERIMENTAL

1°- H$^+$-piperidine (P) and its 2-methyl (2MP), 2,6-dimethyl (2,6
    DMP) ; 3 methyl (3MP) and 4 methyl (4MP) derivatives systems.

    In the study of proton ionization of piperidine and its 2-me-
thyl ; 2,6-dimethyl ; 3-methyl and 4-methyl substituted derivatives
(7) we have found that the stability constants (or free energies of
conjugate acid formation) are not changed by the introduction of a
substituent on the parent molecule (see table I)

Table   I

H$^+$-piperidines systems (μ = 0.5M, KNO$_3$; 25°C)

Amine	log K	$\Delta G°$ kcal.mol^{-1}	$\Delta H°$ kcal.mol^{-1}	$\Delta S°$ cal.deg^{-1}.mol^{-1}
P	11.26	15.38	13.19	7.34
4MP	11.23	15.34	13.48	6.24
3MP	11.20	15.30	13.75	5.20
2MP	11.21	15.31	14.03	4.29
2,6DMP	11.26	15.38	14.53	2.85

    Yet in such a case, the transmission of the σ-inductive ef-
fect exerted by the substituent is supposed to take place by successive polarisation of the carbon-carbon σ-bonds separating this
substituent from the reaction center, the polarisation decreasing
in passing from one bond to the next (8). The inductive effect
should thus modify the stability constants, but no such modifications are observed.

    Nevertheless, it is known (9) that electric effects essential-
ly modify enthalpies of reaction, whereas steric and solvent effects
change both enthalpies and entropies.

    This affirmation allows the interpretation of proton ionization
enthalpies here observed : they decrease (algebraically) in the order P $>$ 4MP $>$ 3MP $>$ 2MP $>$ 2,6DMP , because the inductive effect of
the methyl substituent increases in the order 4MP $<$ 3MP $<$ 2MP $<$
2,6 DMP.

    As regards the entropies, the solvating water molecules, predo-
minantly drawn towards the methyl group (10), are from the position
4 to the position 2 progressively nearer to the nitrogen atom. That
is a possible explanation of the decrease of the entropies values in
the order P $>$ 4MP $>$ 3MP $>$ 2MP $>$ 2,6DMP ; in fact, the site of the

reaction is more and more solvated from P to 2,6DMP (P < 4MP< 3MP < 2MP < 2,6DMP). Moreover, this solvation produces a steric hindrance that exerts a progressively stronger influence on the H_3O^+ ion.

For these systems, the compensation of the variations of enthalpic and entropic terms in the free energy values is almost rigorous. Only the individual determination of each of these has permitted us to find out the influences exerted by the methyl group in different positions.

2°- Ag^+-pyridazine system.

In the comparative thermodynamic study of complexation of silver ion with the heterocyclic amines pyridazine, pyrimidine and pyrazine (11), we have noticed a regular decrease of stability constants β_1 versus β_2 values (see table II), but the examination of the reaction enthalpies and entropies permits us to find an anomalous situation of ΔH°_1 and ΔS°_1 values for Ag^+-pyridazine system.

Table II

Ag^+-diazines systems ($\mu = 0.1M, KNO_3$; 25°C)

	Ag^+-pyridazine	Ag^+-pyrimidine	Ag^+-pyrazine
log β_1	1.48	1.61	1.38
log β_2	2.82	2.98	2.41
$- \Delta G^\circ_1$ (kcal.mol^{-1})	2.02	2.20	1.88
$- \Delta G^\circ_2$ (kcal.mol^{-1})	3.85	4.07	3.29
$- \Delta H^\circ_1$ (kcal.mol^{-1})	_7.39_	4.21	4.07
$- \Delta H^\circ_2$ (kcal.mol^{-1})	8.06	8.35	8.10
$- \Delta S^\circ_1$ (cal.deg^{-1}.mol^{-1})	_18.0_	6.74	7.34
$- \Delta S^\circ_2$ (cal.deg^{-1}.mol^{-1})	14.1	14.35	16.1

In fact, the $\Delta H°_1$ value relative to the first complex formation (Ag^+ + pyridazine $\rightleftharpoons$ $Ag(pyridazine)^+$) is very low (algebraically) and almost equal to $\Delta H°_2$, which is relative to the overall
formation of the second complex (Ag^+ + 2 pyridazine $\rightleftharpoons$ Ag(pyridazine)$_2^+$).

It appears that the most suitable interpretation of this irre-
gularity is the simultaneous coordination of the two nitrogen atoms
of the pyridazine molecule with the Ag^+ ion in the first complex (I)
formation.

(I)   (II)

Therefore, the very constrainted structure of the form (I) may
explain the unusually low value of the corresponding entropy $\Delta S°_1$.

In practice, the second complex formation corresponds with a
steric reorganisation, the structure (II) allowing a more important
number of degrees of freedom. The entropy of the reaction (I)$\rightarrow$(II)
is effectively positive (about + 4 cal.deg^{-1}.mol^{-1}), the enthalpic
term (-0.67 kcal.mol^{-1}) representing a very small part of the free
energy (-1.83 kcal.mol^{-1}).

In this case, the type of binding here evidenced was completely
unforeseeable by examining only stability constants.

3°- Ag^+-thiourea system.

In our initial study of the Ag^+-thiourea system (12), we had
primarily characterised the simultaneous existence of four complexes
in solution, their successive stability constants

$$k_n = \frac{\left[Ag\,(thiourea)_n^+\right]}{\left[Ag(thiourea)_{n-1}^+\right] . \left[thiourea\right]}$$

regularly decreasing from k_1 to k_4. The determination of these

constants at different temperatures had also permitted an approximate calculation of the corresponding reaction enthalpies and entropies
(see table III), evidencing a spectacular discontinuity of their variations.

Table III

Ag$^+$-thiourea system ($\mu = 0.5M$, KNO$_3$; 25°C)

comple-xation degree	log k$_n$	$- \Delta G_n^°(12)$ kcal.mol^{-1}	$- \Delta H_n^°(12)$ kcal.mol^{-1}	$- \Delta H_n^°(14)$ kcal.mol^{-1}	$\Delta S_n^°(12)$ e.u.	$\Delta S_n^°(14)$ e.u.
1	7.04	9.62	0	2.6	32	23.5
2	3.57	4.88	16.5$\pm$4	19.33	-38.5	-48.5
3	2.10	2.87	0	1.13	9.4	5.8
4	0.78	1.07	7.5$\pm$2	8.93	-21.6	-26.4

By refering to a study of UUSITALO (13) in which thermodynamic
and spectroscopic data are systematically compared, we had deduced
that the first and the third complex had sulphur bonds, but that the
second and the fourth had nitrogen bonds (12).

More recently, we have obtained, by means of direct calorimetry
(14), more accurate results about this system which have confirmed
the first hypothesis (the value $\Delta H = -2$ kcal.mol^{-1} found in littera-
ture (15) relating to Cu-S bond tends also to support this opinion)
and permitted a finer interpretation of the binding of these complexes.

Effectively, by examining in detail our last values (14), it
may noticed that the enthalpy of the second complex formation approximately corresponds with the forming of two Ag-N bonds (16), while
the Ag-S bond of the first complex is broken.

The enthalpy value of the third complex formation corresponds
with the addition of one Ag-S bond, without breaking the previously
extant Ag-N bonds.

Finally, the successive enthalpy $\Delta H_{3,4}^°$ ($\Delta H_4^° - \Delta H_3^°$) may be explained by forming of a supplementary Ag-N bond, without breaking any
bond.

The successive forms suitable for the thermodynamic functions
obtained (14), without taking into account the steric configurations,
are as follows :

```
                              |                    |
                              S                    S
                              |                    |
   Ag-S-       -N-Ag-N-    -N-Ag-N-            -N-Ag-N-
                                                    |
                                                    N
                                                    |

    (I)          (II)         (III)               (IV)
```

CONCLUSION

The three examples here just illustrated show the incontestable utility of knowing individually the enthalpies and the entropies of coordination reactions for discussing the bonding and also, in the case of comparison of different systems, the influences exerted by substituents.

As we have seen, it frequently occurs that the variations of the enthalpic and entropic terms are completely compensated for. In such cases, the determination of only the free energies is slender for interpreting coordination phenomena.

BIBLIOGRAPHY

(1) SILLEN, L.G. ; MARTELL, A.E.   "Stability constants" Special publication n° 17, The Chemical Society, London (1964) ; Supplement n° 1, Special publication 25 (1971).

(2) PERRIN, D.D.   "Dissociation constants of organic bases in aqueous solution" Butterworths, London (1965) ; Supplement (1972).

(3) PAOLETTI, P. et al.   J. Chem. Soc. (1961), 65, 1224 ; J. Phys. Chem. (1965), 69, 3759 ; Helv. Chim. Acta (1971), 54, 243 ; J. Chem. Soc. (1972), 736.

(4) CHRISTENSEN, J.J. et al.   J. Phys. Chem. (1966), 70, 2003 ; (1968), 72, 1208 ; J. Chem. Soc., A (1969), 1212 ; Thermochim. Acta (1972), 3, 219 ; 5, 35.

(5) CABANI, S. ; GIANNI, P.   J. Chem. Soc. (1968), 547 ; Anal. Chem. (1972), 44, 253.

(6) WILLIAMS, D.R. et al.   J. Chem. Soc. (1968), 2965 ; (1970), 1550 ; (1972), 790.

(7) BLAIS, M.J. ; ENEA, O. ; BERTHON, G.   Thermochim. Acta (in press).

(8) DEWAR, M.J.S. ; GRISDALE, P.J.   J. Amer. Chem. Soc. (1962), 84, 3548.

(9) LAIDLER, K.J.   Trans. Farad. Soc. (1959), 55, 1725.

(10) CABANI, S. ; CONTI, G. ; LEPORI, L.   Trans. Farad. Soc. (1971) 67, 1933.

(11) BERTHON, G. ; ENEA, O.   Thermochim. Acta (1973), 6, 57.

(12) BERTHON, G. ; LUCA, C.   Bull. Soc. Chim. (1969), 2, 432.

(13) UUSITALO, E.   Ann. Acad. Sc. Fennicae (1967), 6, 47.

(14) ENEA, O. ; BERTHON, G.   Thermoch. Acta (1973), 6, 47.

(15) ASHCROFT, S.F. ; MORTIMER, C.T.   "Thermochemistry of transition metal complexes", Academic Press, London (1970).

(16) CHRISTENSEN, J.J. ; IZATT, R.M.   "Handbook of metal ligand heats" M. Dekker, New-York (1970).

(17) ENEA, O. ; BERTHON, G.   C.R. Acad. Sc., série C (1972), 274, 1968.

(18) ENEA, O. ; BERTHON, G. ; BOKRA, Y.   Thermochim. Acta (1972), 4, 449.

(19) ENEA, O. ; HOUNGBOSSA, K. ; BERTHON, G.   Electrochim. Acta (1972), 17, 1585.

(20) BERTHON, G. ; ENEA, O. ; HOUNGBOSSA, K.   C.R. Acad. Sc., série C (1972), 273, 1140.

(21) HOUNGBOSSA, K. ; BERTHON, G. ; ENEA, O.   Thermochim. Acta (1973), 6, 215.

THE DETERMINATION OF THE HEAT OF FUSION BY DIFFERENTIAL SCANNING CALORIMETRY AND BY MEASUREMENTS OF THE HEAT OF DISSOLUTION

Erwin E. Marti

Central Research Services Department

Ciba-Geigy Limited, Basle, Switzerland

Abstract

A comparison of the DSC and solution calorimetry methods is presented. The instruments used were a DSC-1B and a LKB-8700 precision calorimeter. The calibration procedures and the reproducibility of caloric measurements are reported. As an example the heat of fusion of p-xylene was determined by both methods and the results are compared. The analytical potential of the methods for the characterization of substances is discussed.

Introduction

The determination of the heat of fusion is possible by the following methods:

- calorimetric methods with adiabatic and nonadiabatic instruments
- measurements of the heat of dissolution for the crystalline and the amorphous form of a substance
- vapor pressure measurements in the melting point region of a substance for the solid and the liquid phase.

The investigations presented here are restricted to the first two methods. The caloric calibration, the measurements and the results of the reproducibility and the discussion of the accuracy by comparison of values obtained by the methods applied as well as by comparison with literature values are reported. The selection of p-xylene in our investigations brought advantages such as: practically no experimental difficulties, and thermodynamic properties which are well described in the literature.

The solid-liquid equilibrium for substances with eutectic impurities is of great importance for any method applied for the determination of the heat of fusion. The theory of the solid-liquid equilibrium will not be discussed because the facts are presented in references (1).

The errors of the mean values are given without additional comment in terms of confidence intervals on a 95 % level.

Experimental

Differential Scanning Calorimetry
The instrument used for these investigations was the DSC-1B of the Perkin-Elmer Corporation. Indium with a purity better than 99.9995 % from NV Hollandse Metallurgische Industrie Billiton was the substance for the caloric calibration of the instrument and for the determination of the reproducibility. The heat of fusion was determined for p-xylene Fluka puriss.

DSC curves were recorded on a Speedomax W recorder with a 10 mV range. The data were collected simultaneously with a Digital Data Acquisition system ERA. The data collection rate of the ERA instrument was chosen as 10 points sec^{-1}. The melting areas and the eutectic impurities were calculated with a computer program from data stored on magnetic tape by the ERA system and from experimental conditions and calibration factors, which were punched on cards. The purity calculations are based on the program described in references (1a). The part of the program for the calculations of the melting areas and the caloric calibration factors from experimental data obtained by the ERA system are outlined below. The evaluation is called ERA standard procedure.

Two temperature intervals are selected visually from the recorder curve in the baseline ranges one below and one above the melting region. The arithmetic mean values and their standard deviations are calculated with the computer program from the x,y co-ordinates of all data points within each of the two intervals. The x-axis represents the temperature and the y-axis the heat flow. The means of the x and y values of the data points within these two intervals are used for the determination of the baseline. The selection of the melting region is discussed as a next step. A distance, calculated by multiplying the mean value of the two standard deviations in the direction of the heat flow axis with the factor of 3, is used as the discrimination level for the determination of the melting range. The number of the data points within the melting range is then reduced by bunching to less than 150. These reduced points represent the melting curve for the calculation of the melting area and the purity value. The melting area is determined by the

summation of the trapezoids formed with these points and with the
baseline.

The caloric calibration factor of the instrument for given
experimental conditions is calculated by the equation

$$(f_c)_{R,\dot{T},RS} = \frac{\Delta H_{f,In} \cdot m}{A_{o,n} \quad A} \qquad (1)$$

where $(f_c)_{R,\dot{T},RS}$ is the caloric calibration factor in mcal cm^{-2} for
a given range R in mcal sec^{-1}, for a given scan speed $\dot{T}$ in ^{o}C min^{-1},
and for a given recorder speed RS in cm min^{-1}, $\Delta H_{f,In}$ is the heat
of fusion of indium in cal mole^{-1}, m is the sample weight in mg,
$A_{o,n}$ is the melting area in cm^2, and A is the relative atomic mass
of indium.

A few melting areas of indium were also evaluated by two other
procedures for a comparison with the melting areas calculated by
computer:

- mechanical planimetry
- weighing of the recorder paper, which was cut along the melting
  curve and along the baseline, and comparing it with the weight of
  a piece of paper with a known area.

For both methods the baseline was drawn as a straight line from
the premelting to the postmelting DSC-curve.

Solution Calorimetry
The measurements of solution calorimetry were performed with the
LKB 8700-1 Precision Calorimetry System.

The heat of solution was determined for p-xylene Fluka puriss
in the solvent toluene p.a. Merck. The 1 ml ampul was filled with
p-xylene and the 100 ml standard vessel of the instrument was filled
with 100.0 ml toluene. The set-up of the instrument and the experimental procedure of the caloric determinations were in accordance with
LKB instructions (2), the only exception being that the output of the
electronic galvanometer was connected to a W + W 1100 recorder. The
temperature-time curves obtained from electrical calibrations or by
heats of solution were evaluated by a graphical procedure (2), which
yields results of high accuracy especially for the observed relaxation
times in order of 1 minute.

Results and Evaluations
Differential Scanning Calorimetry
The caloric calibration factor determined with the heat of fusion of
indium and calculated according to Eqn. (1) is presented in Table I

as the mean value of three or four single measurements. The melting
point of indium is 429.76°K (3), the heat of fusion is 781 cal mole^{-1}
(3,4), and the relative atomic mass is 114.82.

The measurements with the same sample of indium were performed
without removing the sample pan from the sample pan holder.

Table I. Caloric Calibration Factors
of a DSC-1B determined with Indium
(Range 4 mcal sec^{-1}, Recorder speed 10.2 cm min^{-1})

Sample weight mg	Mean values of the caloric calibration factor in mcal cm^{-2}			
	Scan speed °C min^{-1}			
	0.5	1	2	4
8.353	0.923±0.009	0.923±0.003	0.927±0.012*	0.959+0.012*
8.620		0.930±0.003		
12.785	0.932±0.003	0.922±0.003		
26.49	0.926+0.003	0.953±0.009*		

* Heat flow rate at maximum beyond the linear range
of the DSC output.

Investigations on the long time stability of the caloric calibration
factor from 2.2.73 to 10.9.73 with a total of 23 measurements of
the same sample of indium with a weight of 8.353 mg show a mean
value of

$$(f_c)_{R=4,\ \dot{T}=1,\ RS=10.2} = 0.922 \pm 0.004 \text{ mcal cm}^{-2}$$

The lowest value of these measurements is 0.911 and the highest
value 0.946.

The caloric calibration factors determined by mechanical
planimetry of melting curves are about 1.5 % higher than the values
obtained by the ERA standard procedure. Another evaluation of
melting curves - the weighing of the melting areas - shows caloric
calibration factors which are 2.8 % higher than the corresponding
values calculated by our standard method.

The melting point T_o, the eutectic purity n_o, and the heat of
fusion were determined for p-xylene puriss. The heat of fusion was
calculated for p-xylene by three different procedures. The heat
of fusion ΔH_f, uncorr. was calculated by applying the ERA standard
procedure for the calculation of the melting areas. The melting

area, the calibration factor, the sample weight, and the relative
molecular mass of the substance are used for the calculation of
ΔH_f, uncorr. A more accurate value of the heat of fusion $\Delta H_{f,DSC,A}$
was obtained by the normal DSC linearization procedure for purity
calculations (1a). The third procedure corrects the sample weight
for the molten fraction which exists at the temperature where melting
is observed by the ERA standard procedure. The molten fraction is
calculated by a numerical integration of Eqn. (33) from the
references (1). This equation is based on the solubility equilibrium.
The thermodynamic properties and the eutectic impurity inserted into
Eqn. (33) for the numerical integration were obtained from the DSC
linearization procedure for purity calculation. It should be mentioned
that the linearization procedure yielding the values $\Delta H_{f,DSC,A}$ corrects
also the eutectic premelting but with a different approach. The heat of
fusion calculated by the third procedure is indicated as $\Delta H_{f,DSC,B}$.
The results for p-xylene are as follows:

$$T_o = 13.3 \pm 0.1 \ ^{\circ}C$$

$$n_o = 99.61 \pm 0.02 \text{ mole } \%$$

$$\Delta H_{f,uncorr.} = 3956 \pm 86 \text{ cal mole}^{-1}$$

$$\Delta H_{f, DSC, A} = 4126 \pm 70 \text{ cal mole}^{-1}$$

$$\Delta H_{f, DSC, B} = 4084 \pm 70 \text{ cal mole}^{-1}$$

Solution Calorimetry

The internal electrical calibration of the LKB calorimeter showed
the following reproducibility of the measured temperature difference
presented in Table II.

Table II. Solution Calorimetry: Reproducibility of the Temperature
Difference measured for a given Amount of Heat.

Range	Heat generated	Number of measurements	Relative error of the mean value of measured temp. diff.
μV	cal		$\%$
10^3	7.17	5	0.47
10^2	0.717	6	0.88
10^1	0.0717	5	1.36

Two sets of measurements were performed with p-xylene. The crystalline p-xylene was dissolved in a volume of 100 ml of toluene at
6.1 $^{\circ}C$ and the liquid p-xylene at 15.4 $^{\circ}C$ in the same amount of
toluene. The sample weight for p-xylene was 100 mg. The results
obtained with the LKB calorimeter are shown in Table III.

Table III. Heats of Dissolution of p-Xylene in Toluene

Range	Temp.	State *	Number of Measurements	ΔH_s
μV	$^{\circ}C$			cal mole^{-1} of p-xylene
10^2	15.4	liq	4	+ 23 $\pm$ 5
10^3	6.1	c	3	+4051 $\pm$ 20

(* liq: liquid,   c: crystalline)

The heat of dissolution of the crystalline p-xylene in Table III is corrected for the amount of the liquid phase due to the eutectic premelting of p-xylene at 6.1 $^{\circ}$C. The relative amount of this liquid phase is approximated for the given conditions and the known eutectic impurity of the sample to 1.9 $\pm$ 0.25 % [see references (1), Eqns. (29) and (33)]. The error of the amount of the molten fraction at 6.1 $^{\circ}$C is estimated by the standard deviations of the thermodynamic properties and the value of the eutectic impurity introduced into the two equations.

The heat of dissolution of the liquid p-xylene in toluene is an extremely low value: therefore the solution can be regarded as practically ideal. As a consequence the heat of dissolution of liquid p-xylene in toluene is practically independent of temperature. The heat of fusion of p-xylene at 6.1 $^{\circ}$C is by a good approximation given by

$$\Delta H_{f,LKB}^{6.1^{\circ}C} = \Delta H_{s,c} - \Delta H_{s,liq} = 4028 \pm 25 \text{ cal mole}^{-1}$$

The heat of fusion of p-xylene measured with the LKB instrument should be known at the melting point for a comparison with the corresponding value obtained by DSC. Obviously the following equation can be applied for the calculation of the temperature dependence of the heat of fusion

$$\Delta H_{f,LKB}^{T_o} = \Delta H_{f,LKB}^{T_1} + (T_o - T_1)(c_{p,liq} - c_{p,c}) \qquad (2)$$

where $\Delta H_{f,LKB}^{T_o}$ is the heat of fusion at the temperature T_o, $\Delta H_{f,LKB}^{T_1}$ is the heat of fusion at the temperature T_1, $c_{p,liq}$ is the heat capacity of the liquid substance, and $c_{p,c}$ is the heat capacity of the crystalline substance. The mean values of the heat capacities for p-xylene in the temperature interval of 0-13.3 $^{\circ}$C are according to Corruccini and Ginnings (5)

$$c_{p,liq} = 42.2 \text{ cal deg}^{-1} \text{ mole}^{-1}$$
$$c_{p,c} = 36.8 \text{ cal deg}^{-1} \text{ mole}^{-1}$$

The temperatures for our calculation are $T_0 = 13.3$ and $T_1 = 6.1$ °C. These values introduced in Eqn. (2) lead to

$$
\begin{aligned}
\Delta H_{f,LKB}^{13.3\ °C} = \Delta H_{f,LKB} &= 4028 \pm 25 + (13.3 - 6.1)(42.2 - 36.8) \\
&= 4028 \pm 25 + 39 \pm 5 \\
&= 4067 \pm 30 \text{ cal mole}^{-1}
\end{aligned}
$$

Discussion

The caloric calibration factors of the DSC-1B are reproducible under given experimental conditions and with the evaluation called ERA standard procedure to a relative error of the mean values of 1 %. The long time stability of the caloric calibration factor measured during a period of 8 months shows a relative error of 0.4 % for the mean value and 1.9 % for a single measurement. The purity of indium used as calibration substance is practically 100 % and no correction for the eutectic premelting is necessary. The comparison of caloric calibration factors calculated by several evaluation methods for the melting area determinations reveals a good agreement of these factors. A high degree of accuracy in calorimetric measurements can only be achieved by applying the same method for the evaluation of the melting areas in the case of melting curves for calibration work as well as for calorimetric determinations.

The linearization procedure has to be applied for the determination of the heat of fusion in the case of samples containing eutectic impurities (1b, 6). The calculated difference between the uncorrected and the corrected melting area or heat of fusion according to the procedure A is for the sample of p-xylene 4.1 %. This correction can not be neglected in the evaluation of the value of the heat of fusion even in the case of a substance with a rather high purity (1c). The improvement in the reproducibility of the determination of the heat of fusion compared to results presented in (1c) is explained in terms of a better experimental handling: especially more accurate weighing of the sample and more accurate calibration factors and also in terms of an elaborated data evaluation.

The heat of fusion of p-xylene was measured by DSC as well as by solution calorimetry. The following literature values are used for a comparison:

$\Delta H_{f,Lit}$ cal mole^{-1}	Author
4040	Parks, Huffman (7)
4090	Rossini (8)
4100	Nakatsuchi (9)

The mean value of these heats of fusion is $\Delta H_{f,Lit} = 4077 \pm 80$ cal mole^{-1}.

The melting point of the pure p-xylene measured with the DSC is in agreement with the literature value.

$$T_{o,DSC} = 13.3 \pm 0.1 \ ^{o}C$$

$$T_{o,Lit} = 13.23 \pm 0.05 \ ^{o}C \qquad (10, 11, 12)$$

The mean values of the heats of fusion for p-xylene at the melting point are as follows:

$$\Delta H_{f,DSC,A} = 4126 \pm 70 \text{ cal mole}^{-1}$$

$$\Delta H_{f,DSC,B} = 4084 \pm 70$$

$$\Delta H_{f,LKB} = 4067 \pm 30$$

$$\Delta H_{f,Lit} = 4077 \pm 80$$

All these results are consistent within the given error limits. The difference between the heats of fusion by DSC according to the two evaluation procedures A and B is not significant and therefore both procedures are equivalent within the given accuracy.

Conclusions

The heat of fusion of substances of a relatively high purity is determined with a high degree of accuracy by DSC as well as by solution calorimetry. The measurement of the heat of fusion with a reproducibility for the DSC method better than 2 % is close to the limitation of the method. Included in this reproducibility are operations such as weighing, recording of the melting process and evaluation. The solution calorimetry with a value of the reproducibility of 0.8 % can be further improved without great difficulties. It should be pointed out that we compare a microanalytical method, the DSC method, with a macroscopic method, the solution calorimetry.

An accurate and reproducible measurement of the heat of fusion in the case of substances with a rather high purity is important for the characterization of identity, crystal modifications, and the absolute or relative degree of crystallinity.

Acknowledgements

The author wishes to thank the following persons for collaboration in experimental and theoretical work: A. Geoffroy, O. Heiber, W. Huber, and G. Tonn.

References

1.   E.E. Marti, Thermochimica Acta, 5 (1972) 173 - 230
1a   ibid., 5 (1972) 203
1b   ibid., 5 (1972) 185
1c   ibid., 5 (1972) 184

2.   LKB 8700-1, Precision Calorimetry System, Instruction Manual,
     LKB Produkter AB

3.   D.D. Wagman et al., American Institute of Physics Handbook,
     3. Ed., (1972) 4 - 233

4.   Landolt-Börnstein, Zahlenwerte und Funktionen, Band II/4,
     (1961) 252

5.   R.J. Corruccini and D.C. Ginnings, Am. Soc. 69 (1947) 2291

6.   G.L. Driscoll et al., Proc. ACSS on Analytical Calorimetry,
     San-Francisco, (1968) 271

7.   H.M. Huffman, G.S. Parks and A.C. Daniels, Am. Soc. 52 (1930) 1547
     G.S. Pauls and H.M. Huffman, Ind. Eng. Chem. 23 (1931) 1138

8.   K.S. Pitzer and D.W. Scott, Am. Soc. 65 (1943) 803
     F.D. Rossini, Selected values (1953), 779

9.   Nakatsuchi, I. Soc. chem. Int. Japan Spl., 32 (1929) 333

10. A.R. Glasgow et al., J. Res. Bur. Stand., 37 (1964) 141
     F.D. Rossini, Selected values (1953), 71

11. K.S. Pitzer and D.W. Scott, Am. Soc. 65 (1943) 803
     Forziati et al., J. Res. Bur. Stand. 36 (1946) 129

12. White and Rose, J. Res. Bur. Stand. 9 (1932) 711

NOVEL METHODS FOR GAS GENERATING REACTIONS

ALLEN A. DUSWALT

HERCULES INCORPORATED
RESEARCH CENTER
WILMINGTON, DE 19899

INTRODUCTION

A study was made on a Hydrox mixture (1) to determine its reactivity during storage in a sealed container. The reaction, written as

$$NaNO_2 + NH_4Cl \rightarrow NaCl + N_2 + 2H_2O$$

presents some difficulties for normal thermal procedures. The volatility of ammonium chloride interferes with kinetic studies by thermogravimetry. Examination of the reaction in a sealed system by normal DTA techniques is not feasible because of the pressure generated by the reaction gases. However, a sample cell designed to compensate for the volatility of the ammonium salt, enabled weight-loss studies to be made in an open system. A closed system study was made by a novel "bursting capsule" technique. The results from the two methods allowed a satisfactory characterization of the reaction.

BURSTING CAPSULE TECHNIQUE

Theory. Gas producing reactions will cause a sealed DSC sample capsule to burst at a reproducible internal pressure. If the stoichiometry is known, rate constants (k) may be calculated from time-to-burst measurements (t) for the reaction carried out at appropriate temperatures, i.e.,

$$F(C) = kt \qquad [1]$$

F(C) is some function of conversion depending on reaction order.
For example, the first order reaction,

$$-\ln (1-C) = kt \tag{2}$$

where $(1-C)$ is the residual fraction of reactant. If reaction
rates are determined at two temperatures (T_1, T_2) the reaction
activation energy may be calculated from

$$\log k_1/k_2 = (E/2.3R)(1/T_1 - 1/T_2) \tag{3}$$

and the Arrhenius frequency factor (Z) from

$$k = Ze^{-E/RT} \tag{4}$$

To calculate rate constants from [1] the extent of reaction
at the burst pressure must be known. This can be determined from
the known stoichiometry of the reaction and the moles of reaction
gas (n) generated at the burst point. Values of (n) for various
reaction temperatures can be obtained from the following
considerations. The burst pressure (P_B) is the sum of the
reaction gas pressure (P_R) and the pressure exerted by the
environmental gas sealed in the capsule (P_A). Then,

$$P_R = P_B - P_A \tag{5}$$

The pressure P_A at reaction temperature T_1 is

$$P_A = P_o T_1/T_o \tag{6}$$

where P_o and T_o denote ambient pressure and temperature when the
capsule was sealed. The capsule free volume (V_F) is

$$V_F = V_C - W/D \tag{7}$$

where V_C, W and D are respectively the empty capsule volume, and
the weight and density of the sample material. The simple gas
law states that

$$n = P_R V_F/RT_1 \tag{8}$$

Assuming ambient pressure is one atmosphere and combining
[5] to [8] gives

$$n = \frac{(P_B - T_1/T_o)(V_c - W/D)}{0.082\ T_1} \tag{9}$$

For gas producing reactions, in which the stoichiometry is
not known, the extent of reaction and therefore the rate

constants cannot be directly determined. However, if the burst pressure occurs at the same degree of conversion for reactions run at different temperatures, an activation energy can be determined. At constant conversion, F(C) is constant and from [1] we can write

$$k_1/k_2 = t_1/t_2 \qquad [10]$$

and substituting in [3] and rearranging,

$$E = 4.57 \frac{\log (t_1/t_2)}{(1/T_1 - 1/T_2)} \qquad [11]$$

To achieve constant conversion at the capsule burst point, the ratio of moles of reaction gas to sample weight must be the same for all reaction temperatures. This is accomplished by calculating a value of (n) from [9] at one temperature, and using the (n/W) value as a standard. Sample weights at all other temperatures are then adjusted so that

$$(n/W)_s = (n/W)_1 = (n/W)_2 = \ldots (n/W)_x \qquad [12]$$

The weight values for runs at other temperatures are calculated from an expression derived from [9] and [12] as

$$W_x = V_c \Bigg/ \left(\frac{n}{W}\right)_s \frac{0.082\ T_x}{P_B - T_x/T_0} + \frac{1}{D} \qquad [13]$$

Equipment. The apparatus for capsule-burst studies has three requirements. (1) It must hold the sample isothermally within about a degree C., for extended periods. (2) The test temperature must be changeable. (3) It must enable determination of the time-to-burst of the capsule. Current DTA and DSC instruments can be used. A Perkin-Elmer DSC-1 with an effluent gas analyzer was used in this study. The DSC signal for the bursting capsule is generally adequate. Occasionally, however, the signal would be small and its position somewhat uncertain due to effects of capsule distortion. The use of the effluent gas analyzer and a helium sweep gas resulted in a sensitive, unequivocal signal for the capsule burst. A second apparatus, constructed from available laboratory materials, worked very well and is shown in Figure 1. The sample is purged in the cool part of the tube, and then moved into the heated section by the external magnet. The thermal conductivity of the helium sweep gas is monitored by the recorder. Then the sample capsule ruptures, a large signal is observed.

The capsules used were Perkin-Elmer aluminum pans and covers (Part No. 219-0062). The Perkin-Elmer sealer assembly

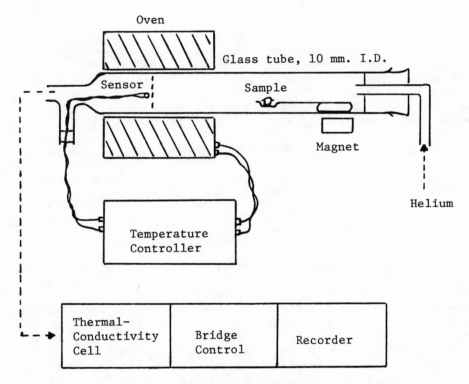

Figure 1.   Apparatus for Capsule-Burst Studies

(No. 219-0061) was used to encapsulate the samples.

Capsule Burst Pressure. The P_B value can be determined from the burst temperature of a sealed droplet of water, when slowly programmed in the instrument. At the burst point, the internal cell pressure is that of the air present plus the vapor pressure of water at that temperature. Table 1 gives the burst temperatures and the contribution of the vapor and air pressure to P_B for a 5 μl. water droplet programmed at 5°C. per minute in a Perkin-Elmer sealed sample capsule.

Table 1.   Determination of Capsule Burst Pressure

Run No.	Burst Temp. °C.	Pressure-Atmospheres		
		Vapor	Air	Burst Pressure (P_B)
1	147	4.3	1.40	5.7
2	146	4.1	1.40	5.5
3	150	4.6	1.41	6.0
4	146	4.1	1.40	5.5
5	141	3.6	1.38	5.0
6	154	5.1	1.42	6.5
7	147	4.3	1.40	5.7
8	148	4.3	1.40	5.7
9	156	5.4	1.43	6.8
10	155	5.3	1.43	6.7
11	150	4.6	1.41	6.0
12	153	5.0	1.42	6.4
13	153	5.0	1.42	6.4
14	155	5.3	1.43	6.7

The average P_B is 6.0 atmospheres $\pm$ 0.5 standard deviation.

Application to the Hydrox Reaction. The study was made to test whether reaction products were influencing the reaction in a sealed system. Since the stoichiometry of the main reaction is known, relationships between reaction time and the extent of reaction can be calculated for any convenient sample size. To simplify calculations, however, equal size samples were used for the study.

Samples were made up of equimolar amounts of pre-dried ammonium chloride and sodium nitrite plus 5% of "bone dry" magnesium oxide stabilizer. 15 mg. portions were sealed in aluminum capsules and exposed to reaction temperatures of 140 to 190°C. either in the DSC-1 or in the apparatus shown in Figure 1. To show the effect of moisture on the reaction, samples were also run which had not been previously dried.

The reaction times of both types of samples are given in Table 2.

Table 2. Burst Times for Hydrox Reactions

Temperature, °C.	Time-to-Burst, Minutes	
	Not Dried	Dried
130	66.8	
140	20.2	327
150	10.1	125
160	4.8	69
170	0.7	27
180		9.0, 12.5
190		2.6

To test whether the reaction behaves differently in a sealed system, the above burst times were compared with those predicted from kinetic values calculated from thermogravimetric (TG) data.

THERMOGRAVIMETRIC ANALYSIS OF HYDROX REACTION

Theory. The expressions for determining the kinetic values from isothermal TG runs are well known. If the weight losses due to reaction are measured at a very early stage, first order kinetics can be used and equations [2] to [4] apply. Specific reaction rate constants (k) can be obtained from [4] for the range of temperatures studied by the bursting capsule method. Then, burst times can be predicted for these temperatures from TG data and compared with the sealed capsule values in Table 2. Capsule burst times (t) can be predicted from

$$t = \left(\frac{n_r}{N_r}\right)\left(\frac{1}{k}\right) \qquad [14]$$

where (n_r/N_r) is the fraction of reactant converted at the burst point. (N_r) is the initial moles of reactant and (n_r) the moles converted. For the Hydrox reaction three moles of gas (n_g) are formed for each mole of reactant converted. Equation [9] calculates (n_g) and predicted burst time then, is

$$t = (n_g/3N_r)\ (\tfrac{1}{k}) \qquad [15]$$

Application. The volatility of ammonium chloride at reaction temperatures posed a problem for the analysis. The weight-loss readings are affected and the stoichiometry of the reaction mixture is changed by the depletion of the chloride. To overcome the first problem, the rate of weight loss of the ammonium salt was determined experimentally in a sample chamber as shown in

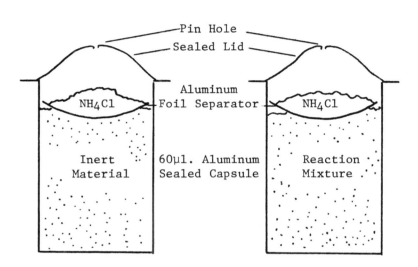

Figure 2. Sample containers for thermogravimetry
(a) for ammonium chloride loss, (b) for Hydrox study

Figure 2a. The chamber is an enlarged version of the Perkin-Elmer sealed capsule with an internal volume of 60 microliters. The hole in the cover is about 10 mils in diameter. Rates of weight loss of ammonium chloride from this container were determined at various temperatures and are given in Table 3.

Table 3.   Volatile Weight Loss of NH_4Cl

Temperature, ($^\circ$C.)	Weight Loss, (mg./min.)
180	7.83×10^{-4}
190	1.32×10^{-3}
200	2.15×10^{-3}
210	3.54×10^{-3}
220	5.43×10^{-3}

The pooled standard deviation for fifteen runs was 1% of the observed weight loss. A plot of log (mg./min.) versus 1/T was linear and allowed extrapolation of the volatilization rate to lower reaction temperatures. Corrections of observed reaction weight losses could then be made for the volatile loss of ammonium chloride. Depletion of the chloride salt from the reaction mixture is minimized by the arrangement in Figure 2b. Excess chloride on top of the reaction mixture saturates the vapor space, minimizing the loss from the mixture.

Reaction mixtures were rapidly brought to temperature and held isothermally in a Perkin-Elmer TGS-1 thermogravimetric instrument. Weight losses were corrected for ammonium chloride volatility and a percent loss rate per hour determined for the reaction at various temperatures. Measurements were made within the first percent or two of reaction and therefore depletion effects are ignored in the rate calculation. Table 4 gives the rate versus temperature results from the thermogravimetric curves.

Table 4.   Rate of Hydrox Reaction

Temperature ($^\circ$C.)	Reaction Rate (% per hr.), (10^{-2})
100	0.15
110	0.39
120	1.31
130	3.52
140	9.13

Comparison of TG and Bursting Capsule Data.   The results from the open (TG) and sealed capsule reactions can be compared by predicting capsule burst-times from TG data. This comparison

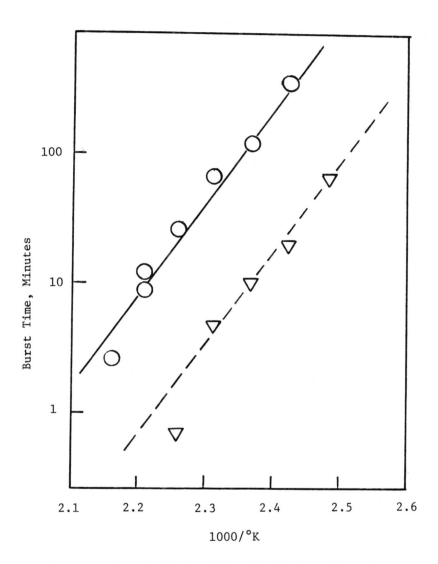

Figure 3.   Arrhenius Plot for Hydrox Reaction

0 - Burst-times for dried sample.   (-) - Predicted
burst-times from TG data. Δ - Burst-times for
undried sample.

can be made by extrapolating TG rate data to appropriate
temperatures and using equation [15] to calculate the times.
Figure 3 shows plots of experimental burst-times from the data in
Table 2 (circles and triangles). The line drawn through the
circles is not a best-fit line. It represents instead, the burst-
times predicted from the TG data for the dried Hydrox reactants.
As shown, the fit of results from the open (TG) and sealed
(capsule-burst) methods is good. Under the conditions studied,
the reaction kinetics which can be calculated for the open system
also apply to the Hydrox reactants in a closed system.

The above study was concerned with the first few percent of
conversion. Later stages of reaction can be examined cumulatively
by sealing in a lower ratio of sample weight-to-cell volume, or
selectively by reacting in an open system for a period and then
sealing for further reaction.

Figure 3 also shows the results of a side study, i.e., the
effect of moisture in an undried sample, on the rate of reaction.
The drier ingredients react at about one-thirteenth the rate of
the other.

CONCLUSIONS

The study of gas producing reactions is an obvious use of
thermogravimetry. By means of the described sample cell arrangement, the method can be extended to two-component systems where
one component is volatile. A complementary technique which uses
a reproducible capsule burst-pressure as a measure of conversion,
can examine the reaction in a sealed system. The effects of
reaction products, moisture, limited oxygen, etc., on the reaction
can then be readily observed. If the reaction and product
stoichiometry are known, specific rate values, the activation
energy and Arrhenius frequency factor, can be obtained. For an
unknown reaction, comparative rate data and an activation energy
may still be calculated.

REFERENCES

(1) Taylor, J., "Solid Propellant and Exothermic
Compositions", 1959, George Newness Limited, pp. 65-74.

DSC: NEW DEVELOPMENTS IN CLINICAL ANALYSIS AND PHYSIOCHEMICAL RESEARCH

Bruce Cassel   and   Tsuyoshi Ohnishi
Perkin-Elmer Corp.   Dept. of Anesthesiology
Norwalk, Conn. 06856   Hahnemann Medical College
                                 and Hospital
                            Philadelphia, Pa.   19102

Over the last decade the field of Thermal Analysis has made significant contributions to a surprising number of fields. In the multifaceted field of polymer science, it has provided a universal tool for ascertaining and predicting the effect of preparation and thermal history on the physical properties of polymers. One reason for this is that a great deal of information about the overall structural state of a complex macromolecular system is revealed in the heat capacity. By establishing correlations between thermal phenomena and viscoelastic properties, these properties can be usefully assessed by differential scanning calorimetry (DSC). Another area where complex systems of macromolecules are of great interest is that of biochemistry and clinical analysis. The question is to what extent can the heat capacity of a clinically accessible sample give useful diagnostic information.

To date there have been a number of interesting DSC studies primarily in the field of biochemical research to gain insight into the mechanisms of physiological processes. For example, Steim, Chapman, Sturtevant and others have investigated the thermal properties of synthetic and/or natural phospholipid membranes (1). These materials display a characteristic thermal adsorption corresponding to an order-disorder transition within the membrane lattice.

A second thermal phenomenon in physiological materials is that of protein denaturation. Chapman, Sturtevant, Privalov and

others have studied scanning calorimetry of protein denaturation
for the purpose of ellucidating the denaturation process (2). As
in the case of the previously cited phospholipid research, much
of the earlier work has been done on hand-fashioned instrumentation. Perhaps, now that second generation DSC instrumentation is
commercially available more laboratories will pursue basic studies of this nature and will progress to the next step, the utilization
of these characteristic thermal transitions for the development of
useful clinical analytical methods.

While some early efforts are now being made to employ
static calorimetry to the practical problems of clinical analysis
(3), there has been to date surprising little attention given to the
possibilities of scanning calorimetry for clinical analysis. This
paper summarizes some of the first steps taken in this direction
toward demonstrating the feasibility of this approach--offered
with the hope that others with expertise in this area will be able to
further pursue this approach.

The first and primary analysis to be considered is that of
blood analysis: to what extent can DSC indicate the concentrations
and types of components in blood from a clinically extracted sample?

The first step is the development of a flow diagram for routine blood separation. Samples from each stage along the flow
diagram would be selected and run with the purpose of ultimately
identifying the various thermal peaks, finding the conditions for
chemically or thermally isolating the components of interest, and,
finally, assaying them. A rudimentary separation scheme can be
seen in figure 1. We have obtained thermal curves of samples
from several of these stages. For comparison, the thermal curves
of several commercially obtained, purified components of blood
have been run. Finally, a more extensive study has been initiated
in one particularly promising vein, that of distinguishing blood
from normal patients from that of those having the sickle cell trait
and from that of those with sickle cell disease.

EXPERIMENTAL

Sample Preparation. Blood samples were prepared with the aid of
Dr. Asakura of the Johnson Research Foundation of the University
of Pennsylvania. Heparinated blood samples from normal and
sickle cell anemic patients were stored at 0°C until shortly before

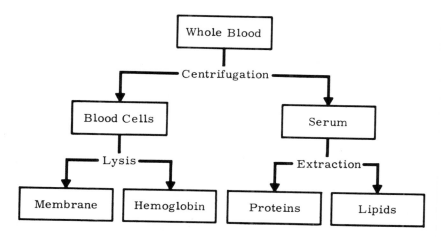

Fig. 1. Blood Separation Scheme

use. Red blood cells were washed thrice with .9% NaCl. These were hemolyzed by water, desalted, concentrated by vacuum dialysis, and buffered at pH 8. Hemoglobin concentrations were determined spectrophotometrically using a Coleman Model 126. Details of the preparative method will appear elsewhere (4).

Reduction of oxyhemoglobin to the deoxy form was accomplished within the DSC sample pans by the addition with stirring of a small fraction of a milligram of sodium dithionite.

Blood proteins and lipids were obtained in a purified form from Sigma Chemical Co. and used directly without further purification. Solutions were prepared directly in the aluminum sample pan by the addition of .01M phosphate buffer to submilligram protein samples weighed to 1 μg using a Perkin-Elmer AM-2 autobalance.

Calorimetry. The thermal curves were obtained on a Perkin-Elmer differential scanning calorimeter (Model DSC-2). A

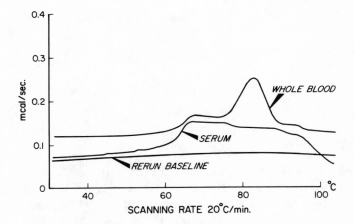

Fig. 2.   Whole Blood, 10 µl

temperature scanning rate of 20°C/min. was found to be most useful for peak resolution. A discussion of recommended procedures for observing low energy transitions in large aqueous samples appears elsewhere (2). A discussion of the careful calibration procedure and the method of data analysis appears in the appendix.

RESULTS AND DISCUSSION

<u>Whole Blood and Serum.</u> The thermal analysis of whole blood and that of serum reveals fairly complex systems with two or more low energy peaks below 60°C and four or more overlapping peaks in the region of 60 to 90°C (See figure 2). The primary difference between the two curves is the hemoglobin which remains in the red blood cells and is centrifuged out of the serum. The only other considerable component present only in the whole blood is fibrinogen, which displays a complex and not too reproducible endotherm (See figure 3).

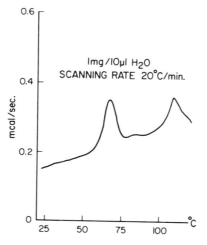

Fig. 3.   Fibrinogen Dispersion

    The primary components of serum and their approximate
distribution (5) are Albumen (60%), α -globulin (12%), β -globulin
(12%), γ-globulin (12%). In order to aid in the identification of
these peaks, thermal curves were made of roughly 10% solutions of Albumen, Hemoglobin, and the γ-globulins. As can be
seen from figure 4, the thermal curves of the pH 7 solution of
freeze-dried hemoglobin results in a peak at 75°C, which is
about ten degrees lower than that observed for hemoglobin in
figure 2. Similarly, the DSC trace of the pH 7 solution of
freeze-dried Albumen gives a peak at 76°C, not in close correspondence to the major endotherm in serum. Apparently,
the freeze-dried solutions do not very closely simulate the
natural system. The purified γ-globulins do appear to give
the sort of broad multistep behavior which could contribute to
the general plateau shape of the serum endotherm. A better
approach to identification of the serum peaks might be afforded
by attempting a further separation by solubility, then thermal
analysis. Also to be explored should be the modification of
the medium, by the addition of urea or NaCl, for instance.

Comparison HbA, HbAS and HbS blood. In 1949 Linus Pauling
demonstrated the existence of electrophoretically distinct molecular forms of hemoglobin, and it is now known that there is a
large variety of hemoglobin molecules which differ from the common HbA (α_2 β_2) form by the substitution of an amino acid group
on the backbones of two of the four polypeptide chains which constitute the protein part of the hemoglobin molecule. The most
common variation is HbS (α_2 β_2^S) wherein the #6 position
glutamic acid side group has been replaced by the neutral valine
group. The result is a molecule which when in the pure state
can polymerize, causing a distortion (sickling) of the red blood
cell and resultant in vivo capillary blockage.

    One clinical problem consists in the identification of patients
having the sickle cell trait (a mixture of HbA and HbS, designated
HbAS). In such a mixture sickling does not ordinarily occur.
However, patients with this trait are adversely affected by
certain anesthesia and other medication and must be identified.
This is presently done somewhat tediously by electrophoresus.
A rapid screening technique to identify even tentatively those
who have this trait would be useful. Moreover, since it has
recently been shown that HbS is less stable with respect to
agitation (5), it was hoped that a thermal difference could be

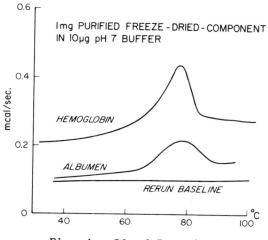

Fig. 4.   Blood Protein

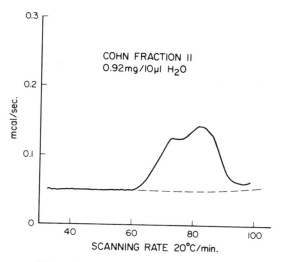

Fig. 5.   γ-Globulins, Bovine

demonstrated. Initial indications are that small but significant differences do exist but that it remains to work out a method for simplifying the differences to the point that an unambiguous determination could be made by a medical technician.

A comparison of thermal curves of red blood cell preparations adjusted to common pH 8 and to common concentrations can be seen in figure 6. The three thermal curves indicate the denaturation of hemoglobin within the red blood cells from normal(RBC A), sickle-cell trait (RBC AS) and sickle-cell diseased (RBC S) patients; the denaturation temperature-- however this may be defined--indicates a stability of HbA > HbS (with HbAS being a mixture of HbA and falling between the two). Taking the denaturation temperature as that of the extreme point of the endothermic peak, we find T_D = 85.0, 84.5 and 83.5 for HbA, HbAS and HbS, respectively, for these conditions. In duplicate experiments the reproducibility was found to be of the order of three or four tenths of a degree. Hence, the denaturation temperatures are only different by an amount slightly greater than the experimental error. In order to amplify this difference, a number of experimental conditions were varied.

The first variation to be considered is that of converting the hemoglobin to the deoxy form by the addition of sodium dithionite. This results in an exothermic reaction, as can be observed if the sample is run on the DSC-2 before the reaction is complete. Curiously, this exotherm often starts dramatically at 66°C, just the point at which the predenaturation peak occurs at this scanning rate. As can be seen in figure 7, the effect of deoxygenation is somewhat surprising. That is, the denaturation enthalpy is almost doubled while the temperatures of denaturation are virtually unchanged.

The next step in the separation was affected, and the thermal curves for hemoglobin were obtained (See figure 8). We note that the 66°C transition has not disappeared and hence it is not a cell membrane transition but rather apparently a structural change within hemoglobin itself. We see also that we have not lost the high temperature endotherm in the case of the HbS sample. This seemed most likely to be the denaturation of HbF, fetal hemoglobin, present in all newborn infants and enduring into childhood in sickle cell patients. Nevertheless, a later sample assayed at 20% HbF did not indicate a discrete peak in this region.

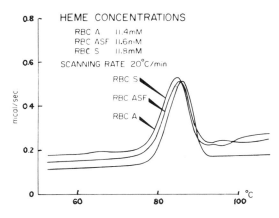

Fig. 6.   Red Blood Cells

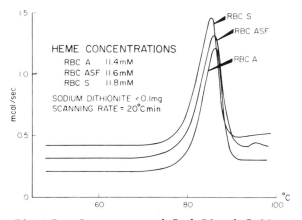

Fig. 7.   Deoxygenated Red Blood Cells

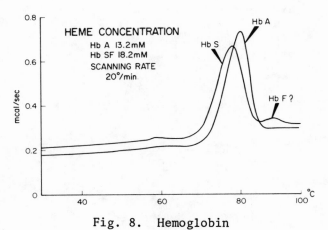

Fig. 8.   Hemoglobin

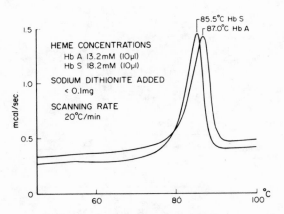

Fig. 9.   Deoxygenated Hemoglobin

A comparison of the HbA and HbS peaks in figure 8 shows predictably that no greater separation exists in the denaturation temperatures of the hemoglobins than in the red blood cells. For this particular pair of samples, however, the concentrations were quite accurately known; and a considerable enthalpy difference appears on a per mole basis between HbA and HbS. Because of uncertainties in the maintenance of the assayed hemoglobin concentrations for some of the data, further study is required to establish this enthalpy difference. With this preparation of samples, the enthalpy difference can be seen also in the denaturation thermogram of deoxygenated hemoglobin in figure 9.

In summary, the thermal denaturation of hemoglobin shows small but experimentally meaningful differences in the temperatures of denaturation of HbA and HbS which could provide a method for distinguishing patients having HbA, HbAS or HbS blood. It is hoped that further research will establish conditions for which the differences in denaturation temperature can be further separated or that differences in the enthalpy can be substantiated and exploited.

OTHER POSSIBLE CLINICAL ANALYSES

Another area where DSC has an opportunity to be a clinical tool is in the denaturation of phospholipids in body fluids. It has been well established that the ratio of lecithin to sphingomyelin (L/S) in amniotic fluid correlates with the occurence of respiratory distress syndrome in newborns. The presently used tests for this ratio require several steps and take more than an hour (6). As can be seen in figure 10, sphingomyelin undergoes a thermal transition at 41°C when totally dispersed and 80°C when dry. Lecithin undergoes a similar transition between 41°C and 105°C, depending on the amount of water present. It was hoped that by taking up these hydrophobic solutes from amniotic fluid into a hydrophobic solvent and drying that a DSC thermal curve would give two peaks the relative areas of which would give the L/S ratio. In fact, what is obtained is one peak having the shape of a melting peak of a very impure material. After shock cooling, two peaks can be obtained but the position and shape of these is greatly dependent on the conditions of cooling.

It is hoped that a van't Hoff analysis of this curve or a

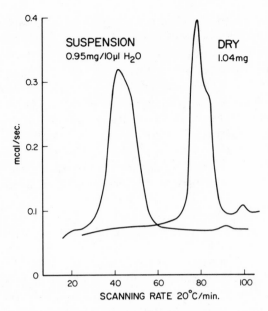

Fig. 10.   Sphingomyelin

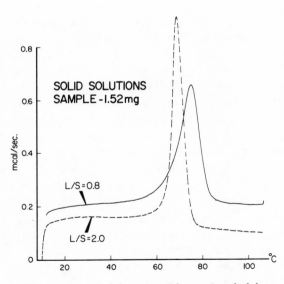

Fig. 11.   Sphingomyelin - Lecithin

comparison with standard curves will allow an evaluation of the L/S ratio. Examples of two L/S ratio mixtures prepared as they would have been extracted by chloroform from 10 cc of amniotic fluid appear in figure 11. It seems likely that a DSC test for L/S could be performed in a matter of a few minutes using a standard one or two milliliter sample.

In conclusion, it is clear from this initial investigation that possibilities exist for the use of DSC in clinical determinations and that laboratories exploring new avenues of clinical analysis should consider thermal analysis as a promising tool.

APPENDIX

When scanning large samples for small heat effects by DSC (or by quantitative DTA) the treatment of the data requires certain corrections if maximum accuracy in temperature and enthalpy is to be strived for. For those interested in availing themselves of the high accuracy of which the method is capable and would like to better understand the theory behind these corrections, several references may be useful (See refs. 7, 8, 9).

Temperature Determination. The sample is exposed to a pro-grammed temperature environment within the DSC cell which can be accurately calibrated to correspond to the linear T- or t-axis of the recorded output. However, if the sample has a non-zero heat capacity, there will be temperature lag between the sample temperature T_S and the programmed temperature T_p caused by, and proportioned to, the thermal resistance R_0 that the heat must flow through from sample holder to sample.

$$T_S = T_p - R_0\dot{Q} \qquad \text{(Newton's Law)} \ldots \ldots (1)$$

where $\dot{Q}$ is the differential heat flow to the sample relative to that for no sample (See figure 12). Graphically this amounts to dropping a line from the denaturation peak to the no-sample baseline parallel to the slope of a thermal curve of a first-order transition such as the melting of high purity indium. Even if the heat capacity of the sample is counterbalanced by a similar heat capacity on the reference side, this temperature lag must be corrected for. In this case the correction amounts to roughly

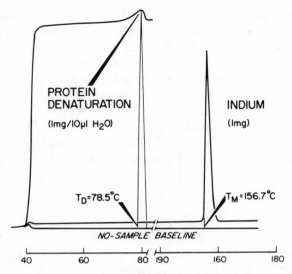

Fig. 12.   Instrument Calibration

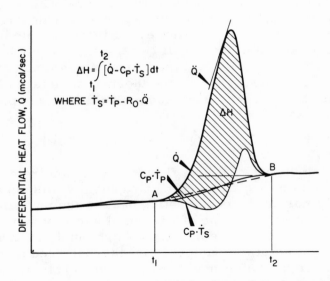

Fig. 13.   Accurate Calculation of Enthalpy

0.7°C for the 10 μl aqueous solution run at 20°C/min., and the temperature axis on the figures have been drawn to correct for this effect.

Enthalpy Determination. One particular advantage of a true differential scanning calorimeter such as the DSC-2 is that the output is directly in units of heat-flow to sample (mcal/sec) and the x-axis is time so the area under a peak is in units of calories. Thus when a baseline is drawn under the peak the area enclosed is the enthalpy of the transition. If there is no heat capacity change in the sample as a result of the transition, then a straight line can be drawn under the peak connecting the pretransition and post-transition baselines. If, however, a shift in the heat capacity does accompany the transition, as is often the case with protein denaturation, a better approximation than a straight line (dashed line $\overline{AB}$, figure 13) can be made. First, the heat capacity during the transition interval can be assumed to take the form of

$$C_p = C_{p_1} (1-x) + x\, C_{p2} \quad \dots\dots\dots\dots (2)$$

where C_{p_1}, C_{p_2} are the linear, sample heat capacities functions before and after the transition, respectively; and x is the fraction of converted material at any given time, t, as determined by the fraction of the total transition enthalpy at that time. In figure 12 this corresponds to the solid designated $C_p \cdot \dot{T}_p$, since on power-time axes this indicates the form of the baseline if the heating rate experienced by the sample were $\dot{T}_p$, the programming rate. In fact, however, since the sample is absorbing energy rapidly during the transition, its scanning rate lags behind during the first part of the transition and catches up during the latter part. Quantitatively, this is expressed by the derivation of Newton's law (9):

$$\dot{T}_p - \dot{T}_s = \ddot{Q}\, R_o \quad \dots\dots\dots\dots (3)$$

where $\ddot{Q}$ is the derivative of the heat flow; namely, the slope of the DSC curve at time t. Graphically, then, the corrected area is the shaded area in figure 12. For most systems the heat capacity change is small and the shaded area below segment $\overline{AB}$

exactly equals and cancels the unshaded area above $\overline{AB}$ so that
the area can be obtained in the conventional way--without going
through the iterative process required to calculate $C_p \cdot \dot{T}s$
exactly (8). In the case of the hemoglobin denaturation, the
difference in the heat of denaturation calculated this way exceeds
that calculated in the conventional way by an amount between
one and two percent. Since the concentration of hemoglobin
was not known to better than this accuracy, the conventional
method was used to compare enthalpies of denaturation of HbA
and HbS. Nevertheless, the DSC-2 is capable of accuracy
better than this (10) and in subsequent experiments where the
concentrations are better established it may be necessary to make
this correction.

One note of caution in reporting enthalpy measurements such
as this is in order. That is, while the enthalpy of transition
for the first-order transitions; that is, whose occurring at a
single temperature with no super-heating, can be reported
unambiguously, those occurring over a temperature range will
be a function of the particular range over which the transition
has occurred. Those transitions which exhibit super-heating will
show an enthalpy dependent on the degree of super-heating.
Thus, transitions such as denaturation and polymer melting will
have enthalpies somewhat dependent upon scanning rate. In this
case, the comparison of hemoglobin enthalpies is valid because
the samples were run under identical scanning conditions. For
the most accurate reporting of a standard enthalpy of denaturation, it would be best to carry out the denaturation experiment
isothermally.

REFERENCES

(1) For brief reveiw of subject with references: TAAS-4,
     Perkin-Elmer Corp., Norwalk, Conn. 06856.

(2) For brief review of subject with references: TAAS-5,
     Perkin-Elmer Corp., Norwalk, Conn. 06856.

(3) Excellent review: Goldberg, R.N., and Armstrong, G.T.,
     "Microcalorimetry: A tool for biochemical Analysis,"
     To Be Published in Medical Instrumentation.

(4) Ohnishi, Asakura, Cassel, R.B., To Be Published.

(5) Ohnishi, T., Asakura, T., and Pisani, R., "Effect of Anesthetics on the Stability of Oxyhemoglobin S," submitted for publication.

(6) Blass, K.G., et al, Clin. Chem. 19/12, 1394 (1973).

(7) Gutman, C.M., and Flynn, J.H., Anal. Chem. 45, 408 (1973)

(8) Heuvel, H.M., and Lind, K.C, J.B., Anal. Chem. 42 9 1044 (1970).

(9) Flynn, J. H., "Thermodynamic Properties from Differential Scanning Calorimetry by Calorimetric Methods," To Be Published.

(10) TAAS-9, Perkin-Elmer Corp., Norwalk, Conn. 06856

THE USE OF THERMAL AND ULTRASONIC DATA TO CALCULATE THE PRESSURE DEPENDENCE OF THE GRUNEISEN PARAMETER

S.R. Urzendowski and A.H. Guenther

Air Force Weapons Laboratory

Kirtland Air Force Base, New Mexico

INTRODUCTION

A knowledge of the Gruneisen parameter is a prerequisite to any practical or theoretical evaluation of the pressure or temperature dependence of thermodynamic properties in equation of state studies. As demonstrated by Gruneisen [1] and others [2,3,4], the Gruneisen ratio may be determined from mechanical properties (such as the bulk modulus and sound velocities), and from thermal properties such as volume coefficient of expansion and specific heat data. This relation is defined by

$$\gamma_g = \frac{\beta \cdot B^T}{C_v \cdot \rho} = \frac{\beta \cdot B^S}{C_p \cdot \rho} = \frac{\beta[\rho(c_\ell^2 - 4/3c_t^2)]}{C_p \cdot \rho} \tag{1}$$

where γ_g is the Gruneisen ratio, β is the volume coefficient of thermal expansion, ρ is the density, C_v and C_p are the heat capacity at constant volume and constant pressure, respectively, B^T and B^S represent the isothermal and adiabatic bulk moduli, and c_ℓ and c_t are the longitudinal and transverse ultrasonic sound velocities, respectively, for isotropic media.

In this study we are interested in the variation of the thermodynamic and elastic property data of Equation (1) as a function of temperature and pressure. Although the temperature dependence of the Gruneisen ratio can be determined from the temperature dependence of the thermo-mechanical properties, this information is generally not available for polymeric materials, alloys, or nonisotropic metals. Difficulties in obtaining temperature derivatives of these materials are associated with material property failure as well as the necessity of multiple thermal expansion and ultrasonic sound velocity measurements to account for contributions along the three coordinate axes of the materials.

It is known that the shock technique has been the most widely applicable technique for the determination of pressure-volume data for equation of state studies. Since the reliability of shock measurements decreases below 100 kbar, most equation of state data has been obtained above 100 kbar. It is necessary to complement the shock data between 1 atm to a pressure of approximately 100 kbar.

Swenson (5) has discussed the difficulties involved in the static determination of equation of state data for certain solids. Due to strong interatomic binding forces, valence and ionic solids (such as diamond and the silver halides) are relatively incompressible. For more complex metals (such as iron or beryllium), very great pressures are necessary to produce significant changes in the cohesive energies or the high temperature thermal properties of these substances. Although molecular solids are the most appropriate for pressure volume studies, further complications arise from material property failure.

In this work, thermal and ultrasonic measurements made at various temperatures were used to calculate the temperature dependence of the Gruneisen ratio according to Equation (1). Although ultrasonic data were obtained for some of the materials at high pressures, the high pressure derivative $(\partial\gamma/\partial P)$ was estimated by applying proper thermodynamic relationships.

DERIVATIONS

The equation of state of solid materials under shock-loaded conditions (6,7) may be expressed as

$$P = au^1 + bu^2 + cu^3 + \cdots + \gamma\partial E/\rho_0 \qquad (2)$$

where u is the compressibility, $u = (\rho/\rho_0)^{-1}$, a, b, and c are constants, E is the internal energy, γ is the Gruneisen ratio, $\gamma=(1/\rho)(\partial P/\partial E)_V$, and ρ is the density.

The constant a may be expressed in terms of the bulk modulus, B, in the Maclaurin expansion as

$$a = (\partial P/\partial u)_0 = \rho_0(c_\ell^2 - 4/3\ c_t^2) = B \qquad (3)$$

where c_ℓ and c_t are longitudinal and transverse velocities, respectively.

As derived previously (8,9) the temperature derivative of the Gruneisen ratio, $(\partial\gamma/\partial T)_0$, was obtained by differentiating each term of Equation (1) to give

$$\left(\frac{\partial\gamma}{\partial T}\right)_0 = \frac{\partial\gamma}{\partial\rho}\left(\frac{\partial\rho}{\partial T}\right)_P + \frac{\partial\gamma}{\partial B^S}\left(\frac{\partial B^S}{\partial T}\right)_P + \frac{\partial\gamma}{\partial\beta}\left(\frac{\partial\beta}{\partial T}\right)_P + \frac{\partial\gamma}{\partial C_p}\left(\frac{\partial C_p}{\partial T}\right)_P$$

$$= \gamma_0\left\{\frac{1}{\beta}\left[\left(\frac{\partial\beta}{\partial T}\right)_P + \beta^2\right] + \frac{1}{B^S}\left(\frac{\partial B^S}{\partial T}\right)_P - \frac{1}{C_p}\left(\frac{\partial C_p}{\partial T}\right)_P\right\} \tag{4}$$

where the zero subscript represents $0°K$ and atmospheric pressure.

The temperature derivative for the specific heat, $(\partial C_p/\partial T)$, for the thermal expansion, $(\partial\beta/\partial T)$, and for the density, $(\partial\rho/\partial T)$, was obtained from linear least square curves of the respective data. The proper specific heat data at constant volume, C_v, was obtained from the following conversion

$$C_p - C_v = TV\beta^2/x^T \tag{5}$$

where x^T is the isothermal compressibility.

Although the experimental measurements yield adiabatic bulk moduli, the data were converted to the isothermal counterparts by

$$B^T = B^S/1 + \beta\gamma T \tag{6}$$

where T is the absolute temperature.

The temperature derivative of the bulk modulus was obtained from the differentiation of Equation (6) to give

$$\left(\frac{\partial B_0^T}{\partial T}\right)_P = \frac{(\partial B^S/\partial T)_P}{1+\beta\gamma T} - \frac{B_p^S\beta\gamma}{(1+\beta\gamma T)^2} - \frac{B_p^S\gamma T(\partial\beta/\partial T)_P}{(1+\beta\gamma T)^2} - \frac{B_p^S\beta T(\partial\gamma/\partial T)_P}{(1+\beta\gamma T)^2} \tag{7}$$

and all of the quantities appearing on the right can be determined from thermal and ultrasonic data.

The pressure derivative, $(\partial\gamma/\partial P)$, may be obtained by differentiating Equation (1) with respect to pressure, so that

$$\left(\frac{\partial\gamma}{\partial P}\right)_T = \frac{\partial\gamma}{\partial\rho}\left(\frac{\partial\rho}{\partial P}\right)_T + \frac{\partial\gamma}{\partial C_p}\left(\frac{\partial C_p}{\partial P}\right)_T + \frac{\partial\gamma}{\partial\beta}\left(\frac{\partial\beta}{\partial P}\right)_T + \frac{\partial\gamma}{\partial B^S}\left(\frac{\partial B^S}{\partial P}\right)_T$$

$$= -\frac{\beta B^S}{\rho^2 C_p}\left(\frac{\partial\rho}{\partial P}\right)_T - \frac{\beta B^S}{\rho C_p^2}\left(\frac{\partial C_p}{\partial P}\right)_T + \frac{B^S}{\rho C_p}\left(\frac{\partial\beta}{\partial P}\right)_T + \frac{\beta}{\rho C_p}\left(\frac{\partial B^S}{\partial P}\right)_T \tag{8}$$

Thurston (10) has shown that the pressure derivatives of β and C_p at constant temperature are related to the mechanical properties of an isotropic solid by

$$\left(\frac{\partial \beta}{\partial P}\right)_T = -\left(\frac{\partial X^T}{\partial T}\right)_P = \frac{1}{(B^T)^2}\left(\frac{\partial B^T}{\partial T}\right)_P = \beta' \tag{9}$$

$$\left(\frac{\partial C_p}{\partial P}\right)_T = -\frac{T}{\rho}\left[\left(\frac{\partial \beta}{\partial T}\right)_P + \beta^2\right] = -T\left(\frac{\partial^2 v}{\partial T^2}\right) = C_p' \tag{10}$$

where the superscript T refers to the isothermal moduli and T is the absolute temperature. If one has the means of measuring the temperature dependence of the bulk modulus and the expansion coefficient, Equations (9) and (10) allow an estimation of the pressure derivatives of β and C_p.

The quantity $(\partial \rho/\partial P)_T$, follows from the relationship of density to volume and from the definition of the isothermal compressibility so that

$$(\partial \rho/\partial P)_T = \rho/B^T \tag{11}$$

where ρ is the density.

Since the adiabatic bulk modulus may be expressed in terms of the longitudinal and transverse sound velocities, Asay et al.(8), and Lamberson (11), have shown that the differentiation of Equation (1) at constant temperature and zero pressure yields

$$\left(\frac{\partial B^S}{\partial P}\right)_{o,T} = 2\rho_o\left[c_\ell c_\ell' - 4/3\, c_t c_t'\right]_{P=o} + (1 + \beta\gamma T) = B^{S'} \tag{12}$$

where $(\partial B^S/\partial P)_{o,T}$ is the pressure derivative of the adiabatic bulk modulus.

Anderson (12) has shown that the conversion to the isothermal pressure derivative results from

$$\left(\frac{\partial B^T}{\partial P}\right)_{o,T} = B_{o,T}^{S'} + \beta\gamma T\left(\frac{B_o^T}{B_o^S}\right)\left[1 - \frac{2}{\beta B_o^T}\left(\frac{\partial B_o^T}{\partial T}\right)_P - 2B_{o,T}^{S'}\right]$$

$$+ \left[\beta\gamma T\left(\frac{B_o^T}{B_o^S}\right)\right]^2\left[B_{o,T}^{S'} - 1 - \frac{1}{\beta^2}\left(\frac{\partial \beta}{\partial T}\right)_P\right] = B^{T'} \tag{13}$$

where the temperature derivative of B_0^T was obtained from Equation (7).

The above derivation leads to

$$\left(\frac{\partial \gamma}{\partial P}\right)_{0,T} = \gamma_0 \left[-\frac{1}{B_0^T} - \frac{1}{C_p}\left(\frac{\partial C_p}{\partial P}\right)_{0,T} + \frac{1}{\beta}\left(\frac{\partial \beta}{\partial P}\right)_{0,T} + \frac{1}{B_0^S}\left(\frac{\partial B^S}{\partial P}\right)_{0,T} \right]$$

$$(14)$$

where the zero subscripts refer to atmospheric pressure and $(\partial \gamma/\partial P)_{0,T}$ is γ'.

Anderson (12) has shown that ultrasonic data taken at relatively low pressures may be used to estimate the pressure-volume isotherm to pressures of the order of the bulk modulus. His assumption that the bulk modulus has a linear pressure dependence resulted in the Murnaghan or logarithmic equation of state (13) defined by

$$\ell n \left(\frac{V_0}{V}\right) = \frac{1}{B_0'} \ell n \left\{ B_0' \frac{P}{B_0} + 1 \right\} . \qquad (15)$$

From the similarity of definitions for the adiabatic and isothermal bulk moduli (Equation 6) the use of B_0^S and $B_0^{S'}$ in Equation (15) results in an adiabat while B_0^T and $B_0^{T'}$ results in the isotherm. Furthermore, Dugdale and MacDonald (14) have shown that the derivative $B_{0,s}'$ may be estimated from

$$B_{0,s}' = 2\gamma + 1 \qquad (16)$$

which is applicable for all pressures lower than B_0.

Since ultrasonic data yield adiabatic bulk moduli and shock wave measurements define the isothermal quantities, the proper conversions are made by

$$B_{0,T}^{S'} = B_{0,s}^{S'} - \left(\frac{\partial B^S}{\partial T}\right)_P \cdot \frac{T\gamma}{B^S} \qquad (17)$$

which utilizes the pressure and temperature derivative of the adiabatic bulk modulus.

Equation (14) may be expressed by a more convenient estimation by

$$\left(\frac{\partial \gamma}{\partial P}\right)_{o,T} = \frac{\gamma}{B^S}\left[\left(\frac{\partial B^S}{\partial P}\right)_T + \left(\frac{1 + \beta \gamma T}{\beta B^S}\right)\left(\frac{\partial B^S}{\partial T}\right)_P - 1 - \gamma - T\left(\frac{\partial \gamma}{\partial T}\right)_P\right] \quad (18)$$

where all the quantities are determined experimentally except perhaps $(\partial B^S/\partial P)_T$.

EXPERIMENTAL TECHNIQUE

Thermal Expansion. A DuPont 940 Thermomechanical Analyzer was used to measure linear thermal expansions between -100 to 200°C. Details of the experimental procedure were previously described (8,9,15). After proper application of chromel alumel thermocouple corrections, individual temperature determinations agreed to within 0.2°C and to within 2.0 to 4.0°C of values reported in the literature. The total probable error of the expansion measurements was ±2.5 percent. This was calculated as the square root of the summation of all errors associated with each component of the instrument.

Heat Capacity. The heat capacity data were obtained with appropriately calibrated differential scanning calorimeters. Both the DuPont DSC and the Perkin Elmer DSC-1B were used for the measurements. The heat capacity values are accurate to approximately ±2.0 percent. For the polymeric materials, transitional regions were eliminated by extrapolating before and after said transitions.

Ultrasonic Sound Velocities. Sound velocity measurements of longitudinal and shear waves at approximately 1 and 3 Mc/sec were made at this laboratory by Asay, et al., (8), and Lamberson (11) and were used to calculate the adiabatic bulk moduli and related ultrasonic data. For most of the velocity-temperature data, a quadratic function was found to fit the data to an accuracy of approximately 1 percent.

Materials Studied. The materials studied included high purity metals, alloys and polymeric materials. Tungsten rod (99.9995 percent pure) was obtained from Alfa Inorganics. The sample density was 19.265 g/cm^3.

The steel sample No. 314 contained the following constituents: 51.2 percent iron, 0.3 percent carbon, 2.0 percent manganese, 0.4 percent phosphorous, 1.5 to 3.0 percent silicon, 23 to 26 percent chromium and 19 to 22 percent nickel. The measured sample density was 7.62 g/cm^3.

Aluminum 2024 was chosen because of its availability, its good machining properties and also because a vast amount of shock wave data is available for comparative purposes. The alloy (density 2.785 g/cm^3) contained 93.4 percent aluminum, 4.5 percent copper, 1.5 percent magnesium and 0.6 percent manganese. Before heat treatment, the porous sample had a density of 2.65 g/cm^3.

The solid polymeric materials studied were teflon, density 2.186 g/cm^3 (E.I. DuPont de Nemours Co.), polystyrene, density 1.046 g/cm^3 (Cadco Corporation), and an epoxy resin, density 1.242 g/cm^3 (Shell Chemical Co.).

RESULTS AND DISCUSSION

Specific heat data in the form of linear equations are presented in Table I. For the polymeric materials the data represent values extrapolated from nontransitional regions, however, for Teflon-1 and Epoxy-1 the data were obtained along transitional regions. The data tabulated for Epoxy-2 represents data extrapolated at higher temperatures.

Table I
Specific Heat and Debye Temperatures
for the Materials Studied

$$C_p = a + bT, \; cal/g^{\circ}K$$

Material	a	b x 10^3	θ,$^{\circ}$K	Temp. Range, $^{\circ}$K
Epoxy	-0.0922	1.166	54.2	243 - 353
Epoxy-1	-0.0732	1.099	56.1	263 - 423
Epoxy-2	0.0026	0.887	49.2	343 - 388
Polystyrene	-0.0846	1.270	70.4	233 - 353
Teflon	0.1544	0.250	52.7	243 - 313
Teflon-1	0.1481	0.304	52.7	253 - 413
Tungsten	0.0284	0.012	321.1	233 - 383
Steel No. 314	0.0281	0.258	454.0	253 - 373
Al 2024	0.1788	0.126	408.5	243 - 373

The Debye characteristic temperatures (θ), of Table I were obtained from the ultrasonic calculations as defined previously (8,9,15). The θ values agree well with previously reported values calculated from low temperature specific heat data. Kok, et al. (16) and Phillips (17) report Debye temperatures for aluminum of θ = 423 ±5 at 0°K and θ = 390 at 298 K.

For teflon, Wunderlich (18) fitted low temperature specific heat data to a Tarasov-type model and obtained a θ value of 47.8 K which agrees with our value obtained at 298 K.

For polystyrene θ = 70.4 at 273 K agrees well with the temperature of 68 K reported by Reese (19) and Choy, et al. (20).

In general, the θ values reported here represent the vibrational frequencies of the materials as determined ultrasonically, however, more information concerning the vibrational characteristics of the polymers is necessary before the observed θ values can be fully understood. The values presented here were calculated from the defined molecular weight of each repeating unit.

Volume coefficient of expansion versus temperature data are presented in Table II. Since the steel and the aluminum alloys were expected to exhibit some anisotropy, expansivity measurements were made parallel and perpendicular to the stratification layers. Expansivity data in the "c" direction, parallel to the axis ($\alpha_{||}$) was 0.8 percent higher than that observed in the perpendicular direction ($\alpha_{\perp}$) for the temperature range studied. Since this increase was within the experimental limitations of the instrument, the volume coefficient of expansion data of Table II was taken as $3\alpha_{||}$.

Table II

Volume Coefficient of Expansion Data for the Materials Studied

$$\beta = a + bT, /^{o}K$$

Material	a x 10^5	b x 10^7	Temp. Range, ^{o}K
Epoxy	5.65	5.49	243 - 353
Epoxy-1	-128.40	53.30	263 - 423
Epoxy-2	-125.65	43.17	343 - 388
Polystyrene	10.46	3.92	233 - 353
Teflon	-72.59	35.87	243 - 413
Teflon-1	-65.33	34.17	253 - 413
Tungsten	0.45	0.31	233 - 383
Steel No. 314	-5.40	2.99	253 - 373
Al 2024	-1.44	2.79	243 - 373

The adiabatic bulk moduli as a function of temperature are given in Table III. The ultrasonic longitudinal and transverse velocities used to calculate the moduli were measured at this laboratory for all the materials listed except tungsten. For this material the elastic data of Featherston and Neighbours (21) were used to calculate the bulk moduli.

The temperature variation of the Gruneisen ratio is given in Table IV. This parameter was calculated from the previously defined thermodynamic relationship of Equation 1.

Table III

Adiabatic Bulk Moduli For The
Materials Studied

$$B^S = a + bT, \text{ dyne/cm}^2$$

MATERIAL	$a \times 10^{-11}$	$b \times 10^{-8}$	Temp. Range, $^{\circ}K$
Epoxy	0.98	-1.47	243 - 353
Epoxy-1	87.32	-1.12	263 - 423
Epoxy-2	2.02	-4.52	343 - 388
Polystyrene	0.58	-0.70	233 - 353
Teflon	0.81	-1.50	243 - 313
Teflon-1	0.79	-1.44	243 - 413
Tungsten	32.02	-3.14	233 - 383
Steel No. 314	16.45	-3.37	253 - 373
Al 2024	8.33	-2.34	243 - 373

Table IV

Gruneisen Parameters For
The Materials Studied

$$\gamma = a + bT$$

MATERIAL	a	$b \times 10^3$	Temp. Range, $^{\circ}K$
Epoxy	2.21	-4.35	243 - 353
Epoxy-1	-1.73	9.89	263 - 423
Epoxy-2	1.07	-1.00	243 - 388
Polystyrene	1.50	-2.84	233 - 353
Teflon	0.13	1.49	243 - 413
Teflon-1	0.34	0.86	243 - 413
Tungsten	0.51	2.91	233 - 383
Steel No. 314	-1.11	9.11	253 - 373
Al 2024	0.11	6.61	243 - 373

Table V

Experimental Thermodynamic Data
Used to Calculate the Pressure Derivatives
of the Gruneisen Ratio

FUNCTION*	UNITS	EPOXY	EPOXY-1	EPOXY-2
ρ_0	g/cm^3	1.25	1.25	1.20
$(\partial \rho_0/\partial P)_T \times 10^{11}$	g/dyne cm	2.29	2.35	2.76
$\beta_0 \times 10^4$	/°K	2.04	2.04	2.46
$(\partial \beta_0/\partial P)_T \times 10^{14}$	cm^2/°K dyne	-5.23	-4.51	-19.43
$C_{p_0} \times 10^2$	cal/g°K	22.27	22.27	30.58
$(\partial C_{p_0}/\partial P)_T \times 10^5$	cm^3/g°K	-20.40	-47.15	-68.16
$B_0^S \times 10^{-10}$	dyne/cm^2	5.77	5.59	20.21
$B_0^T \times 10^{-10}$	dyne/cm^2	5.44	5.28	20.13
$B_{0,S}^{S'}$		3.03	2.98	2.48
$B_{0,T}^{S'}$		3.73	3.51	2.94
$B_{0,T}^{T'}$		3.53	3.33	2.80
γ_0		1.01	0.99	0.74
$(\partial \gamma_0/\partial P)_T \times 10^{11}$	/°K cm^2/dyne	-17.99	-20.35	-2.97

*Zero subscript denotes T = 273 K

The data necessary to calculate the Gruneisen ratio as a function of pressure are given in Tables V, VI, and VII, (Zero subscripts denote values obtained at 273°K). The pressure derivative of the density, the volume coefficient of expansion and the heat capacity, $(\partial \rho/\partial P)_T$, $(\partial \beta/\partial P)_T$ and $(\partial C_p/\partial P)_T$, respectively,

Table VI

Experimental Thermodynamic Data
Used to Calculate the Pressure Derivatives
of the Gruneisen Ratio

FUNCTION*	UNITS	POLYSTYRENE	TEFLON	TEFLON-1
ρ_0	g/cm^3	1.05	2.19	2.19
$(\partial\rho_0/\partial P)_T \times 10^{11}$	g/dyne cm	2.84	5.99	5.78
$\beta_0 \times 10^4$	$/^{\circ}K$	2.36	2.97	3.18
$(\partial\beta_0/\partial P)_T \times 10^{14}$	$cm^2/^{\circ}K$ dyne	-1.92	12.38	-19.42
$C_{p_0} \times 10^2$	$cal/g^{\circ}K$	26.23	22.27	23.42
$(\partial C_{p_0}/\partial P)_T \times 10^5$	$cm^3/g^{\circ}K$	-58.87	-9.57	-9.44
$B_0^S \times 10^{-10}$	$dyne/cm^2$	3.88	4.06	4.00
$B_0^T \times 10^{-10}$	$dyne/cm^2$	3.68	3.87	3.79
$B_{0,S}^{S'}$		2.59	2.18	2.18
$B_{0,T}^{S'}$		2.99	2.75	2.77
$B_{0,T}^{T'}$		2.84	2.65	2.63
γ_0		0.80	0.59	0.59
$(\partial\gamma_0/\partial P)_T \times 10^{11}$	$/^{\circ}K$ $cm^2/dyne$	-12.42	-17.81	-16.25

*Zero subscript denotes T = 273 K

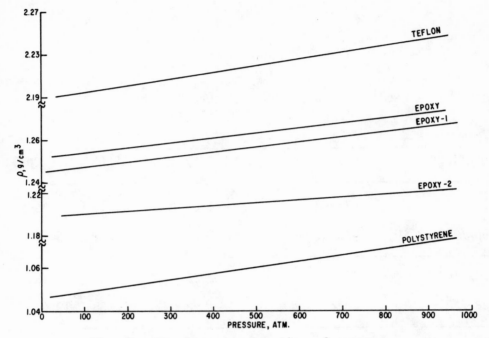

Fig. 1.   Density as a function of pressure.

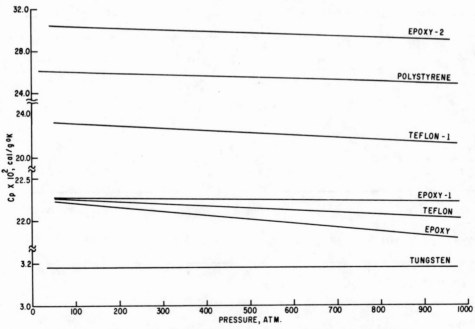

Fig. 2.   The specific heat as a function of pressure.

were obtained by the proper substitution of experimental data into
Equations 9 through 11. The pressure derivative of the bulk
modulus, $(\partial B^S/\partial P)_T = B^{S'}_{0,T}$, was estimated from the Dugdale Mac-
Donald formula, $B^{S'}_{0,S}$ as defined by Equation 16. The proper
conversion to the adiabatic derivative, $B^S_{0,T}$, was made by Equation
17. The function $B^{T'}_{0,T}$, defines the corresponding isothermal
pressure derivative. The pressure derivative of the Gruneisen
ratio was obtained by the proper substitution of data into
Equations (8) and/or (18) where all quantities were determined
experimentally except $(\partial B^S/\partial P)_T$.

For polystyrene, the ultrasonic data of Lamberson (11) were
used to calculate the adiabatic bulk moduli data of Tables III and
VI. The variation of sound velocities with pressure were also
experimentally determined for this material to the 10 kbar region.
Both Lamberson (11) and Asay, et al. (8) have shown that for
polystyrene the validity of the Murnaghan equation was apparent
by the linearity of the bulk modulus to the 10 kbar region with
a pressure derivative of 8.89. This study indicates that the
Dugdale MacDonald relation estimated as 2.59 was valid only to a
pressure of 950 atmospheres. There was a significant difference
between the actual derivative and that estimated through the
Dugdale MacDonald relation, however, for this plastic the compressibility data of Bridgman (22), and the dynamic data of
Wagner, et al. (23) and Hauver, et al. (24) were not in complete
agreement from the 3 to the 20 kbar region. This is probably due
to the transition which occurs within this region.

Ku (25) compared the specific volume of teflon at $30^\circ C$ with
the calculated values obtained by a 12-parameter polynomial, and
by the Murnaghan, Birch and Tait equations. The accepted value at
1000 atm was 0.4367 cm^3/g or a density of 2.289 g/cm^3. This value
compares well with the value of 2.251 g/cm^3 at $0^\circ C$ and a pressure
of 100 atm obtained in this study (Table VI and Figure 1).

The bulk modulus reported by Ku (25) for teflon for the 1 to
5000 atm region was 2.97 x 10^{10} dyne/cm^2. The values obtained in
this study for the actual and extrapolated teflon samples were
4.28 x 10^{10} dyne/cm^2 and 4.34 x 10^{10} dyne/cm^2, respectively.

The bulk moduli versus pressure data for teflon agree well
with data for teflon extrapolated through its room temperature
transition as presented by Weir (26) and Ku (25). For the actual
data (Teflon-1) Ku has shown a more pronounced drop in the bulk
modulus with a transitional shift appearing in the 3 x 10^3 atm
region.

The C_p versus pressure data (Figure 2) and the volume coefficient of expansion versus pressure data (Figure 3) are in

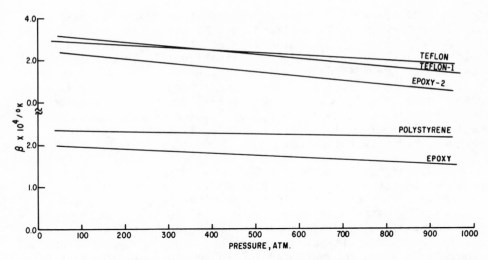

Fig. 3.   The volume coefficient of thermal expansion as a function of pressure.

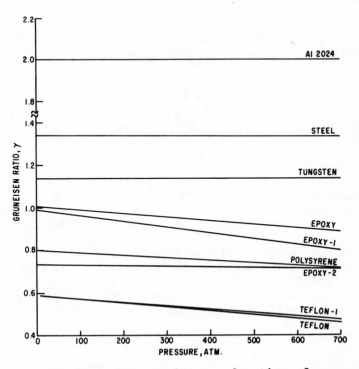

Fig. 4.   The Gruneisen ratio as a function of pressure.

agreement with previously reported data obtained from thermodynamic measurements (2,8,15,18, and 25). Since some of the
metallic materials exhibited only slight changes for the pressure
ranges reported, they were not included on some of the graphs.

The Gruneisen ratio as a function of pressure is given in
Figure 4. Previously reported data for teflon and polystyrene
(25,28,29) list a 1.54 percent decrease in polystyrene and a
30.4 percent decrease for teflon. Our studies indicate a 15.7
percent decrease in γ for polystrene and a 3.06 percent decrease
in teflon for similar temperature and pressure ranges. Differences
in reportes inivial γ values are due to extrapolations through
transition temperatures.

For Al 2024 the ultrasonic sound velocity data to calculate
the bulk moduli (Table III) were compiled from single valued data
presented by Kohn (29) and McQueen, et al. (30). The Hugoniot
curves, the Mie-Gruneisen equation of state, and the Dugdale -
MacDonald relation were employed by McQueen et al., to calculate
the complete thermodynamic description of the alloy. The relation
between shock and particle velocity was applied in order to obtain
the ultrasonic data. In terms of the expansion of the shock
velocity, μ_s, to quadratic terms in the particle velocity, μ_p,

$$\mu_s = a + b\mu_p^2 \tag{19}$$

Ruoff (31) and Pastine (32) give the coefficients a, b, and c as

$$a = \mu_s \ (P = 0) = \sqrt{B_0^S/\rho} \tag{20}$$

$$b = 1/4 \ (1 + B_{0,s}^{S'}) \tag{21}$$

$$c = b/6a \ (2 - b + B_{0,s}^{S'}/2b + \gamma) \tag{22}$$

where the terms have been defined previously

For Al 2024, McQueen, et al. (30) reported a value for b of
1.54 for a sample with a density similar to the one reported in
this paper. Substitution of this value into Equation (21) gave
a shock velocity value of $B_{0,s}^{S'} = 5.161$ and $\gamma = 2.08$. This value
agrees to within 2.3 percent of our value ($B_{0,s}^{S'} = 5.28$, Table VII
which was estimated by the Dugdale-MacDonald relationship. The
$\gamma = 2.14$ of Tables IV and VII, calculated by the thermodynamic
equation agrees to within 2 to 3 percent. Values for γ previously
obtained for two similar alloys, Al 1060 and Al 6061 (8), were
2.06 and 2.13, respectively. It is thus clearly seen that the
accuracy with which γ is obtained will determine the accuracy of
the estimated pressure derivative.

Table VII

Experimental Thermodynamic Data
Used to Calculate the Pressure Derivative
of the Gruneisen Ratio

FUNCTION*	UNITS	TUNGSTEN	STEEL # 314	AL 2024
ρ_0	g/cm^3	19.27	7.62	2.79
$(\partial \rho_0 / \partial P)_T \times 10^4$	g/dyne cm	6.20	4.91	3.77
$\beta_0 \times 10^5$	/$^\circ$K	1.30	2.78	6.92
$(\partial \beta_0 / \partial P)_T \times 10^{17}$	cm^2/$^\circ$K dyne	-0.67	-12.48	-121.56
$C_{p_0} \times 10^2$	cal/g$^\circ$K	3.18	10.10	21.31
$(\partial C_{p_0} / \partial P)_T \times 10^7$	cm^3/g$^\circ$K	-3.43	-92.77	275.01
$B_0^S \times 10^{-11}$	dyne/cm^2	31.15	15.52	7.69
$B_0^T \times 10^{-11}$	dyne/cm^2	31.06	15.36	7.38
$B_{0,s}^{S'}$		3.61	3.68	5.28
$B_{0,T}^{S'}$		3.65	3.76	5.46
$B_{0,T}^{T'}$		3.63	3.72	5.25
γ_0		1.30	1.34	2.14
$(\partial \gamma_0 / \partial P)_T \times 10^{12}$	cm^2/dyne	-3.02	-7.73	-6.60

*Zero subscript denotes T = 273 K

As previously stated, the elastic constants of Featherston and Neighbours (21) were used to calculate the bulk moduli data (Tabel III) for tungsten. Since the elastic wave propagation was directly along a principle direction, the elastic constants were said to be accurate to ± 0.5 percent. The value of B = 3.08 x 10^{12} dyne/cm^2 calculated from Bridgman's dynamic compressibility data agrees well with the calculated difference between the room temperature adiabatic and isothermal value of this study.

No ultrasonic data illustrating the change of elastic constants with pressure were available for tungsten, however, the ($\partial B/\partial P$) value derived from Bridgman's dynamic data was 3.67. This value is in good agreement with the Dugdale-MacDonald estimation of $B_{0,s}^{S'}$ = 3.61 (Table VII) of this study. The corresponding ($\partial\gamma/\partial P)_T$ is listed in Table VII and γ versus pressure is shown in Figure 4. Since no significant change with pressure was noted, the ($\partial\gamma/\partial P$) value should apply to the 10,000 kg/cm^2 pressure range limit given by Bridgman (22).

For the oxidation resistant steel studied, the ultrasonic transverse and longitudinal velocity data of Kohn (29) and McQueen, et al. (30) were used to determine the bulk moduli given in Tables III and VII. The $B_{0,s}^{S'}$ value estimated from the shock velocity data of Kohn was 4.94. The room temperature value of this study ($B_{0,s}^{S'}$ = 4.54, Table VII) agreed to within 8 percent of the value obtained from the shock data. Values for $B_{0,s}^{S'}$ of Tables V, VI and VII were calculated for T = 273 K.

In order to understand the specific heat and thermal expansion data for the epoxy polymer, the differential thermogram of the sample (Figure 5) was compared with Maraset epoxy previously studied (9 and 33). The thermograms show distinct differences with a pronounced transition at $46^{\circ}C$ for the Maraset epoxy and a much broader transition approaching $125^{\circ}C$ for the Shell epoxy. Both thermograms show decomposition beyond $200^{\circ}C$, however, an exothermal double peak was noted for the Shell epoxy as opposed to the broad endothermal peak observed for the Maraset sample. The double peak may be associated with substituents which were added to stabilize the material. A comparison of the thermal expansion data (9 and 33) also showed a slight increase in stability for the Shell epoxy as noted by the shift of the $46^{\circ}C$ transition to slightly higher temperatures. Since the transition is evident, three sets of data were derived for this polymer. Table V, column 1 (labeled Epoxy) illustrates the data obtained as extrapolated through transitional regions. Column 2 (labeled Epoxy-1) defines the actual polymeric thermal and ultrasonic measurements along the transitions and Epoxy-2 defines the high temperature region.

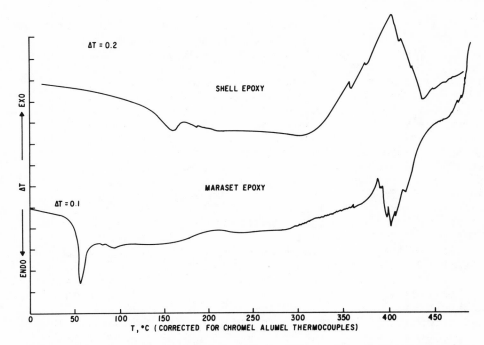

Fig. 5. Differential thermograms for Maraset and Shell epoxies.

The ultrasonic transverse and longitudinal velocities used
to calculate the adiabatic bulk moduli (Table III) were obtained
at this laboratory by Benson, et al. (34). No dynamic shock wave
data are available, however, Benson did fit ultrasonic pressure
data to the Birch equation and found that at 1 kbar pressure, the
adiabatic bulk modulus was 6.2×10^{10} dyne/cm^2. The value calculated from this study is 5.95×10^{10} dyne/cm^2, therefore, the
bulk modulus pressure derivative is applicable to approximately
1 kbar. The γ versus pressure data are illustrated in Figure 4.

CONCLUSIONS

In this paper a relationship between the adiabatic and isothermal pressure derivatives were presented which involves only
the temperature dependent thermodynamic properties of the volume
coefficient of expansion, the heat capacity, volume or density,
and the bulk moduli. The pressure derivatives were estimated
by a Dugdale-MacDonald relation so that the variation of the
Gruneisen ratio with pressure could be determined. Although the
pressure derivatives of the polymers are accurate only for small
pressure ranges, the data will be useful to propose theoretical
models for these materials. Moreover, in high pressure experiments,
the constants of the equation of state are affected by imperfections
in the solid, therefore, it is possible that the ultrasonic
measurements are more reliable.

REFERENCES

1. E. Gruneisen, Handbuch der Phys., $\underline{10}$, 1 (1926).

2. S.R. Urzendowski and A.H. Guenther, Thermal Analysis, Vol. 1, p. 493, Edited by R.F. Schwenker, Jr. and P.D. Garn, Academic Press, New York, 1969.

3. C. Kittel, Introduction to Solid State Physics, John Wiley & Sons, Inc., New York, 1968, page 182.

4. R.E. Barker, Jr., J. Appl. Phys., $\underline{38}$, 4234 (1967).

5. C.A. Swenson, The Physics and Chemistry of High Pressures, Editor, A.R. Ubbelohde, Society of Chemical Industry, London, England, 1963, p. 39.

6. A.H. Guenther, Symposium on Dynamic Behavior of Materials, Special Technical Publication No. 336, Am. Soc. for Testing Materials (1962).

7. J.C. Slater, Introduction to Chemical Physics, McGraw-Hill, New York, 1963.

8. J.R. Asay, S.R. Urzendowski, and A.H. Guenther, Air Force Weapons Laboratory, Tech Rept. No. 67-91, Kirtland Air Force Base, New Mexico, 1968.

9. S.R. Urzendowski and A.H. Guenther, Air Force Weapons Laboratory Tech. Rept. 71-6, Kirtland Air Force Base, New Mexico, 1971.

10. R.N. Thurston, Proceedings of the IEEE $\underline{53}$, 1320 (1950).

11. D.L. Lamberson, Dissertation, Air Force Institute of Technology, Wright-Patterson AFB, Ohio.

12. O.L. Anderson, J. Phys. Chem. Solids $\underline{27}$, 547 (1966).

13. F.D. Murnaghan, Proc. Nat'l. Acad. Sci., $\underline{30}$, 244 (1944).

14. J.S. Dugdale and D.K.C. MacDonald, Phys. Rev., $\underline{89}$, 832 (1953).

15. S.R. Urzendowski, D.A. Benson and A.H. Guenther, Thermal Analysis, Vol. 3, 365 (1971), Editor H.G Wiedemann, Birkhauser Verlag, Basel und Stuttgart, Davos, Switzerland.

16. J.A. Kok, Physics, $\underline{24}$, 1045 (1958).

17. N.E. Phillips, Phys. Rev., $\underline{118}$, 664 (1960).

18. B. Wunderlich, Heat Capacities of Linear High Polymers, Office of Naval Research, Tech. Rept. No. 17, Rensselaer Polytechnic Institute, Troy, New York, 1968.

19. W. Reese, J. Appl. Phys., 37, 3959 (1966).

20. C.L. Choy, G.L. Salinger, Y.C. Chiang, and J.I. Treu, Bull. Am. Phys. Soc., 12, 1063 (1967).

21. F.H. Featherston and J.R. Neighbours, Phys. Rev., 1324 (1963).

22. P.W. Bridgman, Collected Experimental Papers, Vol. VI, Harvard University Press, p. 3846, 1964.

23. M.H. Wagner, W.F. Waldorf, and N.A. Louie, AFSWC-TDR-62-66, Vol. I, Aerojet-General Corp., 1962.

24. G.E. Hauver and A. Melani, Shock Compression of Plexiglas and Polystyrene, BRL Rept. No. 1259, Aberdeen Proving Ground, Maryland, 1964.

25. P.S. Ku, Equation of State of Organic High Polymers, AD678-887, General Electric, Phil., Penn. 1968.

26. C.E. Weir, J. Res. Nat'l. Bur. Std., 53, 245 (1957).

27. R.S. Spencer and G.D. Gilmore, J. Appl. Phys., 20, 502 (1949).

28. J. Brandup and E.H. Immergut (Editors), Polymer Handbook, John Wiley and Sons, New York, 1972.

29. B.J. Kohn, Air Force Weapons Laboratory, Tech. Rept. No. 69-38, Kirtland Air Force Base, New Mexico, 1969.

30. R.G. McQueen, S.P. Marsh, S.W. Taylor, J.N. Fritz and N.J. Carter, High Velocity Impact Phenomena, R. Kinslow, Editor, Academic Press, New York, 1970.

31. A.L. Ruoff, Sandia Corp., Tech. Rept., SC-RR-66-676, 1966.

32. D.J. Pastine and D.J. Piacesi, J. Phys. Chem. Solids, 27, 1783, (1966).

33. S.R. Urzendowski and A.H. Guenther, "The Combination of Thermal and Ultrasonic Data to Calculate Gruneisen Ratios and Various Thermodynamic Functions", paper presented at International Symposium on Thermal Expansion, Lake of the Ozarks, Mo., Nov. 1972. (Paper is in the process of publication).

34. D.A. Benson, R.N. Junck and J.A. Klosterbuer, Air Force Weapons Laboratory Tech. Rept., No. 70-169, Kirtland, New Mexico, 1971.

THE USE OF THERMAL EVOLUTION ANALYSIS (TEA) FOR THE

DETERMINATION OF VAPOR PRESSURE OF AGRICULTURAL CHEMICALS

Roger L. Blaine and Paul F. Levy

E.I. DuPont de Nemours & Co., (Inc.)

Wilmington, Delaware 19898

INTRODUCTION

The title of this paper is listed as " The Use Of Thermal Evolution Analysis for the Determination of Vapor Pressures of Agricultural Chemicals ". The phrase "agricultural chemicals" includes an extremely broad range of chemicals including insect repellents and attractants, herbicides, pesticides, fertilizers, and many other products. Such a broad field would be extremely difficult to cover in the space of this paper and so I will, of necessity, limit my remarks to one, much smaller area, that of insect repellents. The techniques used in this study are perfectly general, however, and I leave it to the imagination of the reader to extend this work to the general case.

Insect repellents are compounds which prevent the approach, the settling and/or the biting of insects. Repellents differ from insecticides in that they do not kill the offending insect. Indeed, the LD_{50} for most suitable insect repellents is in the grams or tens of grams per kilogram of body weight range (1). Repellents are used to protect crops and animals as well as human beings from the damage of insects.

An ideal repellent for use by humans has a number of requirements unrelated to the repellent effectiveness of the compound including such criteria as long lastingness, universality, lack of toxicity and irrita-

tion to the skin and mucous membranes, and lack of an
unpleasant odor (2). Experience during World War II
showed that cosmetic acceptability is the most important
criteria. No matter how effective a particular repellent is, it is worthless if it is not used.

Probably the oldest of man-made insect repellents
is smoke. Since prehistoric times, man has stoked his
fires with smoldering and strong smelling materials as
a defense against the bites of insects. The ancient
Egyptians are said to have smeared themselves with
strongly smelling substances to protect their bodies a-
gainst the bites of gnats and mosquitoes. In more recent times, plant extracts, such as the juice of green,
raw tomatoes, garlic, or olive oil were attributed repellent effects(1). Their use has led to the identification of a number of essential oils known to possess
repellent effects. In 1903 Smith recommended the use of
oil of citronella still used today (3). Some other examples of essential oil repellents are oil of thyme,
lemon oil, orange blossom oil, peppermint oil, coconut
oil, and oil of nutmeg. Citronellol, one of the ingredients in oil of citronella,is such a good insect repellent that it remained for many years the standard
material against whose repellent effects other compounds
were measured.

The work on plant extracts as insect repellents led
to the study of natural insect secretions and their repellent effects. In this way several classes of compounds possessing repellent effects were uncovered. Aldehydes, acids, and quinones were found to be widely used
by insects to repel their enemies (4).

The search for highly effective, long lasting, universal insect repellents was greatly intensified during
the second World War. It became imperative to protect
soldiers in tropical war theaters from diseases such as
yellow fever, malaria, and spotted fever, which are transmitted by insects. More than 4300 chemical compounds
were tested by the Entomological Research Branch of the
U.S. Department of Agriculture at Orlando, Florida,be-
tween 1942 and 1947 as mosquito repellents (5). From
this work, as well as work done at Rutgers University, a
number of quite effective repellents became available.

The exact mechanism of the protection afforded by
repellents is not well understood. Their smell or taste
to humans has no relationship to their repellent ability.

Most repellents are volatile, however, and appear to be effective in the vapor phase rather than being tactilly or gustatorially offensive to the insect.

For this reason, the vapor pressure of repellents is of interest to the theory of repellency (that is,"how do they work") as well as to the practically of repellents (that is,"how long will they last"). Moreover, the vapor pressure of insect repellents, like all agricultural chemicals, is of interest for safety and environmental considerations. While insect repellents are almost never toxic, some of the related compounds such as insect attractants, insecticides, herbicides and pesticides are. Indeed,a number of European countries now require data on the vapor pressure of all agricultural chemicals at ambient temperatures (that is,25°C) before they are released for public use. Where the European countries go in this area, can the U.S. be far behind?

The accurate measurement of very low vapor pressures is not a trivial task. Vapor pressure data is usually obtained by raising the temperature of the sample to some temperature where a measurable vapor pressure is available and then by a series of such measurements and the use of the Clausius-Clapeyron equation, the data is extrapolated to low pressures and temperatures.

Thus for dimethyl phthalate, a widely used insect repellent, there is little reliable data at pressures and temperatures below 1.0 mm Hg at 100.3°C. The picture is even more complicated, in that a number of workers have observed that for some insect repellents, such as dibutyl phthalate, the plot of log P versus 1/T is not a straight line but is better fitted by the addition of a $1/T^2$ term (6,7). Straight line extrapolation of medium temperature-pressure data to low pressures is therefore tenuous.

Thermal Evolution Analysis (TEA) comes to the rescue of investigators examining low vapor pressure at temperatures only slightly above ambient. Commercially available TEA equipment was introduced by Don Carle and coworkers at the Pittsburg Conference in 1969 (8), and is now marketed by DuPont Instruments. Eggertsen and his coworkers at Shell Development Co., who developed the concept (9), have used the TEA to measure the vapor pressure of a number of hydrocarbons (10).

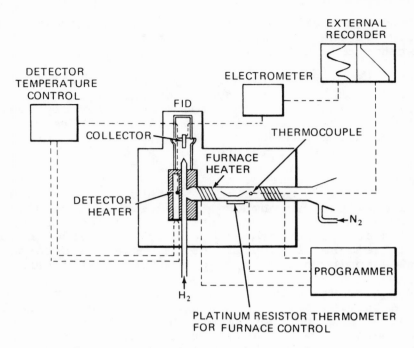

Fig. 1.   Block Diagram of TEA

A block diagram of the TEA is shown in Figure 1. A
sample oven whose temperature is programmable from near
ambient to as high as 500°C is closely coupled to a flame
ionization detector (FID). A nitrogen purge gas passes
over the sample carrying any evolved material to the extremely sensitive FID. The output from the FID is fed
to one pen of a two pen recorder while the sample temperature is indicated by the second pen.

Figure 2 illustrates the vapor pressure probe used
in this work. The material whose vapor pressure is to
be measured is intimately mixed with 20-40 mesh silver
granules and packed into the left end of the probe. The
silver granules provide a high wetted surface area to
the sweeping gas as well as insuring uniform sample temperature. A thermocouple embedded in this mixture acc-

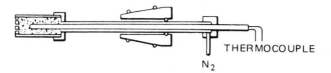

Fig. 2. Vapor Pressure Probe

urately reads the sample temperature. In operation a
stream of nitrogen is slowly fed through the sample/sil-
ver mixture where it is equilibrated with the sample to
the level dictated by the vapor pressure of the sample.
Once the sample containing nitrogen leaves the probe it
is combined with a much larger nitrogen purge gas and
fed to the FID. The rate of nitrogen purge through the
sample is selected by a valve on the front of the TEA.
Settings of 1/30, 2/30, and 5/30 of the total nitrogen
stream are selectable.

Shown in Figure 3 is a photograph of the TEA and its
accompanying two pen strip chart recorder. This particular recorder is fitted with a Disc Integrator for ease
of calibration. The TEA is 18 inches high and 26 inches
long. The TEA and recorder will fit easily in 4 feet of
bench space.

In this work, a Newport Laboratories Series 2600
Digital Thermocouple Indicator was used to read the sample temperature to 0.1°C. This is a far more accurate
temperature measurement than is called for. The recorder
pen can be easily read to within $\pm$ 1°C which is less than
0.5% of the absolute temperature measurement used in the
Clausius-Clapeyron equation.

If one were to temperature program the sample oven
at a linear rate and a constant sample purge gas rate,
one obtains a scan similar to that shown in Figure 4.
The exponential increase in the vapor pressure of the
sample as the temperature increased, reflects the logarithmic relationship between vapor pressure and temperature. One could, of course, pick the vapor pressure

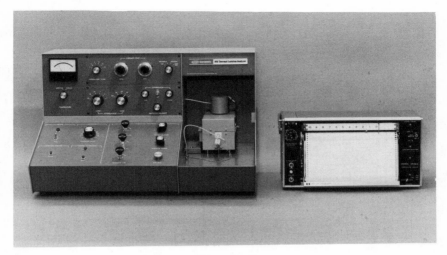

Figure 3

data off this type of curve directly but such a procedure is not as precise as the method to follow. By the
way, the di-(2-ethylhexyl)phthalateshown in this figure
is not in itself an insect repellent but is related to
such repellents as dimethyl-and dibutyl-phthalate.

PROCEDURE

The relative proportion of the sample in the vapor
phase issuing from the vapor pressure probe is proportional to the ratio of sample vapor pressure (VP) to
the atmospheric pressure (P).

$$\frac{M_S}{M_S + M_N} = \frac{VP}{P} \tag{1}$$

Since the number of moles of the sample (M_S) is very
much smaller than the number of moles of the nitrogen
carrier gas (M_N), equation 1 reduces to:

$$VP = \frac{M_S}{M_N} \times P \tag{2}$$

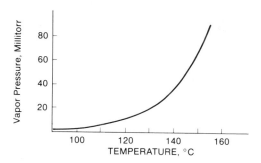

Fig. 4.   Octyl Phthalate Vapor Pressure

M_N is just equal to the carrier gas flow rate (F) divided by the gram molecular volume. M_S is given by:

$$M_S = \frac{\Delta Y \ x \ R \ x \ C}{MW}$$

where: ΔY = vertical chart deflection
        R = instrument range setting
        C = FID response to the sample
       MW = sample molecular weight

Substituting these values into equation 2 yields:

$$VP = \frac{\Delta Y \ x \ R \ x \ C \ x \ P \ x \ 22.4}{F \ x \ MW} \qquad (3)$$

For a given set of experimental conditions, the constants in equation 3 can be lumped into one calibration coefficient K.

$$VP = \frac{\Delta Y}{F} \ x \ K$$

This is the working relationship for each measurement.

In this work, the temperature of the sample was adjusted to the desired value and allowed to equilibrate for several minutes with no nitrogen flow over the sample. This established the baseline. A nitrogen flow

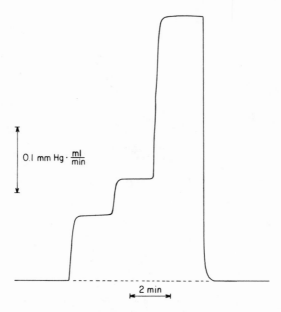

Fig. 5.   Vapor Pressure Measurement – Electrometer Response

of 0.4 ml/min was then passed over the sample for a few minutes. This was followed in succession by 0.7 and 1.8 ml/min flow rates then again by no flow conditions. Such a procedure produces a series of stair steps as shown in Figure 5. The vertical deflection when divided by the purge gas flow rate, yields the vapor pressure.

The vertical axis deflection calibration is accomplished by integrating the area under the total evolution of a known weight of sample. This yields FID response to the particular compound. Range and attenuation selectors on the TEA permits the vertical axis scale to be properly selected for a given vapor pressure measurement.

Dimethyl phthalate is the modern day standard for insect repellents. It is one of the most active of fly and mosquito repellents and was widely used in World War II by itself and in mixture 622 with ethyl hexanediol

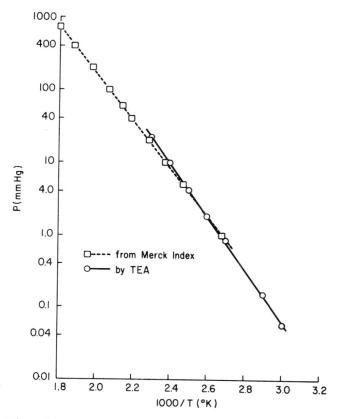

Fig. 6. Vapor Pressure of Dimethyl Phthalate

and indalone. It is also widely used in the plastics industry as a plasticizer. Because of its use as a standard repellent and because its physical properties have been thoroughly investigated, it would seem a natural compound to use as a comparison standard for this TEA work.

Shown in Figure 6 is the Clausius-Clapeyron plot for dimethyl phthalate obtained by this technique. The dimethyl phthalate used was obtained from Aldrich Chemical Co. and was better than 99% pure. The temperatures of interest were selected such that the data fell at even values on the 1/T plot. Each data point was taken in random order and represents an average of three determinations (that is, the three sample gas flow rates).

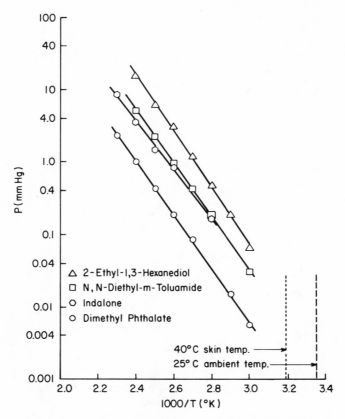

Fig. 7. Vapor Pressure of Insect Repellents

The upper dotted line represents the vapor pressure data
taken from standard tables (11). The lower solid line
represents the values obtained by TEA. In the range of
overlap, between 1.0 and 20mm Hg, the two methods are in
agreement. At lower pressures, the TEA data is somewhat lower than would be expected by a linear extrapolation of the semi-log plot of the standard. This is in
agreement with the trends observed in other phthalic
acid esters at low pressures (6,7). Good agreement is
found for vapor pressures obtained by TEA and those
found by more conventional methods.

Figure 7 illustrates the vapor pressures obtained
by TEA for several common insect repellents.

2-Ethyl-1,3-hexanediol is the familiar insect repellent Rutger's 612. It is one of the oldest, most widely used and most effective of the synthetic insect repellents. It is widely used by itself and in mixtures with other repellents. Ethyl hexandediol is the active ingredient in such commercial repellents as 6-12 marketed by Union Carbide. The ethyl hexandediol used in this experiment was obtained from Aldrich Chemical Co. and was better than 97% pure.

N,N-Diethyl-m-toluamide is a more modern synthetic repellent. It is one of the most effective repellents commercially available. In mixtures with ethyl hexanediol it has no equal. Diethyl toluamide is one of the active ingredients in almost every available off-the-shelf insect repellent. It is marketed under such trade names as Off (by Johnson's Wax), 6-12 Plus (by Union Carbide) and many others. The diethyl toluamide used here was obtained from Aldrich Chemical Co. and was better than 98% pure.

Inadalone is a trivial name for 3,4-dihydro-2,2-di-methyl-4-oxo-2H-pyran-6-carboxylic acid-n-butyl ester. Indalone is one of the few good essential oil insect repellents although almost all of the indalone used for this purpose is synthetic. Indalone finds few uses today having been replaced with diethyl toluamide and ethyl hexanediol. Large quantities of it were used in World War II in the mixture 622. (The 6.2.2 designation corresponds to the relative proportions of the dimethyl phthalate, ethyl hexanediol and indalone used in it). The indalone used in this experiment was insect repellent grade provided by FMC Corporation.

Dimethyl phthalate is also shown in Figure 7 for comparison purposes.

DISCUSSION

The lack of data at lower pressures does not reflect a lack of instrument sensitivity. Indeed, the instrument has the capability to determine vapor pressures at levels three orders of magnitude lower than the lowest point in this figure. The lack of lower pressure data is due to a lower temperature limit of approximately 60°C. The FID, which normally operates at 550°C, is so close to the sample oven that sample temperatures less than 60°C are hard to maintain.

Compound	ave.dev. (%)	Vapor Pressure at 40°C (mm Hg)	at 25°C (mm Hg)
Ethyl Hexanediol	5.8	0.0144	0.00351
Indalone	7.8	0.00750	0.00217
Diethyl Toluamide	3.0	0.00644	0.00167
Dimethyl Phthalate	2.6	0.00120	0.000311

Fig. 8. Vapor Pressure of Insect Repellents

The temperatures of interest for insect repellents are skin temperature, considered to be 40°C, and ambient temperature, 25°C. Figure 7 illustrates that the lower useful temperature of the TEA is not unreasonably far from these temperatures.

Shown in Figure 8 are the extrapolated vapor pressures at 40°C and 25°C for these four insect repellents. We can have confidence in these values to at least two significants figures and possibly three. The average deviation for the pressure measurements shown on the previous Figure are shown in the second column. This average deviation is for at least 5 data points of three measurements each. It should be pointed out that this is the average deviation for the pressure measurement and not the log of the pressure measurement as the data is plotted.

The last Figure 9 represents vapor pressure work done by Dr. O.L. Carter at Rohm and Haas on TOK Herbicide (12). His work illustrates the vapor pressure measurement concept of the TEA applied to another class of agricultural chemicals, that of herbicides. It also illustrates the extreme sensitivity of the TEA measuring vapor pressures as low as 6×10^{-7} mm Hg.

In conclusion, then, the DuPont 916 Thermal Evolution Analyzer is shown to be an effective means of ob-

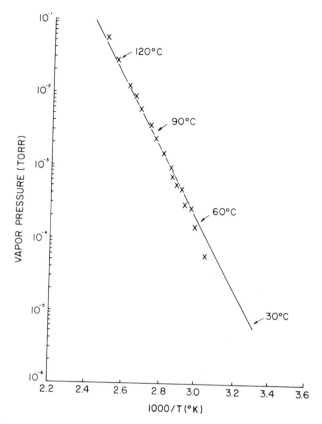

Fig. 9. Vapor Pressure Data for TOK Herbicide

taining vapor pressure on high boiling materials. Vapor pressures as low as 10^{-6}mm Hg can be easily measured. The TEA's dyamic range of 7 orders of magnitude makes it applicable to a wide variety of liquid and solid organic materials.

In this study non-toxic insect repellents were investigated. It takes little imagination, however to extend such work to the area of insect attractants, herbicides, pesticides and other agricultural chemicals.

REFERENCES

1) K.H. Buchel, "Insekten Repellents" in Chemie der Pflanzenschutz- und Schädlings-bekämpfungsmittel,

Richard Wegler (ed.), 1970 pages 487-496.

2) V.G. Dethier, Chemical Insect Attractants and Repellents, Blakiston Co., 1947, page 214.

3) Smith, "Report of the New Jersey State Agricultural Experimental Station Upon Moskitos", Trenton, N.J. 1904.

4) L.M. Roth and T. Eisner, "Chemical Defenses of Arthropodes", Ann. Rev. Entomol., 7, 107-136 (1962).

5) B.V. Travis, F.A. Morton, H.A. Jones and J.H. Robinson, "The More Effective Mosquito Repellents", J. Econ. Entom., 42, 686-694 (1949).

6) E. Hammer and A.L. Lydersen, "Vapour Pressure of Di-n-butylphthalate, Di-n-butylsebacate, Lauric Acid and Myristic Acid", Chem. Eng. Sci., 7, 66-72 (1957).

7) P.A. Small, K.W. Small, and P. Cowley, "The Vapour Pressures of Some High Boiling Esters", Trans. Faraday Soc., 44, 810-816 (1948).

8) A.C. Stapp and D.W. Carle, "A New Thermal Analysis Instrument", presented at the Pittsburgh Conference on Analytical Chemistry and Applied Spectroscopy. Cleveland, Ohio, March 1969.

9) F.T. Eggertsen, H.M. Joke and F.H. Stross, in Thermal Analysis, Vol.1, R.F. Schwenker, Jr. and P.D. Garns (eds), Academic Press, 1969, pages 341-351.

10) F.T. Eggertsen, E.E. Seibert, and F.H. Stross. "Volatility of High Boiling Organic Materials by a Flame Ionization Detection Method", Anal. Chem., 41, 1175-1179 (1969).

11) The Merck Index, Eighth Edition, Merck & Co., Rahway, N.J., 1968 page 378.

12) O.L. Carter, Rohm and Haas Co., Bristol, PA 19007. private communication.

THERMAL EVOLUTION ANALYSIS OF SOME ORGANIC MATERIALS

Edward W. Kifer and Lee H. Leiner

Koppers Company, Inc., Research Department
440 College Park Drive
Monroeville, Pennsylvania 15146

There has long been considerable interest in the search for effective chemicals which make organic building materials resistant to fire, to investigate the basic mechanism of ignition of these materials, and to develop the relevant technique or equipment for fire prevention and protection.

A flame-retardant additive operates by interfering with at least one of the stages of the burning process. Hilado[1] has presented a concise picture of the burning process involving a series of time resolved stages from the initial transfer of heat to the surface of a flammable material to the final step of a self-sustained fire. These stages include: a) the heating of the flammable substrate, b) its subsequent degradation and decomposition, c) the ignition of the flammable gases involved, and d) their continued combustion with sufficient net heat from the combustion sustaining the flame propagation. Normally, an efficient flame-retardant additive will affect more than one of these steps, either in a physical or chemical way. The end result of the presence of the flame-retardant additive is the retardation and final elimination of the burning process mechanism.

In this paper the results of studies of fire retardants in wood and self-extinguishing agents in polystyrene, polyethylene, and latices are reported.

Experimental Procedures

The instrument used in evaluating the flame-retardant additives was the model 916 TEA plug-in module for the DuPont 900
Thermal Analysis System. The system measures the evolution of
organic carbon caused by temperature induced increases in volatility
and thermal decomposition of the sample. The heart of the instrument consists of a temperature-programmable quartz furnace, directly
coupled to a high temperature flame ionization detector whose response is specific to organic carbon and directly proportional to
the gram atoms of carbon evolved from the sample. Samples placed
in the furnace are rapidly swept into the detector by nitrogen
purge gas as they are volatilized or thermally decomposed during
the sample analysis, giving a measure of the rate of organic carbon
evolution as a function of temperature as well as total organic
carbon content of the sample. The total fuel value of any organic
system can thus be correlated to the response of the instrument.

All of the organic systems used in this study were either commercially prepared products currently on the market (being sold as
fire-retardant or self-extinguishing preparations) or in the case
of wood, proposed fire-retardant chemicals still in the development
stage. The compositions of the various fire-retardant chemicals
were not known, and, in many cases, this data was not available
from the manufacturer.

The thermograms were all obtained at a heating rate of 20°C
per minute, and the areas under the curves were measured by planimetry. The peak area is then related to the fuel value of the sample
by the equation

$$F.V. = \frac{A/W}{c}$$

where c is the area per milligram of untreated sample, A is the
area under the curve for the sample, and W is the weight of the
sample in milligrams. Duplicate runs were made for each system and
reported values were within 3% relative in all cases.

Fire-Retardant Wood Systems

Typical thermograms for treated and untreated woods are shown
in Figure 1. The results of a series of experiments using commercially available preparations and fire-retardant systems still in
the development stages are listed in Table I. The standard used in
the evaluation of these systems was ponderosa pine treated with
mono-ammonium phosphate. This has long been used by industry as a
standard to evaluate the effectiveness of fire retardants.

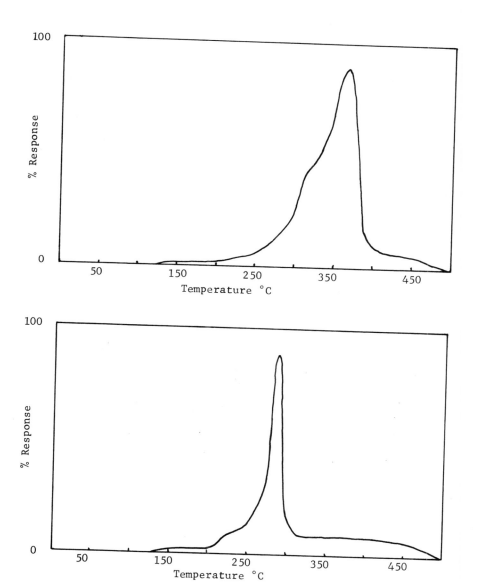

Fig. 1 Thermograms for untreated (top) and treated wood (bottom).

TABLE I

System	% Fuel Value	ΔT Peak, °C	% Char
Untreated Wood	100.00	0	17.07
A-P - 2.1	26.67	103	43.75
A-P - 0.7	62.32	76	31.20
C-101	28.34	111	43.72
C-101L	50.78	87	37.88
C-102	31.28	73	41.66
D-108	43.65	50	38.39
D-114	60.06	56	31.79
D-137	32.67	101	43.06

A-P - Ammonium Phosphate
C - Commercially available.
D - Currently in developmental stage.
Sample C-101L is a leached sample of C-101.

1. Untreated Wood. The pyrolysis of untreated samples of
ponderosa pine using the TEA yields a curve which can be separated
into three distinct regions. The decomposition of the wood to
yield combustible products begins at ≈ 125°C. The amount of combustibles appears to remain constant from this temperature up to
≈ 225°C where the curve begins to increase gradually. A shoulder
appears on the curve at ≈ 320°C, followed by a sharp peak at
≈ 370°C. The descending portion of the curve occupies a narrow
temperature range from 370°C ≈ 390°C.

The curve does not fall back to the baseline at this point,
however, which results in a fourth region on the curve probably
due to a lignin reaction. This reaction yields a straight line to
500°C where the curve falls back to the baseline.

2. Treated Samples. The ammonium phosphate treated wood
samples yielded curves very similar to curves for untreated wood
except that the peak temperatures were all significantly lower.
The temperature of the peak appears to be directly related to the
amount of phosphorus in the wood sample. In all cases a sharp peak
appeared which was narrow compared to the untreated samples.

The other curves of commercially prepared samples and developmental stage samples cannot be interpreted without knowing what
the fire-retardant chemical is. These were run to determine the
effectiveness of the system concerned.

TABLE II

System	% Fuel Value	ΔT Peak, °C
Polystyrene	100.00	0
Expandable Polystyrene	116.70	+ 12
SE Polystyrene	100.08	0
SE-2 Polystyrene	101.07	0

TABLE III

System	% Fuel Value	ΔT Peak, °C
Polyester Resin	100.00	0
Fire-retardant Polyester Resin	73.32	20

TABLE IV

System	% Fuel Value	ΔT Peak, °C
Polyethylene	100.00	0
SE Polyethylene	78.21	18

TABLE V

System	% Fuel Value
Latex	100.00
SE Latex 107	98.22
SE Latex 114	79.63
SE Latex 115	53.64

Polystyrene

The results for polystyrene and self-extinguishing polystyrene are shown in Table II.

As these results show, self-extinguishing agents have no effect in reducing the amount of combustible material available in polystyrene.

It appears from these results that the fire retardant does have a slight effect in reducing the amount of material available for combustion.

In the latex system, a peak temperature could not be determined as the shape of the curve was too complicated to assign a single peak temperature. Two of the self-extinguishing agents do show a significant decrease in the amount of combustibles available. SE-107, however, shows practically no decrease in fuel value.

Discussion

The total area under the curve is the most important parameter for consideration of fire-retardant efficiency. This area, as noted, represents the fuel value of the sample since the whole sample is heated uniformly and completely decomposed in the temperature range used (25-500°C), and quantitative comparisons of total fuel values can be made between different samples. Since the total area under the curve represents a finite, quantitative property of the sample, it does not depend highly upon the experimental conditions, and can be considered (approximately) as an intensive property of the system. As long as the decomposition temperature of the reaction is reached or surpassed, the area under the curve per milligram of sample should be constant for any particular sample. This has been demonstrated by pyrolyzing samples of treated and untreated systems with varying rates of heating.

The reaction rate and the peak temperature of the reaction, on the other hand, can be considered as extensive properties, as these do depend on the experimental conditions. At any point in the reaction the portion of the sample not yet decomposed is represented by the area under the curve to the right of that point. The reaction rate at any point can be found by dividing the curve height at that point by the amount of sample not yet decomposed.

It appears that the Thermal Evolution Analyzer is a simple, fast and accurate method for comparing the effectiveness of fire-retardant treatments on organic materials. This apparatus measures the maximum fuel that would be available in a sample if it were completely pyrolyzed.

References

1. Hilado, C. J., Flammability Handbook for Plastics, Stamford,
    Connecticut, Technomic Publishing Co., 1969.

HIGH SENSITIVITY ENTHALPIMETRIC DETERMINATION OF OLEFINS

L.A. Williams, B. Howard and D. W. Rogers

Chemistry Department, Long Island University

Brooklyn, New York 11201

INTRODUCTION

We have investigated the use of an enforced linear resistance bridge to increase the sensitivity of direct injection enthalpimetry of olefins. Enthalpimetry of olefins, as we have developed it, depends on the exothermic nature of the addition of hydrogen across a double bond in the presence of 5% palladium catalyst on charcoal. The term "direct injection enthalpimetry" indicates that the unknown sample is injected directly into the calorimeter at constant pressure, making the observed temperature rise proportional to the enthalpy change for hydrogenation. If the olefin in question is one component of a mixture containing only saturated or aromatic compounds as the remaining components, the temperature rise, as measured with a thermistor-Wheatstone bridge circuit, is also proportional to the amount of olefin present in the mixture. A knowledge of the molar heat of hydrogenation for the compound, or comparison with a calibration curve of bridge output as a function of olefin concentration, leads directly to the amount of olefin present in an unknown mixture. As implied above, aromatic compounds are not hydrogenated under the mild conditions described here.

The present work was undertaken to extend the useful concentration range covered by hydrogen enthalpimetry (1-4) into the nanomolar region.

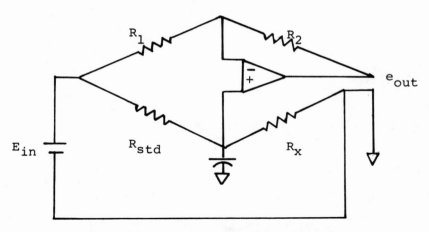

Fig. 1. Deviation Measuring Circuit

EXPERIMENTAL

Apparatus

In order to attain this increase in sensitivity, we replaced
the Wheatstone bridge described in earlier work (1-4) with an enforced linear resistance bridge using operational amplifiers. Current was constant through the measuring arms, giving an output voltage which was linear with resistance change.

The first configuration tried is shown in Figure 1.
Although there are many advantages to this circuit, such as the
grounding of R_x, which aids in guarding and shielding and the large
current one can drive through R_x since R_{std} is involved in that
path, reproducibility was less than expected and the linearity was
not good. Another configuration was to place R_x in R_2, which should
give a linear calibration, reproducibility however, was again not
good. Preliminary testing showed that the configuration in Figure
2 was the best of the three because it gives reproducible results
and has linear response.

If analysis is made of this bridge circuit, Figure 3,

$$e_{out_1} = - e_{in} \frac{R_b}{R_1} \tag{1}$$

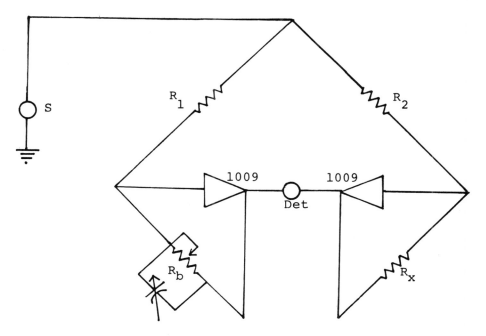

Figure 2. Linear Resistance Bridge

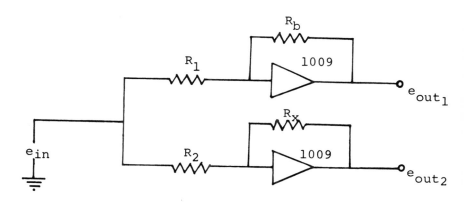

Figure 3. Equivalent Bridge Circuit

$$e_{out_2} = - e_{in} \frac{R_x}{R_2} \tag{2}$$

$$\Delta e_{out} = - e_{in} (R_b/R_1 - R_x/R_2) \tag{3}$$

When $R_1 = R_2 = R$

$$\Delta e = - \frac{e_{in}}{R} (R_b - R_x) \tag{4}$$

$$\Delta e = - \frac{e_{in}}{R} (ERROR) \tag{5}$$

we will observe that the bridge actually measures the "ERROR" or the difference between R_b and R_x once it is balanced. Eq. 5 also tells us that by increasing e_{in} and decreasing R_1 the sensitivity can be made to be much greater.

The Philbrick/Nexus Operational Amflifier Model 1009 (5) was used because it features a FET (Field-Effect Transistor) input to provide high input impedance and low bias currents. The FET input performance is an essential requirement, because it is a "unipolar" device, that is, it contains only one type of current carrier, either holes or electrons, but not both. This has an advantage for very small input signals (6). Other features of the Model 1009 are low noise characteristics, moderate full output frequency, good open-loop gain and gain-bandwith specification, and a 6 db per octave roll-off with large tolerance to capacitive loading. The output will swing ± 10 volts into a 2,000 ohm load without limiting. It is encapsulated as a solid epoxy block for protection which ensures an almost completely isothermal environment.

The entire circuit in Figure 2 was built on the Philbrick/ Nexus Operational Manifold, Model 5001. The heart of it is its plug-in Function Boards, which can accomodate various types of operational amplifiers, linear and non-linear function modules. Circuit wiring is performed using banana plug patch cables, shorting bars, and plug in elements (7). To test the linearity of the bridge circuit, R_x was set at 2000 ohms and R_b balanced the bridge also at approximately 2000 ohms. Then R_b was varied and e_{obs} measured in the forward and reverse directions. The recorder was set at maximum sensitivity (.5 mv) and gave full-scale deflection with a 0.7 ohm change.

Voltage, e_{obs} , vs. resistance is linear between 1999.3 and

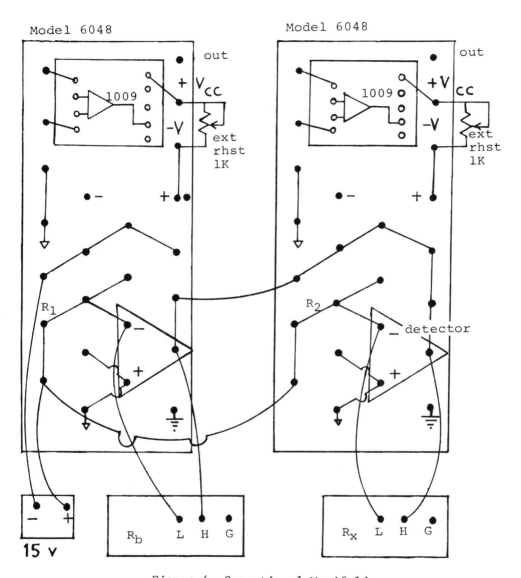

Figure 4. Operational Manifold

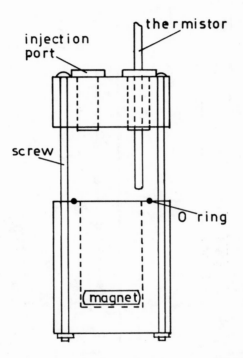

Figure 5. Teflon Calorimeter, outside diameter,
7 cm, cover thickness, 1.2 cm, inside diameter, 3.8 cm
length, 4.8 cm, depth 3.8 cm

2000.6 ohms (e_{obs} : - 0.987 to 0.964 mv) with a slope of 14.3 mv ohm^{-1} and a standard deviation from linearity of 32.5 μv as determined by a linear least squares fit to 14 experimental points at 0.1 ohm intervals.

The calorimeter is a modification of calorimeters previously described (1-4) except that the thin plastic container has been replaced by a heavy Teflon cylinder drilled out and fitted with a cover and "O" ring as shown in Figure 5. The cap of the calorimeter was fitted with a hard septum through which protruded a thermistor probe (8) and a hypodermic needle which served as the hydrogen inlet. Opposite the hard septum was a soft septum through which samples were injected with a microliter syringe of the kind used in gas chromatography. The calorimeter contained 30-35 ml of n-hexane and 0.6 g of 5% Pd catalyst adsorbed on powdered charcoal. The resulting slurry was stirred magnetically.

Procedure

The apparatus was warmed up for approximately five minutes and, since there is a change of offset voltage, less than 1 mv/v and a change of drift with time, less than 50 μv/24 hours, the apparatus was trimmed or balanced to zero with a voltmeter every 48 hours before the data was taken.

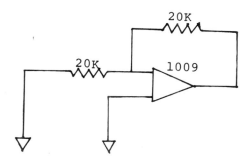

Figure 6. Balancing Circuit

For each series of runs, a new rubber septum was placed in the injection port of the calorimeter and the thermistor was thrust through a hole in the hard septum so that it would be immersed in the reaction slurry but would not come in contact with the magnetic stirring bar.

Thirty-five milliliters of hexane and 0.6 grams of catalyst (5% palladium on charcoal) along with the stirring bar, were placed in the calorimeter, and it was fastened shut by four screws the length of the calorimeter. The entire operation of balancing the circuit and preparing the calorimeter took 15 minutes. The thermistor was now connected to the bridge circuit and R_b set at 2000 + 300 which balanced the bridge. The H_2 tank was turned on, an inlet tube and needle were bled, and the needle was inserted through the hard septum into the calorimeter. H_2 was admitted into the calorimeter up to a pressure of 2 atmospheres. Since there is heat evolved due to catalyst activation, the recorder went off scale. The recorder can be brought back on scale by varying R_b or by cooling the calorimeter with ice (1). The apparatus was left standing for 45 minutes to reestablish temperature equilibrium with the surroundings. As soon as an acceptable baseline was obtained, a sample of the olefin was injected and the scale deflection noted. One can wait until the baseline is reestablished, less than a minute, or one can immediately bring it back on scale by varying R_b . When an acceptable baseline was obtained again, a new sample was injected. Replicate samples of the same olefin or different samples can be hydrogenated on the same charge of catalyst without disassembling the apparatus. It takes about 45 minutes to do 20 replicate samples. All samples were injected with a Hamilton syringe (9) with a Shandon Repro-Jector (10) which gives a reproducibility of better than 0.6%.

The degree of imbalance of the bridge was indicated by a potentiometric recorder (11) which showed full 10 in. scale deflection for a voltage change of 0.5 m_v Any change in temperature within the reaction chamber caused a deflection of the recorder pen. When a sample of olefin was hydrogenated, the exothermic reaction caused a sharp resistance drop of the thermistor probe resulting in the sigmoid curve of temperature vs. time familiar from combustion calorimetry and illustrated in reference 1.By suitable extrapolation of the baseline and the temperature curve after reaction has taken place, one can draw a vertical, intersecting both extrapolated curves, which represents ΔT for the hypothetical instantaneous reaction.

RESULTS AND DISCUSSION

Synthetic unknowns of eleven different olefins in solution with n-hexane were analyzed with the amount of olefin ranging from 70 nanomoles to 3.04 μmoles. Standard deviation of a typical set of 15 - 20 injections from the "best" straight line representing bridge output as a function of olefin concentration was about 5% of the average bridge output for the set leading to an expected average error of about the same percentage in this concentration range. Detailed results are given in Table I.

Table I. Enthalpimetric Determination of Various Olefins

Compound	Amount μ moles	Replicate Samples	Standard Deviation (%)
1 - Hexene	0.62 - 2.48	30	4.9
Cyclohexene	0.91 - 3.04	14	3.7
Cyclooctene	0.47 - 1.06	15	7.8
Cyclododecene	0.43 - 1.36	18	9.8
1,3-Cycloocta-diene	0.72 - 2.41	18	6.5
1,5-Cycloocta-diene	0.65 - 1.98	17	6.5
1,5,9-Cyclodo-decatriene	0.33 - 1.64	17	3.9
1,5,9-Trimethyl-cyclododecatriene	0.25 - 1.98	16	7.8
Bicyclo [4,3,0] nona-3,7-diene	0.75 - 2.26	14	3.0
5-Methylene-2-norbornene	0.50 - 1.49	14	4.2
Dicyclopenta-diene	0.07 - 0.52	12	6.4

The data were treated by a linear least squares method and σ was calculated from the deviation of the set of results from the "best" straight line representing ΔT vs. olefin concentration. The lower limit in concentration (for dicyclopentadiene) represents a 20 fold increase in sensitivity for this method over the most recently published results (2).

ACKNOWLEDGEMENT

We acknowledge the financial assistance of the National Institute of Health in this work.

REFERENCES

1. D. W. Rogers, Anal, Chem., 43, 1468 (1971).
2. D. W. Rogers and R. J. Sasiela, Talanta, 20, 232 (1973).
3. D. W. Rogers and R. J. Sasiela, Microchim. Acta (1973), p.33.
4. D. W. Rogers and R. J. Sasiela, Analytical Biochemistry, in press.
5. Philbrick/Nexus Research, Operational Amplifiers Model 1009, Electronic Engineers Master, Sec. 1100, Dedham, Massachusetts.
6. H. V. Malmstadt, C. G. Enke, E. C. Torens, Jr., Electronics for Scientists, W. A. Benjamin, Inc., New York, 1963.
7. Philbrick/Nexus Research, Operational Manifold Model 5001, Dedham, Massachusetts.
8. YSI #406, Cole Parmer Inst. Co. #8436.
9. Hamilton Syringe Co., P. O. Box 307, Whittier, Calif. 90608.
10. Shandon Instruments Co., Siwickley, Pa.
11. Microcord 44, Photovolt Corp., 1115 Broadway, New York, N.Y. 10010.

BIOCHEMICAL AND CLINICAL APPLICATIONS OF TITRATION CALORIMETRY AND ENTHALPIMETRIC ANALYSIS

Albert C. Censullo, John A. Lynch, Dan H. Waugh and
Joseph Jordan
Department of Chemistry, 152 Davey Laboratory
The Pennsylvania State University
University Park, Pennsylvania 16802

INTRODUCTION

Several analytical methods based on adiabatic reaction calorimetry have been developed in recent years. Authoritative reviews of thermometric enthalpy titrations (including kinetic titrimetry), direct injection enthalpimetry and thermokinetic analysis are available.[1-8] Applications to biochemical and clinical analysis[8-20] have thus far been sporadic notwithstanding the fact that these fields are in dire need of instrumental methods. The relevant fundamental phenomenon (enthalpy change) is universal for all chemical reactions including biochemically and clinically important processes. Successful studies of biologically significant oxidation-reductions by titration calorimetry are reported in this paper. A new method, called "peak enthalpimetry," is described and assessed in the context of contemporary require- ments for instrumental approaches to clinical analysis. The use of novel Curie point temperature sensors for immunological determinations is discussed.

THEORETICAL BASIS OF PEAK ENTHALPIMETRY[21,22]

The principle of a typical peak enthalpimetric experiment is illustrated in Fig. 1. Two isothermal liquids of temperature T_o were made to flow through concentric tubes. The external tube contained an aqueous reagent solution, $\underline{R}$. A "carrier" (the pure water solvent) flowed through the internal tube. A discrete volume (on the order of 0.1 ml. or less) of an "unknown" sample solution, $\underline{X}$ (the concentration of which was small relative to that of $\underline{R}$ in the reagent stream), was injected into the carrier stream

217

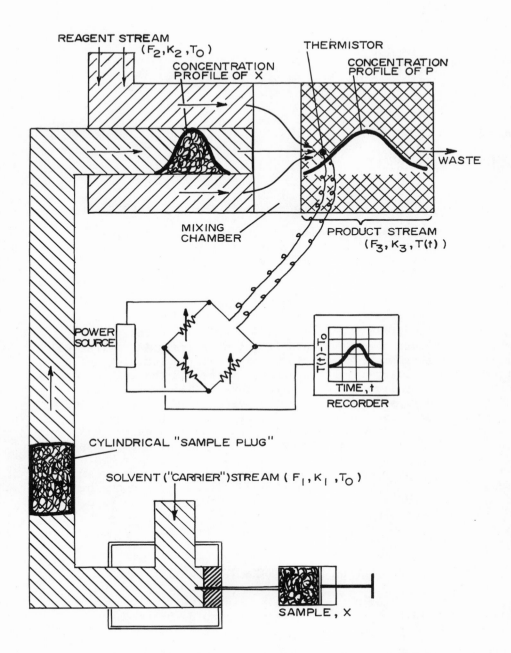

FIG. 1. Schematic Drawing of Peak Enthalpimetry Apparatus.
F--flow rates, K--specific heats, T--temperature,
t--time

as is apparent in the figure. This produced an initially "cylindrical plug" of sample solution bounded by the carrier. As this plug traveled through the carrier tubing the sample $\underline{X}$ spread by diffusion, yielding an approximately Gaussian concentration profile. Upon reaching the mixing chamber, $\underline{X}$ and $\underline{R}$ reacted instantaneously and virtually completely yielding

$$xX + rR = pP \tag{1}$$

This reaction dissipated an amount of heat

$$Q = -n_p\Delta H = -C_p V \cdot \Delta H \tag{2}$$

where n_p denotes the number of moles of product in $\underline{V}$ liters of solution, C_p denotes the corresponding concentration, and $\underline{\Delta H}$ is the heat of the reaction. Considering the volume element whose temperature was sensed by a thermistor positioned as shown in Fig. 1, evidently:

$$T(t) - T_o \equiv \Delta T(t) = \frac{Q(t)}{VK} = -\frac{\Delta H}{K} C_p(t) \tag{3}$$

where $\underline{T(t)}$ denotes the instantaneous temperature "seen" by the thermistor, $\underline{Q(t)}$ is the instantaneous heat dissipated in the relevant volume element $\underline{V}$, $C_p(t)$ is the corresponding instantaneous concentration of product formed, and $\underline{K}$ is the specific heat per unit volume. Because of the assumed complete conversion of $\underline{X}$ to $\underline{P}$ via Reaction 1 the temperature-time profile "seen" by the thermistor reflects the (Gaussian-like) concentration profile of $\underline{X}$ entering the reaction (mixing) chamber, i.e.:

$$C_p(t) = ab \frac{p}{x} C_X(t) \tag{4}$$

where $\underline{a}$ and $\underline{b}$ are dimensionless proportionality constants, which account respectively for the dilution effects associated with mixing the two streams and for the subsequent broadening of the concentration profile of $\underline{P}$ enroute to the thermistor. Combination of Eq. 3 and 4 yields:

$$\Delta T(t) = ab\left(\frac{-\Delta H}{K}\right) \frac{p}{x} C_X(t) \tag{5}$$

where $\Delta T(t)$ is the temperature-time profile recorded as the unbalance potential output of the thermistor.

Further insight into the quantitative significance of $\Delta T(t)$ transpires from the following heat balance considerations.[23-26] Obviously, assumption 6 holds for the mixing chamber where Reaction 1 occurs:

$$\text{Heat flow in = Heat flow out} \tag{6}$$

or, stated mathematically:

$$F_1 T_o K_1 + F_2 T_o K_2 - F_1 \Delta H (\frac{p}{x}) C_X(t) = F_3 T(t) \ K_3 \tag{7}$$

In Eq. 7, T_o denotes the common temperature of the solvent stream (which carries the sample) and of the isothermal reagent stream (which contains R in the same solvent); F and K are flow rates and specific heats respectively and the subscripts 1,2,3 refer to the "streams" (carrier, reagent and product) identified in Fig. 1. The formulation of Eq. 7 implies that heats of dilution are negligible and that ideal adiabaticity prevails. Since

$$F_3 = F_1 + F_2 \tag{8}$$

and

$$K_3 = (K_1 F_1 + K_2 F_2)/(F_1 + F_2) \tag{9}$$

Eq. 7 simplifies to

$$T(t) - T_o = \frac{F_1}{F_1 K_1 + F_2 K_2}(-\Delta H) \ \frac{p}{x} \ C_X(t) \tag{10}$$

Whenever $K_1 = K_2 = K$:

$$\Delta T(t) = (\frac{F_1}{F_1 + F_2})(\frac{-\Delta H}{K}) \ \frac{p}{x} \ C_X(t) \tag{11}$$

Comparison of Eq. 11 and 5 reveals that the two expressions are identical when a $= F_1/(F_1 + F_2)$ and b = 1. Thus, a represents the dilution factor determined solely by the flow rates of the reagent and carrier streams which merge and react in the mixing chamber. The factor b can be visualized as a "correction" for changes of $C_p(t)$ occurring downstream between the locus of the reaction and the thermistor. Under judiciously optimized experimental conditions (as were maintained throughout the investigation described later in this paper) of constant geometry and invariant flow rates, b = const. Furthermore, for a given reaction in a given solvent ΔH = const.; $\frac{p}{x}$ = const.; K = const. Accordingly, Eq. 5 can be reformulated as follows:

$$\Delta T(t) = mC_X(t) \tag{12}$$

where m = $-abp\Delta H/Kx$. It is apparent from Eq. 12 that the thermistor bridge functions in peak enthalpimetry as a differential detector whose output represents a measure of instantaneous concentration. This is similar to analogous detectors in gas chromatography (GC). As in GC the integral under the Gaussian detector output is proportional to sample size. Indeed:

$$C_X(t) = \frac{n_X(t)}{V} \tag{13}$$

Substituting into Eq. 12 and integrating over the entire peak makes it evident that

$$\int \Delta T(t)dt = \frac{m}{V}\int n_X(t)dt = gN_X \tag{14}$$

where N_X is the total number of moles in the sample and $g = m/V$. For a given determination the constant g can conveniently be determined by calibration with a known sample (e.g., N_1 moles) of X: the analytical significance of Eq. 14 is inherent in this simple proportional relationship. The physical significance of Eq. 14 is that the recorded integrals represent a measure of the total heat evolved (or absorbed) in Reaction 1, when N_X moles of sample X are converted to

$$N_P = \frac{p}{x}N_X = \frac{-Q}{\Delta H} \tag{15}$$

moles of product P. Indeed, substitution of N_X from Eq. 15 into Eq. 14 gives

$$\int \Delta T(t)dt = -\frac{gx}{p} \cdot \frac{1}{\Delta H} \cdot Q = (const.)Q \tag{16}$$

Combining Eq. 14 and 16 yields the expression

$$\int \Delta T(t)dt = gN_X = (const.)Q \tag{17}$$

which is the overall "working equation" of peak enthalpimetry: it correlates experimental integrals, analytical sample sizes and heat outputs.

EXPERIMENTAL

Materials. Iron hematoporphyrin was prepared from hematoporphyrin dihydrochloride (Nutritional Biochemicals, Cleveland OH) and ferrous acetate (Alfa Inorganics, Beverly MA) by the method of Montalvo and Davis.[27] Unreacted porphyrin was removed by acid extraction from ether. Sodium dithionite (Fisher Scientific, Pittsburgh PA) was obtained as the purified grade. All other chemicals used in this investigation were reagent grade and used without further purification.

Instrumentation. Essential details of the experimental setup for peak enthalpimetry (schematized in Fig. 1) are illustrated in Fig. 2. The reagent stream inlet port and carrier inlet port were connected to constant pressure reservoirs. The total pressure drop across the system, and thus the flow rate of

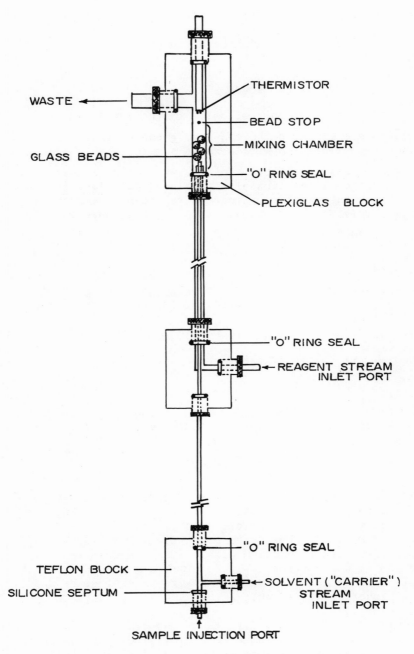

WASTE

THERMISTOR

BEAD STOP

MIXING CHAMBER

GLASS BEADS

"O" RING SEAL

PLEXIGLAS BLOCK

"O" RING SEAL

REAGENT STREAM
INLET PORT

"O" RING SEAL

TEFLON BLOCK

SILICONE SEPTUM

SOLVENT ("CARRIER")
STREAM
INLET PORT

SAMPLE INJECTION PORT

FIG. 2. Experimental Details of Peak Enthalpimetry Apparatus

solutions, was then adjusted by changing the height of one or both of these reservoirs. Constant flow rates were maintained in this manner. The two streams were made coaxial in order to enhance temperature equalization. This was accomplished with the aid of 2 meters of concentric teflon tubing immersed into a 30°C water bath thermostated to ±0.001°C by a proportional temperature controller (Model PTC-40, supplied by Tronac, Orem, UT). The two discrete streams merged in the mixing chamber which contained glass beads. The mixing chamber was attached to a 60 Hz. electromagnetic vibrator. The amplitude of the vibration was approximately 0.5 mm., which provided efficient agitation of the glass beads and mixing. Upon emerging from the mixing chamber, the temperature of the product stream was monitored via a thermistor (#K2109, Fenwal Electronics, Framingham, MA) wired as one arm of a dc Wheatstone bridge whose unbalance potential was proportional to temperature. The bridge output was fed to an adjustable range potentiometric stripchart recorder (Model #601-61-175-3089-6-55 Speedomax XL, supplied by Leeds and Northrup Co., Philadelphia, PA) equipped with a retransmitting slidewire. A potential drop of 10.00 volts was applied to that slidewire and the retransmitted output was fed, in turn, to an operational amplifier integrator (supplied by G. A. Philbrick Researchers, Inc., Boston, MA). The integrated signal was recorded on a second strip chart unit. By using this type of instrumentation both the enthalpimetric peaks and their integrated areas were recorded simultaneously with separate (electrically isolated) measuring and integrating circuits.

The apparatus for thermometric enthalpy titrations has been described elsewhere.[8]

Procedures. Thermometric redox titrations of 75 ml. solution samples were performed at 25°C in a Dewar flask whose capacity was approximately 125 ml. Corresponding potentiometric titrations were recorded as the output of a digital pH meter (Model 801, Orion Research, Inc., Cambridge, MA). The indicator electrode was bright platinum, while a saturated calomel electrode served as reference. Iron hematoporphyrin complexes were prepared by adding an excess of ligand (imidazole, pyridine, etc.) to solutions of iron hematoporphyrin IX, the required excess having been previously determined by spectrophotometric means.[28] Due to the extreme sensitivity of the reduced (ferrous) iron hematoporphyrin complexes to air oxidation, all titrations (both potentiometric and thermometric) involving these species were performed in a glove box filled with dry nitrogen.

The International Union of Pure and Applied Chemistry (IUPAC) sign convention [29] for electrode potentials was used throughout this investigation.

RESULTS AND DISCUSSION

Redox Thermodynamics of Iron Hematoporphyrin Complexes

The reduction of several iron hematoporphyrin complexes [from the Fe(III) to the Fe(II) state] with dithionite was investigated by thermometric enthalpy titrations. The relevant reactions were of the type:

$$(18)$$

which may be equivalently formulated as:

$$L_2hm(III) + \tfrac{1}{2}S_2O_4^{--} + 2OH^- = L_2hm(II) + SO_3^{--} + H_2O$$

$$(18')$$

In Eq. 18 the tetrapyrrole ring (and the symbol "hm") denote the equatorial hematoporphyrin ligand. Fig. 3 illustrates the structure of iron hematoporphyrin. The relevant iron complexes were octahedral in both the oxidized (ferric) and reduced (ferrous) forms, with the axial ligands (cyanide, histidine, imidazole and pyridine) denoted by L. Experimental conditions (concentration of excess axial ligand, pH, etc.) were judiciously selected in such a manner that "bis-complexation" (with respect to L) prevailed to the virtual exclusion of all other species in both the oxidized and reduced forms. A representative thermometric enthalpy titration curve is shown on the right side of Fig. 4. The experimental end-points in Fig. 4 corresponded to the theoretical equivalence point [½ mole of dithionite per mole of $L_2hm(III)$] predicted by Eq. 18. Heats of reaction could readily be evaluated from the temperature increment, ΔT (see Fig. 4 right) via the equation

$$\Delta T = -(N_P \Delta H)/(VK) \qquad (19)$$

The sharpness of the thermometric end-point in Fig. 4 (right side plot) indicates that the reaction described by Eq. 18 was virtually complete at the equivalence point. Consequently, the equilibrium constant of Eq. 18 could not be obtained from the

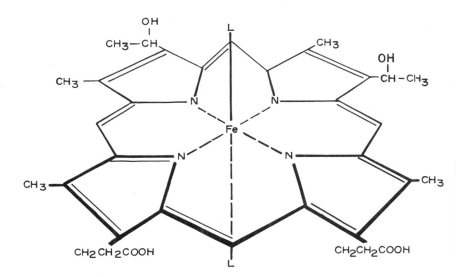

FIG. 3. Structure of Iron Hematoporphyrin Complexes.
L--axial ligands

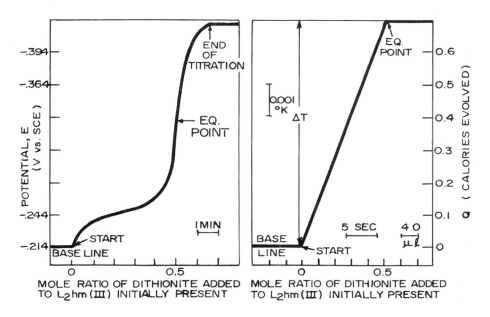

FIG. 4. Potentiometric and Thermometric Titration Curves of
Ferric Hematoporphyrin Complexes with Dithionite,
pH = 9.50.

TET curves. Instead, the relevant K_{eq} (and the corresponding Gibbs Free Energies) were calculated from the formal potentials using the well-known equations:

$$\Delta G = -RT \ln K_{eq} \tag{20}$$

and

$$E_1 - E_2 = \frac{RT}{nF} \ln K_{eq} \tag{21}$$

The value for the sulfite/dithionite couple, E_2, was taken from the literature.[30] The values for the hematoporphyrin complex couples, E_1, were not available, and were determined ad-hoc by potentiometric titrations (with dithionite) of the appropriate bix-axial hematoporphyrin complex. A typical titration curve of this type is shown on the left side of Fig. 4. The values for E_1 were computed by the method of Reed and Berkson.[31] This mathematical treatment of the data involves the rectification of sigmoidal potentiometric titration curves and yields formal potential assignments based on the analytic geometry of the entire plot. Thermodynamic data obtained in this manner are listed in Table I.

TABLE I

THERMODYNAMICS* OF THE REACTION:

$$L_2hm(III) + \tfrac{1}{2}S_2O_4^{--} + 2OH^- = L_2hm(II) + SO_3^{--} + H_2O$$

L	ΔH(KCAL/MOLE)	ΔG(KCAL/MOLE)	ΔS(CAL/°C·MOLE)
CYANIDE	-20.8 ± 0.2	-14.37	-21.5
HISTIDINE	-23.2 ± 0.3	-16.44	-21.8
IMIDAZOLE	-14.5 ± 0.2	-7.47	-23.4
PYRIDINE	-17.9 ± 0.3	-11.65	-21.1

*Formal reaction heat, free energy and entropy assignments (per mole of iron)

TABLE II

REDOX THERMODYNAMICS* OF THE HALF-REACTION

$$L_2hm(III) + e = L_2hm(II)$$

L	ΔH(KCAL/MOLE)	$-nFE = \Delta G$(KCAL/MOLE)	ΔS(CAL/°C·MOLE)
CYANIDE	-18.9	-0.76	-60.8
HISTIDINE	-21.4	-2.84	-62.3
IMIDAZOLE	-12.6	6.13	-62.9
PYRIDINE	-16.0	1.96	-60.2

*Formal assignments per mole of iron referred to the H^+/H_2 half-reaction.

From the experimental findings listed in Table I, we computed assignments for the equilibria between the relevant half-reaction and the H^+/H_2 reference half-reaction (See Table II). This was accomplished by combining known ΔG and ΔS assignments[32] for the reaction

$$Fe(CN)_6^{-3} + \tfrac{1}{2}H_2 = Fe(CN)_6^{-4} + H^+ \tag{22}$$

with data obtained in the present investigation from thermometric and potentiometric titrations, viz.:

$$Fe(CN)_6^{-3} + \tfrac{1}{2}S_2O_4^{-2} + 2OH^- = Fe(CN)_6^{-4} + SO_3^{-2} + H_2O \tag{23}$$

ΔH = -31.0 Kcal/mole, ΔG = -23.5 Kcal/mole, ΔS = -24 e.u.

The virtually identical, highly negative entropies in the last column of Table II are remarkable. In contradistinction, it should be noted that the comparable entropy of the aquo-ferric/ferrous reduction is positive (+24 e.u.). Work is in progress to assess the significance of these findings.

Chloride Determination By Peak Enthalpimetry

A modified version of the classical Schales and Schales method[33] for serum chloride determination was selected for a pilot study to explore potentialities and limitations of peak enthalpimetry in clinical analysis. Solutions of pure sodium

chloride in triply distilled water served to simulate typical
saline concentrations in serum. The carrier stream (See Fig. 1)
consisted accordingly of pure water. The determinative step made
use of the heat of the association process

$$2 \; Cl^- + Hg^{+2} = HgCl_2 \tag{24}$$

the thermodynamic parameters of which are as follows:[34]

ΔH = -13.4 Kcal/mole, ΔG = -18.2 Kcal/mole, ΔS = 16.0 e.u.

The reagent stream (Fig. 1) consisted of 0.25 M. aqueous mercuric
nitrate. "Unknown samples," 100 µl. in volume, of the simulated
serum were introduced via the injection port (Fig. 2).
Representative experimental results are illustrated in Fig. 5.
A more extensive set of numerical data is presented in Table III.
Fig. 5a represents a recording of temperature versus time with
amounts of "chloride taken" as the curve index. Fig. 5b is the
corresponding integral plot of cumulative heat evolution (Q) as a
function of time. Amounts of "chloride found" are listed in
Table III, computed from Eq. 17. The linear correlation between
experimental peak integrals of total heat evolved and amounts of
"chloride taken" is illustrated in Fig. 6. The correlation
coefficient was 0.997. As is apparent from Table III, the precision
of our peak enthalpimetric chloride determination was on the order
of 5%. Though the conventional Schales and Schales procedure,
using diphenylcarbazone as an indicator for the excess mercury
titrant, shows a precision of better than 2% it is anticipated
that our methodology may show comparable precision with further
development.

Differentiation between normal, low and high serum chloride
levels (103 meq/1, <95 meq/1, >110 meq/1, respectively) is
presently possible. By recycling a sufficiently concentrated
mercuric nitrate reagent stream the seriatim performance of
several hundred chloride analyses is evidently feasible. The
technology for automatic injection and digital printout is
readily available.

CONCLUSIONS AND OUTLOOK

The wealth of thermodynamic and thermochemical information
obtainable by judicious use of titration calorimetry is apparent
from Tables I and II. The iron hematoporphyrin complexes
studied in this investigation are convenient models of the so-called
"hemichromes," which are analogous complexes derived from heme
(iron protoporphyrin IX), the prosthetic group of hemoglobin,
myoglobin and cytochrome c. The only difference between the
macrocyclic equatorial ligand of heme (protoporphyrin) and
hematoporphyrin (Fig. 3) is that the former has two vinyl

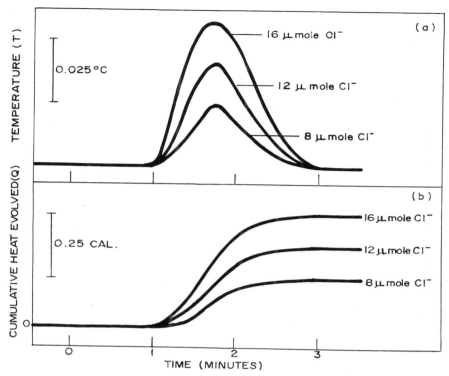

FIG. 5. Determination of Chloride in Simulated Serum
        a. Experimental Peak Enthalpograms   b. Corresponding
                                                  Integral Plots

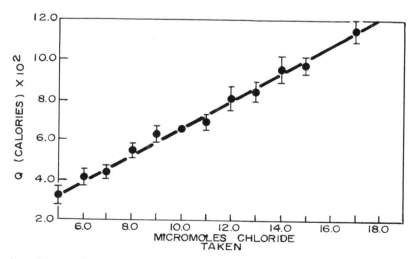

FIG. 6. Plot of Total Heat Evolved as Function of Amount of
        Chloride. Ordinate represents limiting Q-s where Fig. 5b
        levels off.

TABLE III

DETERMINATION OF CHLORIDE IN SIMULATED SERUM BY PEAK ENTHALPIMETRY

TAKEN		FOUND[a]		
Micromoles in total sample[b,c]	Q Calories	Micromoles in total sample[c,d]	Precision % Std. Dev. of Mean	Accuracy % Error
5.0	3.35×10^{-2}	5.0	8	Zero
6.0	4.22×10^{-2}	6.3	7	+5
7.0	4.49×10^{-2}	6.7	5	-4
8.0	5.25×10^{-2}	7.8	4	-3
9.0	6.25×10^{-2}	9.3	4	+3
10.0[e]	6.70×10^{-2}			
11.0	7.10×10^{-2}	10.6	4	-3
12.0	8.25×10^{-2}	12.3	5	+3
13.0	8.45×10^{-2}	12.6	5	-3
14.0	9.65×10^{-2}	14.4	4	+3
15.0	9.85×10^{-2}	14.7	2	-2
17.0	1.15×10^{-1}	17.2	4	+1
AVERAGE			5	

a) average of ten replicates
b) samples of 100 μl were used throughout
c) to obtain clinical units (meg/l) multiply by ten
d) computed from equation 17 and calibration runs
e) "calibration run"

substituents in lieu of the hydroxyethyl groups in the latter.
However, heme is known to dimerize and polymerize[35] in aqueous
solution, which interferes with this type of study, while
corresponding iron hematoporphyrin solutions exhibited no such
tendencies. The oxidation-reduction behavior of heme is crucial

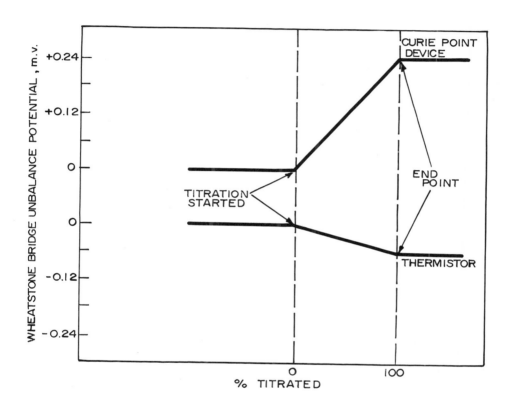

FIG. 7. Comparison of results obtained in a thermometric titration using different temperature sensors. Total temperature change was 0.0024°C.

in its role as a biological transfer agent. In hemoglobin and
myoglobin, reversible oxygen binding occurs while the iron of the
heme is in the ferrous form, this capacity being lost upon
oxidation to the ferric form. In cytochrome c, the heme iron
apparently undergoes reversible oxidation and reduction, while
serving as an electron transfer agent in complex biochemical
oxidations. Recent studies[36-38] seem to indicate that the required
biochemical functions of heme may be mimicked in protein-free
models. Thermometric enthalpy titrations have proven to be a
rapid and convenient method for elucidating relevant thermodynamics
of heme related complexes.

Because of the multi-component nature of samples, clinical
analyses are fraught with difficulties and are often not
quantitative in nature. Many of the newer automated clinical
procedures depend upon spectrophotometry thereby adding the
requirements that a chromogen be present and that there be no
turbidity. This usually leads to numerous sequential manipulations
which are prone to error propagation. Enthalpimetric methods,
which are more direct, do not require color development and are
feasible in turbid solutions, may provide a timely improvement
of instrumental analysis in clinical practice. Immunological
reactions obviously possess the requisite specificity. Indeed,
thermometric titrations have yielded well-defined end-points and
illuminating thermodynamic insights in a recent study involving
reactions of antibodies and haptens with γ-immunoglobulins.[8,39]
However, limitations in available sample sizes and concentrations
create special problems of insufficient temperature sensitivity.
Curie point devices can be used as resistance thermometers in
lieu of thermistors to attain the requisite enhancement in
sensitivity.[40,41] A comparison of typical thermometric titration
curves obtained with a Curie point device and a thermistor is shown
in Fig. 7. In both instances, a Wheatstone bridge circuit yielded
unbalance potentials which were proportional to temperature. These
were opposite in sign because our Curie point device (PTC
#713T25, supplied by Penn. Elec. Tech. Co., Pittsburgh, PA) had a
positive temperature coefficient of resistance, while thermistors
have negative temperature coefficients. It is apparent from Fig. 7
that the capabilities of Curie point temperature sensors are
formidable: the response was on the order of 100 millivolts per
degree which is the equivalent of a five-thousand junction
thermopile.

REFERENCES CITED

1. L.S. Bark, S.M. Bark, Thermometric Titrimetry, Pergamon Press,
        Oxford, 1969.

2. L.S. Bark, P. Bate, J.K. Grime, "Thermometric and Enthalpimetric
    Titrimetry" in Selected Annual Reviews of the Analytical
    Society, Vol. II (L.S. Bark, Ed.) Society for Analytical
    Chemistry, London, 1972, pp. 121-149.

3. P.W. Carr, Crit. Rev. of Anal. Chem., 3, 491 (1972).

4. R. M. Izatt, J.J. Christensen, "Thermochemistry in Inorganic
    Solution Chemistry," in Physical Methods in Advanced
    Inorganic Chemistry, H.A.O. Hill, P. Day, Eds., Interscience,
    New York, 1968, Chap. 11, pp. 538-598.

5. J. Jordan, "Thermometric Enthalpy Titrations" in Treatise in
    Analytical Chemistry, I.M. Kolthoff, P.J. Elving, Eds.,
    Part I, Vol. 8, Interscience, New York, 1968, pp. 5175-5242.

6. J. Jordan, in Topics in Chemical Instrumentation, G. W. Ewing,
    Ed., Chemical Education Publ. Co., Easton, Pa., 1971,
    pp. 193-199.

7. H. J. V. Tyrell, A. E. Beezer, Thermometric Titrimetry, Chapman
    and Hall, London, 1968.

8. J. Jordan, N.D. Jespersen, Coll. Int. Cent. Nat. Rech. Sci.,
    201, 59 (1972).

9. E. B. Smith, P. W. Carr, Anal. Chem., In Press (1974).

10. J.J. Christensen, R.M. Izatt, D.P. Wrathall, L. Hansen,
    J. Chem. Soc. A., 1212 (1969).

11. J.J. Christensen, J.L. Oscarson, R.M. Izatt, J. Am. Chem. Soc.,
    90, 5949 (1968).

12. J.J. Christensen, J.H. Rytting, R.M. Izatt, Biochemistry, 9,
    4907 (1970).

13. J.J. Christensen, J.H. Rytting, R.M. Izatt, J. Chem. Soc. B.,
    1643 (1970).

14. J.J. Christensen, J.H. Rytting, R.M. Izatt, J. Chem. Soc. B.,
    1646 (1970).

15. J.J. Christensen, M.D. Slade, D.M. Smith, R.M. Izatt, J. Tsang,
    J. Am. Chem. Soc., 92, 4164 (1970).

16. R.M. Izatt, J.J. Christensen, J.H. Rytting, Chem. Rev., 71,
    439 (1971).

17. R.M. Izatt, J.J. Christensen, Handbook of Biochemistry, H.A.
     Sober, Ed., Chemical Rubber Co., Cleveland, 1968, p.J-49.

18. N.D. Jespersen, J. Jordan, Anal. Letters, 3, 323 (1970).

19. M.A. Marini, R.L. Berger, D.P. Lan, C.J. Martin, Anal. Biochem.,
     43, 188 (1971).

20. B. Sen, W.C. Wu, Anal. Chim. Acta, 2, 457 (1971).

21. R.J. Thompson, Thesis in Progress, Pennsylvania State
     University.

22. D.H. Waugh, Thesis in Progress, Pennsylvania State University.

23. P. Priestley, W. Sebborn, R. Selman, Analyst, 90, 598 (1965).

24. W. McLean, G. Penketh, Talanta, 15, 1185 (1968).

25. T. Crompton, B. Cope, Anal. Chem., 40, 274 (1968).

26. H. Yoshida, M. Taga, S. Hikime, Japan Analyst, 20, 361 (1971).

27. J.G. Montalvo, D.G. Davis, J. Electroanal. Chem., 23, 164
     (1969).

28. A.C. Censullo, Thesis in Progress, Pennsylvania State
     University

29. M.L. McGlashan, Pure and Appl. Chem., 21, 3 (1970).

30. W. Mansfield Clark, Oxidation-Reduction Potentials of Organic
     Systems, Williams and Wilkins Co., Baltimore, 1960, p. 318.

31. Lowell J. Reed, J. Berkson, J. Phys. Chem., 33, 760 (1929).

32. S.J. Ashcroft, C.T. Mortimer, Thermochemistry of Transition
     Metal Complexes, Academic Press, New York, 1970, p. 276.

33. O. Schales, S. Schales, J. Biol. Chem., 140, 879 (1941).

34. R.J.P. Williams, J. Phys. Chem., 58, 121 (1954).

35. T.M. Bednarski, J. Jordan, J. Am. Chem. Soc., 89, 1552 (1967).

36. C. K. Chang, T.G. Traylor, J. Am. Chem. Soc., 95, 5810 (1973).

37. C.K. Chang, T.G. Traylor, Proc. Nat. Acad. Sci. U.S.A., 70,
     2647 (1973).

38. J.E. Baldwin, Joel Huff, J. Am. Chem. Soc., 95, 5757 (1973).

39. N.D. Jespersen, Thesis, Pennsylvania State University, 1971.

40. J.A. Lynch, Thesis in Progress, Pennsylvania State University.

41. J.A. Lynch, J. Jordan, Abstracts, 25th Pittsburgh Conference
    on Analytical Chemistry and Applied Spectroscopy,
    Cleveland, 1974, Paper #54.

ACKNOWLEDGMENTS

The work described in this paper was supported in part by
Research Grant GP 38478X from the National Science Foundation and
NATO Research Grant No. 794 from the Scientific Affairs Division
of the North Atlantic Council.

RECENT ANALYTICAL APPLICATIONS OF TITRATION CALORIMETRY

Reed M. Izatt, Lee D. Hansen, Delbert J. Eatough,
Trescott E. Jensen, and James J. Christensen

Departments of Chemistry and Chemical Engineering,
and Contribution 58 from the Center for Thermochemical
Studies, Brigham Young University, Provo, Utah 84602

INTRODUCTION

Many chemical systems cannot be satisfactorily studied with
common analytical methods based on spectrophotometric and potentiometric measurements. The difficulties encountered in using these
techniques have led to an active interest in developing new analytical methods. One method which has proven to be extremely useful
as an analytical tool for the study of many systems is titration
calorimetry. A common property of all chemical reactions is
enthalpy change and it is the ability to quantitatively measure
this property that makes calorimetry useful for the study of many
chemical systems. Indeed, the general nature of the measured
parameter, temperature or heat change, is the major asset in applying titration calorimetry to a wide range of problems.

Initial analytical applications of thermometric titrimetry
were based primarily on the analysis of the thermogram for end
points. The development of titration calorimeters capable of
measuring heat changes accurately has made possible the determination of additional information from the thermogram. The
additional information which can be obtained includes reaction
enthalpies, stoichimetries and sometimes equilibrium constants
for the reaction(s) of interest.[1,2,3]

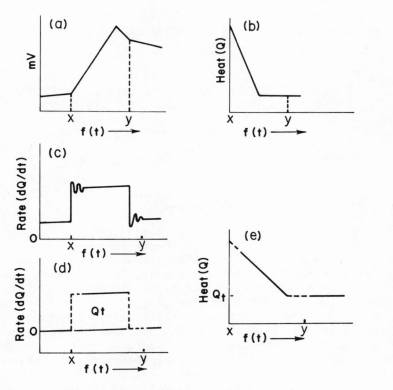

Figure 1. Forms of calorimetric data presentation: a) direct isoperibol data, b) isoperibol data plotted as Q_{corr} vs. time, c) direct isothermal calorimetric data, d) corrected isothermal data where Q_T is the total heat evolved during the reaction, and e) isothermal data converted to a Q_{corr} vs. time plot. In these plots f (t) is some time dependent function, i.e., volume, moles, etc. and x and y designate the start and end of titrant addition.

Typical data presentation capabilities of the isoperibol and isothermal calorimeters are illustrated in Figure 1. Isoperibol calorimeters generally give a thermogram of millivolts (temperature) vs. time (Figure 1a) which can then be converted to the more desirable Q_{corr} vs. time thermogram (Figure 1b). In order to convert Figure 1a to Figure 1b corrections must be made for the heat exchanged with the surroundings during the run. This correction along with heat of dilution corrections make it possible to obtain Q_{corr}. These corrections for heat exchange can be calculated accurately only over a limited time period. This limits the isoperibol method to experiments which are complete within one to two hours. On the other hand, the isothermal calorimeter baseline varies only with solution volume in the reaction vessel and can be predicted accurately. Therefore, experiments not complete in two hours may be done using an isothermal instrument. Isothermal calorimeters directly produce the time derivative of the thermogram (dQ/dt/time) (Figure 1c). The area under the corrected curve (enclosed area, Q_t of Figure 1d) is the total change in the heat content of the system studied.[4,5] A portion of the curve may be integrated to give the heat produced or absorbed during the time interval contained in that portion (Figure 1e).

The dashed lines in the thermograms in Figures 1d and 1e show the major disadvantage of the isothermal compared to the isoperibol calorimetric method. The response time of the system to sudden changes in the rate of heat production caused by the start or cessation of reactions is two to three minutes. The actual response of the instrument to a sudden change in the rate of heat production is a damped sine wave as shown in Figure 1c. The total area under the curve and above the baseline during the period of oscillation gives the heat produced during the time period covered by the oscillation, but the areas of increments of the curve do not give correct values for the heat produced in this portion of the curve. Thus, after any reaction begins or ends the isothermal method does not give usable information for a period of two to three minutes. End points which occur within the first two minutes of a titration cannot be observed directly, however, the amount of heat produced during this period becomes insignificant if runs are sufficiently long. Isothermal experiments lasting 48 hours have been run successfully in our laboratory.

Continuous improvement of instrumentation has resulted in a steady increase in the accuracy achievable on both the time (buret delivery rate) and temperature dependent axes of thermograms. Also, the miniaturization of reaction vessels now makes titration calorimetry competitive with other microanalytical techniques for the study of molecular species. Since calorimetric data are linearly

related to concentrations for a given equilibrium constant, sharper
end points are observed at lower concentrations than is possible
using log methods such as pH and EMF.[6,7] Because of the intrinsic
nature of the quantity being measured, homogeneity of the titrate
solution is unimportant. Whether the studied system contains
small molecules, ions, polymers, living cells or rocks is imma-
terial because only the net heat change is observed and this is
unaffected by particle size. Other notable advantages are the
speed of data collection, the quantity of data which can be
accumulated, and the convenience of automated data collection.

This method is, unfortunately, not without disadvantages.
First, heat changes are non-specific. In simple systems data
can usually be interpreted unambiguously (i.e. simple acid-base
titrations, metal-complex reactions). However, in more complex
systems ambiguities may arise which create the need for additional
information before the data can be interpreted. Second, specialized
and sophisticated equipment and specialized personnel are
required to perform calorimetric titrations. The increasing
commercial availability and reliability of this type of equipment

TABLE 1

Summary of Information Obtained Using
Several Calorimetric Methods

Type of Information	Isoperibol Titration Calorimetry	Isothermal Titration Calorimetry
End points	yes	yes
Stoichiometry	yes	yes
Equilibrium constant	yes, $K < 10^4$	yes, $K < 10^4$
ΔH Values (kind of group)	yes	yes
Kinetics	yes $t_{1/2} < 9$ min	yes $t_{1/2} > 9$ min

and of the number of people using it to study increasingly diverse systems are reducing the seriousness of this latter disadvantage. A summary of the information that may be obtained by analytical methods which utilize the thermal effects of reactions in solution is given in Table 1.

The use of titration calorimetric equipment for the determination of the number and type of ionizable groups in proteins, the measurement of the adsorption capacities of zeolites, and the study of the response of bacterial cells to cytotoxic agents will now be discussed as examples of the flexibility and utility of this equipment.

APPLICATIONS

Proton Ionization from Proteins

A recent titration calorimetric study[8] of insulin using a 3 ml isoperibol titration calorimeter shows that the technique is a direct, rapid, and capable method of producing information on the acid-base chemistry of proteins. A typical thermogram for the titration of aqueous NaOH into an aqueous insulin solution is shown in Figure 2. By the method of linear least squares, intrinsic ΔH values are obtained for each of the regions (I-IV) in the calorimetric titration curve. Assignment of the type of group ionizing in each region (Table 2) is accomplished using the ΔH value calculated for the observed region and comparing it to ΔH values obtained for model compounds.[9] (Data for small molecules show that ΔH is usually dependent on the type of functional group from which the proton ionizes and is only slightly dependent on the structure of the rest of the molecule.) Inspection of the curve (Figure 2) shows several distinct end points which correspond stoichiometrically to the successive titration of protons from the functional groups indicated in Table 2. The number of groups titrated in each of the regions of the curve can be obtained from these end points. These numbers (Table 2) agree well with those calculated from pH titration data and amino acid analysis.

The pH titration of insulin does not provide exact information concerning the number and identity of ionizable groups, however, this information is obtained directly from the thermogram. Assignment of the pK values for the ionizing groups from insulin may be made using pH titration data obtained under identical conditions as with the thermogram.[8] The midpoint of each calorimetric region is assumed to have 1 : 1 stoichiometry of protonated and non-protonated groups. From the pH titration curve obtained using the same solutions of NaOH and insulin used in the calorimetric titration the pH at 1:1 stoichiometry is taken to be the pK value

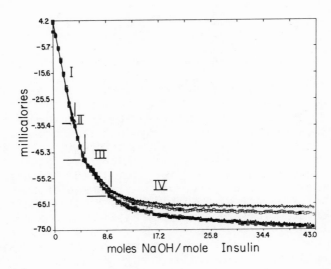

Figure 2:   The addition of 0.17 M NaOH to Insulin (2mg/ml) in a 2.76 ml reaction vessel using a Tronac isoperibol calorimeter at 298 °K. Data for four separate runs were recorded at ten second intervals using a buret delivery rate of 26 μl/second. A linear least squares method was used to obtain the lines through the data points.

TABLE 2

Summary of Insulin Data

Region on Curve	Group	Intrinsic ΔH[a] (kcal/mole)	Intrinsic pK Cal[b]	pH[c]	No. of groups Cal[a]	pH[c]	Anal[d]
I	β-, γ- carboxyl	+ 1.8	4.7	4.7	8.5	8.5	8
II	imidazole	+ 5.6	6.1	6.0-6.4	4.1	4	4
III	phenolic	+10.4	8.4	9.6	8.5	8	8
IVa	α-, ε- amino	+12.1[e]	10.8[f] α7.5 ε9.6		6.5[f]	6	6
IVb	guanidinium		11.4[f]	11.9	2[f]	2	2

[a] Values determined from calorimetric data[8]. [b] Values are obtained using data from a parallel pH titration where the pK is assigned as the pH at the midpoint of the region (Figure 2). [c] Values given are from Tanford and Epstein, see reference 10. [d] Amino Acid analysis, see reference 9, p. C-263. [e] ΔH values from calorimetric titration with base are not separable. [f] The values given were assigned from calorimetric data for insulin modified with formaldehyde. [8]

for that group. Thus, a pK value is obtained which is not dependent on assigning end points from the pH titration curve.

Protein denaturation has also been studied by isoperibol calorimetry. The curves in Figures 3a and 3b show the calorimetric titration of HCl into native and heat denatured human serum albumin (HSA) solutions (4 mg/ml) respectively. The initial pH values of the native and heat-denatured protein solutions are 5.2 and 6.8, respectively. In both cases the final pH at the end of the titration is about 2.0. The endothermic region of the curve in Figure 3a corresponds to the same pH region (5.0 to 3.8)[11] where denaturation of this protein occurs. The heat of denaturation of the protein determined from this endothermic region is + 20 ± 2 kcal/mole protein. The heat liberated in the titration of the native protein is similar to that observed for the denatured protein. The sensitivity of the instrument is apparent from the fact that a total of only 10 mcal of heat were liberated during the entire titration shown in Figure 3a. The precision obtainable with the isoperibol calorimeter is evident by the good agreement of the separate runs in Figures 3a and 3b.

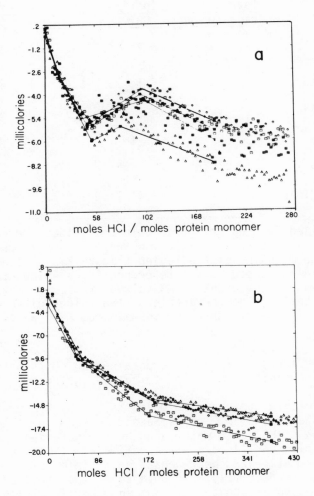

Figure 3:   Calorimetric titration of HCl (0.16 M) into; (a) native human serum albumin (4 mg/ml) (4 runs) and (b) heat-denatured human serum albumin (4 mg/ml solution which was previously placed in boiling water for 10 minutes) (3 runs).

In summary, these examples demonstrate the usefulness of isoperibol titration calorimetry in studying proteins. The study of insulin shows the applicability of the method to the identification of functional groups, the determination of the number of such groups, and the assignment of pK values to them. The experiments with HSA demonstrate that continuous isoperibol titration calorimetry can be used as a tool to study protein conformation changes.

Adsorption of Aromatic Compounds by Zeolite (LMS 13X)

The adsorption of aniline, nitrobenzene, and toluene by Zeolite (ground rock) is an interesting example of the use of isothermal calorimetry to determine a concentration when no end point can be observed directly.[12] Results from several runs for the addition of aniline, nitrobenzene, or toluene to a hexane suspension of Zeolite are given in Figure 4. An equation has been derived from which the number of binding sites and log K and ΔH values resulting from the interactions are obtained by fitting the equation to the thermogram.[13,14] (Figure 4). Representative of the results obtained is the log K value for the adsorption of aniline by Zeolite in hexane, 3.08 ± 0.14. This value compares favorably with the value, 3.05 ± 0.03, obtained independently by a direct analytical method.

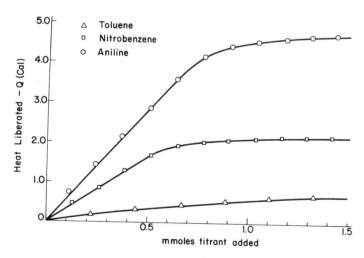

Figure 4: Typical calorimetric titration curves for the interaction of toluene, nitrobenzene, or aniline with 0.5 grams of Zeolite LMS 13X in hexane. (Reproduced with permission from reference 12).

The adsorption capacity of Zeolite for aniline in hexane
has also been determined by calorimetric and analytical methods
to be 1.64 ± 0.10 and 1.67 ± 0.01, respectively. It is interesting
that the accuracy of the calorimetrically determined values is good
even though the thermogram has no definite end point. These
data indicate that reliable stoichiometry and thermodynamic values
may be obtained from the calorimetric data without the requirement
for quantitative reactions or visible end points if the reaction
is known. Although the method outlined above is applicable to any
system of chemical reactions, the particulate nature of the system
studied imposes certain restrictions. The rate of reactions
generally decreases as particle size increases above small colloid
dimensions. Thus, incremental isothermal, rather than continuous
isoperibol titration, calorimetry must often be used in studying
these systems.

Bacterial Cell Response to Cytotoxic Agents

The time derivatives of the thermograms for the addition of
cytotoxic materials to $\underline{S}$. faecalis cells in an isothermal calorimeter[15] are shown in Figure 5. The thermogram provides information
about the cell metabolism and thus becomes an indicator of the
extent of cytotoxicity of the added agent. The degree of cell
response and the character of that response appear to be a
property of the concentration and type of cytotoxic agent used.

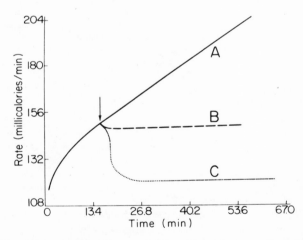

Figure 5: The time derivative of the thermogram for the addition
of cytotoxic material to $\underline{S}$. faecalis (4% DMSO broth), A) standard
curve. The arrow indicates injection of: B) 0.1 ml of 12.6 mg
penicillin 'G'/ml (4% DMSO broth), and C) 0.1 ml 11.8 mg tetra-
cycline HCl/ml (4% DMSO broth).

One complete determination of cell response can be accomplished in less than one hour. The range of possible uses
of this technique include: 1) the study of the types of response
of growing cells to cytotoxic agents, 2) the determination of the
concentrations of cytotoxic agents, and 3) the study of other
factors that influence cell metabolism, i.e. serum factors such
as antibody, complement, and β-lysin.

Summary

Solution calorimetry is a developing analytical tool that
will continue to be used for new applications. Development of
more sensitive and dependable instruments using small volumes
(see previous paper) and their greater availability will continue
to increase the usefulness of this technique. Along with improved
instrumentation has come a significant step in miniaturization
($<$ 5 ml) (see previous paper) making possible the study of systems
whose investigation was previously not practicable because
of the amount of material required.

The capability of collecting large amounts of information
using automated instruments with the output data in a form
easily used for analysis makes it possible for the investigator to
obtain the maximum benefit from the data collected and the time
expended. Due to the nature of isoperibol and isothermal calorimetry, the methods are complementary and are best suited to
different types of investigative applications. Generally it can
be stated that isoperibol calorimetry is applicable to systems
where fast kinetics and reasonably large quantities of heat are
involved and run times of between 20 and 60 minutes are used.
On the other hand, isothermal calorimetry is particularly well
suited to systems where slow reaction kinetics or very small heat
changes are encountered and where run times of hours or days are
needed.

During the past several years significant improvements in
instrumentation and data handling have taken place resulting in
many diverse applications. The future should bring further
advancements in calorimetry and increasing application to
additional fields of research.[16]

ACKNOWLEDGMENT

This work was supported by NIH Grants GM-18816-03 and AM-
15615-02 from the U.S. Public Health Service.

REFERENCES

1. Christensen, J. J., Eatough, D. J., Ruckman, J., and Izatt,
    R. M., Thermochimica Acta, 3, 203 (1972).
2. Eatough, D. J., Christensen, J. J., and Izatt, R. M.,
    Thermochimica Acta, 3, 219 (1972).
3. Eatough, D. J., Izatt, R. M., and Christensen, J. J.,
    Thermochimica Acta, 3, 233 (1972).
4. Christensen, J. J., Johnston, H. D., and Izatt, R. M., Rev.
    Sci. Instrum., 39, 1356 (1968).
5. Christensen, J. J., Gardner, J. W., Eatough, D. J., Izatt,
    R. M., Watts, P. J., and Hart, R. M., Rev. Sci. Instrum.,
    44, 481 (1973).
6. Hansen, L. D., Izatt, R. M., and Christensen, J. J., "Appli-
    cation of Thermometric Titrimetry to Analytical Chemistry,"
    In "Treatise on Titrimetry. Vol. 2: New Developments in
    Titrimetry," edited by Joseph Jordan, Marcel Dekker, in press.
7. Tyrrell, H. J. V., and Beezer, A. E., "Thermometric Titrimetry,"
    Chapman and Hall, Ltd., London (1968).
8. Eatough, D. J., Kassierer, E. F., Hansen, L. D., Jensen, T. E.,
    Izatt, R. M., and Christensen, J. J., Biochemistry, submitted
    for publication.
9. Christensen, J. J., and Izatt, R. M., Table on "Heats of
    Proton Ionization and Related Thermodynamic Quantities," in
    Handbook of Biochemistry With Selected Data for Molecular
    Biology, The Chemical Rubber Publishing Co., Cleveland,
    Ohio, 2nd Ed., 1970, J-58-J-172.
10. Tanford, C., and Epstein, J., J. Amer. Chem. Soc., 76,
    2163 (1959).
11. Wallevik, K. J., Biol. Chem., 248, 2650 (1973).
12. Eatough, D. J., Salim, S., Izatt, R. M., Christensen, J. J.,
    and Hansen, L. D., Anal. Chem., 46, 126 (1974).
13. Eatough, D. J., Izatt, R. M., and Christensen, J. J.,
    Thermochimica Acta, 3, 233 (1972).
14. Wrathall, D. P., and Gardner, W., "Temperature, Its Measure-
    ment and Control in Science and Industry," H. Plumb (Ed.),
    Vol. 4, part 3, 1973, p. 2223.
15. Jensen, T. E., Hansen, L. D., Eatough, D. J., Sagers, R. D.,
    Izatt, R. M., and Christensen, J. J., Science, submitted for
    publication.
16. Ross, P. D., Science, 183, 229 (1974).

CALORIMETRIC STUDIES OF PI-MOLECULAR COMPLEXES

William C. Herndon, Jerold Feuer, and Roy E. Mitchell

Departments of Chemistry, University of Texas at

El Paso, and Texas Tech University, Lubbock

INTRODUCTION

The work reported here is the result of an experimental calorimetric investigation of a series of intermolecular complexes
which fall into the category labeled "charge transfer" complexes
(1-5) or "donor-acceptor" complexes (4). Specifically, the systems studied consist of two compounds which are postulated to
have an electron donor-acceptor relationship to one another (1).
The donor molecules used were benzene, the methyl substituted
benzenes, thianthrene, naphthalene, acenaphthalene, phenanthrene,
and pyrene. The acceptor molecules were tetracyanoethylene
(TCNE), 1,3,5-trinitrobenzene (TNB), and picric acid (1,3,5-
trinitrophenol or PA). The calorimetric experiments were carried
out in methylene chloride solution, all possible donor-acceptor
combinations were used, and a 1:1 acceptor to donor complex con-
stitution was assumed. These calorimetric investigations are the
first studies of this type.

Several books have been published which deal almost exclusively with this type of molecular complex. Among these are the
recent books by Mulliken and Person (1) and Foster (5) and the
older books by Andrews and Keefer (2), Rose (3), and Briegleb (4).
These books give an excellent history of charge transfer complexes
and go into greater detail on many points than will the present
discussion.

Past studies (6-34) of this type of complex have focused upon
the determination of the equilibrium quotients and the thermodynamic values associated with the complex formation reaction shown
in equation 1

$$A + D \rightleftharpoons C \tag{1}$$

where A is the acceptor molecule, D is the donor molecule, and C is the complex. Many methods (1-5) have been devised for the determination of these values utilizing such properties as dielectric constants of complex containing solutions, NMR chemical shifts, vapor pressure, etc. The most frequently used technique has been to measure the intensity of an absorption band which sometimes (6) appears and is considered to be the direct result of the presence of a complex (1). This band is called the "charge transfer band" and cannot be attributed to either of the complex's partners. The main reason for the initiation of our calorimetric studies was a desire to compare the spectroscopic thermodynamic parameters with results determined by a completely different technique. We chose to employ the methods of standard reaction calorimetry in the belief that they would lead to accurate values of the complexation ethalpies.

Past Methods for Obtaining Thermodynamic Values

By far, the most widely used methods for the determination of association constants (thermodynamic values) for molecular complexes have been those employing the charge transfer band. One of the first and best known of these methods is that of Benesi and Hildebrand (7). For a 1:1 complex, the equilibrium constant is defined by equation 2

$$K = \frac{(C)}{(A^0 - C)(D^0 - C)} \tag{2}$$

where parentheses indicate molar concentrations and the superscript zero indicates an initial concentration. This equation is valid thermodynamically only if all activity coefficients are assumed to be unity. The standard state is an infinitely dilute solution. If Beer's Law, as shown in equation 3, is assumed

$$A = \varepsilon b(C) \tag{3}$$

where A is the absorbance, ε is the extinction coefficient and b is the cell length set equal to unity, and the experimental conditions are set such that $(D^0) \gg (A^0)$, a linear equation results as shown in equation 4.

$$\frac{(A^0)}{A} = \frac{1}{K\varepsilon} \frac{1}{(D^0)} + \frac{1}{\varepsilon} \tag{4}$$

This equation is the Benesi-Hildebrand (7) equation. A plot of
$(A^0)/A$ versus $1/(D^0)$ thus should yield a straight line with a
slope equal to the reciprocal product of the equilibrium constant
and the extinction coefficient and an intercept which is the reciprocal of the extinction coefficient. The optimum implementation
of the experimental procedure for this and similar methods requires
a number of different points in each experiment and a statistical
data analysis such as a least squares method in order to obtain
the desired constants. Since this method only allows for the
obtaining of the equilibrium constant, separate experiments must
be done at other temperatures in order to obtain the enthalpy
changes associated with the reaction.

Scott (27) has argued that equation 4 is incomplete in that
it requires extrapolation to more concentrated solutions, and
suggested equation 5

$$\frac{A}{(D^0)} = -KA + K(A^0)\varepsilon \qquad (5)$$

as more correct (31). Equations 4 and 5 reduce to the same equation. However, a statistical analysis of the data could lead to
different results based on the two equations. If one were to use
an unrestricted least squares method there should be agreement
between the two equations.

Ketelaar, et al. (28), have developed an equation similar to
equation 4 for the analysis of the absorption band data which may
be a sum of the true charge transfer band and a component absorption band. DeMaine (29,30) has suggested a rather complicated
data analysis method which does not assume that the acceptor or
donor extinction coefficients, εA and εD, respectively, are
unchanged with the addition of the other component. Rose and
Drago (33) modified the Ketelaar equations and made their application more practical.

An important criticism of Benesi and Hildebrand type methods
has been put forth by Dewar and Thompson (19). In their work,
Dewar and Thompson measured the equilibrium constants of the complexes of the acceptor TCNE with a number of non-substituted
aromatic compounds in order to compare to Dewar and Thompson's
theoretical model. The results showed that the equilibrium constants varied with the wavelength of measurement for a number of
individual complexes. Dewar and Thompson discounted the explanation put forth by Johnson and Bowen (21) (that ternary complexes
were the cause of this variation) by reason of the fact that Dewar
and Thompson observed a decreasing equilibrium constant with
increasing concentration for the TCNE hexamethylbenzene complex.
This observation is contrary to that expected if Johnson and Bowen
were correct.

Other methods employ such properties as infrared absorption, NMR absorption, polarography and partition methods. In Foster's book these methods are discussed in detail. Most of these methods employ equations of the form of the Benesi-Hildebrand equation. For example, equation 6 is used in nuclear resonance techniques applied to the obtaining of data on complexes.

$$\frac{1}{\Delta} = \frac{1}{K}\frac{1}{(D^0)} + \frac{1}{\Delta^0} \tag{6}$$

Δ^0 is the chemical shift of the pure complex in solution relative to the shift of the pure acceptor in solution. The fraction of the complex, P_c, is then given by equation 6a, where Δ is the chemical shift of the complex-containing solution.

$$P_c = \frac{\Delta}{\Delta^0} \tag{6a}$$

Foster (5) has tabulated some values which compare NMR and optical studies for a few complexes. As can be seen from Table I, there is some agreement between the two methods, and an interesting con-

Table I

A Comparison of Complex Equilibrium Constants

Complex	Solvent	Method	Relative Concentration	K
A	$CHCl_3$	Optical	$(A^0) \gg (D^0)$	1.0
A	$CHCl_3$	Optical	$(A^0) \ll (D^0)$	0.6
A	$CHCl_3$	Optical	$(A^0) = (D^0)$	0.4
A	$CHCl_3$	NMR	$(A^0) \ll (D^0)$	0.4
B	$CHCl_3$	Optical	$(A^0) \gg (D^0)$	1.0
B	$CHCl_3$	Optical	$(A^0) \ll (D^0)$	1.4
B	$CHCl_3$	Optical	$(A^0) = (D^0)$	1.0
B	$CHCl_3$	NMR	$(A^0) \ll (D^0)$	1.0
C	$CHCl_3$	Optical	$(A^0) \ll (D^0)$	3.3
C	$CHCl_3$	Optical	$(A^0) = (D^0)$	2.5
C	$CHCl_3$	NMR	$(A^0) \ll (D^0)$	2.6

Complex A: N,N-dimethylaniline TNB
Complex B: TNB N,N,N^{1}N^1-tetramethyl-phenylenediamine
Complex C: HMB Fluoranil

centration dependence of the equilibrium constant in the optical results. The concentration dependence could be the result of a solvent effect. If the NMR technique and UV-visible spectral technique gave exactly the same result, the assumption of a single equilibrium including a 1:1 complex would be strongly supported. Although Foster considers the agreement in Table I to be good, one can see that under similar experimental conditions the values of equilibrium constants differ by as much as 40 percent. Therefore, the question of the validity of the numerical results found by means of either technique is still open.

Calorimetry

Calorimetry has been a widely practiced and established general method of gathering thermodynamic data for many decades (35,36). It is, in fact, the most direct method for this purpose, since its specific methods directly measure temperature changes in controlled systems. Though specific instrumentation varies, the general procedure of reaction calorimetry is easily seen from the following discussion of the method employed here.

There have been some past calorimetric investigations into complexed systems (37-45). One method, which was used in the present work, has been called the "entropy titration" method (37, 38,39). This method specifically requires a series of sample introductions into a calorimetric vessel in which the reactions take place. The heats involved in these reactions are noted. If the system involves a simple bimolecular equilibrium to give a complex, it is necessary only to analyze the experimental points consisting of corrected heats and concentrations in order to obtain the equilibrium constant and enthalpy of reaction of the system. As explained below, all thermodynamic parameters for the equilibrium reaction are obtainable in one experiment. As noted before, the spectroscopic methods require a minimum of two series of experiments, each at different temperature, in order to obtain thermodynamic parameters.

The entropy titration method has been successfully applied by Izatt, Christensen, et al. (37,38,45) in the determination of pK values for the proton ionization of bisulfate (HSO_4^-) ion and biphosphate ion (HPO_4^{-2}) (37) and gave excellent agreement with other experimental methods. This method was also used in the determination of the thermodynamic values for the interaction of thiourea with $Hg(CN)_2$ in water-ethanol solutions (38).

The general equation of reaction calorimetry is as shown in equation 7

$$Q_{CORR} = \Delta H C V \qquad (7)$$

where Q_{CORR} is the observed heat corrected for dilution effects, ΔH is the heat of the complex formation reaction in calories/mole, V is the volume of the calorimetric vessel in liters, and the concentration units are in mole/liter.

Solving for C, the molar concentration of the complex, as given in equation 2, gives equation 8.

$$K = \frac{(C)}{(A^0 - C)(D^0 - C)} \tag{2}$$

$$C = 0.5\left[A^0 + D^0 + 1/K \pm \sqrt{(A^0 + D^0 + 1/K)^2 - 4\ D^0 A^0}\right] \tag{8}$$

If the boundary condition that C = 0 if D^0 = 0 is enforced, it is easily seen that the negative sign of the radical is the correct sign. Substituting this expression into equation 7 gives equation 9.

$$Q_{CORR} = \frac{\Delta H V}{2}\left[A^0 + D^0 + 1/K - \sqrt{(A^0 + D^0 + 1/K)^2 - 4\ D^0 A^0}\right] \tag{9}$$

This is a nonlinear equation in two unknowns, K and ΔH. Specific methods of data analysis will be discussed in a later section.

EXPERIMENTAL

The calorimetric vessel, as shown in Figure 1, consisted of a 265 milliliter silvered Dewar flask placed in a cylindrical brass can. The top of this brass can has five pieces of precision bore brass tubing placed through appropriately drilled holes and the junctions sealed with silver solder so as to be waterproof. An electrical heater, an all glass stirrer, a thermistor probe (51,53), and an RGI (Roger Gilman Incorporated) 2 milliliter micrometer syringe precise to 0.001 milliliters are introduced into the calorimetric vessel through four of the tubes and are thus supported by the top of the can. The fifth tube is used as a vent. This top is secured to the rest of the can by six 1/4 inch bolts and the two parts are sealed by a teflon gasket so that the seal is watertight. The calorimeter had a thermal modulus of 6 x 10^{-3} min^{-1}.

The heater is composed of a piece of glass rod with non-inductive windings of Constantane wire around it, coated with an epoxy resin and immersed in a glass tube partially filled with silicone oil for better heat conduction. The top of this tube is

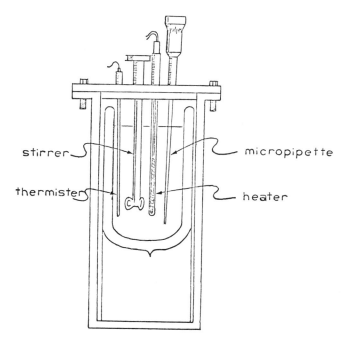

Fig. 1. The calorimeter.

sealed with paraffin so that the two insulated wires come through
the paraffin. The heater is electrically connected to a constant
current source, a Sargeant IV coulometer, and both the coulometer
current and the heater resistance were periodically checked using
a Leeds and Northrup potentiometer, an Eppley standard cell, and
standard resistors. They were found constant to within 0.01 percent over a period of two years.

The thermistor probe is composed of a 10,000 ohm thermistor
placed in the bottom of a thin glass tube and immersed in paraffin
for more sensitivity and stability. The thermistor resistance was
found to be linear over the range of temperature used in this
research. The thermistor probe was connected to a Wheatstone
bridge, composed of two 2.5 K resistors and a 0-10 K variable
resistor box, which is in turn connected to a Kiethley Model 150B
Microvolt ammeter and a Barber-Coleman model 8400-56000-000-1-92
recorder.

The calorimetric vessel rests in a water bath kept at 24.75 ± 0.01°C by a Tronac precision temperature controller, model PT-1000. In an actual experiment, after the calorimetric vessel is allowed to come to approximately bath temperature, the procedure involved sequential calibrations and sample introductions. The calibrations were carried out using a heater and the constant current source. Calibration time intervals were measured by an electrical timer in the constant current source. The samples were introduced by means of the RGI precision micrometer syringe, which is constructed of teflon with a pyrex tip. The liquid in the Dewar was constantly stirred with a glass stirrer and an external motor so that the sample was immediately well distributed and the thermal effect was almost immediately detectable. The recorder output consisted of a flat line with a number of resistance jumps, a sample of which is shown in Figure 2. The number of these jumps is equal to the sum of the number of calibrations and the number of sample injections. The method used for obtaining the numerical data from the graph paper output of the recorded was suggested by Wadso (53). This method requires that the base lines at either side of a jump be extended past the jump and a line be drawn perpendicular to the direction of the graph paper's ejection from the recorder such that the shaded areas in Figure 2 are equal. The measured distance in recorder paper units between the two intersection points A and B corresponds to the actual heat involved in the step corresponding to the jump. If this was a calibration, the heat involved would be known and the heat capacity of the system in units of calories/ graph paper division could be calculated. The average heat capacity as obtained from two calibrations, one before and one after the sample introduction, was used to calculate the heat involved per sample introduction.

This experimental procedure was used for the blanks corresponding to the introduction of donor solutions to pure solvent and the actual complexation reactions carried out for complexes of the

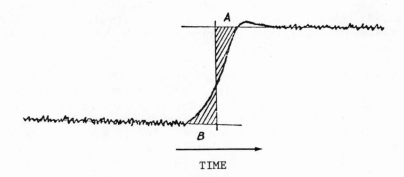

Fig. 2. A sample of recorder output.

acceptors TCNE, 1,3,5-trinitrobenzene (TNB) and picric acid (PA), each with the donors benzene, all of the methyl substituted benzenes through hexamethylbenzene (with the exception of 1,2,3-trimethylbenzene which was not available), naphthalene, phenanthrene, pyrene, thianthrene, and acenaphthalene.

The calorimetric system was always calibrated electrically, but it was also checked by measurement of a standard reaction. Various standards have been published for calorimeters (54-56). The method chosen in this study was that of titrating aqueous 1 molar hydrochloric acid into an excess of aqueous 1 molar sodium hydroxide. Reported ΔH values (57,58) are -13.49 and -13.51 kilocalories/mole; this system gave a ΔH value of -13.60 ± 0.08 for HCl concentrations of 0.707 and 0.702 molar.

The heat measured in a sample injection may be considered to be a sum of three terms, as shown in equation 10. The first term is heat due to the complex forming reaction, equation 11. The second term, $Q_{MECHANICAL}$, is heat due to the introduction of the sample at bath temperature into the calorimeter at a different temperature. The calorimeter equilibrium temperature is higher than bath temperature due to stirring. The third term is heat due to the dilution of the donor solution in the solvent methylene chloride. The reaction heat is exothermic but the other two terms may be either endothermic or exothermic.

$$Q_{TOTAL} = Q_{REACTION} + Q_{MECHANICAL} + Q_{DILUTION} \qquad (10)$$

$$Q_{REACTION} = Q_{CORR} \qquad (11)$$

$Q_{MECHANICAL}$ can be calculated from knowing the heat capacity of the calorimeter, C_P^C, and the heat capacity of the solution in the syringe, C_P^S, in calories/milliliter degree. The resistance change of the thermistor is a measure of the temperature change due to sample introduction. The final resistance of the thermistor R is used in a conservation of heat requirement, equation 12,

$$(C_P^C)(R - R_c^O) = (C_P^S)(R_{BATH} - R) \qquad (12)$$

where R_c^O is the thermistor resistance before the sample introduction. Rearrangement gives equations 13 and 14 where P is the number of ohms/degree (P = 396 $\Omega/°C$).

$$R = (1 + C_P^S/C_P^C) = R_{BATH}(C_P^S) + R_c^O = (R_{BATH})(C_P^S)R_c^O/[1 + C_P^S/C_P^C] \qquad (13)$$

$$Q_{MECHANICAL} = (R - R_{BATH})c_P^S/P \tag{14}$$

The calorimeter contains 200 milliliters of solution so $c_P^C \gg c_P^S$ and the term $R_{BATH}(c_P^S)/(c_P^{CAL})$ is small compared to R_C^O. Also $R \cong R_C^O$ since only very small resistance changes (1 ohm) were observed. Therefore, equation 14 can be written as equation 15.

$$Q_{MECHANICAL} = (R_c^O - R_{BATH})(c_P^S)/P \tag{15}$$

The $Q_{DILUTION}$ for every reactant was measured separately over the entire concentration range used in each series of experiments. The actual experimental data are listed and summarized graphically in a dissertation by one of us (J.F.)*. The experimentally observed Q_{TOTAL} and the corrections for $Q_{MECHANICAL}$ and $Q_{DILUTION}$ are also listed in the same dissertation. Examples of the data collected are shown in Table II for the system TCNE-benzene in methylene

Table II

TCNE-Benzene Calorimetric Data
(TCNE = 0.0760 M (200 ml), Benzene = 7.481 M)

Volume (ml)	Total Volume (ml)	$-Q_{OBSERVED}$	$-Q_{TOTAL}$	$-Q_{TOTAL}^{CORRECTED}$
0.45	0.45	0.123825	0.044508	0.0832123
0.34	0.79	0.09452	0.0893191	0.1472336
0.16	0.95	0.066258	0.1313871	0.180936
0.52	1.47	0.1242955	0.2121464	0.2907635
0.48	1.95	0.1159614	0.296381	0.3816953
0.50	2.45	0.134212	0.378561	0.4681556
0.50	2.95	0.1215456	0.4666167	0.5597886
0.30	3.25	0.0583060	0.5096791	0.6048841
0.50	3.75	0.97759	0.5915727	0.689277
0.71	4.46	0.1406681	0.7172235	0.8118150

*Jerold Feuer, A Theoretical and Calorimetric Study of Tetracyano-ethylene-Aromatic Complexes, Dissertation Abstracts International, 32B, 2093B (1971), Order No. 71-17, 896, 336 pages, University Microfilms, A Zerox Company, P.O. Box 1764, Ann Arbor, Michigan 48106.

chloride solution. In this experiment the calorimeter was loaded
with 200 milliliters of 0.0760 molar TCNE solution. Pure benzene,
7.481 molar, was added; the volume increment and total volume
added are listed in the first and second columns of Table II.
$Q_{OBSERVED}$ for the increment are listed in column three, column
four lists the $Q_{MECHANICAL}$ for the total volume addition. The
corrected total heats evolved are listed in the last column.

Data Analysis

Various methods of data analysis are possible and these
methods have been summarized and applied to entropy titration data
by Izatt, Christansen, et al. (46).

Assume that the unknowns, K and ΔH, and the observed reaction
heats define a three dimensional surface. If one assumes specific
values of K and ΔH one can use equation 9 to calculate the heats
of reaction for each sample introduction. The set of values given
by the sum of the squares of the calculated heats minus the
observed heats, equation 16, should be a minimum at the correct
values of K and ΔH assuming no systematic experimental errors.
The value W in equation 16 is a statistical weighing factor which
was set equal to 1.0 in the present work. The problem is to find
the absolute minimum value of U and its corresponding values of K
and ΔH. One method, called "schematic mapping" (46), involves

$$U(K,\Delta H) = \sum_{i=1}^{N} W_i (Q_{CORR}^{CALC} - Q_{CORR}^{OBS})^2 \tag{16}$$

variation of the K values with a constant ΔH until the best minimum
U value is obtained by trial and error or by an absolute minimization procedure (47,48). This procedure has an advantage in that
it can be applied to a multiple equilibria problem. According to
Izatt, Christensen, et al. (46), the method is "thorough and
accurate" (46) but requires much computer time for multiple equilibria problems. A second method, called "pit mapping" (46), has
been used by Sillen (49) and Paelette, et al. (50). This method
assumes a quadratic form for $U(K,\Delta H)$ and iteration methods are
used for self consistency. This method is described (46) as
advantageous in that little computer time is used but has the disadvantage of locating saddle points and decreasing the area of
convergence. A third method is the simultaneous solution of the
calorimetric equations. This method, according to Izatt,
Christensen, et al. (46), has no advantages, is difficult to use
and is not applicable to multi-equilibria problems. The fourth
method was the "variable metric method of minimization" (46) and
this method uses a non-linear least squares method, a method which

calculates values $\partial U/\partial K_i$ and $\partial U/\partial \Delta H_i$ for a given K_i and ΔH_i, and approximates the Hessian matrix values.

$$\left|\left|\frac{\partial^2 U}{\partial K_i \partial K_j}\right|\right| \quad .$$

This method locates only relative minima and uses more computer time than the pit mapping method (46).

The data from all runs for a particular complex were placed together in one set and analyzed by a method very similar to schematic mapping (46). Values of K and ΔH were chosen over wide ranges and the U values were printed in the squares corresponding to the particular values of K and ΔH as shown in Figure 3. When a possible minima was located the particular square was expanded by narrowing the ranges of K and ΔH. This process was repeated to a desired number of significant figures which was limited by the computing system used (IBM 360-50). All possible minima were expanded and the best values of K and ΔH corresponding to the smallest U value were used. A disadvantage in the entropy titration method is that an error in one of the first steps would be carried over in all proceeding steps. This would effect the ΔH value more than the K value. All data were examined and compared to the calculated values and, if the last value was out of agreement with the general agreement, this value was discarded and the whole procedure was again carried out. This datum discarding was, however, a rare occurrence.

One can see from the values reproduced in Figure 3, that the method is highly sensitive. The best values of the equilibrium constant and the enthalpy of reaction are obtained rapidly by this

$-\Delta H$ (calories/mole)	K = 0.070	0.090	0.110	0.130	0.15
2800	6.56557	3.93378	1.32051	1.32274	3.91537
2900	6.23357	3.50813	0.81034	1.93123	4.62083
3000	5.90150	3.08265	0.33209	2.54203	5.32650
3100	5.56950	2.65742	0.19967	3.15379	6.03230
3200	5.23752	2.23259	0.79982	3.76605	6.73819

K_{BEST} = 0.1076 ΔH_{BEST} = -3158 calories/mole

Fig. 3. A sample of the modified schematic mapping computer output. The numbers represent the values U x 100 for TCNE-benzene.

procedure. The same best-fit values were obtained in experiments carried out after a time interval of 18 months.

Reagents

Columbia and Aldrich reagent grade tetracyanoethylene (TCNE) was recrystallized from chlorobenzene and doubly sublimed in vacuum. The sealed tube melting point range was 198-200°C in agreement with the literature (34).

Mallinckrodt analytical reagent grade dichloromethane (methylene chloride) was doubly distilled and checked for purity by gas chromatography.

Baker picric acid containing ten percent water for safe transport was recrystallized twice from absolute ethanol and carefully sublimed in vacuum. The melting point range was 123-124°C (59).

Eastman reagent grade 1,3,5-trinitrobenzene (TNB) was recrystallized twice from absolute ethanol and very carefully sublimed in vacuum with a resulting melting point of 121-123°C (59).

Benzene, toluene, para-xylene, ortho-xylene, meta-xylene and mesitylene were all reagent grade, doubly distilled and checked by gas chromatography and NMR for impurities.

Columbia reagent grade 1,2,4-trimethylbenzene, 1,2,3,4-tetramethylbenzene, and 1,2,3,5-tetramethylbenzene were all distilled and checked for purity by gas chromatography and NMR.

Columbia reagent grade durene was recrystallized from absolute ethanol and doubly sublimed in vacuum with a resulting melting point of 78.5-80°C (59).

Pentamethylbenzene was obtained in reagent grade from Aldrich, recrystallized from absolute ethanol and doubly sublimed in vacuum. The melting point was 54-55°C.

Aldrich reagent grade hexamethylbenzen was recrystallized from absolute ethanol and doubly sublimed in vacuum. The resulting point was 165-167°C (59).

Columbia reagent grade naphthalene was recrystallized from absolute ethanol and doubly sublimed in vacuum. The resulting point was 70-81°C (60).

Columbia reagent grade phenanthrene was recrystallized from absolute ethanol and doubly sublimed in vacuum. The resulting melting point was 99-101°C (60).

Aldrich reagent grade pyrene was recrystallized from absolute ethanol and doubly sublimed in vacuum. The resulting melting point was 154–156°C (60).

Aldrich reagent grade acenaphthalene was recrystallized from absolute ethanol and doubly sublimed in vacuum. The melting point was 97–99°C (60).

CALORIMETRIC RESULTS AND COMPARISON WITH SPECTROSCOPIC RESULTS

The thermodynamic values determined for TCNE, TNB, and PA complexes are presented in Tables III, IV, and V, respectively.

Table III

Tetracyanothylene Complexes
Calorimetric Results
($T = 25°C$, Solvent CH_2Cl_2)

Donor	K_c	$-\Delta H°$ (cal/mole)	$\Delta G°$ (cal/mole)	$-\Delta S°$ (cal/mole°K)
benzene	0.1076 ± 0.0003	3158 ± 1	1321	15.02
toluene	0.1320 ± 0.0026	7881 ± 20	1200	30.22
p-xylene	0.2516 ± 0.0003	9335 ± 3	818	36.07
m-xylene	0.2210 ± 0.0005	9428 ± 4	894	34.62
o-xylene	0.2291 ± 0.0023	7910 ± 18	873	29.46
mesitylene	0.5990 ± 0.0010	8385 ± 8	290	30.61
1,2,4-TMB	0.6075 ± 0.0008	9186 ± 7	237	31.60
durene	1.150 ± 0.0005	8953 ± 4	-83	29.75
1,2,3,4-TMB	2.405 ± 0.0019	8711 ± 15	-545	27.39
1,2,3,4,5-TMB	1.220 ± 0.0013	9378 ± 11	-118	31.06
PMB	4.358 ± 0.0029	6981 ± 15	-872	20.49
HMB	11.998 ± 0.0008	7474 ± 4	-1472	20.13
naphthalene	0.2473 ± 0.0022	4364 ± 9	828	17.41
acenaphthalene	0.4254 ± 0.0010	8510 ± 8	507	30.24
phenanthrene	1.874 ± 0.0012	11340 ± 12	-372	36.79
pyrene	0.8880 ± 0	4530 ± 0	70	15.43
thianthrene	0.9340 ± 0.0029	4930 ± 13	40	16.67

Table IV

Trinitrobenzene Complexes
Calorimetric Results
(T = 25°C, Solvent CH_2Cl_2)

Donor	K_c	$-\Delta H°$ (cal/mole)	$\Delta G°$ (cal/mole)	$-\Delta S°$ (cal/mole°K)
benzene	0.0691 ± 0.0018	756 ± 1	1583	7.85
toluene	0.0485 ± 0.0008	1663 ± 1	1793	11.60
p-xylene	0.0726 ± 0.0039	2396 ± 9	1554	13.25
m-xylene	0.0784 ± 0.0013	2417 ± 3	1509	13.17
o-xylene	0.0744 ± 0.0026	2380 ± 6	1540	13.15
mesitylene	0.1123 ± 0.0035	4574 ± 16	1296	19.70
1,2,4-TMB	0.08775 ± 0.0035	5812 ± 20	1442	24.34
durene	0.1165 ± 0.0049	6338 ± 30	1274	25.54
1,2,3,4-TMB	0.1470 ± 0.0028	6680 ± 18	1136	26.23
1,2,3,5-TMB	0.1116 ± 0.0024	7664 ± 18	1299	30.08
PMB	0.348 ± 0.0045	5786 ± 25	625	21.51
HMB	0.4556 ± 0.0033	6346 ± 20	466	22.86
naphthalene	1.436 ± 0.0007	3140 ± 2	-214	9.82
acenaphthalene	0.824 ± 0.0020	8805 ± 16	115	29.93
phenanthrene	0.427 ± 0.0014	13020 ± 37	504	45.38
pyrene	1.688 ± 0.0014	8516 ± 10	-310	27.54
thianthrene	0.987 ± 0.0078	2481 ± 16	7.65	8.35

Literature values, where available for TCNE, TNB, and PA complexes are presented in Tables VI, VII, and VIII, respectively. Litera-ture values were, in some cases, in mole fraction form and have been transformed to molar concentration form in these tables as suggested by Herndon and Goodin (61). The error limits given with the calorimetric data were calculated by formulae presented by Rosslinsky and Kellaini (62) and are standard deviations.

Equilibrium constants of TCNE complexes in CH_2Cl_2 solution are compared graphically in Figure 4. There is a good correlation between the two sets of values along line B which has a slope of approximately 2.5. Since there is no 1:1 correspondence (line A), a systematic error in one or both methods is indicated.

Table V

Picric Acid Complexes
Calorimetric Results
(T = 25°C, Solvent CH_2Cl_2)

Donor	K_c	$-\Delta H°$ (cal/mole)	$\Delta G°$ (cal/mole)	$-\Delta S°$ (cal/mole°K)
benzene	0.0505 ± 0.0030	1004 ± 3	1758	9.27
toluene	0.01416 ± 0.0067	2426 ± 16	2523	16.61
p-xylene	0.0726 ± 0.0050	2280 ± 10	1554	12.87
m-xylene	0.0735 ± 0.0032	2450 ± 7	1547	13.41
o-xylene	0.06265 ± 0.0036	2372 ± 9	1642	13.47
mesitylene	0.4709 ± 0.0102	2474 ± 23	446	9.80
1,2,4-TMB	0.145 ± 0.0089	1951 ± 17	1144	10.39
durene	0.3324 ± 0.0006	2522 ± 1	653	10.65
1,2,3,4-TMB	0.5605 ± 0.0042	2450 ± 9	343	9.37
1,2,3,5-TMB	0.3604 ± 0.0065	3012 ± 18	604	12.13
PMB	0.538 ± 0.0008	1988 ± 1	368	7.91
HMB	1.430 ± 0.0057	2880 ± 14	-212	8.95
naphthalene	1.259 ± 0.0008	1993 ± 1	-137	6.23
acenphthalene	1.077 ± 0.0016	6730 ± 9	-46	22.43
phenanthrene	1.356 ± 0.0005	4608 ± 2	-182	14.85
pyrene	1.886 ± 0.0013	7123 ± 8	-377	22.64

Figures 5 and 6 compare calorimetric and spectroscopic values
for enthalpies and entropies of TCNE complexes in CH_2Cl_2 solution.
There seems to be no correlation between these values.

Various criticisms of spectral methods for obtaining thermodynamic values of complexes have been put forth (13,19,20,65-69).
The Benesi-Hildebrand method (7) has been criticized by Foster,
et al. (67), on account of the dependence of the value of the
equilibrium constant on the ratio between the donor and the acceptor concentrations. Foster, et al. (67), attributed this anomaly
to a deviation of the charge transfer absorption from Beer's Law.
The Benesi-Hildebrand method (7) has also been criticized for the
variation of K with wavelength of measurement by Foster and
Horman (69), Johnson and Bowen (21), and Dewar and Thompson (19),

Table VI

Tetracyanoethylene Complexes
Literature Values
(T = 22°C, Solvent CH_2Cl_2)

Donor	K_c	$-\Delta H°$ (cal/mole)	$\Delta G°$ (cal/mole)	$-\Delta S°$ (cal/mole °K)
benzene (34)	0.1286	2540	1214	+12.59
toluene (34)	0.2377	2960	851	12.79
p-xylene (34)	0.4905	3610	421	13.53
m-xylene (34)	0.3863	–	563	–
o-xylene (34)	0.4498	–	474	–
mesitylene (34)	1.113	4760	-63	15.76
durene (34)	3.485	5320	-740	15.37
PMB (34)	7.913	7210	-1226	20.08
HMB (34)	16.911	–	-1676	–
naphthalene (32)	0.7521	–	169	–
pyrene (34)	1.896	–	-379	–

Table VII

Trinitrobenzene Complexes
Literature Values

Donor	K_c	$-\Delta H°$ (cal/mole)	$\Delta G°$ (cal/mole)	$-\Delta S°$ (cal/mole°K)
*benzene (63)	0.0657	450	1613	6.92
*toluene (63)	0.1459	860	1141	6.71
**m-xylene (63)	0.0839	2160	1468	12.18
*mesitylene (63)	0.2140	1210	914	7.13
**durene (62)	0.2266	4010	880	16.41
**PMB (64)	0.3134	–	688	–
**HMB (63)	0.6846	4710	224	16.56
*naphthalene (63)	1.362	2200	-183	6.77
*acenphthalene (63)	1.940	2400	-393	6.74
*phenanthrene (63)	3.086	2630	-668	6.58

*T = 25°C, Solvent $CHCl_3$
**T = 20°C, Solvent CCl_4

Table VIII

Picric Acid Complexes
Literature Values

Donor	K_c	$-\Delta H°$ (cal/mole)	$\Delta G°$ (cal/mole)	$-\Delta S°$ (cal/mole°K)
*durene (63)	1.91	3310	−383	9.82
*HMB (63)	4.86	4020	−937	10.35
*naphthalene (63)	3.59	4140	−757	11.35
**acenphthalene (63)	4.36	−	−872	−
**phenanthrene (63)	8.04	−	−1235	−

*T = 20°C, Solvent CCl₄
**T = 20°C, Solvent CHCl₃

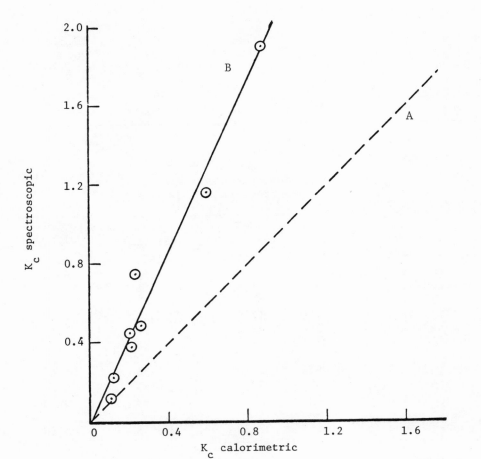

Fig. 4.   Comparison of the calorimetric and spectroscopic equilibrium constants for TCNE complexes.

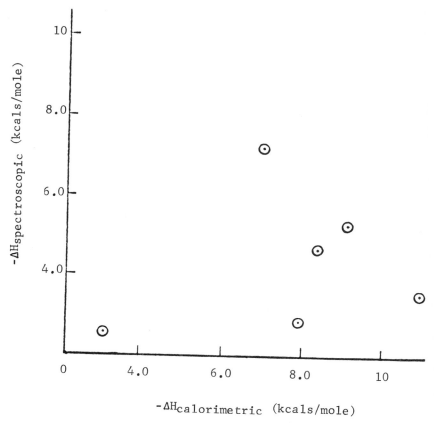

Fig. 5.   Comparison of the calorimetric and spectroscopic enthalpies for TCNE complexes.

whose work has already been discussed. Foster and Horman (69) explained their results on the basis of a deviation from Beer's Law.

Hammond (68) has studied the effects of concentration errors and optical density errors in the analysis of different forms of the Benesi-Hildebrand (7) equation and has found that, for small equilibrium constants, small errors in concentrations lead to large errors in the association constants obtained by these methods. It was concluded by Hammond that for very weak interactions, the Benesi-Hildebrand method and other similar methods, while theoretically valid, may be useless in practice.

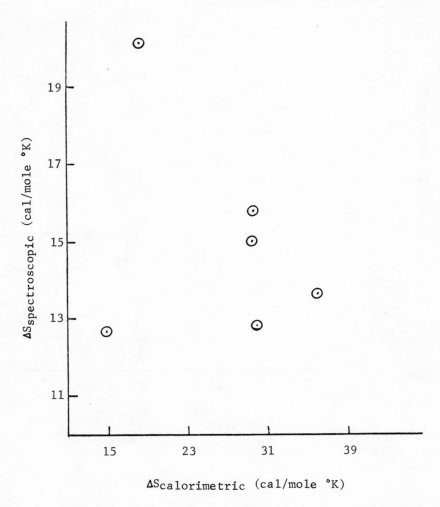

Fig. 6. Comparison of the calorimetric and spectroscopic entropies for TCNE complexes.

Orgel and Mulliken (13) have made two proposals which may
qualitatively explain the discrepancies listed above. In one
model, Orgel and Mulliken (13) postulate a number of isomeric 1:1
complexes. These isomers contribute statistically to the observed
properties of the bulk complex system. Mulliken and Orgel (13)
stated that the Benesi-Hildebrand method (7) would yield a set of
constants which are the averaged extinction coefficient and a

total equilibrium constant. A disturbing point is that the model predicts that the results should be "independent of the wavelength employed in the analysis" (13). This was shown not to hold in some cases by Dewar and Thompson (19) and by others. Another consequence of this model is that, as pointed out by Mulliken and Orgel (13), the spectroscopic methods such as the Benesi-Hildebrand method (7) do not allow differentiation between the presence of only one complex and a family of complexes. It should be mentioned that the calorimetric method and many other analytical techniques also have this drawback.

Orgel and Mulliken also proposed that charge transfer absorption could occur during random collisions of a donor and acceptor pair. They called this model the "contact charge transfer" model (13). This could be considered as a special case of isomeric complexes. The important point is that absorption of light can occur even when no stable complexes are found. The theoretical justification for contact charge transfer is based on previous work of Mulliken (9) on the general theory of complexes and spectra and will not be discussed further in this paper.

Person (66) has criticized the Benesi-Hildebrand method and included contact charge transfer (13) into his discussion. Person (66) has concluded that the experimental conditions must have the initial donor concentration between 0.1 and 9.0 times the inverse of the equilibrium constant in order for the Benesi-Hildebrand linear plot (7) to give a non-zero intercept. Since many complexes which were reported have equilibrium constants of the order of 10^{-2}, this presents experimental problems of solubility and, according to Person (66), makes the accurate measurement of these complexes impossible.

Carter, Murrell, and Rosch (20) have shown that the Benesi-Hildebrand procedure (7) overestimates the value of the equilibrium constant and underestimates the value of the extinction coefficient. Murrell, et al. (20), also criticized the contact charge transfer model (13) on the basis of the failure of the contact complexes to obey the mass action law (equation 2) and the absence of the solvent's role in the model. Murrell, et al. (20), have simply displaced the contact complex hypothesis (13) with a model which assumes competitive complexing involving solvent molecules.

Prue (65), in reviewing the paper by Murrell, et al. (20), pointed out that the extra interactions will lead to a larger equilibrium constant than expected, and that the models of contact charge transfer (13) and solvation complexing (20) are basically the same.

Based on the criticisms, it was not totally unexpected that the calorimetric values for equilibrium constants would be smaller than those obtained spectroscopically.

It may be of some interest to compare the calorimetric values obtained for the complexes of the three acceptors.   Such comparisons could shed some light on the relative sizes of contributing factors to the stabilities of these complexes.

Figure 7 shows a graphic comparison of the equilibrium constants determined by the calorimetric method for TCNE and TNB

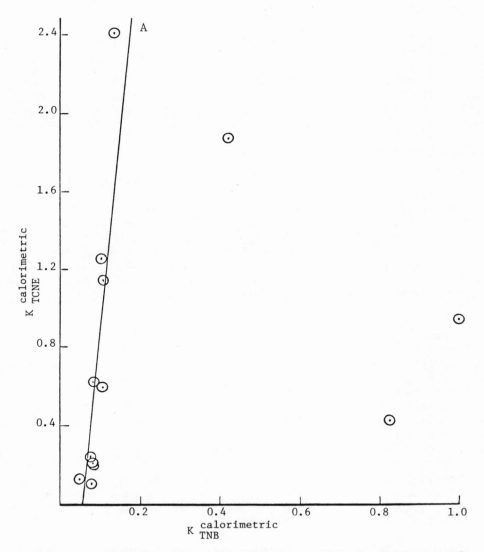

Fig. 7.   Comparison of the calorimetric equilibrium constants for TCNE and TNB complexes.

complexes. There is a definite correspondence inclusive of the points which correspond to benzene and the substituted benzene acting as donors. The donors phenanthrene, pyrene, naphthalene, acenaphthalene and thianthrene are not included in this correlation. Figure 8 is a similar plot using the values of PA and TCNE

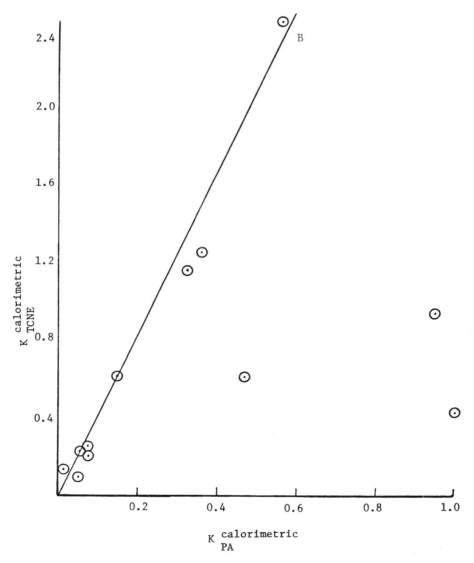

Fig. 8. Comparison of the calorimetric equilibrium constants for TCNE and PA complexes.

complexes. There is also a correspondence present. The same five
donors and 1,2,3,4-tetramethylbenzene do not fit this correspondence. Figure 9 represents the same comparison between the equilibrium constant values for TNB and PA complexes. Due to the
similarity between these two molecules, one might expect a better
correspondence than in the comparisons of the molecules with TCNE
results. This does not seem to be the case. Here, again, the
methyl substituted benzenes seem to follow a correspondence of
their own (line C). The other donors do, however, also seem to

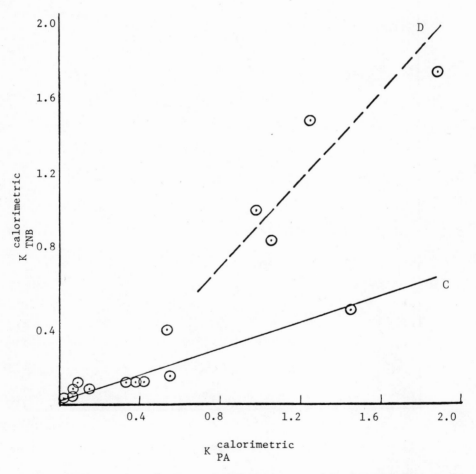

Fig. 9.   Comparison of the calorimetric equilibrium constants for
          TNB and PA complexes.

follow a correspondence (line D). As a possible explanation, it should be remembered that in the donors which are not benzene derivatives the area covered by the TCNE molecule is larger. The cyano groups and their bond dipoles might play a more important role in these complexes. Another possibility is that the greater area of these molecules allows different solvation in the complex which could effect the various contributions to the total stabilization of the complex. Similar comparisons are shown in Figures 10, 11,

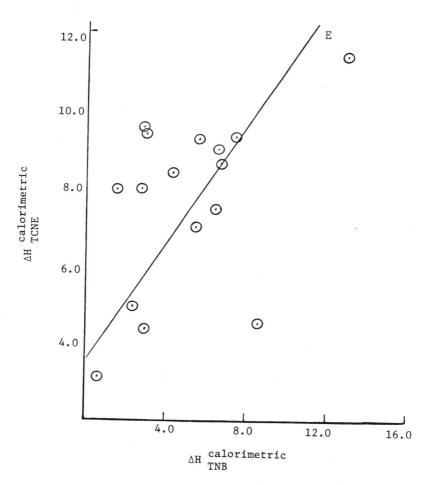

Fig. 10. Comparison of the calorimetric enthalpies for TCNE and TNB complexes.

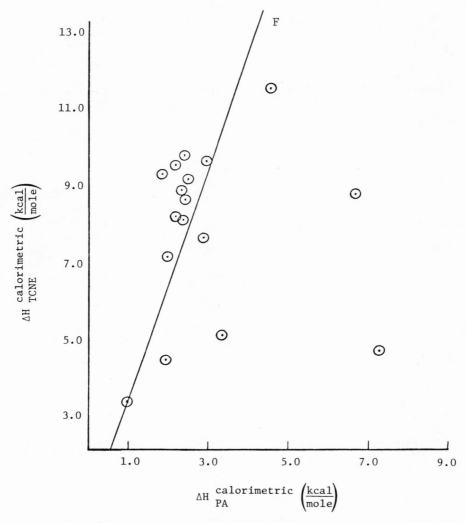

Fig. 11.   Comparison of the calorimetric enthalpies for TCNE and
           PA complexes.

and 12 for the calorimetric enthalpies. Correspondence is also
evident (lines E, F, and G) in each of these comparisons. These
correlations could also be interpreted as an indication that the
same types of forces are involved in the formation of all of these
complexes except for those donors not involved in the correspondence
(pyrene, acenaphthalene, phenanthrene, and naphthalene). A

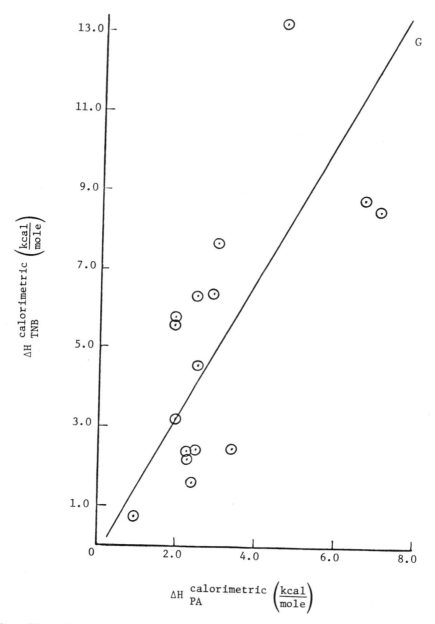

Fig. 12.   Comparison of the calorimetric enthalpies of TNB and
            PA complexes.

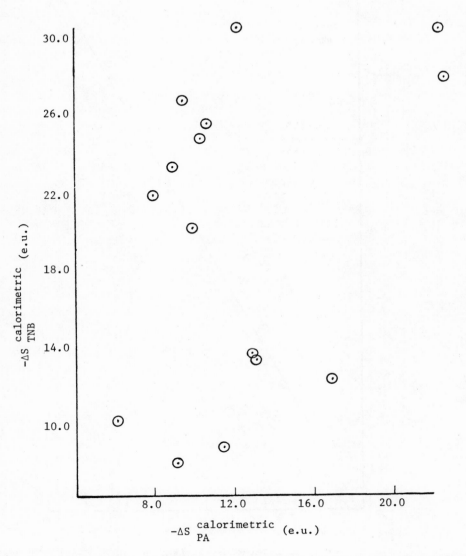

Fig. 13.   Comparison of calorimetric entropies for PA and TNB complexes.

non-correspondence could be considered as due to experimental error, but since there seems to be general correspondence in Figure 12, it seems likely that the reason for the results being as they are could not be caused by a random error. The general correspondence in Figure 12 seems to indicate a balance of forces such that the difference in geometries between TCNE and the other acceptors causes a difference in the importance of the various contributions to the total stabilization. This same effect could affect the entropy in a similar manner thus allowing the correspondence of enthalpies (Figure 12) but not allowing the correspondence of equilibrium constants or free energies (Figure 9). A comparison of the entropy values for the PA and TNB complexes is shown in Figure 13. This figure shows a lack of correlation in the entropy values. No definitive general patterns are evident.

With these comparisons, we wish to end our discussion of this data at the present time. Several different explanations for the discrepancies between the calorimetric and spectroscopic results could be presented, but we have no critical evidence that supports any particular explanation. We believe that we have eliminated sources of systematic errors in the calorimetric experiments, and we think there is a good possibility that the spectroscopic experiments do not exclude such errors as rigorously. The calorimetric method measures heats of a reaction in a more direct manner than the spectroscopic method, which is based on usage of the Van't Hoff equation. Based on the fact that there is a correlation in equilibrium constant plots, cf. Figure 4, it seems reasonable to assume that there is either an error in the spectroscopic method of determining enthalpies or that the data thus obtained has been misinterpreted.

ACKNOWLEDGEMENT

The financial support of the Robert A. Welch Foundation, and the Petroleum Research Fund of the American Chemical Society is gratefully acknowledged. Jerold Feuer also thanks the National Science Foundation for an NSF traineeship. The Texas Tech Computer Center made several generous gifts of computer time available for this work. This work is abstracted from a dissertation submitted by Jerold Feuer to the Graduate Faculty of Texas Tech University in partial fulfillment of the requirements for the Degree of Doctor of Philosophy.

REFERENCES

(1) Mulliken, R.S. and Person, W.B., "Molecular Complexes," Wiley-Interscience, New York (1969).

(2)   Andrews, L.J. and Keefer, R.M., "Molecular Complexes in Organic Chemistry," Holden-Day, New York (1964).

(3)   Rose, J., "Molecular Complexes," Peragon Press, New York (1967).

(4)   Briegleb, G., "Electronen-Donator Complexe," Springer-Verlog, New York (1961).

(5)   Foster, R., "Organic Charge Transfer Complexes," Academic Press, New York (1969).

(6)   Bentley, M.D. and Dewar, J.J.S., Tetrahedron Lett., $\underline{50}$, 5043 (1967).

(7)   Benesi, H.A. and Hildebrand, J.H., J. Amer. Chem. Soc., $\underline{70}$, 2382 (1948); $\underline{71}$, 2703 (1949).

(8)   Mulliken, R.S., J. Amer. Chem. Soc., $\underline{72}$, 600 (1950).

(9)   Mulliken, R.S., J. Amer. Chem. Soc., $\underline{74}$, 811 (1952).

(10) Mulliken, R.S., J. Phys. Chem. $\underline{56}$, 801 (1952).

(11) Mulliken, R.S. and Reid, C., J. Amer. Chem. Soc., $\underline{76}$, 3869 (1954).

(12) Mulliken, R.S., Reciul, $\underline{75}$, 845 (1956).

(13) Orgel, L.E. and Mulliken, R.S., J. Amer. Chem. Soc., $\underline{79}$, 4839 (1957).

(14) Murrell, J.N., J. Amer. Chem. Soc., $\underline{81}$, 5037 (1959).

(15) Subomaira, H.T. and Mulliken, R.S., J. Amer. Chem. Soc., $\underline{82}$ 5966 (1960).

(16) Mulliken, R.S. and Person, W.B., Ann. Rev. Phys. Chem., $\underline{13}$, 107 (1962).

(17) Murrell, J.N., Quart. Rev., $\underline{15}$, 191 (1961).

(18) Lepley, A.R., J. Amer. Chem. Soc., $\underline{86}$ 2545 (1964).

(19) Dewar, J.J.S. and Thompson, C.C., Tetrahedron Suppl., $\underline{7}$, 97 (1966).

(20) Carter, S., Murrell, J.N., and Rosch, E.J., J. Chem. Soc. (London), 2048 (1965).

(21) Johnson, G.D. and Bowen, R.E., J. Amer. Chem. Soc., 87, 1655 (1965).

(22) Dewar, J.J.S. and Lepley, A.R., J. Amer. Chem. Soc., 83, 4560 (1961).

(23) Lepley, A.R., J. Amer. Chem. Soc., 84, 3577 (1962).

(24) McConnel, H., Ham, J.S., and Platt, J.R., J. Chem. Phys., 21, 66 (1953).

(25) Foster, R., Tetrahedron, 10, 95 (1960).

(26) Andrews, L.J. and Keefer, R.M., J. Amer. Chem. Soc., 73, 4169 (1951).

(27) Scott, R.L., Recuil, 75, 787 (1956).

(28) Ketelaar, J.A.A., van de Stolpe, C., Goudsmit, A., and Dzcubas, W., Recuil, 71, 1104 (1952).

(29) de Maine, P.A.D. and Seawaight, R.D., Ind. Eng. Chem., 55, 29 (1963).

(30) de Maine, P.A.D., Spectroschim. Acta, 15, 1051 (1959).

(31) Foster, R., Hammick, D.L., and Wardley, A.A., J. Chem. Soc. (London), 3817 (1953).

(32) Nagakura, S., J. Amer. Chem. Soc., 80, 520 (1958).

(33) Rose, N.J. and Drago, R.S., J. Amer. Chem. Soc., 81, 6138 (1959).

(34) Merrifield, R.E. and Phillips, W.D., J. Amer. Chem. Soc., 80, 2778 (1958).

(35) Weissberger, A., "Physical Methods of Organic Chemistry," Vol. I, Part I, Interscience, New York (1959).

(36) Skinner, H.A., "Experimental Thermochemistry," Interscience, New York (1962).

(37) Christensen, J.J., Izatt, R.M., Hansen, L.D., and Partridge J.A., J. Phys. Chem., 70, 2003 (1966).

(38) Izatt, R.M., Eatough, D., and Christensen, J.J., J. Phys. Chem., 72, 2720 (1968).

(39)   Gol'clstein, I.P., Gur'yanova, E.N., and Karpovich, I.R., Russian J. Chem., 39, 491 (1965).

(40)   Jordan, J. and Alleman, T.G., Anal. Chem., 29, 9 (1957).

(41)   Bolles, T.F. and Drago, R.S., J. Amer. Chem. Soc., 87, 5015 (1965).

(42)   Olofsson, G., Acta Chim. Scandanavia, 20, 2289 (1966).

(43)   Epley, T.D. and Drago, R.S., J. Amer. Chem. Soc., 89, 5770 (1967).

(44)   Nelander, B., Acta Chim. Scand., 20, 2289 (1966).

(45)   Gill, S.J. and Forquhar, E.L., J. Amer. Chem. Soc., 90, 3039 (1968).

(46)   Izatt, R.M., Eatough, D., Snow, R.L., and Christensen, J.J., J. Phys. Chem. 72, 1208 (1968).

(47)   Christensen, J.J., Wrathall, D.P., Izatt, R.M., and Tolman, D.O., J. Phys. Chem., 71, 3001 (1967).

(48)   Christensen, J.J., Wrathall, D.P., and Izatt, R.M., Anal. Chem., 40, 175 (1968).

(49)   Sillen, L.G., Acta Chim. Scand., 16, 159 (1962).

(50)   Pasletti, P., Vacca, A., and Arenare, D., J. Phys. Chem., 70, 193 (1966).

(51)   Somen, G., Coops, Y., and Tolk, M.W., Recuil, 82, 231 (1963).

(52)   Wagner, W., Zeischrift für Physical Chemistry, 218, 392 (1961).

(53)   Wadso, I., Science Tools, 13, 33 (1966).

(54)   Vandersee, C.E. and Swanson, J.S., J. Phys. Chem., 67, 208 (1963).

(55)   Hale, J.P., Izatt, R.M., and Christensen, J.J., J. Phys. Chem., 67, 2605 (1963).

(56)   Gunn, S.R., J. Phys. Chem., 69, 2920 (1965).

(57)   Everett, D.H. and Wynne-Jones, W.F.K., Trans. Faraday Soc., 35, 1380 (1939).

(58) Harned, H.S. and Robinson, R.A., Trans. Faraday Soc., 36, 973 (1940).

(59) Fieser, L.F. and Fieser, M., "Reagents in Organic Synthesis," Wiley, New York (1967).

(60) "Handbook of Chemistry and Physics," Chemical Rubber Company, 50th edition (1969).

(61) Herndon, W.C. and Goodin, R.D., J. Phys. Chem., 73, 2793 (1969).

(62) Roseinsky, D.R. and Kellaivi, H., J. Chem. Soc. (London), 1207 (1969).

(63) Thompson, C.C. and de Maine, P.A.C., J. Phys. Chem., 69, 2766 (1965).

(64) Jureniki, N.B. and de Maine, P.A.D., J. Amer. Chem. Soc., 86, 3217 (1964).

(65) Prue, J.E., J. Chem. Soc. (London), 7534 (1965).

(66) Person, W.B., J. Amer. Chem. Soc., 87, 167 (1965).

(67) Emslie, P.H., Foster, R., Fyfe, C.A., and Norman, I., Tetrahedron, 21, 2843 (1965).

(68) Hammond, P.R., J. Chem. Soc. (London), 481 (1964).

(69) Foster, R. and Horman, I., J. Chem. Soc. (B) (London), 171 (1966).

CALORIMETRIC INVESTIGATION OF THE PYRIDINE - CHLOROFORM COMPLEX

Gary L. Bertrand and Thomas E. Burchfield

Department of Chemistry
University of Missouri-Rolla
Rolla, Missouri 65401

ABSTRACT

Heats of solution of small amounts of pyridine have been measured in mixtures of chloroform and carbon tetrachloride at 25°C. This data has been interpreted in terms of a complex between chloroform and pyridine with simultaneous solution for K_f and $\Delta H_f°$, using both the ideal dilute solution approximation and a second approximation which attempts to account for the effects of nonideality. The "best" values determined with the ideal dilute solution approximation are K_f = 0.28 1/mole, $\Delta H_f°$ = - 3.3 kcal/mole; and K_f = 0.28 1/mole, $\Delta H_f°$ = - 2.8 kcal/mole determined with the second approximation.

INTRODUCTION

The application of calorimetric methods to the investigation of strong complexes in solution has been firmly established (1-4). However, as was noted in a recent paper (4b), the validity of these methods in the investigation of weak complexes has not yet been established. The three basic calorimetric techniques are the binary solution method developed by Dolezalek (5) and recently used by Hepler (6); the "high dilution" method developed by Lamberts (1) and Drago (2); and the "pure base" method developed by Arnett (3). The dilution method used by Hepler and Woolley (7) for the study of self-association is actually a binary solution method, but more closely resembles the high dilution method in both principle and application. The binary solution technique usually involves the possibility of multiple equilibria and has not yet been subjected to the rigorous testing the other methods have received. The relative merits of the latter two methods have been discussed in several papers, in some

283

cases showing excellent agreement between the two and in other cases
showing strong disagreement. Actually, the various methods are limited only by the validity of their inherent approximations. The
binary solution method depends primarily on the assumption that the
unassociated components and the complex form an ideal ternary solution. The basic approximation of the high dilution method is the
assumption of ideal dilute solutions over the range of compositions
studied. In the pure base method, the basic approximation involves
the use of a "model compound" to account for the enthalpy of transfer
of one associating component from an inert solvent to an unassociated
state in the other pure component as solvent. Obviously, the results
of these investigations can be compromised by failure of the basic
assumptions or approximations. The present work was designed to
study the effects of such failure in the high dilution method.

The pyridine - chloroform complex is so weak that the high dilution method requires rather high concentrations of one or both of
the components to obtain precise results, thus possibly violating
the ideal dilute solution approximation. The weakness of this complex also presents a problem to the pure base method in that the
assumption of complete complexation of one component at high dilution
in the other is questionable. The possibility of multiple equilibria
involving pyridine dimers and other species makes the application of
the binary solution method of limited validity. In order to show
where the various approximations are invoked in the different methods
and to describe the approximations we will make, we review here the
basic premises of the high dilution and pure base methods.

The High Dilution Method

Normally, a series of measurements are made as $\Delta H(n_A^\circ, n_B^\circ, n_I^\circ)$
for the process

$$n_A^\circ \ A(\text{pure or concentrated solution}) + [n_B^\circ \ B + n_I^\circ \ I]$$

$$\rightarrow [n_{AB} \ AB + (n_A^\circ - n_{AB}) \ A + (n_B^\circ - n_{AB}) \ B + n_I^\circ \ I] \quad (1)$$

with A and B representing the associating components, AB the complex,
and I the inert solvent; with ratios of (n_B° / n_I°) ranging from zero
to perhaps 0.1 or more. The difference between the enthalpy change
for this process and the same process with no B present is attributed
to the product of the amount of complex formed and the enthalpy of
formation of the complex

$$\Delta H(n_A^\circ, n_B^\circ, n_I^\circ) - \Delta H(n_A^\circ, n_I^\circ) = n_{AB} \Delta H_f \quad , \tag{2}$$

$$K_f = \frac{n_{AB} V}{(n_A^\circ - n_{AB})(n_B^\circ - n_{AB})} \quad , \quad V = V(n_A^\circ, n_B^\circ, n_I^\circ) \ . \tag{3}$$

If K_f is known from independent studies, ΔH_f can be easily determined by combination of Eq. 2 and 3; or simultaneous solution of
these equations for a series of measurements can yield both K_f and
ΔH_f.

    To rigorously describe the enthalpy changes involved, we must
consider the limit of Process 1 as $n_A° \to 0$ for specific values of $n_B°$
and $n_I°$. We define

$$\Delta H_A = \lim_{n_A° \to 0} \frac{\Delta H(n_A°, n_B°, n_I°)}{n_A°} \tag{4}$$

$$\Delta H_A° = \lim_{n_A° \to 0} \frac{\Delta H(n_A°, n_I°)}{n_A°} \tag{5}$$

$$f_{AB} = \lim_{n_A° \to 0} \frac{n_{AB}}{n_A°} \quad . \tag{6}$$

In terms of the partial molar enthalpies of the actual species in
solution,

$$\Delta H_A = f_{AB}(\overline{H}_{AB} - \overline{H}_A - \overline{H}_B) + (\overline{H}_A - \overline{H}_A') \quad , \tag{7}$$

with $\overline{H}_A'$ representing the partial molar enthalpy of $\underline{A}$ in the initial
state of Eq. 1. Representing the partial molar enthalpy of $\underline{A}$ at infinite dilution in the solvent $\underline{I}$ as $\overline{H}_A°$,

$$\Delta H_A° = \overline{H}_A° - \overline{H}_A' \quad , \text{ and} \tag{8}$$

$$\Delta H_A - \Delta H_A° = f_{AB}(\overline{H}_{AB} - \overline{H}_A - \overline{H}_B) + (\overline{H}_A - \overline{H}_A°) \quad , \text{ with} \tag{9}$$

$$K_f = \frac{f_{AB}\gamma_{AB}V}{(1 - f_{AB})n_B°\gamma_A\gamma_B} \simeq \frac{f_{AB}}{(1 - f_{AB})C_B°} \quad . \tag{10}$$

The relative partial molar excess enthalpy of a component is defined

$$\overline{L}_i = \overline{H}_i - \overline{H}_i° \quad . \tag{11}$$

Eq. 9 can then be written

$$\Delta H_A - \Delta H_A° = f_{AB}\Delta H_f° + f_{AB}(\overline{L}_{AB} - \overline{L}_A - \overline{L}_B) + \overline{L}_A \quad . \tag{12}$$

In the ideal dilute solution approximation, all of the relative
partial molar excess enthalpy terms are equated to zero, and Eq. 12
reduces essentially to Eq. 2. Experimental determination of the
relative partial molar excess enthalpy is possible only for component

$\underline{B}$, for which $\overline{L}_B$ can be obtained from heats of mixing of $\underline{B}$ and $\underline{I}$. In order to estimate $\overline{L}_A$ and $\overline{L}_{AB}$, we have made slight modifications in approximations proposed by Arnett and Christian.

The Pure Base Method

In using this technique, Arnett (3) has assumed that virtually all of component $\underline{A}$ exists as the complex $\underline{AB}$ at infinite dilution in pure $\underline{B}$. We consider here the possibility that only some fraction, $f_{AB}*$, is associated. The process of dissolving pure $\underline{A}$ in an infinite amount of $\underline{B}$ can be written

$$A(pure) + f_{AB}* \ B\cdot\infty B + \infty B \rightarrow f_{AB}* \ AB\cdot\infty B + (1 - f_{AB}*) \ A\cdot\infty B \quad . \quad (13)$$

$$\Delta H_A* = f_{AB}*(\overline{H}_{AB}* - \overline{H}_A* - \overline{H}_B*) + (\overline{H}_A* - \overline{H}_A') \ . \quad (14)$$

Combination with Eq. 8 gives

$$\Delta H_A* - \Delta H_A° = f_{AB}*(\Delta H_f*) + (\overline{H}_A* - \overline{H}_A°) \quad . \quad (15)$$

To remove the last term on the RHS of Eq. 15, Arnett has assumed that some model compound, $\underline{M}$, will interact with the solvents $\underline{B}$ and $\underline{I}$ in much the same manner as unassociated $\underline{A}$, such that

$$\overline{H}_A* - \overline{H}_A° = \overline{H}_M* - \overline{H}_M° = \Delta H_M* - \Delta H_M° \quad , \text{ and} \quad (16)$$

$$(\Delta H_A* - \Delta H_A°) - (\Delta H_M* - \Delta H_M°) = f_{AB}*(\Delta H_f*) \quad , \quad (17)$$

in which ΔH_M* and $\Delta H_M°$ are the standard heats of solution of $\underline{M}$ in the solvents $\underline{B}$ and $\underline{I}$, respectively. Arnett has used this approximation for complexes of sufficient strength that $f_{AB}*$ can be approximated as unity, and has shown that in many cases there is very little difference between ΔH_f* and ΔH_f determined by the high dilution method. It should be noted, however, that these values do not refer to the same solvent, since ΔH_f* is for the formation of the complex in solvent $\underline{B}$ rather than $\underline{I}$.

If Eq. 16 is valid for $\underline{B}$ as solvent, it should be reasonably valid for mixtures of $\underline{B}$ and $\underline{I}$,

$$\overline{H}_A - \overline{H}_A° = \overline{H}_M - \overline{H}_M° = \overline{L}_M = \overline{L}_A \quad , \quad (18)$$

since this reduces to an identity in pure $\underline{I}$. $\overline{L}_M$ can be determined as the difference between the heat of solution in pure $\underline{I}$ and in mixtures of $\underline{B}$ and $\underline{I}$, thus providing an approximation for $\overline{L}_A$ in Eq. 12.

The α-Approximation

In order to account for solvent effects on equilibrium constants,

Christian (8) has proposed the approximation

$$(\Delta G_{AB}{}^{\circ})_{R \to S} = \alpha[(\Delta G_{A}{}^{\circ})_{R \to S} + (\Delta G_{B}{}^{\circ})_{R \to S}] \ , \tag{19}$$

in which the terms $(\Delta G_i{}^{\circ})_{R \to S}$ refer to the standard free energy of transfer of a component from a reference solvent $\underline{R}$ to some other solvent $\underline{S}$, and a similar equation for the enthalpy of transfer

$$(\Delta H_{AB}{}^{\circ})_{R \to S} = \alpha[(\Delta H_{A}{}^{\circ})_{R \to S} + (\Delta H_{B}{}^{\circ})_{R \to S}] \ . \tag{20}$$

Taking $\underline{R}$ as the solvent $\underline{I}$, and $\underline{S}$ as a mixture of $\underline{I}$ and $\underline{B}$, this becomes

$$\overline{L}_{AB} = \alpha(\overline{L}_A + \overline{L}_B) = \alpha(\overline{L}_M + \overline{L}_B) \ . \tag{21}$$

Combination of Eq. 12, 18, and 21 gives an equation relating the measureable quantities ΔH_A, $\Delta H_A{}^{\circ}$, $\overline{L}_M$, and $\overline{L}_B$ with the constants K_f, $\Delta H_f{}^{\circ}$, and α:

$$\Delta H_A - \Delta H_A{}^{\circ} = f_{AB} \Delta H_f{}^{\circ} + f_{AB}(\alpha - 1)(\overline{L}_M + \overline{L}_B) + \overline{L}_M \ . \tag{22}$$

Application to the Pyridine - Chloroform Complex

The pyridine - chloroform complex appeared to be the most suitable of the weak complexes we considered for this study, though several difficulties must be noted. Drago (2b) claims that pyridine has unusual interactions with carbon tetrachloride, and that in some solvents pyridine may exist as dimers (2d). The possibility of dimers of pyridine restricts this study to low concentrations of pyridine and thus high concentrations of chloroform. There is the possibility that in addition to the hydrogen bond, pyridine may interact with chloroform through the chlorine atoms in much the same manner as it interacts with carbon tetrachloride. Two types of hydrogen bonds are possible, involving either the nitrogen atom or the π-electrons of pyridine.

We have chosen to work with small amounts of pyridine in mixtures of chloroform and carbon tetrachloride, with benzene as the model compound for pyridine. Thus, in the equations we have developed; $\underline{A}$ = pyridine, $\underline{M}$ = benzene, $\underline{B}$ = chloroform, and $\underline{I}$ = carbon tetrachloride. This choice was based on consideration of a molecule of pyridine at infinite dilution in mixtures of chloroform and carbon tetrachloride. The interactions of pyridine with the chlorine atoms of chloroform should be more similar to the interactions with carbon tetrachloride than with any other solvent. Similarly, the interactions between chloroform and the π-electrons of pyridine should be somewhat similar to the interactions of chloroform with benzene.

EXPERIMENTAL

Materials

All materials were MCB Spectroquality. Chloroform, benzene, and carbon tetrachloride were stored over molecular sieves and were distilled shortly before use. Pyridine was stored over KOH pellets, distilled directly into glass ampoules, and sealed.

Calorimetry

The calorimeter is very similar to one described previously (9), except that the metal lid and Trubore bearings have been replaced by Teflon. The calorimeter contains 300 ml of solution and sample sizes ranged from 0.25 to 5 ml. Observed enthalpy changes were 1 - 10 cal. Heats of solution of successive samples of solute (pyridine, benzene, or chloroform) were measured at 25.0 ± 0.1 °C in an aliquot of chloroform - carbon tetrachloride mixture prepared by weight, and the molar enthalpy effect was extrapolated to zero added solute. These extrapolations involved very little correction of the observed enthalpy effects, in every case less than 10 cal/mole. Formal concentrations of pyridine ranged from 0.01 to 0.05 $\underline{M}$.

RESULTS

Ideal Dilute Solution Approximation

A least squares treatment was used to determine the values of K_f and ΔH_f which gave a minimum standard deviation in the quantity on the LHS of Eq. 2. As a test for failure of the ideal dilute solution approximation on the basis of goodness-of-fit, the parameters were calculated with data for successively higher concentrations of chloroform. Uncertainties in the parameters were calculated in much the same manner as the marginal uncertainties used by Drago (2e). As is shown in Table I, there is not a statistically reliable difference between the parameters calculated for maximum chloroform concentrations between 0.82 and 8.23 $\underline{M}$. However, inclusion of the data measured in pure chloroform (12.4 $\underline{M}$) causes a substantial change in the calculated parameters and in the goodness-of-fit. These results would normally be interpreted as evidence that the ideal dilute solution approximation is valid to chloroform concentrations as high as 8.23 $\underline{M}$, and that the "best" parameters for the pyridine - chloroform complex in carbon tetrachloride are ΔH_f = - 3.3 kcal/mole and K_f = 0.28 1/mole.

Table I.   Thermodynamic Properties of the Pyridine - Chloroform
Complex at 25 °C in Carbon Tetrachloride, Based on the
Ideal Dilute Solution Approximation.

Max. Concentration of Chloroform (mole/l)	N	ΔH_f (kcal/mole)	K_f (1/mole)	$\sigma_r \underline{a}$
0.82	3	- 3.2 ± 1.2	0.29	0.034
3.05	4	- 3.24 ± 0.18	0.28	0.031
6.12	5	- 3.33 ± 0.08	0.28	0.027
8.23	6	- 3.28 ± 0.05	0.28	0.022
12.39	7	- 3.03 ± 0.09	0.32	0.042

$\underline{a}$ See Eq. 23 in text.

Pure Base Method

Since the data point measured in pure chloroform deviates
considerably from the remainder of the set in the previous calculation, the pure base method might be expected to give somewhat different results. Assumption of complete complexation of pyridine leads
to $\Delta H_f^* = - 1.8$ kcal/mole. If the value of $K_f = 0.28$ 1/mole, determined by the high dilution method, is assumed to be valid in pure
chloroform, a value of $\Delta H_f^* = - 2.3$ kcal/mole is calculated, based
on 78% complexation of pyridine at infinite dilution in chloroform.

Possible Effects of Nonideality

Treatment of this data with Eq. 22 requires determination of an
additional parameter, α. In order to determine the "best" set of
parameters, ΔH_f° and K_f were calculated for particular values of α
with a least squares technique. In this case, the parameters were
determined on the basis of a minimum relative standard deviation (σ_r)
in the quantity $(\Delta H_A - \Delta H_A^\circ)$:

$$\sigma_r^2 = (N - 1)^{-1} \sum [\frac{(\Delta H_A)_{obs} - (\Delta H_A)_{calc}}{(\Delta H_A)_{obs} - \Delta H_A^\circ}]^2 \tag{23}$$

The value of α was varied to produce a minimum value of the relative
standard deviation of the data set. For each value of α, calculations were performed at successively higher concentrations of chloroform. In every case, the data point for pure chloroform deviated
greatly from the remainder of the set, so this point was not included
in the final calculations. Table II shows the results of these
calculations.

Table II.   Nonideality Effects on the Calculated Thermodynamic
Properties of the Pyridine - Chloroform Complex.

α	ΔH_f° (kcal/mole)	K_f (1/mole)	σ_r
0	- 3.37	0.24	0.024
0.2	- 3.22	0.25	0.024
0.4	- 3.03	0.27	0.023
0.6	- 2.86	0.28	0.021
0.8	- 2.70	0.30	0.022
1.0	- 2.52	0.32	0.023

Values of ΔH_f° and K_f are somewhat sensitive to the value of α, but
the relative standard deviation is not. The "best" parameters appear
to be approximately $\alpha = 0.6$, $K_f = 0.28$ 1/mole, and $\Delta H_f^\circ = - 2.8$ kcal/
mole. This value of α appears reasonable in comparison to values
between 0.5 and 0.68 reported by Nozari and Drago (2d).

DISCUSSION

The three methods of calculation described here yield different
values for the enthalpy of formation of the pyridine - chloroform
complex. The value determined by the pure base method might be expected to differ from the other two values for two basic reasons:
(a) this value is for complex formation in chloroform rather than in
carbon tetrachloride as determined by the other methods; and (b) this
value is based on an equilibrium constant determined in mixtures of
chloroform and carbon tetrachloride, and one must recognize that the
equilibrium constant for association of a solvent is a poorly defined
quantity. The difference of 0.5 kcal/mole between enthalpies of
formation calculated by the other two methods is not so easily explained. Both data treatments give results which are as statistically precise as the results of investigations of stronger complexes
for which an overall accuracy of 0.2 kcal/mole has been claimed.
At this time, we cannot claim that ΔH_f° determined with the nonideality approximation is any more reliable than the value determined with
the ideal dilute solution approximation, since it is quite possible
that our approximation has compounded errors; nor can we claim that
all of the effects of nonideality have been taken into account.
Since the two methods lead to the same value of K_f, their relative
reliability cannot be tested with an independently determined equilibrium constant. Therefore, we conclude from this study that the
high dilution method of calorimetric investigation can lead to errors
of 0.5 kcal/mole or more in the enthalpy of formation of weak

complexes. We further suggest that the effects of nonideality might lead to errors of this magnitude for much stronger complexes. The chloroform - carbon tetrachloride system is very nearly an ideal solution. The relative partial molar excess enthalpy of chloroform (as defined in Eq. 11) at 8.4 M in carbon tetrachloride is about - 200 cal/mole, which is quite comparable to the relative partial molar excess enthalpy of pyridine at 1 M in carbon tetrachloride (pyridine in cyclohexane is considerably more nonideal) or phenol at 0.03 M in cyclohexane - levels of concentration which have often been used in investigations of much stronger complexes. Eq. 22 provides a means to study the possible effects of nonideality on the thermodynamic properties calculated for these stronger complexes.

Slejko, Drago, and Brown (2e) determined K_f = 1.4 1/mole for the pyridine - chloroform complex in cyclohexane at 9.5 °C, and they estimate ΔH_f = - 4.6 kcal/mole. Assuming that ΔH_f is independent of temperature, we estimate K_f = 0.9 1/mole at 25 °C. This indicates that the complex is much more stable in cyclohexane, as is expected if pyridine interacts strongly with carbon tetrachloride.

ACKNOWLEDGEMENT

We thank the Research Corporation for support of this work

REFERENCES

1. (a) L. Lamberts and T. Zeegers-Huyskins, J. Chim. Phys., <u>60</u>, 435 (1963); (b) L. Lamberts, ibid., <u>62</u>, 1404 (1965).

2. (a) T. F. Bolles and R. S. Drago, J. Amer. Chem. Soc., <u>87</u>, 5015 (1965); (b) T. D. Epley and R. S. Drago, ibid., <u>89</u>, 5770 (1967); (c) R. S. Drago and T. D. Epley, ibid., <u>91</u>, 2883 (1969); (d) M. S. Nozari and R. S. Drago, ibid., <u>94</u>, 6877 (1972); (e) F. L. Slejko, R. S. Drago, and D. G. Brown, ibid., <u>94</u>, 9210 (1972).

3. (a) E. M. Arnett, T. S. S. R. Murty, P. von R. Schleyer, and L. Joris, ibid., <u>89</u>, 5955 (1967); (b) E. M. Arnett, L. Joris, E. Mitchell, T. S. S. R. Murty, T. M. Gorrie, and P. von R. Schleyer, ibid., <u>92</u>, 2365 (1970).

4. (a) W. C. Duer and G. L. Bertrand, ibid., <u>92</u>, 2587 (1970); (b) G. L. Bertrand, D. E. Oyler, U. G. Eichelbaum, and L. G. Hepler, Thermochim. Acta, <u>7</u>, 87 (1973).

5. F. Dolezalek, Z. Phys. Chem., <u>64</u>, 727 (1908).

6. T. Matsui, L. G. Hepler, and D. V. Fenby, J. Phys. Chem., <u>77</u>, 2397 (1973).

7. (a) E. M. Woolley, J. G. Travers, B. P. Erno, and L. G. Hepler, J. Phys. Chem., 75, 3591 (1971); (b) E. M. Woolley and L. G. Hepler, ibid., 76, 3058 (1972); (c) N. S. Zaugg, L. E. Trejo, and E. M. Woolley, Thermochim. Acta, 6, 293 (1973).

8. (a) S. D. Christian, J. Phys. Chem., 70, 3376 (1966); (b) J. Grundnes and S. D. Christian, Acta Chem. Scand., 23, 3583 (1969); (c) S. D. Christian and E. E. Tucker, J. Phys. Chem., 74, 214 (1970).

9. (a) G. L. Bertrand, R. D. Beaty, and H. A. Burns, J. Chem. Eng. Data, 13, 436 (1968); (b) E. L. Taylor, M. Sc. Thesis, University of Missouri - Rolla, 1969.

APPLICATION OF THERMAL ANALYSIS AS A SUBSTITUTE FOR STANDARD ASTM POLYMER CHARACTERIZATION TESTS

P. S. Gill and P. F. Levy

E. I. Du Pont De Nemours & Company
Instrument Products Division
Wilmington, Delaware

INTRODUCTION

Thermal analysis techniques have for some time been used in the field of materials characterization (1). The data obtainable from such procedures can be applied directly to the study of polymer physical properties such as phase changes, stability measurements, mechanical effects, and thermodynamic data. The ability of modern commercially available Thermal Analysis Systems to provide accurate and precise data has been accepted in both the academic and industrial research environments. In the field of polymer quality control, however there are many techniques still widely used which do not adequately define the physical property being monitored. It is the purpose of this paper to present data which illustrates the advantages of the thermal analysis technique of Differential Scanning Calorimetry (DSC) over the more traditional procedures used for determining polymer melting behavior. Correlations are presented between data obtained using the DuPont Thermal Analysis System and that from standard ASTM methods used in the polymer industry. Illustrations are also given of several instances in which typical DSC procedures can simultaneously provide additional data which can also be used to evaluate other parameters significant in the characterization of the product.

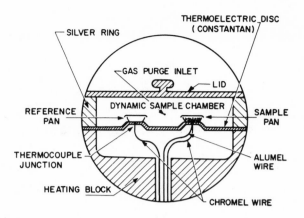

Figure 1. DSC Cell - Cross Section

DETERMINATION OF POLYMER MELTING BEHAVIOR BY DSC

Differential Scanning Calorimetry is rapidly
becoming the most widely used and universally accepted
of all the thermal analysis techniques. It is pertinent
therefore to extend this acceptance into the areas of
polymer quality control, and provide a degree of precision, accuracy, and automation previously not possible.

Differential scanning calorimetry (DSC) is a
thermal analysis technique which quantitatively measures
the differential heat flow (milli-calories per second)
between a sample and its reference material, as a
function of the linearly programmed temperature (^{O}C).
Figure 1 depicts a cross-section of the DuPont DSC cell.
In this configuration the differential thermocouples
are positioned external to the sample, and separated
from the sample by means of a thermoelectric disk (2),
thus providing the advantages:

• The sample thermal conductance is not an integral part of the heat flow path to the sample temperature sensor.

• The disk, having a constant and reproducable thermal resistance, permits quantitative calorimetric measurement of the differential heat flow.

• Sample handling becomes greatly simplified since the sample is remote from the thermocouple.

Since the thermal sensor is external to the sample, a slight time lag between the sample temperature and that monitored at the thermocouple junction is produced. This lag, though minimal at normal program rates (5-20 deg C/min), can be eliminated by temperature programming at a sufficiently slow rate (e.g. 2 deg C/min) such that the time lag produces an insignificant temperature lag. Programming at slow rates provides other improvements in transition temperature measurements of polymeric materials as will be discussed later.

Most commercial polymers consist of amorphous and crystalline fractions and usually have a wide molecular weight distribution. The melting behavior of such polymers (temperature of fusion, temperature range of fusion, and heat of fusion) is very much dependant upon the inherent structural properties which in turn are dependant upon several experimental variables. Before proceeding further, let me define some of the terms frequently used in referring to the melting behavior of polymeric materials. The melting point of a polymer is the temperature at which the last traces of crystallinity disappear. This "point" which is most accurately measured by X-Ray diffraction, has been shown (4) to correspond to the extrapolated completion temperature in DSC (see Figure 2). This point has proved difficult to measure for some polymers by the visual techniques now frequently used as quality control procedures due to the somewhat low degree of crystallinity and wide melting range typical of polymers. The melting temperature referred to by these visual methods is most closely represented by the temperature corresponding to the maximum rate of polymer fusion. This melting temperature is also the endothermic peak temperature as measured by DSC. The "onset temperature" for polymeric fusion can also be measured by DSC, and corresponds to the temperature at which polymer fusion commences. Any complete description of polymer melting behavior should include

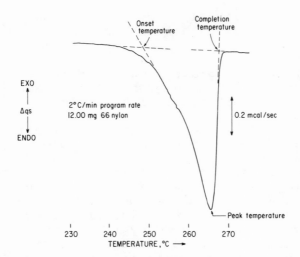

Figure 2.   DSC Fusion Curve for 66 Nylon

each of these three temperatures, and the experimental
conditions under which the data is obtained. Such a
complete thermal description of polymer melting is only
possible using instrumental measuring devices such as
DTA or DSC, which can unambiguously and reproducibly
provide this data. Visual methods suffer from the disadvantages of being subjective and of not providing sufficient control of the experimental variables.

VARIABLES AFFECTING POLYMER MELTING BEHAVIOR

The influence of certain parameters can be significant in determining the melting behavior of polymers,
some to the more cirtical of these factors include:

- heating rate   • polymer thermal history
- sample size   • polymer crystallinity
          • sample preparation

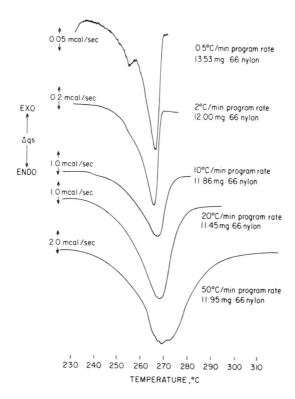

Figure 3. Effect of Heating Rate on 66 Nylon Melting Behaviur

HEATING RATE can affect polymer fusion in two ways; first, by inducing a significant thermal lag at high heating rates, and second, by causing superheating of the polymer crystals at high heating rates, due to the fact that the polymer crystals are often very slow melting (3). Optimum linear programmed heating rates, depending upon the particular polymer, were found to be between 0.5 and 2.0°C/minute. Figure 3 depicts graphically the effects of heating rate upon the melting behavior of a 66-nylon polymer. Two notable features are; the shifting of the peak to higher temperatures with increased heating rate, and the appearance of a second peak at lower temperature in the case of very slow heating rates. The appearance of the small endotherm has been explained (4) as being due to a secondary crystallization or annealing process rather than from the

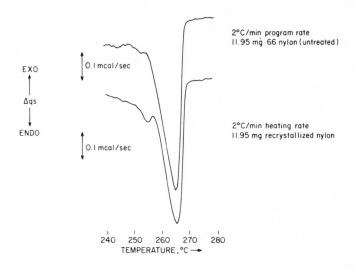

Figure 4.   Effect of Thermal History on 66 Nylon Melting

primary crystallization.   Its manifestation at slow
heating rates being due to resolution enhancement (fast
heating rates are shown to cause peak broadening).   The
small endotherm can be shown to be dependent upon
crystallization since its contribution to the total
melting endotherm increases when the 66-nylon is slowly
recrystallized from the melt (Figure 4).   This effect
illustrates another of the dependent variables which
influence polymer melting, that of the THERMAL HISTORY
of the material.   It is very important, in making thermal measurements of polymeric materials, to ensure that
the effects of thermal history are taken into consideration.   Alternatively, one can use this phenomenon to
obtain data directly relating to the polymer thermal
history (7,8).   Such thermal history effects may be removed from the polymer by reproducible recrystallization of the polymer from the melt after holding above
the melt temperature for a period of time (5-10 min.).
Such an annealing process can be successfully achieved
by means of the DuPont DSC cell, which thus results in
a reproducible polymer crystal morphology, and also more
precisely measured melting temperatures.   Figure 4 depicts the effect of annealing on the melting behavior
of 66 Nylon.

The effects of SAMPLE SIZE on the melting behavior
of polymers are primarily due to:

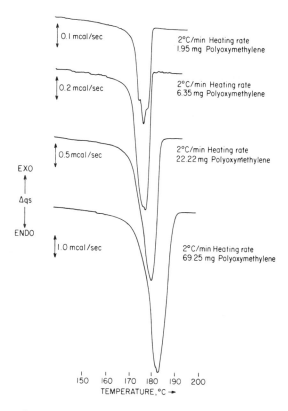

Figure 5.   Effect of Sample Size on Polyoxymethylene Fusion

•   peak broadening at increased sample size causing loss of resolution.

•   increased thermal resistance between sample and
thermocouple with larger sample sizes, resulting in
larger time and temperature lags.

Illustration of sample size effects upon the melting of a polyoxymethylene is shown in Figure 5.   Each
of the two phenomena are clearly depicted.   The appearance of multiple peaks at small sample sizes is again
due to improved resolution, permitting observation of
more than one type of crystalline form.   Annealing of
the polymer, as described previously, removes this
thermal history effect, permitting re-orientation of

the polymer crystals, resulting in the appearance of only one peak. Optimum sample size for determination of
polymer melting behavior is found to be in the range of
2 to 10 mg depending upon the particular polymer, its
molecular weight range and its degree of crystallinity.

POLYMER CRYSTALLINITY is a very critical variable
upon which the thermal characteristics are very dependent. Most polymers consist of crystalline and amorphous fractions. The amorphous fraction affects the crystalline melting behavior by lowering the fusion temperature and broadening the melting range. This phenomenon provides two major advantages of using DSC for
polymer melting behavior determinations.

  ● Polymers of low crystallinity have very broad
melting ranges which prove extremely difficult to define by visual techniques. DSC measurements on such
systems can be made with far greater objectivity and
precision.

  ● DSC, in addition to defining polymer melting
from qualitative temperature measurements can also be
used to quantitatively determine the degree of polymer
crystallinity.

The area of the DSC polymer fusion endotherm gives
a measure of the heat of fusion which when compared to
the heat of fusion of the 100% crystalline material
can yield the degree of polymer crystallinity. Figure
6 shows such a measurement for a commercial polyethylene
sample (5) which agreed closely with the X-ray diffraction crystallinity determination. The lowering of the
melting temperature can also be used to determine polymer crystallinity, but, since this is also affected by
other phenomena (particulary impurity levels), it is
not as accurate as the heat of fusion determination.
This same type of heat of fusion measurement can also
be used in determining compositional information on
polymer blends or polymer filled compositions. The
data reported in Table I was obtained from a series of
66 nylon compositions blended with varying amounts of
a specific additive. DSC has the obvious advantages
over the gravimetric acid digestion technique of being
faster, safer and non-destructive, permitting subsequent analysis of other polymer properties.

Recent data for the melting and crystallization behavior of Polytetrafluorethylene (PTFE) has also shown

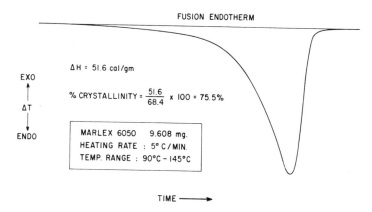

Figure 6. Polymer Crystallinity

Table I

Sample	Wt% 66 Nylon (by DSC heat of fusion)	Wt% 66 Nylon (by acid digestion)
#1	79.0	81.3
#2	78.1	79.5
#3	72.7	74.7
#4	68.7	68.5

a good correlation between the heat of crystallization
and the number average molecular weight for PTFE (6).

SAMPLE PREPARATION is also significant in the determination of polymer melting and crystallization behavior in two ways. First the crystallinity can be
affected by the grinding techniques used for particle
size reduction, and secondly the nucleation during crystallization is very much dependant upon sample treatment prior to melting. These effects can be negated
by either sampling directly, or by annealing in the
same way as described for removal of thermal history.

Having thus evaluated the more critical variables

Table II

Sample	Onset temp. DSC (°C)	Peak temp. DSC (°C)	Completion temp. DSC (°C)	ASTM Fisher-John°C
66 Nylon (Type I)	241	265	268	257
6 Nylon (Type II)	182	221	227	214
6:10 Nylon (Type III)	199	223	227	218
6:12 Nylon (Type VIII)	200	219	221	215
HFP/TFE copolymer	211	264	281	271
Polyoxymethylene	165	179	180	172

we are now in a position to more accurately define the optimum experimental parameters for use with DSC in evaluation of polymer melting behavior. That is:

- slow heating rates ≤ 2°C/minute

- small sample size < 10mg

- removal of thermal history and sample prepara-tion effects (where necessary).

Under these optimized conditions the following DSC data was obtained for a series of polymeric materials. This data together with that for traditional visual techni-ques is produced in Table II.

A statistical evaluation of the precision of the DSC peak temperature measurement showed a sigma (stan-dard deviation) of ± 1°C for 6:6 Nylon (10 runs).

DISCUSSION

This data illustrates how the polymer melting range can vary significantly for different polymeric composi-tions. It is also noted that the visual technique (in this case ASTM method D789) gives a melting temperature which falls within the polymer melting range, and approximates that of the DSC peak temperature. This point corresponds to the maximum rate of fusion of the polymer under the timized conditions (i.e. no time lag) and provides the most precise temperature measure-ment point in the DSC curve. The true melting tempera-ture however corresponds to the extrapolated completion temperature which requires a degree of subjectivity in its measurement. The recommendation would be that both temperatures be used in defining polymer melting behavior. DSC provides the following major advantages over the visual techniques in the determination of polymer melting temperatures

- greater precision - less subjectivity

- better definition of the total melting range

- ease of sample handling

- ability to provide additional quantitative data

- significant reduction in manpower requirements

This final point is particularly important in the qual-ity control areas where large sample loads provide jus-tification for some degree of automation.

REFERENCES

1. Miller, G.W., App. Poly. Symp., 10, 35 (1969).

2. Baxter, R.A., in Thermal Analysis, Vol. 1, R.F.
    Schwenker, Jr. and P.D. Garn (eds,), Academic Press
    (1969) pages 65-84.

3. Wunderlich, B., Thermochim. Acta., 4, 175 (1972).

4. Chiu, Jen, J. Macromol. Sci., Chem. Ed., A8, 3 (1974).

5. DuPont Thermal Analysis Application Brief #12.

6. Takeshi Suwa, et.al., J. App. Poly. Sci., 17, 3253
    (1973).

7. Nakagawa, K., Ishida, Y., J. Poly. Sci., 11, 2153
    (1973).

8. Lord, F.W., Polymer, 15, 42 (1974).

APPLICATION OF SCANNING CALORIMETRY TO

PETROLEUM OIL OXIDATION STUDIES

F. Noel, G. E. Cranton

Imperial Oil Enterprises Ltd, Research Department

Sarnia, Ontario, Canada   N7T 7M1

INTRODUCTION

The purpose of this paper is to indicate the potential in applying Differential Scanning Calorimetry (DSC) to petroleum oil oxidation studies. The practical value of this approach appears to be extensive; but the understanding remains meager, and much more work is required to develop techniques and understanding. Hopefully, this paper will stimulate additional research in this area.

LITERATURE BACKGROUND

Krawetz[1] has described the use of pressurized Differential Thermal Analysis (DTA) to define the thermal stability of lubricants in an inert atmosphere. Vajta and Adonyi[2] applied Thermogravimetric Analysis (TGA) to derive the kinetics for thermal degradation of lube oil and additives at 150 to 500°C. Vaclav[3] used TGA to show that asphaltenes in fuel oils contribute to residue formation in both inert, and oxidizing atmospheres. On this basis it has been suggested that TGA could be used as a quick method of screening fuel oils[4]. The same technique was found to be useful when applied to new and used motor oils[5]. Levy et al[6] used pressurized DTA to characterize the oxidative stability of brake fluids, as a quick screening test. A similar technique was applied by May and Bsharah[7] to characterize the stability of various hydrocarbons under 500 psig oxygen. The effect of antioxidants was also demonstrated. Commichau[8] used DTA to determine the efficiency of antioxidants in motor oil formulations. We have previously reported briefly on the use of DSC at 7 atm. to study the degradation of motor oils[9].

EQUIPMENT

The Perkin-Elmer DSC unit is designed mainly for operation at, or near, atmospheric pressure. The advantages of DSC analysis at elevated pressures were obtained in our laboratory by constructing a simple pressure cell system, which is shown in Figure 1.

The pressure cell consists of two parts held together by three 1/2 in. bolts. The bottom part of the pressure cell is the Perkin-Elmer DSC assembly itself, while the top of the cell consists of the bottom part of a Perkin-Elmer low temperature cover, with the coolant cup removed. For this application, the purge lines beneath the assembly are plugged. Two lines run through the head for pressurizing and purging the system. The cell has been used repeatedly at pressures up to 100 psig. The air or nitrogen is dried by molecular sieves before entering the cell.

TECHNIQUES

For thin film oil oxidation studies, the size of the sample in the pan is critical. A sample size of 1 mg or less produces unambiguous and fairly reproducible degradation exotherms. Since there is no gas-sample agitation, the major oxidation tends to be on the liquid surface. Therefore, larger samples (>1 mg) produce lower degradation energy values.

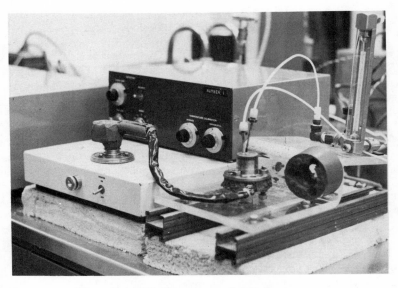

Fig. 1. DSC Pressure Cell Accessory

For routine analysis, the oil sample is smeared on the bottom of the aluminum sample pan, then covered by a pan shaped cover, which is punctured to provide four small holes. The pressure cell is purged at 30 cc/min with air at 100 psig pressure. The temperature program is set at 10°C/min with a range setting of 4 or 8 millical/sec for full scale deflection on a 10 mv recorder.

For isothermal work, the sample is equilibrated at the desired temperature under N_2, then switched to air or O_2. Most of the isothermal work was carried out at temperatures between 180-230°C, since this is the temperature range to which motor oils are frequently exposed during engine operation.

DISCUSSION OF THERMOGRAMS

Figure 2 shows a typical thermogram obtained during the temperature programmed oxidation of 0.5 mg stabilized lube oil in 100 psig air. With the programmed temperature technique the oxidative stability of the oil is defined in terms of three temperature parameters: temperature at onset of degradation, peak temperature and end of peak temperature. The latter temperature does not really define the end of oxidation, since it will obviously continue on to the maximum scanning temperature of the DSC-1B. Table 1 shows the reproducibility of results obtained for temperature programmed data. Reasons for the scatter are not yet clear; however, part of the problem lies in the definition of the onset of degradation.

Figure 3 shows the type of thermogram obtained during isothermal oxidation. Induction time is defined as the time interval between the introduction of air or oxygen and the onset of the oxidation process. Degradation energy, which is related to the extent of oxidation at that temperature, is calculated from the peak area. The end of the exotherm defines the end of the oxidation period, which is followed by a level trace. The latter tracing is usually slightly below that of the level tracing before the oxidation process. This shift seems to be related to the accumulation of degradation products on the underside of the dome. Table 2 shows the reproducibility of the isothermal data to be fair for definition of induction time and degradation energy. For many applications this scatter is not a problem, but certainly needs improvement for more stringent research studies. More development work is required in this area.

EFFECT OF ANTIOXIDANTS

The performance of an antioxidant zinc dialkyldithiophosphate (ZDDP), at various concentrations in lube oil, was studied by DSC

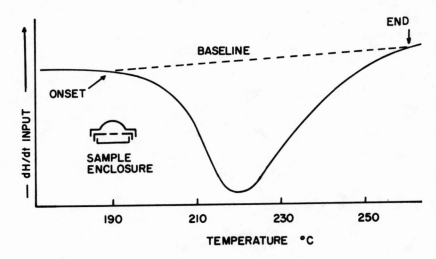

Fig. 2. DSC Degradation Thermogram - 10°C/min., 100 psig air, lube oil.

TABLE 1

REPRODUCIBILITY OF TEMPERATURE

PROGRAMMED OIL DEGRADATION DATA

0.2-0.45 mg unstabilized lube oil
10°C/min, air 100 psig, 5 samples

Degradation Temp, °C

	Onset	Peak	End
	175-194	210-225	257-268
Average	182.9	218.3	260.9
Standard Dev.	6.7	5.8	5.7

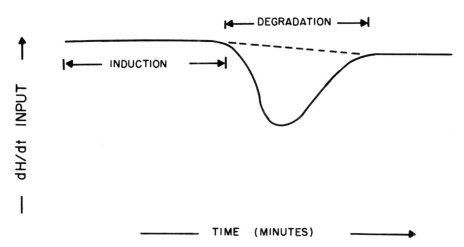

Fig. 3.   DSC Degradation Isotherm - 100 psig air, 210°C

TABLE 2

REPRODUCIBILITY OF ISOTHERMAL OXIDATION DATA

0.4 to 0.5 mg Inhibited Lube Oil, 210°C, 100 psis air, 27 repeats

	Induction Time (min)	Degradation Energy (cal/g)
Range	4.5 to 7.5	344-479
Average	5.9	420
Standard Dev.	1.0	39

analysis. The oil degradation thermograms for two concentrations of ZDDP are shown in Figure 4.

The curves show that in the absence of antioxidant the lube oil starts to degrade at 190°C, and produces a single peaked exotherm. In the presence of antioxidant the degradation starts at a higher temperature, and a double peaked exotherm is produced. The first peak possibly relates to oxidation reactions which include the ZDDP, while the larger second peak relates primarily to the oxidation of the oil after the antioxidant has been consumed. The two peak character is accentuated at higher ZDDP concentrations.

Figure 5 shows the application of isothermal DSC to compare three lube oil antioxidants (A, B, C) of the phenate type. These are complex salts of calcium and substituted phenol, where the substituents are alkyl groups and/or sulphur linkages. In this example the phenates are added at the same calcium level. The DSC results show that a change in the phenate structure produces a change in both the induction time and the peak area, or extent of oxidation.

ANTIOXIDANT STABILITY

Lubricating oils are applied in a wide variety of situations involving time, and temperature, and it is important to choose antioxidants which will function in the desired temperature range. From DTA and TGA studies Kalmutchi et al[11] concluded that the efficiency of ZDDP type antioxidants depended on their thermal stability, which was found to increase with alkyl group chain length[11,12]. In the study described here, we have used three different ZDDP type additives, which are briefly described below:

ZDDP	MOL WT*
A (alkyl)	578
B (alkyl)	817
C (aryl)	2043

* calculated from Zn content

The ZDDP samples were of the commercial grade and were not purified. DSC scans on the additives alone were carried out under N_2 and air each at 100 psig and heating rate of 10°C/min. Thermograms are shown in Figure 6.

The antioxidant ZDDP-A has an irreversible endothermic process at 208°C which is not altered by changing from N_2 to air. The antioxidant ZDDP-B starts to degrade thermally at 220°C, and

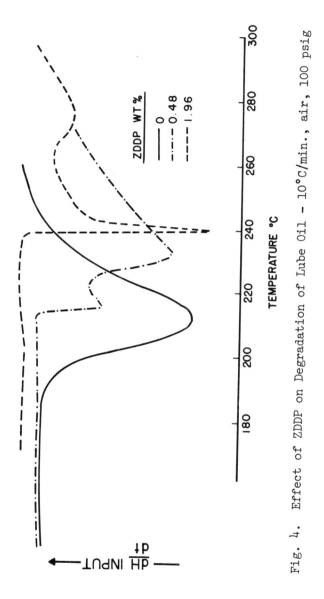

Fig. 4. Effect of ZDDP on Degradation of Lube Oil – 10°C/min., air, 100 psig

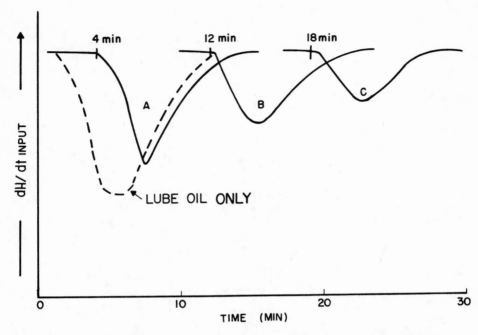

Fig. 5. Comparison of Various Phenate Salts in Lube Oil –
DSC Analysis, 100 psig air, 210°C

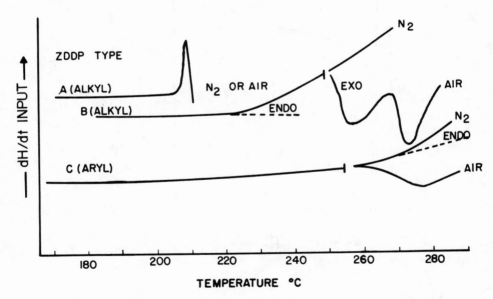

Fig. 6. Stability of Various ZDDP Antioxidants - 10°C/min.,
100 psig air

oxidation then starts at 248°C. Antioxidant ZDDP-C did not degrade until 254°C either in nitrogen or air. These results show that the stability of the ZDDP type antioxidants increases in the order A, B, C.

EFFECT OF ANTIOXIDANT CONCENTRATION

In the formulation of motor oil additive packages the objective is to achieve optimum concentration for each of a large number of additive types based on performance. In connection with new antioxidants it is important to be able to define quickly the useful concentration range to be considered in screening tests.

Figure 7 summarizes a study of the effect of antioxidant concentration on the stability of a lube oil, as defined by DSC analysis. The antioxidants used in this study were three different ZDDP type antioxidants, dibenzyl disulphide (DBDS), and a phenol type antioxidant. Each solution contained only lube oil and one antioxidant. The curves show that for each antioxidant there is a maximum additive concentration, above which there is little improvement in stability. These maximum concentrations are different on a weight basis, but, as shown in Table 3, are similar on a molar basis, namely 0.04 mole/1000 g oil. The free radical phenol type antioxidant has a different maximum concentration value of 0.1 mole/1000 g oil. It is evident from this data that a few short DSC scans will define the useful range of antioxidant concentration to be considered for further conventional tests.

EFFECT OF METAL SURFACE

While in use, the lube oil is subjected to elevated temperature, atmospheric oxygen, and the presence of metal surfaces, which are known to affect the hydrocarbon oxidation process. In order to study the metal surface effect, special metal discs were prepared, which had a diameter of 0.25 in. and were 7 mils thick. These discs were prepared from a forged rod with the same composition as SAE bearing steel #52,100. A sample of oil (0.5 μl) was then smeared on the disc. The oxidative degradation of oil samples was compared on the discs and the normal aluminum pans. The results are summarized in Table 4.

The presence of the bearing metal lowers the temperature of the onset of degradation, both in absence and presence of water. The addition of the ZDDP removes this metal surface effect, since the degradation temperatures for the aluminum and bearing metal surfaces appear to be the same.

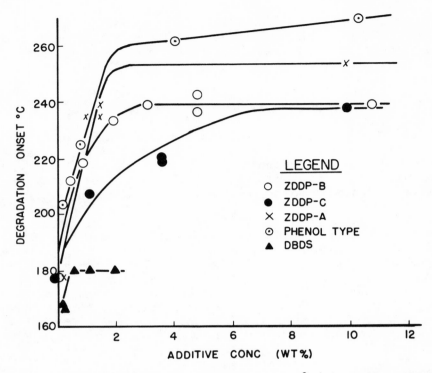

Fig. 7.   Antioxidant Effects Level Out - 10°C/min, 100 psig, Air

TABLE 3

ADDITIVES SHOW SIMILAR CONCENTRATION EFFECTS

Additive	Mol Wt[1]	Oil	Air Press (atm)	Conc. at Plateau ~Wt %	Moles/1000 g oil
DBDS[2]	218	A	1	0.7	0.032
ZDDP-A	578	B	7	2.1	0.036
ZDDP-B	817	B	7	3.0	0.037
ZDDP-C	2043	B	7	7.5	0.037
Phenol Type	208	C	7	2.0	0.1

(1)   calculated
(2)   dibenzyl disulfide

TABLE 4

METAL SURFACE AFFECTS DEGRADATION TEMPERATURE

DSC, 10°C/min, air

		Degradation Onset, °C	
Sample	Metal:	Aluminum	Bearing*
lube oil A		184, 186, 190	151, 161
A + 1% H_2O		172	146, 155
A + ZDDP		260, 256	252

* SAE 52,100 untempered

OXIDATION KINETICS

The oxidative degradation of hydrocarbon oils is a complex process, since at any one temperature during oxidation numerous components are being oxidized. During DSC analysis, therefore, each dH/dt value is the net sum of concurrent reactions at a particular time and temperature. Nevertheless, an overall pattern presumably exists, which can be used to characterize the lube oil oxidation process. All DSC oxidation studies discussed in this paper involved relatively high temperatures (150-230°C) and O_2 concentrations (100 psig Air). Consequently the oxidation process is one of the short kinetic chain length, and the exotherms probably depict mainly peroxide decomposition. The application of DSC analysis to derive kinetic data from exothermic processes has already been applied to peroxides[13] and explosive compounds[14].

We have analyzed several inhibited and uninhibited lube oils which provided a number of DSC thermograms with information relating time, temperature, and degradation energy. One particularly important point noted is that many exotherms were approximately symmetrical about (dH/dt) max as shown below:

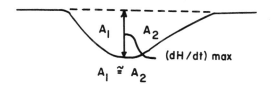

$$A_1 \cong A_2$$

These thermograms lend themselves to a simple treatment for kinetics. Assuming an overall first order process, the half-life form might be usable with these near symmetrical DSC traces.

(1) $kt_{\frac{1}{2}} = 0.693$

In the case of lube oil oxidation by DSC, k is a composite of initiation, propagation, and termination constants, which would be a similar situation for the derived activation energy value. Nevertheless, these values are useful in characterizing the oxidation of lubricating oils.

The integrated form of a first order expression is:

(2) $-\ln(1 - x) = kt$

Assuming that heat evolved "H" is proportional to moles (N) oxidized then:

(3) $x = N/N_o = H/H_o$ where H_o = total heat

also $H/H_o = A/A_o$ where A_o = total peak area

Substituting in (2) we obtain:

(4) $-\ln\dfrac{H_o - H}{H_o}$ or $-\ln\dfrac{A_o - A}{A_o} = kt$

At the condition of (dH/dt)max, H approximates 0.5 H_o, and equation (5) applies conditionally:

(5) $[-\ln(0.5) = kt_{\frac{1}{2}} = 0.693]$ dH/dt max

In equation (5.) "t" is the time interval from the onset of oxidation to (dH/dt)max, or half oxidation of the oil at one temperature. This is half of the oxidation which can occur at that temperature, which is not equivalent to oxidation of half of the total oil.

The DSC data from the oxidation of inhibited and uninhibited oils were used to calculate oxidation kinetics by equation (5). Results, given in Table 5, were then used in Arrhenius plots shown in Figure 8. Both oils appear to follow the same kinetics; however, it is evident that from the change in slope above 190°C that there is a change in the oxidation process. Slope 1 depicts an oxidation process below 190°C with an activation energy (Ea) of 26.5 k cal/g mole. Slope 2 depicts an oxidation process above 190°C with an Ea 18.3 k cal/g mole. Both of these values agree with reported range of values for oxidation of simple hydrocarbons under similar conditions[15]. These Ea values are nevertheless composites of simultaneous reactions during oxidation.

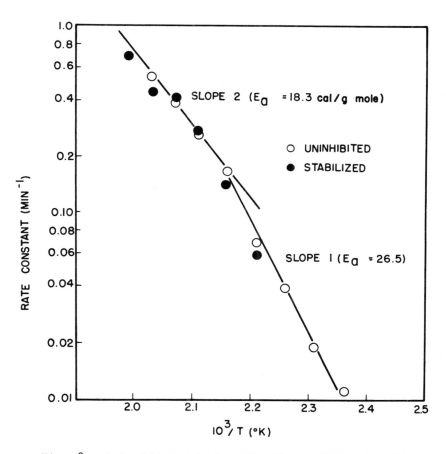

Fig. 8. Lube Oil Oxidation Kinetics - 100 psig air

TABLE 5

CALCULATION OF OXIDATION RATE CONSTANTS
obtained from isothermal DSC at 100 psig air

Sample	Oxidation Temp. °C	°K	t ½(min)	k(min^{-1})
Lube Oil	220	493	1.3	0.533
	210	483	1.8	0.385
	200	473	2.6	0.266
	190	463	4.1	
	180	453	10, 10.0	0.069
	170	443	17.6	0.039
	160	433	36.0	0.019
	150	423	63.0	0.011
Stabilized Lube Oil	230	503	1.0	0.693
	220	493	1.6	0.433
	210	483	1.7	0.408
	200	473	2.5	0.277
	190	463	5.0	0.138
	180	453	12.0	0.058

Since these oxidation kinetics appear to be the same for the two oils, the presence of the antioxidants did not affect the overall oxidation rates once the antioxidant is consumed. This is in agreement with known behaviour of some inhibited oxidation systems[15].

EFFECT OF LUBE OIL COMPOSITION

Table 6 shows the DSC degradation energies for lube oils of various boiling ranges and from two different crudes. It is apparent that for the same lube oil boiling range different crudes will provide oils of different degradation energies. For lube oils from crude A, as the boiling temperature increases (as composition changes) there is a corresponding decrease in the degradation energy. This suggests that the degradation energies at 210°C are related to a particular class of components whose concentration changes with increasing lube oil boiling temperatures. This point is supported by the results in Figure 9. For this work the lube oil was fractionated into saturates and aromatics, and the fractions were then subjected to DSC oxidations at various temperatures.

TABLE 6

EFFECT OF OIL COMPOSITION

ON DEGRADATION ENERGY

DSC 210°C, 100 psig air

Crude	Boiling Range, °C	Degradation Energy, cal/g
A	300-450	847
	350-500	650
	450-550	497
	500-600	144
B	300-450	1360
	350-500	1220

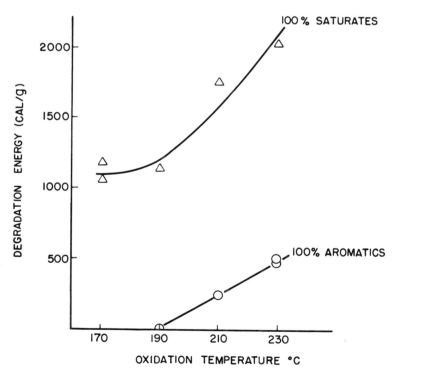

Fig. 9. Composition and Temperature Affect Degradation Energy

The results show that between 170-230°C the saturates contribute more to the degradation energy than the aromatics. Near 190°C, the aromatics start oxidizing and there is a rather abrupt change in the saturates oxidation energy. These reults support the kinetic change depicted in Figure 8. This particular point will be elaborated on in a future paper[16].

REFERENCES

(1)   A. A. Krawetz, I and EC Prod R-58 Dev, 1966, (5) 2, 191-198.

(2)   L. Vajta, et al, Acta Chim (Budapets) 1968, 58 92), 207-229.

(3)   M. Vaclav, Erdel Kohle, Erdgas Petrochem, 1968, 21 (9), 546-549.

(4)   M. Vaclav, Hutn, Listy, 1968, 23 (1), 43-46.

(5)   M. Vaclav, Wiad Naft, 1969 (5), 105-109.

(6)   F. P. Levy, et al, Thermochim Acta, 1970, 1, 429.

(7)   W. R. May, L. Bsharsh, Ind. Eng. Chem. Prod Res. Div., 1971, 10 (1), 66.

(8)   A. Commichau, Erdoel Kohle, Erdgas, Petrochem. Brennst. - Chem, 1972, 25 (6), 322-327.

(9)   F. Noel, J. Inst. Pet., 1971, 55 (558), 354-357.

(10) F. Noel, Thermochimica Acta, 1972, 4, 377-392.

(11) G. Kalmutchi et al, Lucr. Conf. Nat. Chim Anal, 3rd, 1971, 4, 317-321.

(12) A. D. Brazier, J. Inst. Pet., 1967, 53, 518.

(13) K. E. J. Barrett, J. Appl. Poly Sci. 11, 1617-1626, (1967).

(14) A. A. Duswalt P. 313, Anal. Calorimetry, Ed. Porter and Johnson, Plenum Press, 1968.

(15) Autoxidation of Hydrocarbons and Polymers, L. Reich, S. S. Stivala, Marcel Dekker, N.Y., 1969.

(16) G. E. Cranton, F. Noel, Aromatics and Base Oil Oxidation (to be published in J. Inst. Petroleum (London)).

THE DETECTION OF IMPURITIES BY THERMAL ANALYSIS

H. J. Ferrari and N. J. Passarello

Lederle Laboratories Division, American Cyanamid Co.

Pearl River, New York 10965

Thermal analysis is used in our laboratory to check material which is intended for use in clinical trials or toxicological studies. Invariably the quality of such samples will be of high purity, 98% or better. This immediately gives some idea of the test which confronts thermal analysis. Generally, we are dealing with samples containing a single impurity of <2% or several impurities of <2%. We feel that DTA has met this challenge with plenty to spare. It has earned its place as one of the techniques employed to evaluate the final purity of material intended for clinical and toxicological use.

Previous publications from this laboratory have discussed many interesting aspects of thermal analysis relative to compounds from Lederle Laboratories.[1-5] This report will discuss additional examples in which thermal analysis has continued to contribute useful analytical data toward the profile of potential pharmaceutical products.

Although most of the data we have obtained from thermal analysis is concerned with qualitative aspects, in some cases it has been possible to obtain semi-quantitative data.[2] A few additional semi-quantitative examples are included in this publication. The scope of thermal analysis has been increased by obtaining a cooling curve, a reheat cycle, a variation in the atmosphere in which the compound is heated, and the monitoring of the effluent vapors. The use of ancillary techniques such as hot-stage microscopy, vapor phase chromatography, mass spectrometry, and chemical examination in conjunction with thermal analysis has added new dimensions to thermal studies.

All temperatures have been corrected for thermocouple non-
linearity. Heating rates are 10°C/min. unless otherwise specified.

A METHYLAMINO COMPOUND

Several batches of a methylamino compound were checked by
thermal analysis and the resulting thermograms generally displayed
the usual patterns; a small exotherm @ 330°C immediately preceding
the melting decomposition endotherm around 336-342°C (Fig. 1).

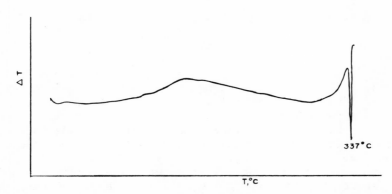

FIG. 1. A typical endotherm for a methylamino compound.
Exothermic region precedes the 337° melting-decomposition endotherm.

A batch of the methylamino compound was received for purity evaluation and the thermogram of this batch was different from previous
batches (Fig. 2). Now there is a tremendous exotherm in the 330-
340°C area and no melting decomposition endotherm.

Additional attempts at purification did not alter the picture
of the thermogram. It was decided to investigate the possibility
of the presence of either or both, the chloro compound and the dimethylamino compound. A thermogram was obtained on a sample of
each (Fig. 3,4). Since it appeared more likely that the exothermic area was the result of the dimethylamino compound, a synthetic mixture (not co-crystallized) was prepared using the dimethylamino and the chloro (3:1). A thermogram of the mixture is
shown in Fig. 5. A co-crystallized mixture of methylamino, chloro
and the dimethylamino (5:1:1) was prepared by dissolving these

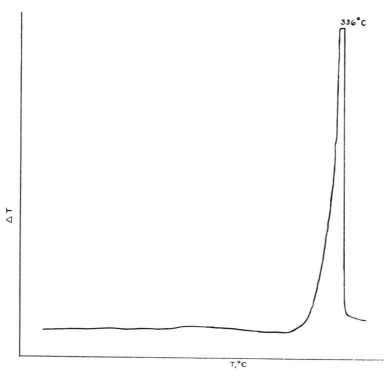

FIG. 2. A similar batch of a methylamino compound – only a 336°C exotherm is present. The melting decomposition endotherm is absent.

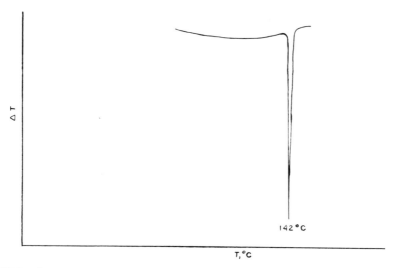

FIG. 3. Melting endotherm 142°C for the methylaminochloro compound.

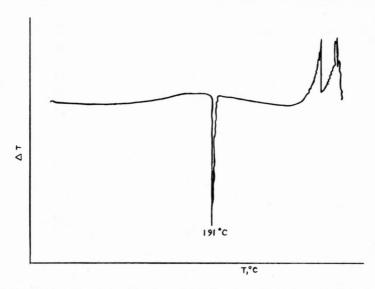

FIG. 4. Melting endotherm 191°C for the dimethylamino compound followed by decomposition at 310°C.

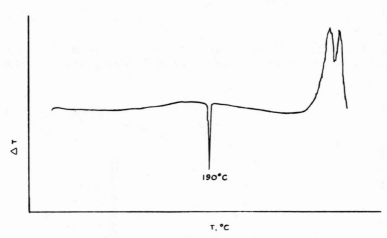

FIG. 5. Synthetic mixture of the methylamino compound, methylaminochloro and the dimethylamino.

compounds in DMF. After evaporating to dryness, a thermogram was obtained on the mixture (Fig. 6). The endotherm at 131° could be the chloro compound, the endotherm at 149° was a solvent (DMF) and the exotherm in the 310° area probably was the dimethylamino compound. Additional purification attempts to remove the

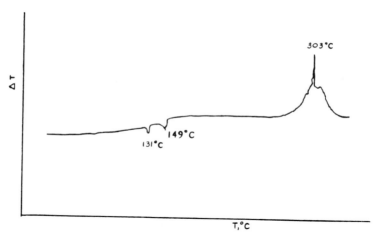

FIG. 6. Co-crystallized mixture of the methylamino compound, methylaminochloro and dimethylamino. Chloro compound 131°C endotherm; solvent endotherm 149°C, decomposition of the dimethylamino compound 303°C exotherm.

dimethylamino compound were apparently successful, judging from the disappearance of the 310-340°C exotherm and reappearance of a sharp melting-decomposition endotherm at 338°C (Fig. 7). It is also significant that there is not any indication of an exotherm prior to the melting-decomposition endotherm as had been the case in previous batches. (Fig. 1).

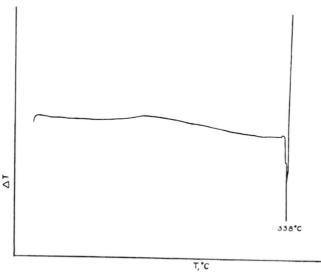

FIG. 7. Thermogram after recrystallization of the methylamino compound. No exotherm or exothermic area is present.

Shortly thereafter another batch was submitted and the story repeated itself. The thermogram once again indicated a hugh exotherm in the 300-340°C area and the absence of the expected melting-decomposition endotherm @ 338°C. Additional purification treatment resulted in the disappearance of the 338°C melting-decomposition endotherm. However an additional problem appeared with the presence of the 193°C endotherm not seen until this sample (Fig. 8). The endotherm was suspected as being the result

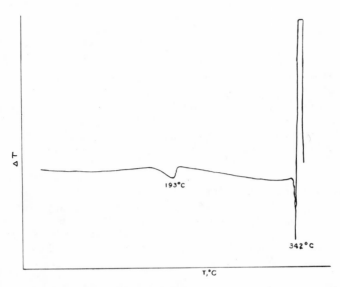

FIG. 8. Thermogram of a sample of methylamino compound recrystallized from HCl. Additional endotherm at 193°C results from loss of HCl.

of the loss of HCl since this was the final recrystallization solvent. The presence of HCl was confirmed by a chlorine analysis (6% found) and TGA weight loss of approximately 5% (Fig. 9). The effluent vapors from TGA were acid as determined by placing moistened alkacid paper in the exit port of the furnace tube. The vapors were also bubbled into an acidified solution of $AgNO_3$ resulting in the formation of a white precipitate. Thermal analysis was extremely sensitive to the presence of the dimethylamino compound (<0.5%) but much less sensitive to the presence of the chloro compound (1-2%).

AN IMIDAZOLIDINONE COMPOUND

The thermogram for a standard sample of an imidazolidinone compound has a single melting endotherm at 172°C. A batch was submitted for analysis and the resulting thermogram had an additional endotherm at 151°C (Fig. 10). At first glance this type

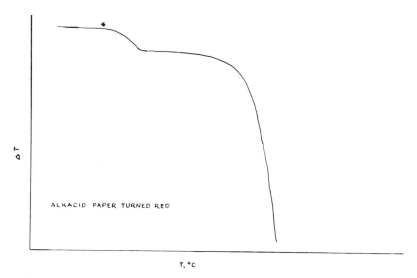

FIG. 9. Thermogram (TG) showing the weight loss (HCl) from the recrystallized sample of the methylamino compound.

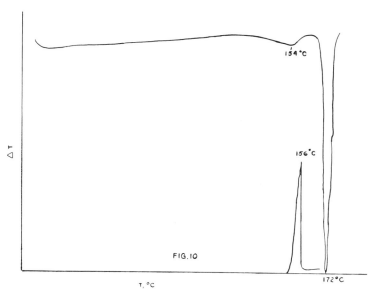

FIG. 10. Thermogram for a sample of an imidazolidinone compound. An additional endotherm is present at 154°C.

of endotherm appeared to be typical of a solvent but the thermo-gram using reduced pressure failed to confirm this. It now seemed to be the result of a possible impurity. Both TLC and HSLC indicated that two additional minor components were present. With the aid of mass spec, these two components were identified

as the rearranged compound and the hydantoin. A thermogram was
obtained for the rearranged compound (Fig. 11) and the hydantoin
(Fig. 12). The melting endotherms were 207°C and 184.5°C respectively. Generally, when the melting endotherm of the impurity
is at a higher temperature than the desired compound the chance
for detecting the impurity is slim. A thermogram was obtained on

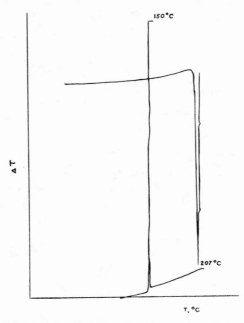

FIG. 11. Thermogram for the rearranged imidazolidinone compound melting endotherm 207°C.

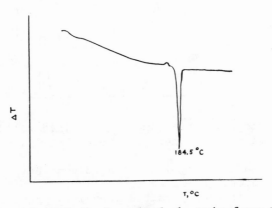

FIG. 12. Thermogram for the hydantoin from the imidazolidinone compound. Melting endotherm 184.5°C.

a mixture of the crystals of the imidazolidinone and the rearranged compound, prepared by grinding the two compounds in a mortar. A completely different pattern of endotherms is displayed in Fig. 13. Endotherms are now present at 147°, 151° and 155°C and 172°C endotherm for the imidazolidinone is absent.

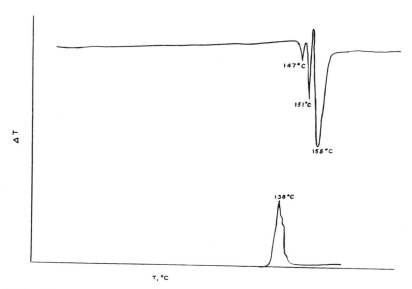

FIG. 13. Synthetic mixture: imidazolidinone compound and a rearranged compound therefrom.

A similar mixture was prepared for the imidazolidinone and the hydantoin. The thermogram (Fig. 14) shows two endotherms, one at 159°C and a second at 184°C. Obviously, the 184°C endotherm represents the melting of the hydantoin and the 173°C melting endotherm for the imidazolidinone is absent. The explanation of this thermogram was resolved by the hot-stage microscopy work of D. Grabar at the Stamford Central Research Labs of American Cyanamid Company. Grabar observed that both the imidazolidinone and the rearranged compounds have polymorphs and behave similarly when recrystallized from the melt. Mixed fusion of the imidazolidinone and the rearranged compound shows that the phase diagram is a simple eutectic between form I (imidazolidinone) and form I' (rearranged) melting at 154°C. Form II (imidazolidinone) and II' (rearranged) form a solid solution with a eutectic temperature at 150°C. A similar explanation would also follow for the mixture of the imidazolidinone and the hydantoin compounds.

BIPHENYL COMPOUND

The thermogram for a purity standard of the biphenyl compound indicates two endotherms, 148° and 164°C (Fig. 15). The

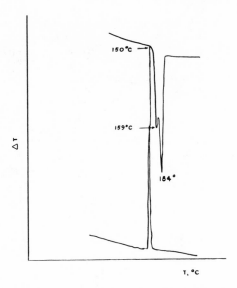

FIG. 14.   Synthetic mixture: imidazolidinone compound and
hydantoin compound.

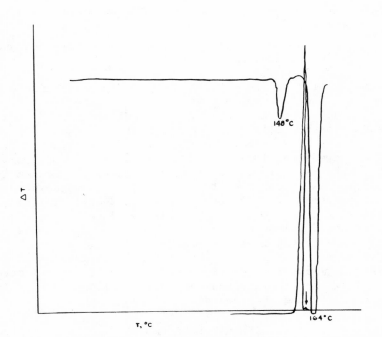

FIG. 15.   Thermograms for the purity standard of a biphenyl
compound.   Endotherms at 148° and 164°C.

first step in the absence of any other information is to rule out
the possibility of solvation by repeating the thermogram using
reduced pressure (Fig. 16). The thermogram obtained using reduced

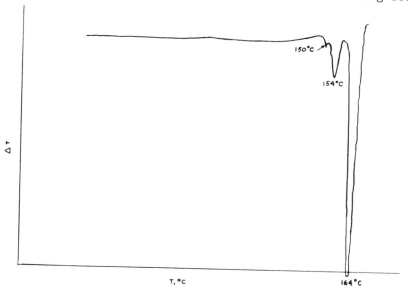

FIG. 16. Thermogram for a biphenyl compound using reduced
pressure. Three endotherms are present 150°, 154° and 164°C.

pressure appears to effect the separation of an additional endotherm at the 150°-154°C region. There are now three endotherms
with the following temperatures; 150°, 154° and 164°C. Previous
experience would suggest that the appearance of an additional
endotherm with this particular change in environment (reduced
pressure vs. air), might be the result of a nonoxidative effect.[4]
The sample was forwarded to D. Grabar since no other analytical
technique reported any significant impurity to account for this
larger 154°C endotherm (reduced pressure). Grabar observed that a
phase change occurs at about 153°C and some crystals appear to
undergo two to three transformations in rapid succession. Now
that the history of this compound is known there is no cause for
alarm when this phenomenon appears.

NYSTATIN

Various batches of Nystatin were forwarded for comparison by
thermal analysis (Fig. 17, 18, 19). The information supplied to
the interested persons would be as follows: the first endotherm
in the 110°-115°C region is the result of solvent since the endotherm shifted to a lower temperature using reduced pressure
(Fig. 20). The actual percent weight loss of the solvent was

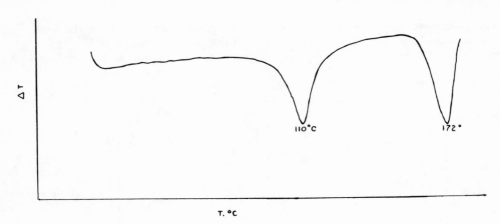

FIG. 17. Thermogram for Nystatin (#1) with two endotherms 110° (solvent) and melting decomposition at 173°C.

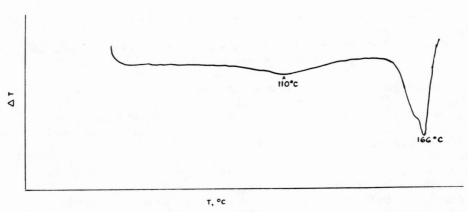

FIG. 18. Thermogram for Nystatin (#2) with an endothermic area at 110°C (solvent) melting-decomposition at 166°C.

determined by thermogravimetric analysis. Additional information for the evaluation of differences among similar samples can be obtained by comparing the following: the temperature of the melting endotherm, the onset of melting Fig. 19 (A); the area in the extrapolated onset Fig. 19 (B); the width of the endotherm Fig. 19 (C).

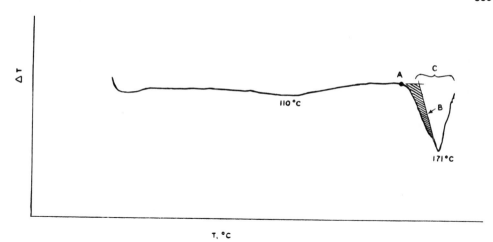

FIG. 19. Thermogram for Nystatin (#3) with an endothermic area at 110°C (solvent) and melting-decomposition at 171°C.

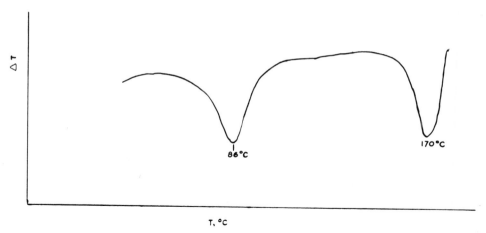

FIG. 20. Thermogram for Nystatin (#4) with an endotherm at 88°C (solvent) and a melting-decomposition endotherm at 170°C.

NITROFURANTOIN

A situation similar to Nystatin was demonstrated with various batches of nitrofurantoin. Only a few samples were selected as an example (Fig. 21, 22, 23). Once again the first endotherm (110-120°C) was explained by repeating the run using reduced pressure.

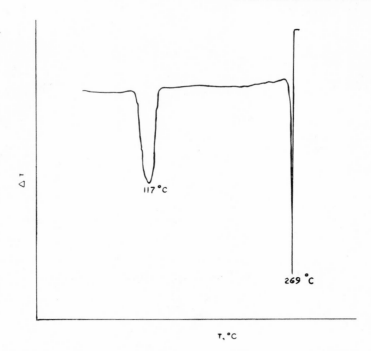

FIG. 21.   Thermogram for Nitrofurantoin (#1) with two endotherms 117°C (solvent) and melting-decomposition at 269°C.

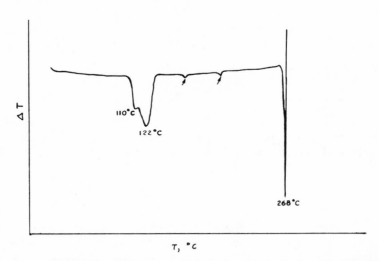

FIG. 22.   Thermogram for Nitrofurantoin (#2) with three endotherms 110°, 122°C (solvents) and melting-decomposition at 268°C.
Plus two minor impurities indicated by arrows.

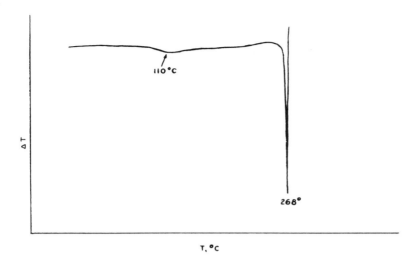

FIG. 23.   Thermogram for nitrofurantoin (#3) with an endo
thermic area at 110°C and a melting-decomposition endotherm at
268°C.

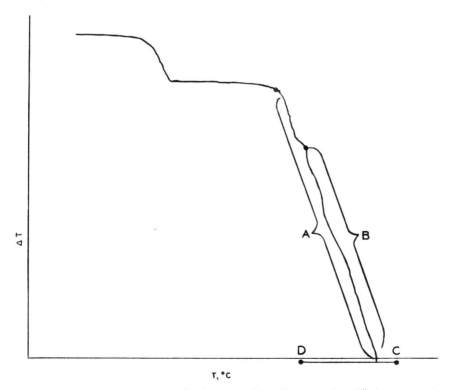

FIG. 24.   Thermogram (TG) for nitrofurantoin (#1) with the
first weight loss for solvent and the second weight loss a multistep violent decomposition.

The shifting of the endotherm to a lower temperature indicates solvation. A thermogravimetric analysis gave the actual weight loss
(Fig. 24). In this case the melting decomposition (A), including
a violent endothermic decomposition (B), with rapid spontaneous
super heating (C) (approx. 30°C), followed by a rapid super cooling (D) (approx. 40°C). This should alert those handling such a
compound to use extreme caution in handling this compound at
elevated temperatures.

STEROID COMPOUND

Thermal analysis was helpful in resolving the discrepancy
between the VPC analysis for solvents (1.2% methanol, 0.3% H_2O,
0.5% undetermined) in samples of the steroid. The values obtained
for total solvents by TGA were generally in the range of 0.3%.
The thermogram (DTA) indicates a broad endothermic region 167–193°C
emphasized by two apparent endotherms at 167° and 190°C (Fig. 25).
A thermogram obtained by TGA indicated a two step decomposition;
one at 150°–187°C and a second at 187°C to complete decomposition
(Fig. 26). It seems possible from the DTA and TGA data that degradation or pyrolysis may have contributed to the anomalous VPC
results since the injection port of the VPC was set at 250°C, the
temperature at which the steroid would be decomposed.

THIADIAZOLE COMPOUND

The usual thermograms for the thiadiazole show one melting
decomposition endotherm at 270°C (Fig. 27). It was desirous to
know if the chloro compound could be detected in the thiadiazole
by thermal analysis. A thermogram for the chloro compound is
shown in Fig. 28. A synthetic mixture of the dry powder was

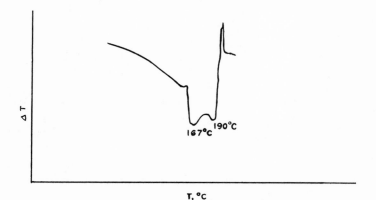

FIG. 25. Thermogram for a steroid compound with two decomposition endotherms at 167° and 190°C.

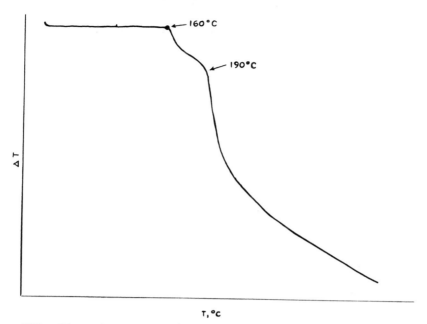

FIG. 26. Thermogram (TG) for the steroid compound in Fig. 25 with a two step decompostion at 160° and 190°C.

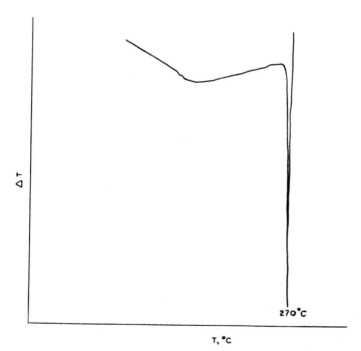

FIG. 27. Thermogram (TG) for a thiadiazole compound with a decomposition endotherm at 270°C.

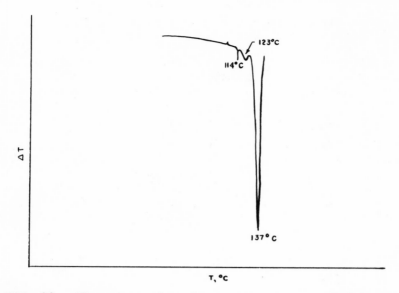

FIG, 28.   Thermogram for the chlorothiadiazole compound with
the melting endotherm at 137°C and two impurities melting at 114°
and 123°C.

prepared to contain 2% of the chloro in a mixture with the thiadiazole. The thermogram of this mixture indicated a small endotherm at 137°C for the chloro compound (Fig. 29). By increasing
the sensitivity setting of the instrument, approximately 0.5-1%
of the chloro compound can be detected by thermal analysis.

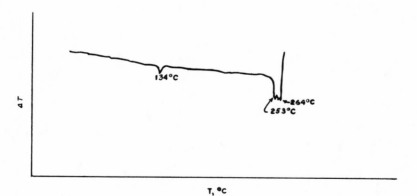

FIG. 29.   Thermogram for a mixture of 98 mg. of thiadiazole
compound, 2 mg. of chlorothiadiazole compound. A 134°C endotherm
represents the chloro and 253-254°C endotherm the final decomposition of the thiadiazole compound.

AN OXAZEPINE COMPOUND

The thermogram for this compound prepared at another location usually showed a melting endotherm at approximately 97°C (Fig. 30) vs the Lederle prep melting at 94°C (Fig. 31). Attempts to obtain a higher melting point by recrystallization were unsuccessful (Fig. 32). The additional endotherm in (Fig. 32) at 84°C is caused by residual solvent. Finally a thermogram was obtained which contained two endotherms and led to the suspicion that polymorphs may be responsible for this behavior (Fig. 33). The cooling curve was of no value in this situation since the compound did not crystallize under these conditions. The existence of polymorphs was confirmed by Grabar using hot-stage microscopy. With this information available the continued efforts toward purification were no longer necessary.

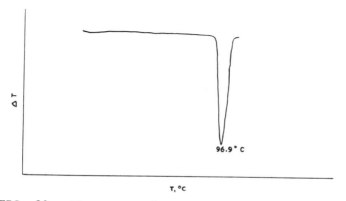

FIG. 30. Thermogram for an oxazepine compound (#16A) melting endotherm at 96.5°C.

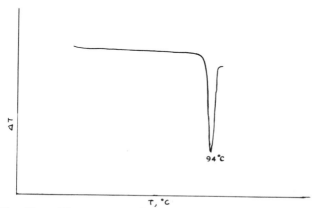

FIG. 31. Thermogram for an oxazepine compound (#1A) melting endotherm at 94°C.

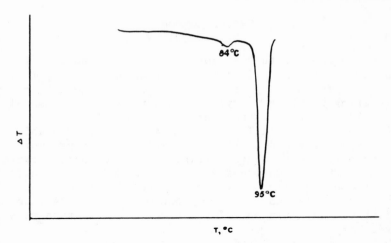

FIG. 32. Thermogram for an oxazepine compound (#8B). Two endotherms 84°C (solvent) and melting 95°C.

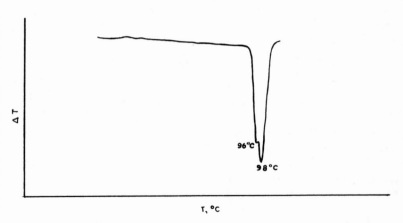

FIG. 33. Thermogram for an oxazepine compound (#68) with two endotherms 96° and 98°C.

A TRIAZOLE COMPOUND

A sample of the triazole compound (X) had a single melting endotherm at 167°C. A sample from the retention file (Y) was run for comparison and its melting endotherm was recorded at 178°C. An examination of the powder under the microscope indicated long thin needle-like crystals for sample (X) and short stubby clumps of crystals for the sample (Y) sample. The purity standard then became available and it had a final melting endotherm at 175.5°C.

(Fig. 34). Microscopically it consisted of very small oval crystalline aggregates. In addition, the thermogram indicated a
possible impurity in the 162-165°C area (Fig. 34). It was obvious
from the microscopic examination and the thermograms that we had
three separate crystalline forms with corresponding different
melting points and an impure purity standard.

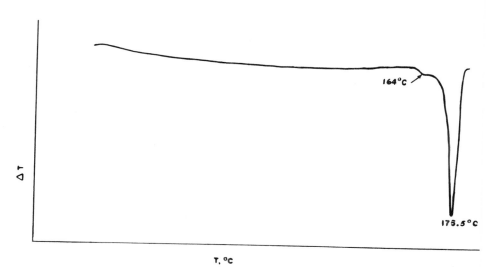

FIG. 34. Thermogram for a purity standard of a triazole with
two endotherms 164°C (impurity) and melting 175°C.

RESEARCH COMPOUND "D"

A sample of research compound DOP was received for purity
evaluation prior to the availability of a purity standard (Fig.35).
Although a purity standard was not available, it appeared as if
we were dealing with at least two impurities (A and B) that are
visible in the thermogram (Fig. 35). Other indications of the
presence of impurities were the broadness of the endotherm and the
onset of melting which is 19°C lower than the melting endotherm.
The less difference between the onset temperature and the melting
endotherm the purer the compound. Analysis by vpc, nmr, and ms
indicated considerable amounts of the trans isomer and OP compound as contaminants.

When the purity standard became available thermograms were
obtained for the purity standard (Fig. 36) and OP compound (Fig.
37). A mixture was prepared to contain 5% OP compound and 95%

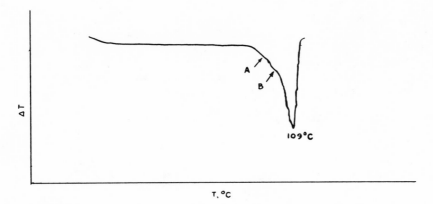

FIG. 35. Thermogram for research compound "D" with a broad
endotherm at 109°C. Indications of at least two impurities by
arrows.

research compound DOP. The resulting thermogram of the mixture
showed a 4°C lowering of the melting endotherm, and a broadening
of the endotherm (Fig. 38). A sample rich in trans isomer could
not be handled by DTA because of its semi-solid consistency. A
sample of this type would have to be handled by DSC which utilizes
sample pans rather than capillary tubes used for DTA.

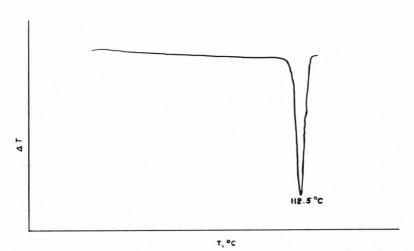

FIG. 36. Thermogram for a purity standard of research compound "D" with one endotherm at 112.5°C.

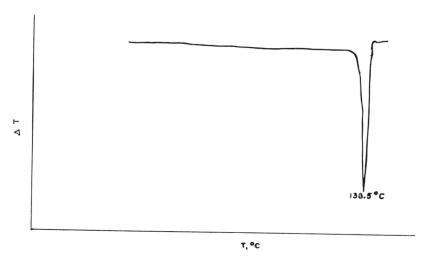

FIG. 37. Thermogram for the oxy compound of research compound "D" with one melting endotherm at 138.5°C.

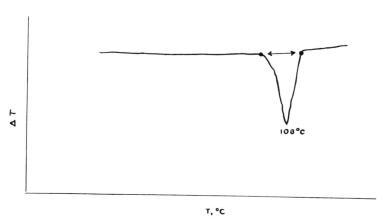

FIG. 38. Thermogram for a synthetic mixture of the purity standard for research compound "D" and its oxy derivative. A broad endotherm is shown and a 4°C lowering of the melting endotherm.

AMINOBENZOIC ACID COMPOUND

When one observes a thermogram (Fig. 39) containing six endotherms for an organic compound, one should forward the sample for

hot-stage microscopy studies. It is obvious that there will be a
possibility of solid solutions, eutectics and polymorphs. The six
endotherms mentioned above were explained by Grabar's hot-stage
microscopy as follows: the sample exists as a mixture of at least
two polymorphs. Transformations were noted at 56°, 90° and 108°C
(to liquid crystals) and melting at 129°C, the other two endotherms
are undoubtedly metastable forms.

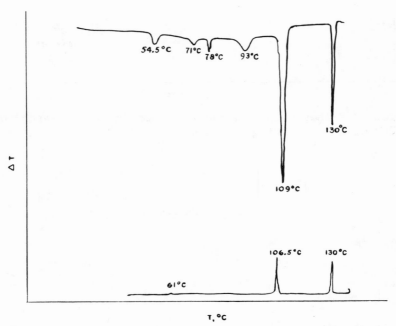

FIG. 39. Thermogram of an aminobenzoic acid compound.
Melting endotherm at 130°C and an assortment of eutectic, solid-
solid, liquid crystals. Phase changes at 54.5°, 71°, 78°, 93°
and 109°C.

AN IMIDAZO COMPOUND

The thermogram in Fig. 40 had eight recrystallization exotherms when first run; the most prominent exotherm was at 187°C.
It was assumed that these exotherms were the result of polymorphs. A year later when the same sample was run a completely
different cooling curve was observed (Fig. 41). Apparently during
the period of storage all of the metastable polymorphs had been
converted to the stable form recrystallizing at 193°C. Storage
conditions such as moisture and temperature are factors which can
induce such changes.

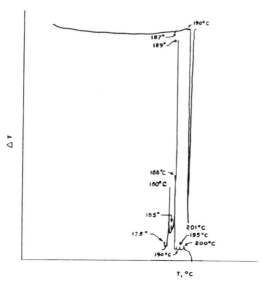

FIG. 40. Thermogram for an imidazo compound with a melting endotherm at 201°C and extremely small indications of phase changes at 187° and 190°C. The cooling curve shows exotherms at 200°, 195°, 190°, 188°, 185°, 180° and 178°C.

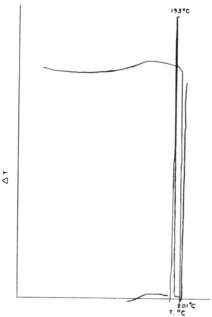

FIG. 41. Thermogram for the imidazo compound in Fig. 40 one year later. Indications of only one recrystallization exotherm at 198°C (stable polymorph).

ANTIBIOTIC "A"

In the preliminary work on Antibiotic "A" a discrepancy
arose between K.F. and TGA methods for the determination of the
degree of solvation. Antibiotic "A" was known to rapidly absorb
moisture on standing. It was decided to equilibrate all samples
of Antibiotic "A" under standardized conditions prior to analysis
in order to eliminate errors due to varying amounts of solvation
at different times of analysis. An analysis of the equilibrated
Antibiotic "A" for water by the Karl Fisher method resulted in a
value of 10.2% and the TGA (wt. loss) was 12.5%. This discrepancy led to further investigation. Eventually nmr showed that
both acetone and methanol were present in addition to water.

MINOCYCLINE

The first TGA attempts at trying to resolve whether minocycline samples were mono or dihydrates were unsuccessful. The
usual TGA operating conditions using a heating rate of 10°C/min.
and a nitrogen purge resulted in weight loss of 3-4%. This would
indicate that all of the samples were monohydrates. More vigorous
conditions were decided upon since it was known that minocycline
bound water very strongly. With a slower heating rate of 1°C/min.
and reduced pressure, 0.2 mm Hg, a monohydrate could be distinguished from a dihydrate. Fig. 42 shows the thermogram for mino-
cycline·HCl using the normal operating conditions of 10°C/min.
heating rate and a nitrogen purge vs Fig. 43 with the modified
conditions of 1°C/min. heating rate and reduced pressure. The
calculated value for the monohydrate is 3.3% H_2O and 6.36% H_2O
for the dihydrate. The data shows that we can now obtain satisfactory values for the dihydrate with a modification of the normal
operating conditions.

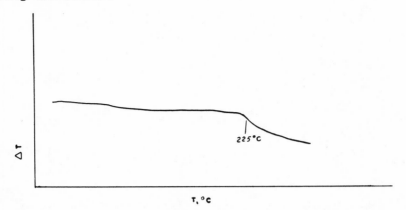

FIG. 42. Thermogram for minocycline using a purge of nitrogen and a heating rate of 10°C/min. (3-4% wt. loss).

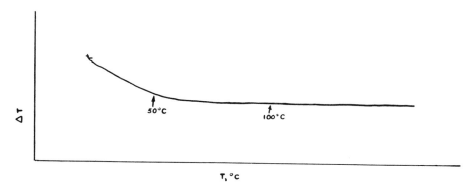

FIG. 43.   Thermogram for the same minocycline sample in Fig. 42 using reduced pressure (0.2 mm Hg) and a heating rate of 1°C/min. (6.4% wt. loss).

TGA has also been useful in monitoring weight losses with time at various fixed temperature ranges. It provides information relative to the appropriate temperatures required to remove solvents and also to serve as a warning for compounds that might be unsafe because of their thermal instability.

ACKNOWLEDGEMENTS

We wish to thank D. Grabar of the Central Research Laboratories, American Cyanamid Company, Stamford, Connecticut for all the helpful hot-stage microscopy information.

REFERENCES

1. Brancone, L. M., Ferrari, H. J., Dept. 912, Vol. III, p. 246-303 (1965).

2. Brancone, L. M., Ferrari, H. J., Microchemical J., 10, No. 1-4, p. 370-392 (1966).

3. Ferrari, H. J., Grabar, G. D., Microchemical J., 16, No. 1, p. 5-13 (1971).

4. Ferrari, H. J., in "Thermal Analysis" (R. F. Schwenker, Jr. and Paul D. Garn) Academic Press, New York, Vol. 1, p. 41-64 (1969).

5. Ferrari, H. J., Inoue, M., in "Differential Thermal Analysis" (R. C. MacKenzie) Academic Press, London, Vol. 2, p. 453-472 (1972).

THERMOMECHANICAL ANALYSIS OF DENTAL WAXES IN THE PENETRATION MODE

John M. Powers and Robert G. Craig

School of Dentistry, The University of Michigan

Ann Arbor, Michigan 48104

Differential thermal analysis (DTA) of commercial and dental waxes has established the presence of solid-solid as well as melting transitions[1]. The phase changes associated with these transition temperatures can be expected to influence the mechanical behavior of wax. Flow under load is an important property of waxes used in dentistry and is known to vary with time and temperature[2,3]; however, the method used in specification testing[4] for measuring flow by loading a specimen at a constant temperature for a given interval of time does not directly indicate the relationship between flow and the phase(s) present in the wax.

In the penetration mode of thermomechanical analysis (TMA), a force can be placed on a sample, and penetration can be recorded continuously as a function of time or temperature. Comparison of TMA with DTA curves can be made to determine the effect of thermal transitions on the mechanical behavior of a material. This study examines the relationship between penetration and solid-solid and melting transitions of dental waxes and their components and serves to provide fundamental information on the formulation and testing of dental waxes.

MATERIALS AND METHODS

The waxes were subjected to thermomechanical analysis in the penetration mode as obtained from the manufacturer (see Table 1) or blended as indicated. Cylindrical wafers (6.0 mm in diameter and 1.0 mm in height) of a wax were made by allowing molten wax to solidify in a warmed stainless steel die. The temperature of the

349

Table 1. Description, Lot Number and Annealing Temperature of
          Dental and Commercial Waxes

Wax	Description	Lot Number	Annealing Temperature
Acrawax C Powdered[*]	synthetic wax	B-1685	--
Barnsdahl[+]	hydrocarbon wax	--	--
Baseplate wax[†]	pink no. 7	7407010	--
Beeswax[§]	refined yellow USP	2247	40°C
Candellila[+]	ester wax	--	--
Carnauba[§]	refined pure #1 yellow	B-2318	60°C
Ceresin[+]	hydrocarbon wax	--	--
Inlay casting wax[+]	blue-hard	0824A865	50°C
Litene[+]	hydrocarbon wax	--	--
Montan[§]	domestic	--	--
Paraffin[§]	fully refined	319	40°C
Paraffin-Beeswax blends[§]	(2.5-75% Beeswax)	--	40°C
Paraffin-Carnauba blends[§]	(2.5-25% Carnauba)	--	40°C
Paraffin-Carnauba blends[§]	(50-75% Carnauba)	--	60°C
Polyethylene wax[∞]	synthetic wax	--	--
Sticky wax[†]	Peeso wax	0746801	--

[*] Glyco Chemicals, Chas. L. Huisking & Co., Inc. Williamsport, Pa.
[+] Kerr Sybron Corporation, Romulus, Michigan 48174.
[†] S.S. White Company, Philadelphia, Pa.
[§] Frank B. Ross Co., Inc., Jersey City, New Jersey.
[∞] International Wax Refining Company, Valley Stream, New York.

die was that chosen as the annealing temperature for a particular
wax or wax blend.

Some waxes were studied in annealed and unannealed conditions
to test the effect of heat-treatment on penetration. Annealed
specimens were prepared in a manner similar to the unannealed
specimens; however, the annealed specimens were heated in an oven
for 24 hours at a temperature 20°C below the melting temperature
of the wax prior to testing. The temperatures at which the
various waxes were annealed are given in Table 1.

Thermomechanical analysis[*] was carried out on three specimens

[*] Du Pont 941 Thermomechanical Analyzer, E.I. du Pont de Nemours &
    Co., (Inc.), Instrument Products Division, Wilmington, Del. 19898

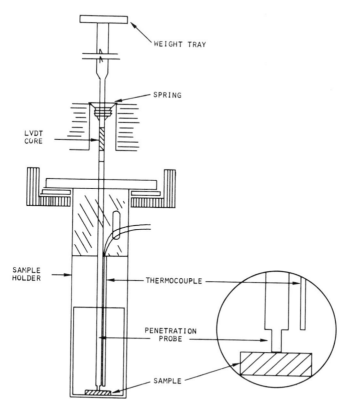

Figure 1.   Schematic Diagram of TMA with Penetration Probe

for each condition from 25°C to the temperature at which maximum
(100 percent) penetration occurred for the various waxes. The
analysis was conducted in air, and a linear heating rate of 5 °C
per minute was used. Differential thermal analysis[+] was carried
out from 25°C to melting in air using a heating rate of 20°C per
minute.

A schematic sketch of the penetration mode of the thermomechanical analysis cell is shown in Figure 1. A quartz probe,
which is attached to a metallic sleeve that serves as the core of

+ Du Pont 900 Differential Thermal Analyzer, E.I. du Pont de
  Nemours & Co., (Inc.), Instrument Products Division, Wilmington,
  Del. 19898

a LVDT, was mechanically positioned by a spring on the surface of a wax sample. The output of the LVDT was then electrically balanced and zeroed such that subsequent vertical displacement could be calibrated and recorded continuously as a function temperature. The cylindrical probe tip shown in the enlarged portion of Figure 1 has a diameter of 0.92 mm and a length of 0.72 mm. One hundred percent penetration thus refers to a vertical displacement of 0.72 mm. Weights were placed on the weight tray to allow stresses of 1.52 and 25.5 g/mm^2 to be applied to the wax. All TMA curves began at 25°C with zero percent penetration. When several curves were plotted on the same figure, the curves were sometimes offset for convenience.

Statistical analysis of the data was performed using analysis of variance[5]. A hypothesis testing the equivalence of means was rejected if the computed value of α was less than or equal to 0.05.

RESULTS

The relationship between DTA and TMA in the penetration mode for unannealed hydrocarbon waxes is shown in Figure 2. At the higher stress level (25.5 g/mm^2), penetration was dominated by the influence of the premelting transitions of paraffin, litene and ceresin waxes. Penetration of barnsdahl at this stress level reflected its broad melting range beginning at 40°C. At the lower stress level (1.52 g/mm^2), major penetration was associated with the melting peak of the wax; although the initial penetration of paraffin (M.P. 52°C) that was associated with the solid-solid transition occurring from 36 to 37°C accounted for approximately 56 percent of the total penetration.

DTA and TMA curves of four ester waxes commonly used in the formulation of dental waxes are compared in Figure 3. Penetration at both stress levels was related to the major melting transition for montan, candellila and carnauba waxes. The difference in temperature between 100 percent penetration at each stress level was 7, 5 and 2°C, respectively. For the hydrocarbon waxes, this value was between 9 and 17°C for the waxes studied. Penetration of the ester waxes was affected to a much lesser extent by the presence of solid-solid transitions. The penetration of beeswax, however, was dominated at the higher stress level by the melting of its hydrocarbon component.

DTA and TMA curves of two synthetic waxes are compared in Figure 4. At the lower stress level, penetration occurred between 75 and 80°C for the polyethylene wax and between 142 and 145°C for the Acrawax C. Transitions occurring at lower temperatures had

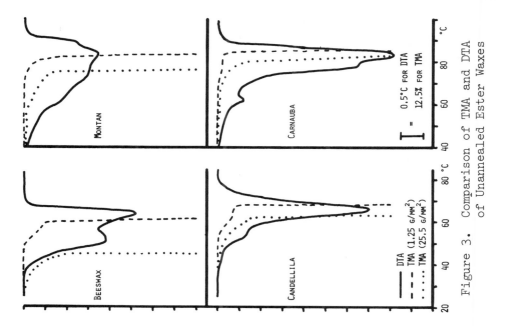

Figure 3. Comparison of TMA and DTA of Unannealed Ester Waxes

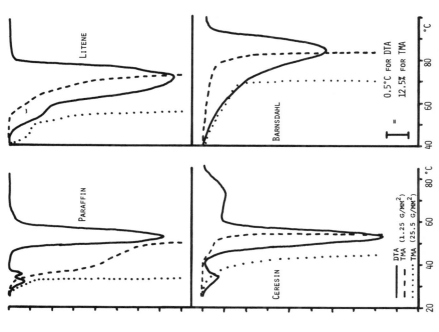

Figure 2. Comparison of TMA and DTA Curves of Unannealed Hydrocarbon Waxes

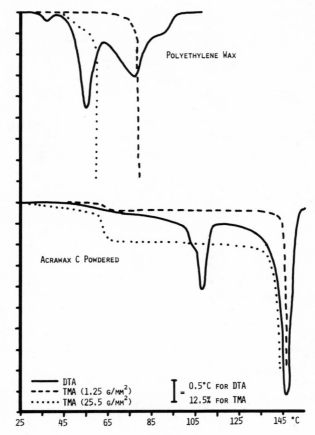

Figure 4.   Comparison of TMA and DTA Curves of Synthetic Waxes

little effect on the penetration of either wax. At the higher
stress level, penetration was complete at 58°C for the polyethylene wax. Approximately, 25 percent penetration occurred
between 55 and 65°C for the Acrawax C; however, further penetration did not occur until a temperature of about 135°C was reached.
The transition occurring at 105°C did not appear to influence the
penetration of Acrawax C.

The relationship between DTA and TMA for annealed and unannealed paraffin is shown in Figure 5. The curve for annealed
paraffin at the lower stress level had an initial penetration
that was associated with the endothermic transition observed for
paraffin from 33 to 37°C. This initial penetration accounted for
approximately 45 percent of the total penetration. The final
penetration was associated with melting transformation. The curve

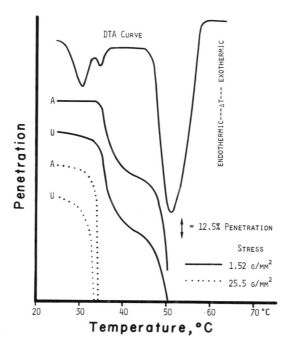

Figure 5.   Relationship Between TMA and DTA Curves for Annealed
            (A) and Unannealed (U) Paraffin Wax

for annealed paraffin at the higher stress level was associated
entirely with initial endothermic transitions of paraffin. The
mean penetration temperatures for paraffin reported in Table 2
were significantly different at the 0.05 level of α when compared
between stress levels and between annealed and unannealed conditions at the lower stress level.

    In order to examine a possible explanation for the
differences observed between the annealed and unanneal conditions
of paraffin, a limited study was made of the influence of annealing on the magnitude of the endothermic transitions observed
from 26 to 37°C for paraffin. The results of this DTA study are
shown in Figure 6. Trial 1 is the DTA curve of an unannealed
sample of paraffin run from 25 to 40°C. Trial 2 is the DTA curve
of the same sample of paraffin that was tested immediately after
cooling from trial 1. Trial 3 is the DTA curve of the sample of
paraffin from trial 2 that was annealed at 40°C for 24 hours prior
to testing. The values of ΔT observed for the endothermic

Table 2.   Penetration Temperatures for Annealed and Unannealed
           Paraffin at Two Stress Levels

Stress	Treatment	Penetration		
		10%	50%	90%
1.52 g/mm²	Unannealed	34.2 (0.2)	38.4 (0.9)	49.1 (0.3)
	Annealed	35.0 (0.5)	46.8 (0.3)	50.0 (0.2)
25.5 g/mm²	Unannealed	29.1 (1.7)	33.8 (0.3)	34.4 (0.4)
	Annealed	33.1 (0.3)	34.2 (0.5)	34.4 (0.5)

* Mean (Standard Deviation) with Sample Size of 3 - Units in °C

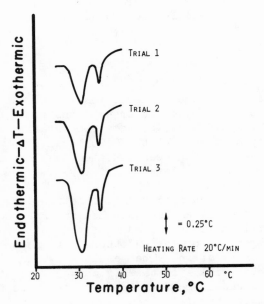

Figure 6.   Influence of Annealing on ΔT of Solid-Solid Tran-
            sitions of Paraffin Wax

transition with a maximum at 31°C were 4.5, 6.0 and 8.5°C for trial 1, 2, and 3, respectively. The values of ΔT observed for the endothermic transition with a maximum at 34.5°C were 2.0, 2.25, and 2.5°C, respectively.

In Figure 7 DTA and TMA curves are compared for four binary mixtures of paraffin with carnauba wax. Temperatures at which ten percent penetration occurred for binary mixtures of paraffin with carnauba or beeswax are plotted versus percent paraffin in Figure 8. As the carnauba content increased, the premelting and melting transitions of paraffin were observed to have a rapidly decreasing effect on the resistance to penetration of the mixture. Beeswax, on the other hand, had only a small effect on resistance to penetration. For both mixtures, the resistance to penetration

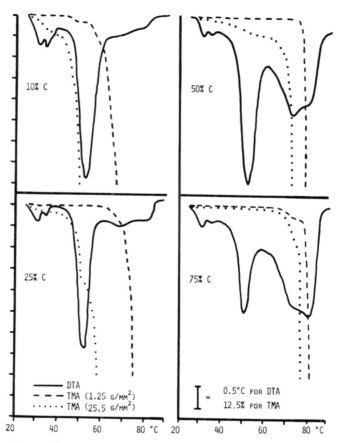

Figure 7. Comparison of TMA and DTA Curves of Binary Mixtures of Paraffin and Carnauba (C) Waxes

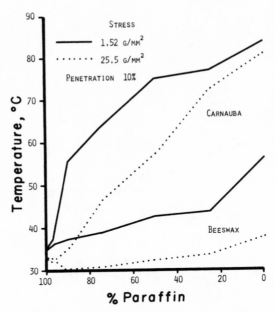

Figure 8.   Penetration Temperature Versus Percent Paraffin for
Binary Mixtures with Beeswax or Carnauba

was improved to a greater extent at the lower stress level. At
the higher stress level for the binary mixture with less than 75
percent beeswax, a small decrease in the resistance to penetration
was observed.

The relationship between DTA and TMA curves for annealed and
unannealed dental inlay wax is shown in Figure 9. The curves at
the lower stress level had an initial penetration that was associated with the onset of the major endothermic transition observed
for the wax from 50 to 68°C. The initial penetration was
approximately 10 percent of the total penetration for the annealed
wax and 21 percent for the unannealed wax. The final penetration
was associated with completion of this transition and the onset
of the endothermic transition observed from 68 to 85°C. The
curves at the higher stress level were associated entirely with
the premelting transition observed from 35 to 50°C. The annealed
wax was observed to be more resistant to penetration than the
unannealed wax at both stress levels.

A comparison of DTA and TMA curves for a dental base plate

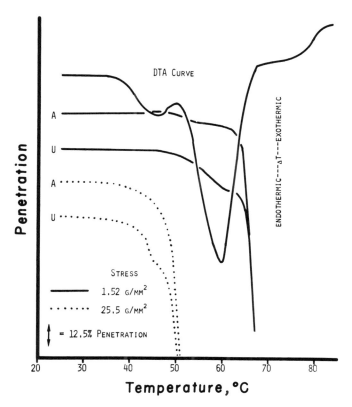

Figure 9. Relationship Between TMA and DTA for Annealed (A) and
           Unannealed (U) Dental Inlay Wax

wax and a dental sticky wax is shown in Figure 10. For both
waxes penetration at the lower stress level was associated with
the higher temperature endotherm, whereas at the higher stress
level penetration was associated with the lower temperature
endotherm.

DISCUSSION

The resistance to penetration of waxes is closely related to
the presence of solid-solid and melting transformations, to the
temperature range of these transitions and to the extent to which
solid-solid transitions occur. Experimentally, penetration is
dependent upon factors such as time, temperature, load, and rate
of heating. The latter factor is especially critical for materials that are poor thermal conductors (such as waxes).

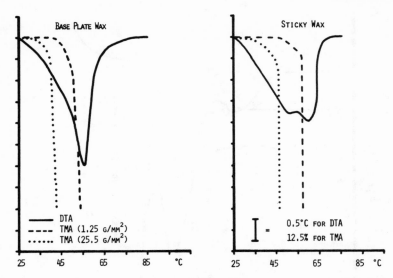

Figure 10. Comparison of TMA and DTA Curves for Dental Baseplate
and Sticky Waxes

The hydrocarbon waxes used in the formulation of dental waxes
exhibit poor resistance to penetration at temperatures corresponding to solid-solid phase transitions, particularly at higher
stresses. The melting temperature of the wax by itself has little
effect on penetration except at lower stresses and, therefore,
may be a poor criterion is the selection of a hydrocarbon wax
that is resistant to flow. Melting temperature is important,
however, in the selection of hard ester waxes (montan, candellila
and carnauba) used in the formulation of dental waxes. The
presence of solid-solid transitions or melting transitions of
lower melting components has a comparatively small effect on the
resistance of these ester waxes to penetration. Beeswax, however,
which has a large hydrocarbon component[1], shows a considerable
difference in resistance to penetration at each stress level. At
the lower stress level, penetration was dominated by the melting
of the ester component, whereas at the higher stress level, penetration was dominated by melting of the hydrocarbon component.

Annealing paraffin at 40°C for 24 hours was shown to increase
the resistance to penetration. A corresponding increase in the
magnitude of the heat of transition of primarily the lower
temperature solid-solid phase transformation was observed as a
result of this heat-treatment. It would appear that the heat-
treatment of paraffin increases the crystallinity of the lattice

primarily with respect to the odd-numbered paraffin hydrocarbons and that this increased crystallinity offers more resistance to penetration.

Carnauba and beeswax are frequently added to paraffin wax in dental compounding to alter the properties of the paraffin. The pronounced effects of small additions of carnauba on the increased resistance to penetration of paraffin in binary mixtures is consistent with flow data[3]; moreover, the importance of the solid-solid transition on the resistance to penetration of a binary wax blend has been shown to decrease noticeably as increasing amounts of a high melting ester wax are added. Addition of beeswax, however, does not produce hardening; rather, small additions of beeswax reduce the resistance of paraffin to penetration probably by interferring with crystallinity of the paraffin.

Dental inlay waxes are mixtures of hydrocarbon and ester waxes[1,6]. The effect of the high-melting ester component on the resistance to penetration of a dental inlay wax at low stress levels is to increase the temperature at which the major penetration occurs beyond the melting transformation of the hydrocarbon component. At high stress levels, however, the temperature of the solid-solid transformation associated with the hydrocarbon is the determining factor of the resistance to penetration. With few exceptions these temperatures occur above mouth temperature (37°C)[7] at which high resistance to flow is desirable. Formulation of waxes where resistance to flow is important should be based more on the temperature at which the solid-solid phase transformation occurs than on the melting temperature of the hydrocarbon wax.

Annealing a dental inlay wax at 50°C for 24 hours improves its resistance to penetration. Because of this improvement, specification testing of an annealed wax for recommended properties may be misleading if the conclusions are applied to the wax used clinically in the unannealed condition.

REFERENCES

1. R.G. Craig, J.M. Powers and F.A. Peyton, J. Dent. Res., 46, 1090 (1967).
2. R.G. Craig, J.D. Eick and F.A. Peyton, J. Dent. Res., 44, 1308 (1965).
3. R.G. Craig, J.D. Eick and F.A. Peyton, J. Dent. Res., 45, 397 (1966).
4. Guide to Dental Materials and Devices, American Dental Association, Chicago, Illinois, 1972-73.

5. Midas-Mean Test.   Statistical Research Laboratory, The
    University of Michigan, Ann Arbor, 1972.
6. R.G. Craig, J.M. Powers and F.A. Peyton, J. Dent. Res., <u>50</u>,
    450 (1971).
7. J.M. Powers and R.G. Craig, J. Dent. Res., scheduled for
    publication in March-April, 1974.

HOT-STAGE ELECTRON MICROSCOPY OF CLAY MINERALS

Donald L. Jernigan[1] and James L. McAtee, Jr.[2]

1. Mary Hardin Baylor College, Belton, Texas   76513

2. Baylor University, Waco, Texas   76703

INTRODUCTION

Clay minerals have been studied[1-12] by a variety of techniques after heat treatment in order to gain information about their thermal transformations. Various clay minerals have been fired to temperatures of interest and then examined by X-Ray diffraction techniques or by electron microscopy. More recently, clay minerals have been examined under continuous heating conditions using heating-oscillating X-Ray diffraction methods.[13-17] However, the examination of clay minerals under continuous heating conditions in the electron microscope (hot-stage electron microscopy) has been very limited[18,19] and subject to difficulties in interpretation of results.[20,21] In the study presented here, two clay minerals and their structural prototypes have been investigated by hot-stage electron microscopy to gain new information about their thermal transformations.

Although the focus of attention in this study has been on a well crystallized kaolinite sample; sodium montmorillonite, mica, talc, and a 1:1 mixture (by weight) of sodium montmorillonite and hydrous chromium oxide have been studied. The thermal behavior of kaolinite has been described in detail by Grim.[22] Considerable disagreement exists among previous studies as to the behavior of kaolinite above its DTA dehydroxylation temperature of 600°C and in the region of its DTA exotherm at 950°C. Brindley and Nakahira[7,8] studied the nature of kaolinite heated above its dehydroxylation temperature and termed the resulting structural type metakaolin. According to Brindley and Nakahira, the essential structural reorganization taking place upon formation of metakaolin from

363

kaolinite is the change in the Al-O network with aluminum in tetrahedral coordination rather than octahedral. In their proposed structure the tetrahedra share corners or edges. Brindley and Nakahira stated that metakaolin was disordered along the c-axis although other workers have not agreed with this conclusion.

Taylor[12]interpreted the thermal behavior of kaolinite as shown by DTA techniques differently than did Brindley and Nakahira. According to Taylor, the 600°C peak is the result of non-homogeneous loss of water from selected sites known as donor regions which not only lose water but also lose aluminum ions due to migration. The cations migrate into acceptor regions and take up tetrahedral coordination while the oxygen network remains unchanged in acceptor regions according to Taylor's mechanism. The kaolinite DTA exotherm at 950°C was thought to be the result of a reorganization in oxygen packing to give a spinel phase which later transformed into mullite.

Differences have been reported among workers as to the nature of the new phases formed upon structural collapse of the metakaolin normally associated with the region from 900-1000°C in DTA work. The formation of γ-alumina[11]or mullite[2-4]alone have both been postulated in the region of this exotherm. Still other investigators[1,6]have reported the formation of both mullite andγ-alumina in kaolinite samples heated to 1000°C and Grim[22]states that the formation of either or both of these phases could be associated with the exotherm at 950°C. This possibility has been supported by investigations which have shown that the products formed upon structural collapse of metakaolin depend upon the crystallinity of the starting material and upon experimental conditions. Both Glass[6]and Wahl and Grim[15]found that there was a difference between a well crystallized and poorly crystallized kaolinite as to the nature of the product formed. Wahl and Grim also found that the presence of certain cations mixed with the kaolinite sample could strongly enhance or inhibit the formation of mullite.

Brindley and Nakahira[7]analyzed heated samples of kaolinite by X-Ray diffraction methods and concluded that at 950°C metakaolin converts toγ-alumina having a spinel-type structure and containing some silicon. Using their proposed structure for metakaolin, they showed that only a simple structural change was involved in this transformation. At 1050°C, they proposed that theγ-alumina was converted to mullite. They also proposed that there was continual and progressive elimination of silica from the crystals as metakaolin was transformed, and that the amorphous silica was converted into cristobalite above 1100°C. Comer[9,10]used electron diffraction methods and came to the same conclusions as Brindley and Nakahira.

Wahl and Grim[15]have pointed out that one difficulty in correlating X-Ray diffraction analysis (particularly continuous heating studies) with DTA work is the poor crystallinity of the products formed at 950°C. These workers found that mullite formation indicated by the DTA exotherm was not indicated on the diffraction tracing. Mullite lines could be found at 950°C only on powder camera film after long exposure. Mullite lines could not be registered on the diffraction tracing until approximately 1050°C.

Kaolinite samples heated to elevated temperatures have been examined by electron microscopy,[3,4,9-11]but none by the continuous heating technique within the microscope. Furlong[18]carried out heating studies by hot-stage electron microscopy on montmorillonite and illite samples and observed the formation of "liquid spots" within the sample at temperatures in the vicinity of 600°C which he attributed to amorphous silica expelled from the clay mineral. He used molybdenum grids with carbon support films to carry his samples. It has been shown[20,21]that molybdenum and other metals react with the carbon film under such circumstances to form carbide particles which are revealed on the carbon film. The possibility of this occurring would make interpretation of results difficult when carbon is used as the support film. Investigations where previously heated kaolinite samples have been examined by electron microscopy have given results consistent with other studies in that they have shown the formation of new phases in fired samples near the 950°C DTA exotherm. As before, both mullite[3,4]andα - alumina[11]have been reported to be formed. In two cases[11,24]the appearance of a new phase as either small particles or as a grain structure has been observed in samples fired to only 800°C.

Recently, Range et al[23]have reported new findings for the thermal behavior of kaolinite. These workers proposed that meta- kaolin cannot be represented by a single structural type but actually consists of several forms they label I, II, and III. These forms may occur together as a mixture. In one investigation, Range and co-workers[23]analyzed heating studies on singly kaolinite particles by X-Ray diffraction methods and reported that a phase change occurred at 385°C without dehydration. They also gave evidence which indicated total structural collapse of kaolinite at 600°C upon prolonged heating. Upon heating, they found that the 001 reflection and the (hkl) reflections disappeared first and then the (hk0) reflections disappeared in the 400-600°C region. After heating kaolinite for several days at 600°C, the particles were amorphous to x-rays. Previously, it had been assumed that complete structural collapse did not occur until the DTA exotherm at approximately 950°C, but in previous studies the heating conditions were considerably different from those used by Range and co-workers. Evidence will be given here which strongly suggests that further structural reorganization is in progress immediately following the dehydroxylation at 600°C, and that metakaolin is unstable upon prolonged heating in the vicinity of this temperature.

EXPERIMENTAL

All heating studies were performed on an Hitachi HU-11A
electron microscope equipped with a special specimen stage serving
as a microfurnace. The samples could be heated to elevated
temperatures with this device and observed continuously at any
temperature of interest up to 1000°C. The heating procedure was
identical to that described previously[20],[21] where photomicrographs
were recorded only after the sample had been allowed to remain
at any temperature of interest for a minimum time of thirty minutes.
The temperature was monitored using a Pt-Pt.Rh(13%) thermocouple
with the temperatures being certain to ± 10°C as found previously.

The samples were dispersed in water using a wrist-type shaker
for the sodium bentonite, kaolinite, and talc. The mica was dispersed in water ultrasonically. One or two drops of these dispersions
were placed on a nickel microscope grid containing a SiO supporting
film and the samples were allowed to air dry. Silicon monoxide
was used as the support film to prevent complex transition metal
carbides from being formed when carbon support films were used.

After preliminary heating studies, the focus of attention was
on changes in the samples near to and above the dehydroxylation
temperature of the clay minerals, i.e. in the vicinity of 600°C.
Consequently, the normal heating procedure was to heat the sample
from room temperature to 600°C in approximately one hour and hold
the sample at that point for a minimum time of thirty minutes or
longer. After observation, the sample was then heated to progressively higher temperatures as desired. Under no circumstances,
were any photomicrographs recorded until after a minimum of thirty
minutes equilibration time, although in most cases the samples
were studied for periods much longer than this at the temperature
of interest. Prior to commencement of heating, a particular area
of interest on a grid was chosen and electron micrographs and
selected area electron diffraction patterns of this area were made.
After equilibration at the temperature of interest, a re-examination
of the chosen area was made by taking identical micrographs and
diffraction patterns as made before heating. Once this examination
was completed, the entire sample was examined at the elevated
temperature in order to insure that all observations were representative of the total sample. Also, repetitive studies were made at a
temperature of interest to guarantee the reproducibility of results--
particularly the selected area electron diffraction data.

A number of heated samples were cooled and re-examined at
room temperature. Some were confined to the low vacuum of the

electron microscope and examined while others were allowed to age
in air for as long as one month and then re-examined with the
electron microscope. The experimental details of performing
selected area electron diffraction measurements under elevated
temperatures has been mentioned elsewhere.[20,21] The d-spacings
given are accurate to at least ±0.01 Å under heating conditions.

RESULTS AND DISCUSSION

The figures which follow in this section were chosen to
give a representative sampling of results as revealed in the
electron micrographs and selected-area electron diffraction patterns.
In all cases, the samples were heated in situ, and in the majority
of cases, the pictures were taken while the sample was being
heated. In some cases, samples were observed after cooling to
ascertain any differences between observations at elevated temperatures and at room temperature. Without exception, there were no
changes to be observed in the samples studied when they were
allowed to cool either in vacuum or in the air.

Figs. 1 and 2 show the results of heating a well crystallized
kaolinite sample. The areas shown in both micrographs are the
same with Fig. 1 showing the area before heating and Fig. 2 showing
the same area after the sample had been heated first to 740°C for
three hours and then to 800°C for three hours. In comparing the
two figures, the appearance of a second phase consisting of small,
dense, and angular particles is observed. Beyond this, though,
a comparison shows the disappearance of sharp edges for the kaolinite
particles as seen before heating even though a rough outline of the
original clay platelets still remains. This suggests that a basic
reorganization of the metakaolin lattice is already in progress
at this point. The particles shown in Fig. 2 were first observed
in these studies after kaolinite samples had been heated for three
hours at 650°C. The extent of particle formation increases rapidly
with both an increase in temperature or a lengthening in time of
heating at a specified temperature.

The selected-area electron diffraction data obtained from
studies on samples heated to 800°C is shown in Table 1 along with
data obtained at 1000°C. The general behavior in terms of electron
diffraction data is as follows. At room temperature the kaolinite
particles give a very sharp and distinct pattern with measured
d-spacings being identical to those reported[22]for a well crystallized kaolinite. As the sample is heated to progressively higher
temperatures, the pattern is characterized by weaker and broader
lines. The disordering of the sample reaches its climax when
after heating for approximately three hours at 600°C the sample
is essentially amorphous to the electron beam. This amorphous

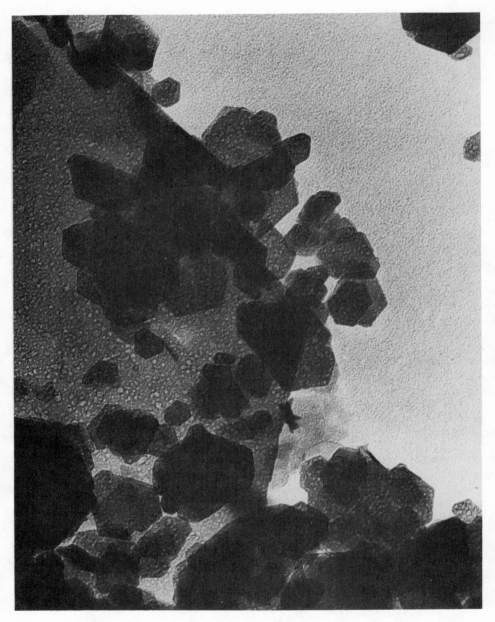

Figure 1.   Electron photomicrograph of kaolinite sample before
            heating.   Fig. 1 is the same area as Fig. 2.   Magnifi-
            cation--33,400X.

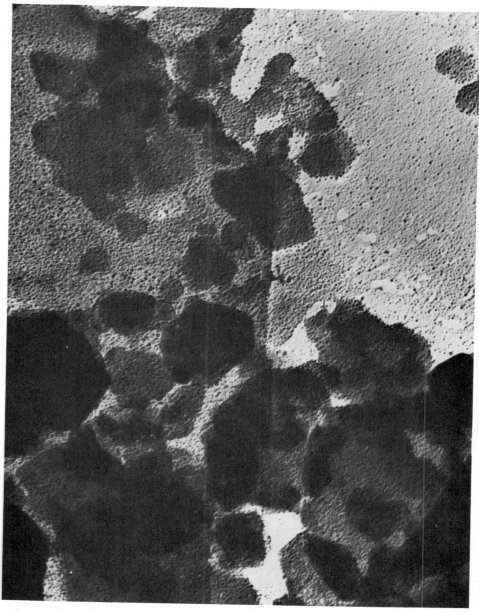

Figure 2.   Electron photomicrograph of kaolinite sample after
            heating to 740°C for 3 hours and then to 800°C for 3
            hours. Fig. 2 is the same area as Fig. 1. Magnifica-
            tion--33,400X.

Figure 3.　Electron photomicrograph of kaolinite sample after heating to 1000°C for 1 hour.　Magnification--19,300X.

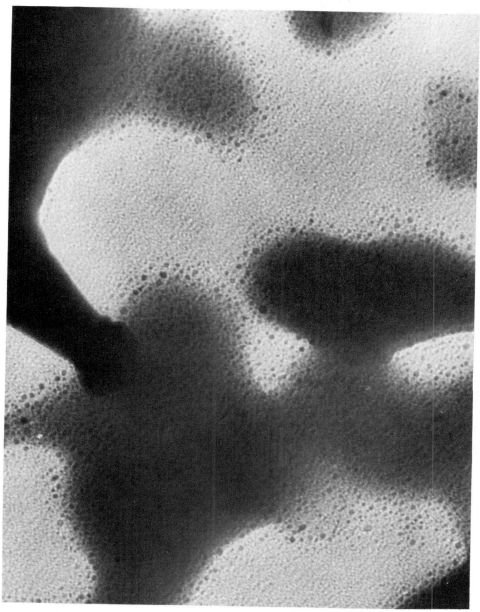

Figure 4. Electron photomicrograph of kaolinite sample after
heating to 1000°C for 1 hour. Magnification--64,400X.

Table 1

Selected-Area Electron Diffraction Data
from Heating Studies on Kaolinite

d-spacings ($\overset{o}{A}$)

800°C	1000°C
4.53	3.12
3.42	2.23
2.91	1.91
2.55	1.64
1.99	1.56
1.53	1.39
1.41	1.28
1.29	1.09
	1.00

period extends until approximately 750°C when a weak pattern
was obtained on a sample heated for three hours. The pattern at
this point is apparently the same as that obtained at 800°C for
which the d-spacings are given in Table 1. In all cases, the
electron diffraction patterns obtained above 750°C were quite weak.

Figs. 3 and 4 show areas of a kaolinite sample heated to
1000°C for one hour. At this point, the original kaolinite
particles have been disrupted and in their place are many small
angular particles. Fig. 4 clearly shows that no large platy
kaolinite particles remain. The selected-area electron diffraction
data given by the sample at 1000°C is shown in Table 1.

The interpretation of the electron diffraction data obtained
from heated kaolinite samples is complicated by the fact that only
a small number of lines were observed. It is obvious that the
small angular particles shown in Figs. 2-4 are not highly ordered.
It is even possible that some of the lines shown in Table 1 at
800°C could be attributed to some remnant of the kaolinite lattice
not disrupted. However, there is reason to doubt this for at least
three reasons: (1) the fact that a region was observed from
approximately 600°C to 700°C where the sample was amorphous to the
electron beam, (2) the electron micrographs (Fig. 2) indicate
considerable disruption in the original particles, and (3) the
recent work by Range, et al[23] indicated that prolonged heating at
just above 600°C resulted in appreciable disorder and eventually
produced a sample amorphous to X-Rays. The amorphous region found
here does not prove, of course, that the sample was totally devoid
of order, but does suggest that the original material was passing

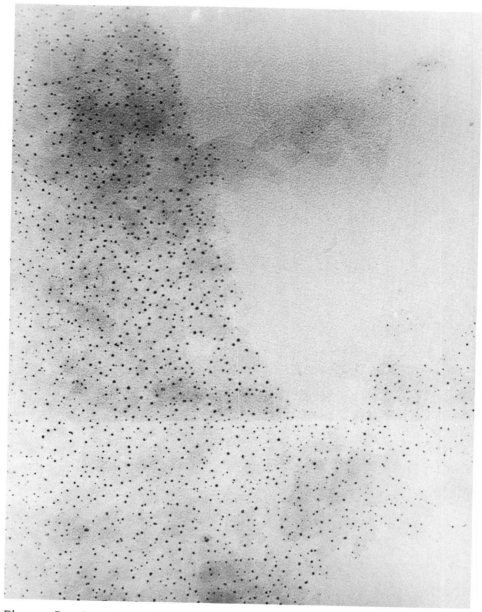

Figure 5. Electron photomicrograph of sodium exchanged bentonite sample heated to 675°C for 3 hours. Magnification-- 29,200X.

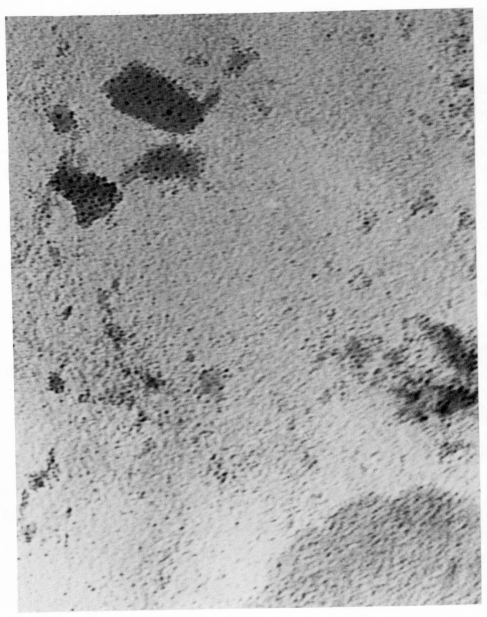

Figure 6.   Electron photomicrograph of mica sample heated to 750°C
for 3 hours. Magnification--61,000X.

Figure 7. Selected-area electron diffraction pattern of a mica sample before heating. Fig. 7 is the same area as Fig. 8.

Figure 8.   Selected-area electron diffraction pattern of a mica
            sample after heating to 675°C for 4 hours.   Fig. 8 is
            the same area as Fig. 7.

through a region where reorganization was taking place.

As has been mentioned, numerous investigators have proposed that mullite or γ-alumina or a mixture of the two is produced upon the structural collapse of metakaolin. The diffraction data obtained here is not of sufficient quality to identify conclusively any new phases formed, although the data could best be accounted for by assuming that a mixture of mullite and γ-alumina exists at 800°C. At 1000°C, there is little evidence in the measured d-spacings that γ-alumina is present. Again, strong conclusions as to the nature of new phases are not possible.

Although the more extensive study has been made on kaolinite, heating studies were run on mica, talc, and a sodium exchanged bentonite at 750°C and below. No heatings studies at higher temperature or detailed analysis of diffraction data was made. Fig. 5 is an example of the formation of new particulate matter within Na-bentonite upon heating for three hours at 675°C. The behavior of Na-bentonite paralleled that of kaolinite concerning electron diffraction to the extent it was studied. The extremely sharp and intense pattern at room temperature progressively deteriorates until at a temperature of approximately 650°C after several hours of heating the sample is amorphous to the electron beam. As mentioned, Furlong[18] observed particulate matter forming in two natural montmorillonites heated to 600°C, but the use of carbon support films made interpretation of his results difficult.

Fig. 6 shows the formation of particulate matter in mica particles heated to 750°C for three hours. Figs. 7 and 8 show the change in the selected-area electron diffraction pattern upon heating for a typical area such as that shown in Fig. 6. The patterns in Figs. 7 and 8 (before and after heating to 675°C respectively) were taken and developed under identical conditions both before and after heating. The intensities are therefore completely comparable. In contrast to kaolinite and the Na-bentonite, the mica samples did not become totally amorphous when heated just above 600°C, although there is evidently a drastic loss in crystallinity upon heating. It appears that the electron diffraction pattern recorded at 675°C is essentially the same as that at room temperature but indicating considerably less order in the sample. Talc samples behaved like the mica samples in that there was an obvious deterioration in the electron diffraction patterns above 600°C but total loss of crystallinity did not occur up to temperatures as high as 700°C.

CONCLUSIONS

The results obtained here along with other recent work[23,24] give evidence for the fact that metakaolin may suffer extensive

disruption and even collapse at a temperature much lower than that
found by DTA techniques and heating-oscillating X-Ray diffraction
when prolonged heating conditions are used. The primary difference
seems to be that in the study reported here a sample was heated
continuously at lower temperatures (e.g. 650°C) while in DTA work
and many other techniques the sample is constantly being raised
in temperature and the total heating time may not be of long duration. The preliminary results obtained on mica, talc, and Na-
bentonite also suggest that they may be capable of beginning a
reorganization in the 600°C region upon prolonged heating. The
results also suggest that all of the materials may be qualitatively
similar as to the temperature at which structural changes in the
mineral lattice can occur.

ACKNOWLEDGEMENT

     The authors express appreciation to the Robert A. Welch
Foundation for support of the research recorded herein.

REFERENCES

1. McVay, T. M., and Thompson, C. L., J. Am. Ceram. Soc., 11,
    829-41 (1928).

2. Insley, H., and Ewell, R.H., J. Res. Natl. Bur. Std., 14,
    615-27 (1935).

3. Eitel, W. and Kedesdy, H., Abhandl. Preuss. Akad. Wiss., 5,
    5-45 (1943).

4. Comeforo, J.E., Fischer, R.B., and Bradley, W.F., J. Am.
    Ceram. Soc., 31, 254-59 (1948).

5. Johns, W.D., Mineral. Mag., 30, 186-98 (1953).

6. Glass, H.D., Amer. Mineral., 39, 193-207 (1954).

7. Brindley, G.W., and Nakahira, M., J. Am. Ceram. Soc., 42,
    311-14 (1959).

8. Brindley, G.W., and Nakahira, M., J. Am. Ceram. Soc., 42,
    314-18 (1959).

9. Comer, J.J., J. Am. Ceram. Soc., 43, 378-84 (1960).

10. Comer, J.J., J. Am. Ceram. Soc., 44, 561-63 (1961).

11. Tsuzuki, Y., J. Earth Sci., Nagoya Univ., 9, 305-44 (1961).

12. Taylor, H.W.F., Clay Minerals Bull., 5, 45-55 (1962).

13. Rowland, R.A., Weiss, E.J., and Lewis, D.R., J. Am. Ceram. Soc., 42, 133-38 (1959).

14. Wahl, F.M., "Effect of Impurities on Kaolinite Transformations as Examined by High-Temperature X-Ray Diffraction", Advances in X-Ray Analysis, 10th Conference, pp. 264-275, Plenum Press, New York (1962).

15. Wahl, F.M., and Grim. R.E., Clays and Clay Minerals, Pergamon Press, 1964, pp. 69-81.

16. McAtee, J.L., Jr., J. Catal., 9, 289-94 (1967).

17. McAtee, J.L., Jr., Proc. Third Toronto Symposium on Thermal Analysis, pp. 13-28, H.G. McAdie, Ed., 1969. Toronto Section, Chemical Institute of Canada.

18. Furlong, R.B., Clays and Clay Minerals, Pergamon Press, 1967, pp. 97-101.

19. Daw, J.D., Nicholson, P.S., and Embury, J.D., J. Am. Ceram. Soc., 55, 149-51 (1972).

20. Jernigan, D.L., and McAtee, J.L., Jr., Thermochimica Acta, 4, 393-404 (1972).

21. Jernigan, D.L., and McAtee, J.L., Jr., Thermochimica Acta, in press.

22. Grim, R.E., Clay Mineralogy, McGraw-Hill, New York, 1968.

23. Range, K.J., Russow. J., Oehlinger, G., and Weiss, A., Ber. Deut. Keram. Ges., 47, 545-49 (1970).

24. McConnell, J.D.C., and Fleet, S.G., Clay Mineral., 8, 279-90 (1970).

HEATS OF IMMERSION OF HYDROXYAPATITIES IN WATER

Hillar M. Rootare and Robert G. Craig

University of Michigan School of Dentistry

Ann Arbor, Michigan 48104

Heats of immersion have been reported for many polar solids of metal oxides and silicates both in polar and nonpolar liquids. These heats have been determined as a function of outgassing temperatures, specific surface areas, and particle size. Usually they are made on the "bare" sample only. However, with most of the hydrophilic metal oxide surfaces, it is not always known whether the surface was "bare" (i.e. completely dehydroxylated) or had some residual hydroxyl groups still on the surface. How "bare" the surface was, would depend greatly upon the outgassing temperature and pressure at which the sample was prepared. Although the heats measured were relatively insensitive to these factors for a nonpolar liquid such as hexane on silica, titanium dioxide, or alumina[1], for a polar molecule such as water on these solids, large and complex changes have been observed[2]. Wade and Hackerman, et al. measured heats of immersion for water on TiO_2[3], SiO_2[4,5], and on Al_2O_3[6,7] as a function of the outgassing temperature and specific surface area. Whalen[8] and Kiselev, et al.[9,10] measured heats of immersion of different silica gel preparations and quartz. Holmes, Fuller and Secoy[11-14] did an extensive study on the thoria-water system, investigating the effect of outgassing temperature on heat of immersion and relating it to surface hydration of ThO_2. Morimoto, et al.[15-18], who investigated the ZnO-water system quite extensively by the water vapor adsorption method, have also conducted calorimetric measurements of heats of immersion on several metal oxides. The effect of outgassing temperature on loss of surface water and the heats of immersion were measured on both α- and γ-alumina[15], ZnO[16], TiO_2[17], and $α-Fe_2O_3$[18]. The heats of immersion were related to change in surface area on heating and to decrease in water content with

increasing outgassing temperature. In general, it can be observed
from these results that the heat of immersion of these oxides
increased with increasing outgassing temperature. Most heat of
immersion curves appear to pass through a maximum at a certain
temperature, e.g. Al_2O_3 at 600°C, ZnO at 500°C, TiO_2 (both rutile
and anatase) at 400°C, and $\alpha-Fe_2O_3$ at 600°C. Two out of four
hematite samples, however, showed almost no change in the heat of
immersion between 600° and 800°C. Also, for some samples like
quartz[8] the increases in the specific surface area appeared to
decrease the heat of immersion, which was attributed to semi-
amorphous character of the smallest particles. In the case of
thoria (ThO_2)[13,14] and hydroxyapatite samples, the opposite trend
was found; the highest surface area VIC-HAP* had the highest heat
of immersion.

It was evident that the increase in outgassing temperature
causes strongly bound surface water to desorb or dehydrate, and
that the increased heat of immersion is therefore attributable to
large amounts of energy involved in rehydration or adsorption on
the strong adsorption sites. It was the objective of this
research to determine the heat of immersion of various hydroxyapatites as a function of outgassing temperature and the extent
of surface coverage in order to understand the reaction of water
at the hydroxyapatite surface.

MATERIALS AND METHODS

Hydroxyapatite Samples

Three hydroxyapatite (HAP) samples that were different in
their preparation were used for heat of immersion measurements.
Victor (VIC-HAP) was a finely powdered sample with 50% of the
particles less than 7.5 μm in size. It was commercially prepared
by precipitation from hot water. The sample had a low bulk
density (0.50 g/cc) and true density (2.70 g/cc), but had a high
surface area of 70 m^2/g (Table 1).

The second sample (NBS-HAP) was originally prepared at the
National Bureau of Standards†. This sample was prepared by
precipitation and by the addition at a very slow rate (∿1 ml/min)
of 2 molar phosphoric acid solution to a bath of boiling water
and calcium oxide.

The third sample (TVA-HAP) was a high temperature solid-state

---

*   Victor hydroxyapatite, Victor Division, Stauffer Chemical Co.,
    380 Madison Ave., New York, N.Y. 10017.
†   National Bureau of Standards, Dental Research Section,
    Washington, D.C. 20234.

Table 1.   Properties of Hydroxyapatites

HAP	Preparation	Ca/P Ratio	Surface Area m^2/g	Median Particle Size, μm
VIC	Comm. Prep. Hot HOH	$1.656 \pm .024$	$70.03 \pm .14$	7.5
NBS	Lab. Prep. Ppt., $100^\circ C$	1.672 ---	$22.50 \pm .26$	6.0
TVA	Lab. Prep. S.S., $1200^\circ C$	$1.665 \pm .001$	$3.04 \pm .03$	16.0

reaction product. It consisted of heating a stoichiometric mixture of monocalcium phosphate monohydrate $Ca(H_2PO_4)_2 \cdot H_2O$ and
calcium carbonate at $1200^\circ C$ in an atmosphere of equal volumes of
steam and nitrogen for three hours. The steam supplied the
required hydroxyl groups and the carbon dioxide was liberated in
the reaction and swept away by the nitrogen gas. The TVA-HAP
sample prepared this way had a low surface area of $3.0 \ m^2/g$ but a
higher bulk (0.84 g/cc) and true densities (3.06 g/cc). The Ca/P
molar ratio was 1.665 which was close to the theoretical value of
1.667.

Of the above three HAP samples the NBS- and the TVA-HAP represented two stoichiometric samples by different preparations.
They also differed greatly in surface area and particle size
(Table 1). The VIC-HAP sample had the highest surface area of
the three, but it had the lowest Ca/P molar ratio which may
indicate Ca deficiency. VIC-HAP also contained 0.5% carbonate as
determined by thermogravimetric analysis.

From the infrared spectra (Figure 1) it was observed that
VIC-HAP sample had lower hydroxyl content than the other two
samples. The sharp absorption peak at 635 cm^{-1} frequency,
assigned to the OH^- vibration mode, present in the TVA- and
NBS-HAP samples, appeared only as a shoulder on the P-O bending
mode (ν_4) peak at 603 cm^{-1}. Also, the reflection at 3530 cm^{-1},
from the internal stretching of the OH^- ion, present in the TVA-
and NBS-HAP as a sharp peak, appeared as a weak small pip riding
on the broad water band (3400 cm^{-1}). The relative size and
sharpness of these absorption peaks gave a good qualitative indication of the crystallinity of the HAP samples. The presence
of carbonate in VIC-HAP was also shown by the reflections at
870 cm^{-1} and a doublet at 1463 cm^{-1} and 1528 cm^{-1}.

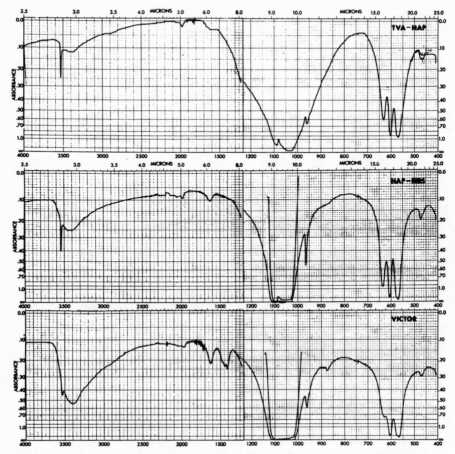

Figure 1.   Infrared Spectrograms of Apatites: TVA, NBS and VIC
            Hydroxyapatites.

Precision Solution Calorimeter

        A precision solution calorimeter* was used and the method
involved measuring of the amount of electrical energy necessary
to duplicate (or nullify, in case of an endothermic process), the
thermal effect accompanying a physical or a chemical process.

        The calorimeter was designed to enable a precise comparison
between an electrical experiment (calibration run), and an actual
experiment (test run). The temperature was measured accurately

*   LKB Model 8721-1. LKB Instruments, Inc., 12221 Parklawn Drive,
    Rockville, Maryland 20852. LKB-Producer AB, S-161 25 Bromma
    1, Sweden.

as a function of time, and the conditions were chosen to give as identical a temperature change as possible in both experiments. Knowing the amount of electrical energy added in the calibration run, and the temperature changes in the two experiments, and by taking the stoichiometry and the amounts of the reactants into account, a precise calculation could be obtained of the change in energy.

The calorimeter cell consisted of a gold reaction vessel fitted with a 2000 ohm thermistor, a 50 ohm calibration heater, a sapphire-tipped rod on the bottom of the reaction cell for breaking the ampule, and an 18-carat gold stirrer (600 rpm) and a holder for a 1 ml spherical ampule. The cell enclosed in a chrome-plated brass vessel was submerged in a thermostated water bath, where the temperature was sensed by a thermistor and regulated by a proportional temperature controller. The bath normally provided temperature control of better than 0.01°C. An electronic galvanometer* allowed temperature change of 5×10^{-5}°C to be detected. Details of the experimental method have been described by Rootare[19].

Heats of Immersion of Hydroxyapatites in Water at 20°C

The heat of immersion, also known as the heat of wetting, is the heat evolved or absorbed in the wetting process of a clean solid in a pure liquid. Since heat is almost always evolved, the enthalpy change ($-\Delta H_I$) is negative. Since the wetting phenomena is surface dependent, and is not affected by the bulk properties of the solid, the heat effect can be reduced to the more practical basis of heat evolved per unit area. The area was measured by the B.E.T. method of nitrogen adsorption. The heat of immersion was expressed in terms of ergs/cm^2.

Pyrex ampules were used for hydroxyapatite samples outgassed up to 400°C, and quartz ampules were made for the use from 400° to 1000°C. The amount of heat generated by breaking an evacuated, empty, Pyrex ampule in 100 ml of double distilled, deionized water was 0.026±0.001 cal. This heat was determined by averaging five tests, and it was used to correct the experimental heats yielded prior to normalizing the values.

Samples were outgassed at 300°C under evacuation of about 10^{-3} torr. The time of outgassing was investigated in order to determine the optimum outgassing times at a given temperature. The VIC-HAP samples were also investigated at outgassing

---

* 419-A DC Null Voltmeter. Hewlett-Packard Co., P.O. Box 301, Loveland, Colorado 80537.

temperatures ranging from 100° to 1000°C, to determine the outgassing or the sintering temperature effect on the heats of
immersion. All sample bulbs were sealed under vacuum with a
microflame* hand torch.

Because low amounts of heat were measured, in the range of
one calorie or less, it was found convenient and more precise to
record+ the change in the thermistor resistance with time. The
10 inch scale on the recorder represented 0.1 ohm or less, thus
making the reading of the resistance change more precise than
with the Wheatstone bridge which had a sensitivity of 0.01 ohm.

RESULTS AND DISCUSSION

Heats of immersion of hydroxyapatite samples in water are
summarized in Table 2 for TVA-, NBS-, and VIC-HAP outgassed at
300°C, and for VIC-HAP at temperatures up to 1000°C. The heat of
immersion for VIC-HAP became essentially constant above 600°C.
The slight drop in the heat at the 1000°C outgassing temperature
may be caused by several factors such as loss of carbonate and a
slightly inaccurate surface area value for the sample. The
surface areas of the sintered VIC-HAP powders were not measured
directly before the immersional experiments but were determined
on other samples and the values are shown in Figure 2. Figure 2
also shows the weight loss data of VIC-HAP plotted as a function
of outgassing temperature for purposes of comparison.

Figures 3 through 6 present heat of immersion values for
VIC-HAP samples outgassed at temperatures of 100° to 1000°C and

Table 2. Heats of Immersion of Hydroxyapatite Powders

HAP	Surface Area m^2/g	Outgassing Temperature °C	$-\Delta H_I$ ergs/cm^2
TVA	3.04	300	458.±14.
NBS	22.5	300	648.± 7.
VIC	70.4	300	696.± 1.
VIC	63.0	400	777.± 4.
VIC	40.0	600	1112.±12.
VIC	20.0	800	1112.±39.
VIC	5.0	1000	1083.±14.

* Microflame, Inc., 7800 Computer Ave., Minneapolis, Minn. 55435.
+ Sargent SRC Recorder. Sargent-Welch Scientific Co., 8560 W.
  Chicago, Detroit, Mich. 48204.

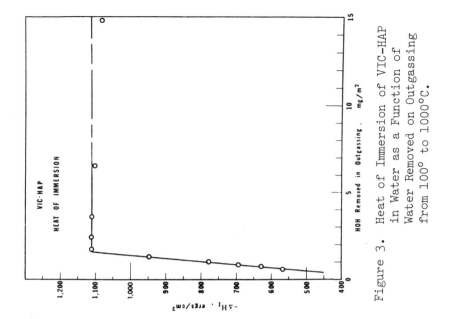

Figure 3. Heat of Immersion of VIC-HAP in Water as a Function of Water Removed on Outgassing from 100° to 1000°C.

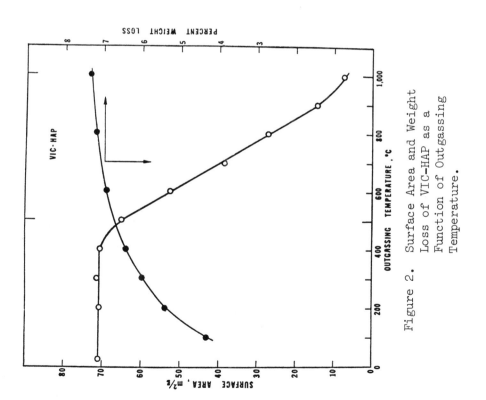

Figure 2. Surface Area and Weight Loss of VIC-HAP as a Function of Outgassing Temperature.

plotted as a function of the water removed from or bound to the
surface. The heat of immersion is plotted as a function of the
amount of water removed on outgassing in Figure 3 and the data
was linear between 100° and 600°C, and the heat of immersion
plateau covered a range of dehydration from >2 to 15 mg/m².
The heat of immersion data are plotted in Figure 4 as a function
of water removed on outgassing in millimoles and a different
shaped curve is obtained. Before attaining a plateau, three different sloped curves were observed for increasing amounts of
water removed. Each linear portion of the curve represents the
binding energy of the water removed from the VIC-HAP surface under
given conditions. In order to relate this energy to that involved in rehydration of the partially dehydrated surfaces on
outgassing, the data is presented as a function of bound water
remaining on the surface in Figures 5 and 6. This treatment is
equivalent to having preadsorbed water covering part of the
surface. It was observed in Figure 5 that there were at least
three segments to the heat of immersion curve for water as a

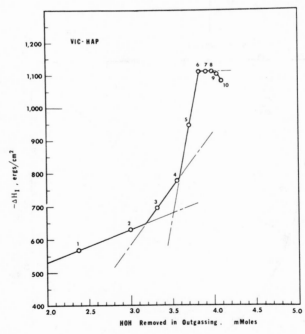

Figure 4.   Heat of Immersion of VIC-HAP in Water as a Function
            of Millimoles of Water Removed From the Surface on
            Outgassing from 100° to 1000°C; Points Numbered 1
            Through 10, Respectively Represent Values at 100°C
            Intervals.

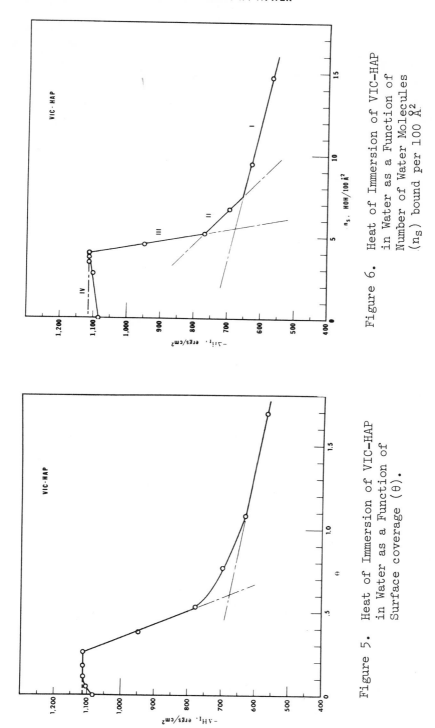

Figure 6. Heat of Immersion of VIC–HAP in Water as a Function of Number of Water Molecules (n_S) bound per 100 Å^2

Figure 5. Heat of Immersion of VIC–HAP in Water as a Function of Surface coverage (θ).

function of the fraction of the surface covered (θ). The first
segment was horizontal from θ = 0 to 0.25. The second segment had
a steep negative slope from θ = 0.25 to $\sim$0.6, which indicated a
large energy change with increased coverage of water. The third
segment had a low negative slope at monolayer and higher surface
coverage, indicating only small energy changes with coverage. At
monolayer coverage the heat of immersion was 643 ergs/cm^2 and
from a vapor phase adsorption study[19], the amount of water in a
monolayer was determined to be 18.34 mg/g, or 0.260 mg/m^2. The
energy of rehydration of this surface covered with a monolayer of
water, which is essentially equivalent to the heat of adsorption
on that surface, can be calculated by the relationship,

$$\text{Energy of Rehydration} = \frac{(-\Delta H_I) \times MW \times K}{(WM/A)} = q$$

where MW is the molecular weight of water, WM/A is (0.260 mg/m^2)
the amount of water adsorbed at monolayer coverage, and K is a
constant for conversion of units. This calculation yielded
10.6 Kcal/mole of water for the energy of rehydration of the
VIC-HAP surface covered with a monolayer of water, which is very
close to the value of 10.5 Kcal/mole for the heat of liquifaction
(E_L) of water.

Only VIC-HAP was used in this study of heat of immersion as
a function of outgassing temperature. Future investigations are
needed to determine the effect of absence of carbonate above
900°C on the heat of immersion measurements of NBS- and TVA-HAP.

If the outgassing removed only physisorbed water from the
HAP surface, the heat of immersion should be independent of the
temperature, because there should be no physically adsorbed water
remaining on the HAP surface under these conditions. Although the
surface area remained constant up to 300°C, the heat of immersion
increased steadily. From 400° to 600°C, where the surface area
decreased about 43% (70 m^2/g to 40 m^2/g), the heat of immersion
per unit area increased even more rapidly, until above 600°C it
leveled off, even though the surface area continued to decrease
as a result of sintering. A plausible explanation for this
phenomena is that chemisorbed water is being progressively removed from the HAP surface which is more strongly bound at the
higher outgassing temperatures. On subsequent immersion of the
sample in water the re-chemisorbed water results in a large net
heat of adsorption.

The heat of immersion curve for VIC-HAP as a function of the
number of molecules (n_s) of bound water per 100 Å^2 is plotted in
Figure 6. Four linear segments, labeled in Roman numerals I
through IV, represent various stages of hydration of the surface

and the energies associated with rehydration of these surfaces. From the slopes the average rehydration energy was calculated. The rehydration of the VIC-HAP surface is confined to a small number of water molecules/100 $\overset{o}{A}{}^2$, about 5 to 6. Only the slopes I through III can be used for rehydration energy calculations.

The initial horizontal portion (IV) represents rehydration of the surface outgassed from 600° to 1000°C, and corresponds to 4 HOH molecules per 100 $\overset{o}{A}{}^2$ of surface and yielded a relatively constant heat of immersion. This value may represent stabilized rehydration. The slightly decreasing values for the outgassing temperatures of 900° and 1000°C may be caused by either competing reactions from decomposition products of carbonate apatite*, or to inaccurate surface area values taken for these sintered samples. The rehydration energy associated with segment III is 43 Kcal/mole of water (Table 3). This energy represents the the completion of hydroxylation of VIC-HAP surface and is confined to the addition of the last water molecule going from 4 to 5 HOH/100 $\overset{o}{A}{}^2$ surface. Each water molecule chemisorbed on the "bare" oxygen-rich surface, will yeild two surface hydroxyl (OH) groups per one surface oxygen. A chemical reaction, therefore, actually occurs between the active surface oxygen and water, where Ca-O-P bonds are broken to form Ca-OH and P-OH bonds for each HOH molecule.

Segment II (Figure 6), involves 5 to 7 HOH/100 $\overset{o}{A}{}^2$ (outgassing temperature of 300° to 400°C) and represents the intermediate rehydration of the surface hydroxyls. The average rehydration energy associated with this reaction is 7.1 Kcal/mole of water (Table 3). Segment I completes the rehydration of the hydroxylated surface and ranges between 8 and 15 HOH/100 $\overset{o}{A}{}^2$, and exhibits an average replacement energy of 1.7 Kcal/mole of water. This value represents full hydrolysis of the partially rehydrated hydroxyl

Table 3. Average Rehydration Energies for Water on VIC-HAP Surface

Slope No.	Temp. Range °C	HOH content/ 100 $\overset{o}{A}$ surface	q av. Kcal Mole
I	100- 200	14.7-9.5	1.7
II	300- 400	6.6-5.2	7.1
III	400- 600	5.2-3.9	43.
IV	600-1000	3.9-0	----

---

* From thermogravimetric analysis about 0.5% of weight loss can be attributed due to carbonate at 1000°C.

surface of VIC-HAP, and is equivalent in energy to that of hydrogen bonding. These rehydration energies were calculated from the increased heat of immersion values as a function of surface coverage (preadsorbed water), when the VIC-HAP surface is completely rehydrated with water (in immersion) to its stable surface state.

Because of the constancy of the segment IV (Figure 6), the average rehydration energy could not be determined in this manner for the heat of immersion data for the outgassing range of 600° to 1000°C. The result, however, suggests the existence of two sintering modes as in the case of thoria (ThO_2)[11-14]. The lower temperature sintering may involve condensation of surface hydroxyl groups, while sintering at higher temperatures of 900° to 1000°C may involve, besides decomposition of carbonate HAP, mass transport of HAP.

Heats of immersion as a function of surface coverage can be used to calculate heats of adsorption (q^*). These calculations were obtained by differentiation of the data that were corrected for the 0.5% weight loss due to carbonate. In order to reduce the energy per unit area $(-\Delta H_I)$ to energy per mole of water (q^*), the former values were divided by the surface area, and as a result are independent of the area change on sintering.

$$q^* = \frac{(-\Delta H_I) \text{ MW} \times 2.389 \times 10^{-11}}{(-\Delta W)\dagger/\text{S.A.}} \text{ Kcal/mole}$$

where MW is the molecular weight of water, 2.38×10^{-11} is the conversion factor from ergs to Kcal, $(-\Delta W)^\dagger$ is the corrected weight loss, and S.A. is the specific surface area in m^2/g.

The net differential heat of adsorption (q^*) curve for water on VIC-HAP is given as a function of the fraction of the surface covered (θ) in Figure 7. This plot was possible because the monolayer capacity was determined from the water adsorption isotherms, and it was assumed that outgassing at 1000°C completely dehydroxylated the surface of VIC-HAP. The very high initial heats of adsorption may be attributed to formation of surface hydroxyl groups by dissociative chemisorption, which is one way of expressing the chemical reaction between a water molecule and a surface oxygen in forming two surface hydroxyl groups. The Ca-O-P bonds are broken in the process of forming a Ca-OH and a P-OH bond on the HAP surface.

The average rehydration energies of VIC-HAP surface (Table 3) calculated from the heat of immersion curve as a function of water content (Figure 6) provide supporting evidence to the proposal that high temperature outgassing (1000°C) produces a completely

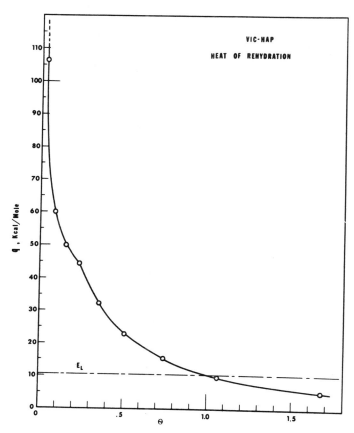

Figure 7. Net Differential Heat of Adsorption of Water Vapor on VIC-HAP.

dehydroxylated, clean surface. If this is so, then the three slopes of diminishing heat of immersion with surface coverage, represent the energies involved in rehydration of the partially covered HAP surface. Thus, a model similar to Fuller and Holmes[13] may be proposed and semi-quantitative enthalpy values may be assigned for the idealized surface reactions for dehydration of the VIC-HAP surface at outgassing temperature ranges given in Table 3. These reactions are shown schematically in equations 1 through 3.

It should be realized that while the above schematic drawings represent a reasonable and probable mechanism, other mechanisms may also be consistent with the experimental results. From the practical point of view, equations 2 and 3 together represent the

$$\begin{array}{l}\text{HOH HOH HOH HOH} \\ \text{OH OH OH OH}\end{array} \xrightarrow[\text{I}]{100\text{–}250°C} \begin{array}{l}\text{HOH \quad HOH} \\ \text{OH OH OH OH}\end{array} + 2\text{HOH}, \quad q_{av} = -1.7\ \frac{\text{Kcal}}{\text{mole}} \qquad (1)$$

$$\begin{array}{l}\text{HOH \quad HOH} \\ \text{OH OH OH OH}\end{array} \xrightarrow[\text{II}]{250\text{–}400°C} \text{OH OH OH OH} + 2\text{HOH}, \quad q_{av} = -7.1\ \frac{\text{Kcal}}{\text{mole}} \qquad (2)$$

$$\text{OH OH OH OH} \xrightarrow[\text{III+(IV)}]{400\text{–}1000°C} \bigwedge\bigwedge + 2\text{HOH}, \quad q_{av} = -43\ \frac{\text{Kcal}}{\text{mole}} \qquad (3)$$

formation of one layer of adsorbed water on the hydroxylated surface of HAP. The quantity of water involved, however, is the weight equivalent of two chemisorbed monolayers. In short, it requires three water molecules to hydrate one surface oxygen site. In the first diagram (equation 1) the dehydration of a fully hydrated hydroxyl surface is being stripped by one HOH molecule to form a bridging hydration. In Figure 6, this is represented approximately by 5 HOH molecules per 100 $\overset{\circ}{A}{}^2$ (slope I, where n_s = 10 to 15). These schematics are intended to reflect the stoichiometry and not particularly the exact configuration, since the configuration of HAP surface is more complex than that of simple metal oxides. The final dehydration of surface hydroxyls takes place at outgassing temperatures of 250° to 400°C, and the temperature above 400°C involves primarily water lost due to dehydroxylation or condensation of hydroxyl groups.

If outgassing temperatures from 100° to 1000°C are considered to correspond to stripping the surface of preadsorbed water according to the proposed model, then the total bound water content (Table 3 and Figure 6) about 14.7 HOH water molecules can be estimated to consist of 5.5 HOH molecules as 11 OH groups forming the hydroxylated surface, and 9.2 HOH molecules chemisorbed on top as a hydrated layer. It is interesting to note that this proposal agrees well with the independently determined value of 9.3 HOH molecules/100 $\overset{\circ}{A}{}^2$ from water adsorbed as a monolayer on the hydroxylated VIC–HAP surface[19]. This latter value was derived from water vapor adsorption isotherms on VIC–HAP compacts outgassed at 300°C. Considering the accuracy of the fractional values of water molecules, these numbers can be rounded-off so that the first 5 HOH molecules to form 10 OH groups per 100 $\overset{\circ}{A}{}^2$ HAP surface and yield a completely hydroxylated surface. The next

Table 4.   Average Heats of Rehydration for HAP Samples Outgassed
            at 300°C

HAP	S.A. m^2/g	$-\Delta W^*$ mg/g	$-\Delta W$ mg/m^2	$-\Delta H_I$ ergs/cm^2	qd Kcal/Mole
TVA	3.04	6.37	2.10	458.	0.94
NBS	22.5	24.5	1.09	648.	2.56
VIC	70.4	59.9	0.851	696.	3.52

* The weight losses ($-\Delta W$) are average values from 3 or more
  replicates.

10 HOH molecules per 100 $\overset{\circ}{A}^2$ chemisorb on the hydroxylated surface
and form the hydrates surface. Thus, the total bound water represents 15 HOH molecules per 100 $\overset{\circ}{A}^2$ of VIC-HAP surface.

The average heats of rehydration of partially dehydrated HAP
samples outgassed at 300°C are given in Table 4. These values
represent samples that include chemisorbed and physisorbed water
under ambient laboratory conditions, and then were stripped of
that water ($-\Delta W$) on outgassing in vacuum at 300°C. It is
interesting to note, that although the amount of water adsorbed
per unit area ($-\Delta W$, mg/m^2) decreased, the corresponding heats of
rehydration increased from about 1 Kcal/mole for TVA- to about
3.5 Kcal/mole for VIC-HAP. It is hoped that in future research
TVA- and NBS-HAP samples are submitted to similar heat treatment
experiments as VIC-HAP so that comparison can be established
among the different HAP surfaces and their rehydration energies.

ACKNOWLEDGMENT

This investigation was supported by the US Public Health
Service Training Grant DE-00181 from the National Institute of
Dental Research, National Institutes of Health, Bethesda, Md.

REFERENCES

1. Wade, W.H. and Hackerman, Norman, J. Phys. Chem., 66, 1823 (1962).
2. Egorov, M.M., Kiselev, V.F., Krasil'Nikov, K.G. and Murina, V.V., Zh. Fiz. Khim., 33, 65 (1959).
3. Wade, W.H. and Hackerman, Norman, J. Phys. Chem., 65, 1681 (1961).
4. Wade, W.H., Cole, H.D., Meyer, D.E. and Hackerman, Norman, J. Phys. Chem., 65, 1968 (1961).
5. Wade, W.H., Every, R.L. and Hackerman, Norman, Advances in Chemistry Series No. 33, Washington, ACS (1961).
6. Venable, R.L., Wade, W.H. and Hackerman, Norman, J. Phys. Chem., 64, 355 (1960).
7. Wade, W.H. and Hackerman, Norman, J. Phys. Chem., 64, 1196 (1960).
8. Whalen, J.W., Advances in Chemistry Series No. 33, Washington, ACS (1961).
9. Egarova, T.S., Zarif'yands, Yu. A., Kiselev, V.F., Krasil'-Nikof, K.G. and Murina, V.V., Russ. J. Phys. Chem., 36, 780 (1962).
10. Egorov, M.M. and Kiselev, V.F., Russ. J. Phys. Chem., 36, 158 (1962).
11. Holmes, H.F. and Secoy, C.H., J. Phys. Chem., 69, 151 (1965).
12. Holmes, H.F., Fuller, E.L., Jr. and Secoy, C.H., J. Phys. Chem., 70, 436 (1966).
13. Fuller, E.L., Jr., Holmes, H.F., Secoy, C.H. and Stuckey, J.E., J. Phys. Chem., 72, 2095 (1968).
14. Holmes, H.F., Fuller, E.L., Jr. and Secoy, C.H., J. Phys. Chem., 72, 2095 (1968).
15. Morimoto, Tetsuo; Shiomi, Koichi and Tanaka, Horoshi, Bull. Chem. Soc. Japan, 37, 392 (1964).
16. Morimoto, Tetsuo; Nagao, Mahiko and Hirata, Miyoshi, Kolloid-Z.Z. Polym., 225, 29 (1968).
17. Morimoto, Tetsuo; Nagao, Mahiko and Omori, Teiji, Bull. Chem. Soc. Japan, 42, 943 (1969).
18. Morimoto, Tetsuo; Katayama, Noriko; Naons, Hiromitu and Nagao, Mahiko, Bull. Chem. Soc. Japan, 42, 1490 (1969).
19. Rootare, H.M., Ph.D. Dissertation, Univ. of Mich. (1973).

HEATS OF SOLUTION OF APATITES, HUMAN ENAMEL AND DICALCIUM-PHOSPHATE IN DILUTE HYDROCHLORIC ACID

Robert G. Craig and Hillar M. Rootare*

University of Michigan, School of Dentistry

Ann Arbor, Michigan 48104

The solubility of hydroxyapatite and dental enamel in water and various buffered solutions has been studied extensively by many investigators and only a few will be listed in the references[1-12]. Heats of wetting and heats of immersion have been measured for water on synthetic hydroxyapatite and ground tooth structure by Tsuchitani, et al.[13,14] and Brauer and Huget[15,16]. The latter authors[17,18], as did Pitt[19], studied the degree of surface modification produced by various ions and functional groups in an aqueous environment by measuring the heats of immersion.

In 1972, Brunetti, Prosen and Brown[20], and Craig and Rootare[21] reported preliminary results for the heat of solution of human enamel and apatite in 0.1 to 1 N aqueous hydrochloric acid. Values by the former authors for the reaction enthalpies for enamel and hydroxyapatite were 71.96 and 69.65 kcal/mole. The latter authors reported heats of solution for highly crystalline apatites from different sources and human enamel to be 73.3 to 80.5 kcal/mole and 66.8 kcal/mole.

The heats of solution of hydroxyapatite, tooth enamel, and fluorapatite are important values in the understanding of the dissolution or decalcification of tooth structure in dilute mineral acid such as occurs in the presence of plaque in contact with tooth structure. This interest is the basis for the present studies of the heats of solution of apatites, human enamel and dicalcium phosphates.

---

* Current address: Research Director, L. D. Caulk Company, Milford, Del. 19963.

EXPERIMENTAL

Preparation and Sources of Apatite Samples

Victor hydroxyapatite (VIC-HAP). It was a fine powder with
50% of the particles <8.5 μm. It was commercially prepared by
precipitation from hot water[*], and has low bulk density and high
surface area. The physical properties are listed in Table 1.

Table 1. Preparation and Physical Properties of Apatite Samples

	Preparation	Mean Size[*] $X_g \pm \sigma_g$ μm	Molar Ratio Ca/PO$_4$	Bulk Density g/cc	Surface Area m^2/g
VIC-HAP	Commercial, hot water ppt.	8.5±2.5	1.658±.046	0.50	70.0±.1
NBS-HAP	Lab. ppt. 100°C water	6.0±2.3	1.672 ----	0.86	22.5±.3
TVA-HAP	Lab. 1200°C Solid state	16.0±1.6	1.665±.008	0.84	3.02±.05
FAP	Lab. 1100°C Solid state	10.2±1.4	1.70 ----	1.26	0.64±.02
CAP	Lab. 1100°C Solid state	---- ---	---- ----	----	0.51 ---
VIC-HAP- 1200	Lab. 1200°C Sintered VIC	---- ---	1.71 ----	1.42	1.34±.08
Enamel	-	---- ---	---- ----	----	3.30 ---

* Coulter Counter data.

NBS-HAP preparation[†]. The method[22] consists of precipitation
by the addition of 2 M phosphoric acid to a bath of boiling water
and calcium oxide at a rate of 1 ml/min. The CaO was prepared from
pure CaCO$_3$ by calcining at 1100°C for 24 hrs. and the resulting CaO
was protected from atmospheric CO$_2$ until used. After the addition
of the acid the mixture was kept boiling for 2 days, the superna-
tent liquid was removed and an equal volume of CO$_2$-free distilled
water was added and the boiling continued for 3 days. The precipitate was rinsed with CO$_2$-free distilled water and acetone, the
resulting material was dried at 105°C for 24 hrs, and the dried
cake was ground to pass a 200 mesh sieve. The sample had a moderate surface area and the properties are listed in Table 1.

*Victor Division, Stauffer Chemical Co., 380 Madison Ave., New York,
N.Y. 10017.
†National Bureau of Standards, Dental Research Section, Washington,
D.C. 20234.

TVA-HAP preparation*. The preparation consisted of heating a stoichiometric mixture of pure $Ca(H_2PO_4)_2 \cdot H_2O$ and $CaCO_3$ at 1200°C in a stream of equal volumes of steam and nitrogen for 3 hrs. This sample had a low surface area and fairly high bulk density as shown in Table 1.

Fluorapatite (FAP) and Chlorapatite (CAP)[†]. The materials were prepared from pure $CaHPO_4$, $CaCO_3$, CaF_2, and $CaCl_2$ by firing at 1100°C, then remixed and refired an additional 2 hrs at 1100°C. No detectable amounts of halide were lost during firing. The properties are listed in Table 1.

VIC-HAP-1200°C. VIC-HAP was sintered in platinum boats at 1200°C in a flow of a 1:1 mixture of steam and nitrogen for 30 min. The sintered material was washed with three 100 ml aliquots of 0.1 M ammonium citrate of neutral pH at 55°C and then washed 3 times with double distilled water before filtering and drying at 110°C. Again the properties are listed in Table 1.

Dicalcium phosphate. The samples were prepared by the method described by Moreno, et al.[11]

Human enamel. The crown of extracted teeth, with pulp and caries removed, were ground in a mill and the 100/200 mesh powder was collected. The enamel and dentin were separated by flotation using bromoform as the suspending medium[23]. The enamel powder was rinsed with acetone, dried overnight at room temperature and then at 105°C for one hour.

Precision Solution Calorimetry

All samples were dried from 105-120°C (overnight) and stored in a desiccator over silica gel until ready for use. Powders were weighed into 1 ml cylindrical glass ampules and sealed. The ampule was placed in the calorimeter[+] containing dilute HCl and the system allowed to reach approximately equilibrium at 20°C. The ampule was broken and the heat of solution was obtained by monitoring the resistance change of the thermistor as a function of time. A typical curve of thermistor resistance versus time is shown in Figure 1. The determination of ΔR is not as straight forward as it might appear since the difference between R_i and R_f must be taken from a line through the half area-point as shown.

---

* Tennessee Valley Authority, Report No. 678, November 6, 1956, Wilson Dam, Ala.

† General Electric Chemical Products Plant, Lamp Metals and Components Dept., 1099 Ivanhoe Road, Cleveland, Ohio 44110.

+ LKB Model 8721-1, LKB Instruments, Inc., 12221 Parklawn Drive, Rockville, Md. 20852.

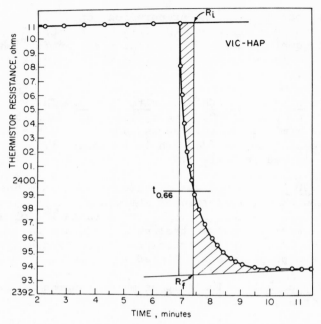

Figure 1. Typical heat of solution curve of VIC-HAP showing the
resistance and time values obtained from the Wheatstone
bridge readings as a function of time. The half areas
under and above the curve are shaded.

The calibration factor was determined by electrical calibration. A
correction for the heat evolved in breaking the ampule was obtained experimentally and the heat amounted to 5×10^{-3} calories. The
reliability of the method was checked by determining the heat of
solution, $-\Delta H_S$, for tris(hydroxymethyl) aminomethane* in 0.1 N HCl
at 25°C; the experimental value of 7.12 kcal/mole compared favorably with the reported value of 7.115 kcal/mole.

A computer program was written[24] to integrate the area under
the reaction curve and to calculate the heat of solution. The
program allows calculation of the electrical calibration, the
actual heat of solution runs, or combinations of both in which the
average calibration factor was used to calculate the heat of solution and also to average the resultant heats from replicate runs.
The program prints out the means, standard deviations, and
coefficient of variations.

* NBS, Certificate of Analysis, Standard Reference Material 724a
  for Solution Calorimetry.

RESULTS AND DISCUSSION

The heats of solution for the various samples are presented
in Table 2. The VIC sample of HAP had the highest heat of solution of 88.59 kcal/mole and it also was the least crystalline of
the HAP samples as shown by the sharpness of -OH peak at 3540 cm^{-1}
in the infrared, the sharpness of the x-ray diffraction patterns
and by scanning electron photomicrographs. When VIC-HAP was
heated at 1200°C in steam and nitrogen for 30 min the crystallinity improved markedly and the surface area decreased from 70.0 to
1.34 m^2/g; this VIC-1200-HAP had the same heat of solution as the
NBS-HAP which had a surface area of 22.5 m^2/g. The TVA-HAP had
the lowest heat of solution of the hydroxyapatites although it had
a low surface area of 3.02 m^2/g. The difference in the $-\Delta H_s$
values for NBS and TVA hydroxyapatites would not be entirely
explained on the differences in surface areas since values of 76.96
and 72.96 kcal/mole, respectively, were obtained when the heats of
solution were corrected for the heats of immersion. It appears,
therefore, that calorimetric heats of solution are useful in
evaluating differences in crystallinity as well as possible
differences in purity and surface enthalpies.

The value reported by Brunetti, et al.[20] of 291.4 J/g was
lower than VIC, NBS, or TVA hydroxyapatites of 368.9, 335.1, and
305.3 J/g which indicates differences in the apatite samples and
again the sensitivity of the method in distinguishing differences.

Table 2. Heats of solution in dilute aqueous hydrochloric acid.

	$-\Delta H_s$, kcal/mole	HCl Conc., N
VIC-HAP	88.59±0.13	0.1
VIC-1200-HAP	80.18±0.01	0.1
NBS-HAP	80.46±0.58	0.1
TVA-HAP	73.30±0.03	0.1
CAP	69.35±0.08	0.1
FAP	36.75±0.01	1
Human Enamel	66.79±0.09	0.1
DCP Dihydrate	−0.20	1
	0.013	0.2
	0.23	0.1
DCP Anhydrous	4.80±0.03	0.1

The value of 36.75 kcal/mole for the $-\Delta H_S$ of FAP was the
lowest of all the apatites while the values for CAP and human
enamel were 69.35 and 66.79 kcal/mole, respectively. The fluorine
content of the human enamel was 133 ppm or 0.0133% compared to
4.15% fluorine in fluorapatite. The substantially lower value of
$-\Delta H_S$ for enamel compared with HAP is not in proportion to the
fluoride concentration and is probably more related to the organization of the crystals than the substitution of $-F$ for some of the
$-OH$ groups. Brunetti, et al.[20] reported 301.1 J/g for the heat of
solution in 1 N HCl for their sample of human enamel which can be
compared with 278.2 J/g in the present study. This difference
could easily be a result of differences in fluorine content and/or
crystal structure.

The heats of solution of FAP in 0.1 and 0.2 N HCl were 37.22
and 35.87 kcal/mole but the reaction was very slow and it was
difficult to determine when the completion of the liberation of
heat occurred since there was always a drift in the temperature of
the calorimeter as shown in Figure 2. This experimental difficulty
caused problems in determining the time for the half area. When
the heat of solution was conducted in 1 N HCl the reaction was
sufficiently rapid so that good repeatability was obtained and the
mean and standard deviation are reported in Table 2. The value of

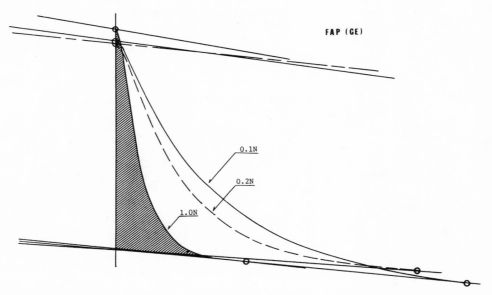

Figure 2.   Heat of solution measurements of FAP in three different
            concentrations of hydrochloric acid, 0.1 N, 0.2 N, and
            1.0 N. The rate of solution increased markedly with
            increasing concentration of the acid.

$-\Delta H_S$ for FAP in 0.1 and 0.2 N HCl were in reasonable agreement with the reported value of 36.75 kcal/mole but the standard deviations were much higher.

Scanning electron photomicrographs showed that both FAP and CAP were highly crystalline as shown in Figure 3.

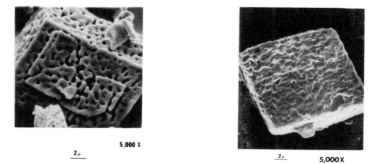

Figure 3.   SEM photomicrographs of FAP X-298 and CAP X-299 at 5000 X.

The heat of solution of DCP dihydrate resulted in some interesting values as can be seen in Figure 4. When $-\Delta H_S$ was determined in 0.1 N HCl the reaction was exothermic, 0.23 kcal/mole, while in 1 N HCl the reaction was endothermic, -0.20 kcal/mole. At the intermediate concentration of 0.2 N HCl the value was only slightly exothermic, 0.013 kcal/mole. No standard deviations are listed since insufficient runs were made. Anhydrous DCP was prepared from the dihydrate by heating at 300°C in flowing nitrogen and the weight loss was stoichiometric and equivalent to 2 moles of water. No evidence of the pyrophosphate was observed in the infrared spectrum. When the $-\Delta H_S$ of DCP was determined in 0.1, 0.2 and 0.3 N HCl the values were 4.80, 4.84, and 4.75 kcal/mole. The rates of solution were adequately rapid so that the reproducibility of the values was good. Since the same ionic species would be obtained with DCP dihydrate and anhydrous the lack of variation of $-\Delta H_S$ of DCP as a function of HCl concentration indicates the variation found with DCP dihydrate is a result of differences in the heats of solution.

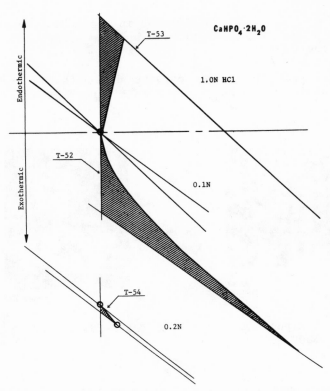

Figure 4. Heat of solution measurements of dicalcium phosphate
           dihydrate in three different concentrations of hydro-
           chloric acid. Note the inversion of the curve from
           exothermic in 0.1 N HCl to endothermic in 1.0 N HCl.

The heat of solution for the various apatites and DCP are of
inherent interest but it should be an important parameter in
characterizing the crystallinity and purity of these biologically
important materials.

ACKNOWLEDGMENT

This investigation was supported by the US Public Health
Service Training Grant DE-00181 from the National Institute of
Dental Research, National Institutes of Health, Bethesda, Md.

REFERENCES

1. Benedict, H.C. and Kanthak, F.F., J. Dent. Res., 12, 227 (1932).
2. Levinskas, G.J. and Neuman, W.F., J. Phys. Chem., 59, 164 (1955).
3. Brown, W.E. and Wallace, B.M., IADR Abstr., paper 669, 220 (1971).
4. Deitz, V.R., Rootare, H.M. and Carpenter, F.G., J. Colloid Sci., 19, 87 (1964).
5. Rootare, H.M., Deitz, V.R. and Carpenter, F.G., J. Colloid Sci., 17, 179 (1962).
6. Gray, J.A., Francis, M.D., and Griebstein, W.J., Chemistry and Prevention of Dental Caries, Springfield, Ill., Thomas, (1962).
7. Francis, M.D., Ann. N. Y. Acad. Sci., 131, 694 (1965).
8. Gray, J.A. and Francis, M.D., Mechanisms of Hard Tissue Destruction, Washington, D.C., Am. Assoc. Advan. Sci. (1963).
9. Brown, W.E., Soil Sci., 90, 51 (1960).
10. Brown, W.E. and Wallace, B.M., Ann. N. Y. Acad. Sci., 131, 690 (1965).
11. Moreno, E.C., Gregory, T.M. and Brown, W.E., J. Res. Nat. Bur. Stand., 72A, 773 (1968).
12. McDowell, Hershel, Wallace, B.M. and Brown, W.E., IADR Abstr., paper 340, 125 (1969).
13. Tsuchitani, Yasuhiko; Sukeno, Shuzo; Inoue, Koyoshi; and Yokomizo, Ichiro, J. Osaka Univ. Dent. Sch., 8, 1 (1968).
14. Tsuchitani, Yasuhiko; Fujisawa, Yoshifumi; Inoui, Koyoshi; and Yokomizo, Ichiro, J. Osaka Univ. Dent. Sch., 8, 9 (1968).
15. Brauer, G.M., Adhesive Restorative Dental Materials, Washington, D.C., U.S. Gov't. Printing Office (1966).
16. Brauer, G.M. and Huget, E.F., J. Dent. Res., 50, 776 (1971).
17. Huget, E.F. and Brauer, G.M., J. Colloid and Interface Sci., 27, 714 (1968).
18. Huget, E.F., Brauer, G.M. and Loebenstein, W.V., J. Dent. Res., 47, 291 (1968).
19. Pitt, D.A., PB 17 578, Springfield, Va., Clearinghouse for Federal Scientific and Tech. Information (1967).
20. Brunetti, A.P., Prosen, E.J. and Brown, W.E., IADR Abstr., paper 115, 77 (1972).
21. Craig, R.G. and Rootare, H.M., IADR Abstr., paper 118, 78 (1972).
22. Avnimelech, Yoran, Israel J. Chem., 6, 375 (1968).
23. Manly, R.S. and Hodge, H.S., J. Dent. Res., 18, 133 (1939).
24. Rootare, H.M., Ph.D. Dissertation, Univ. of Mich., (1973).

VERIFICATION OF THE IONIC CONSTANTS OF PROTEINS BY CALORIMETRY

Mario A. Marini
Department of Biochemistry
Northwestern University Medical and Dental Schools
Chicago, Illinois   60611

Charles J. Martin
Department of Biochemistry
The Chicago Medical School
University of Health Sciences
Chicago, Illinois   60612

Robert L. Berger
Laboratory of Technical Development
National Institutes of Health
National Heart and Lung Institute
Bethesda, Maryland   20014

Luciano Forlani
Centro di Biologia Molecolare
Universita di Roma
Rome, Italy   00185

INTRODUCTION

Dissociation curves of proteins have been extensively used as an analytical tool to provide useful information on the nature and number of the ionic groups in a protein. Changes in the ionic behavior of these groups have been related to structural and functional features of a large number of biologically important molecules. An exact count of the number of groups in any region of a potentiometric titration is difficult because of the overlapping ionizations of the ionizing groups. The residues which ionize in proteins can be grouped into three major sets corresponding to their ionization constants. (In this paper, the ionization constant is referred to by its negative logarithm, i.e., pK'. The concentrations are ex-

pressed as N which is μmoles ionized/μmoles present.) These sets
include the acid region which is composed of C-terminal carboxyl
groups and the γ and δ carboxyls of aspartic and glutamic, the mid-
region which includes the imidazoyl and α-amino group ionizations
and the alkaline region comprising the ionizations of the sulphy-
dryl, phenolic and ε-amino groups from cysteine, tyrosine and lysine
(1), respectively. There is, however, no sharp division between
these regions. Wyman (2) obtained the heats of ionization by titration at different temperatures which when plotted against pH (or
charge) gave an enhanced demarkation of the three major sets of
groups. Although this procedure necessarily gives the average heats
of ionization derived from the groups titrating in any given pH region, it is quite useful for a refined analysis of the ionic dissociation curves of proteins.

It is now possible to obtain directly, the heat liberated during a titration (3). The thermal curve (or thermogram) generated
by this method is a function of the heats of ionization (ΔH_i) of the
groups titrating, the number (N) of these groups and their ionization
constants (pK'). If all three parameters are known, it should be
possible to calculate the thermal curve. We have previously reported (4) a computer assisted evaluation of pK' and N for a mixture of
model compounds. A large number of solutions were possible to describe the potentiometric titration curve of this mixture. Only when
a number of constraints were utilized could the correct solution be
obtained. For proteins, the selection of the proper constraints
requires information from a variety of sources, which, in some cases,
is not known. It occurred to us, that the selection of the appropriate pK' and N from those obtained by computer was possible if
the solutions were utilized to calculate the thermal curve using
known heats of ionization.

This analysis would hopefully require no prior knowledge of the
protein and no constraints. The potentiometric curve would be obtained and evaluated for pK' and N. Each possible solution would
then be utilized and with the heats of ionization to calculate a
thermal curve. That set of values which was more nearly coincident
with the experimentally obtained thermal titration would be the most
acceptable solution. The present paper outlines this approach for
a number of mixtures of model compounds.

EXPERIMENTAL

Materials

Glycylglycine and lactic acid were obtained from Calbiochem.
Imidazole was obtained from Aldrich Chemical Company. β-Mercapto-

ethanol, phenol, pyridine, formic acid and methylamine hydrochloride were reagent grade. All materials were used without purification.

Methods

Stock solutions (approximately 0.05 M) of each model compound were made in water and 4.0 ml of each was titrated from pH 2 to 12.5. From these curves, the actual concentrations (N) and the ionization constants (pK') were calculated. From these solutions, individual samples (0.008 M) were prepared. These samples were also titrated potentiometrically and thermally. On the basis of these titrations, the individual heats of ionization were computed. A variety of mixtures were prepared by diluting the desired compounds (4:50) to obtain solutions of 0.004 M for each component. These were also titrated potentiometrically and thermally. The potentiometric procedure has been described (5).

The thermal titrations were conducted as previously described (6) on 4 ml samples of 0.008 M (32 μmoles) of the individual model compounds. After thermal equilibration, base (1.0 M KOH) was added at the rate of 1.061 μl per second so that the reaction was complete in 30.2 secs. Addition of base continued for 170 seconds. The mixtures were adjusted to pH 4.00 ± .05 prior to titration. The thermal curves were displayed on a recorder trace and recorded on punch paper tape as millivolts/sec.

RESULTS AND DISCUSSION

Since the thermograms of some of the mixtures requires 120-150 seconds to complete, the experimental data must be corrected for heat losses from the calorimeter. As a preliminary approximation it is assumed that all heat losses are a result of Newtonian cooling which may be expressed as:

$$dQ/dt = k(T-T_o) \qquad\qquad (1)$$

where T is the temperature at some time (t) after the addition of base at initial temperature, To. The corrected heat is obtained from:

$$Q_{corr} = Q_{obs} + k\int_{t_o}^{t} (T-T_o)dt \qquad\qquad (2)$$

The constant, k, was empirically determined from the thermal titration curves of the standard HCl. This value was found to be 0.0018 ± 0.006 cal/degree second. A proper cooling correction constant is defined as that value which produced a nearly constant μvolt reading after the reaction has been completed. The use of this correction

is demonstrated for the thermogram of a single run of formic acid
(Fig. 1). The initial temperature base line, $\overline{AB}$, is extrapolated
to point C assuming that the temperature equilibrium would continue
in the absence of a reaction. The cooling after the neutralization
of the formic acid, $\overline{DE}$, is the resultant of the heat loss from the
calorimeter and the cooling produced by the addition of base at the
equilibrium temperature to the warmer reaction solution. Correction
for these known losses and for initial warming (or cooling) gives
the corrected thermogram shown in Figure 2.

The curve shown is an average of seven samples of formic acid
titrated on different days. All the values fall within the limits
of the closed circles (± 5 μvolts) after the cooling correction has
been applied with the exception of the points at D. Because of the
ability to average a number of curves of a given compound, it was
felt that the correction was sufficient. For extreme precision,
however, it appears that a simple Newtonian cooling is an oversimplification and a correction for the excess heat at point D must be
applied. Finally, the simple extrapolation of the initial temperature baseline, $\overline{AB}$, should also be represented as a Newtonian function
for a more accurate assessment of thermodynamic values. Failure to
apply these more precise corrections are estimated to cause errors
of 3% in the calculated values. Since all the curves reported here
have been corrected for cooling, the millivolt readings are referred
to as Q_{obs} rather than Q_{corr}.

Calculation of ΔH_i for Model Compounds

The individual heats of ionization for each thermal curve was
calculated by three procedures.

(A). From the height of the thermogram: The heat observed (Q_x)
expressed as millivolts for any acid is:

$$Q_x = (Q_s C_x / C_{HCl})(1 - \Delta H_i / \Delta H_f) \tag{3}$$

where C_x is the concentration of the unknown in μmoles, and C_{HCl} is
the concentration of the standard HCl in μmoles which gives a heat
rise, Q_s, expressed in millivolts. The heat of ionization of the
unknown is ΔH_i and ΔH_f is the heat of formation of water which is
taken as 13,500 cal/mole at $20°C$ (7). When both standard and unknown
are present in the same concentration, the ΔH_i may be obtained from
the uncorrected (Fig.1) or the corrected (Fig. 2) thermogram by:

$$\Delta H_i = \Delta H_f (1 - Q_f / Q_s) \tag{4}$$

(B). From the slope: Because the addition of base is a constant (1.061 μmoles/sec), the observed heat per second is:

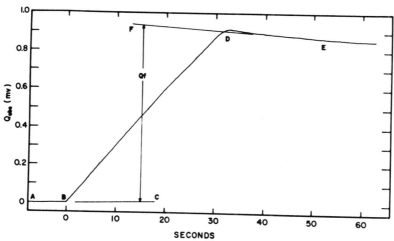

FIGURE 1:   Thermal titration of 32 μmoles formic acid at 20°.   The
segment of the curve (AB) to 0 time represents the thermal equilibration of the system.   This is extrapolated to point C.   Segment $\overline{BD}$
is the result of the heat liberated by titrating the formic acid.
At point D, all the acid is titrated and segment $\overline{DE}$ to point F is a
correction for the cooling.   The heat due to the reaction is represented by Q_f.

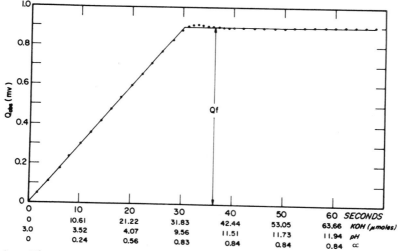

FIGURE 2:   Thermogram for formic acid corrected for thermal equilibrium and cooling according to equation (2).   The closed circles are
the experimental points and the solid line is the thermogram calculated for ΔH_i 388 cals/mole which was found by using equation (4).
The ordinate can be represented as shown.   Seven separate curves
have been corrected (values fall within the points) and averaged.

$$Q_x/\text{sec} = (Q_s C_b/C_{HCl})\,(1 - \Delta H_i/\Delta H_f) \tag{5}$$

where C_b is μmoles titrated/sec. This may be modified to obtain the heat rise for time, t, by:

$$Q_{xt} = (Q_s C_b t/C_{HCl})\,(1 - \Delta H_i/\Delta H_f) \tag{6}$$

At 30.2 seconds, 32 μmoles have been titrated and the calculated curve in Figure 2 can be obtained.

(C). From the amount titrated as a function of pH. When the titration curve of the sample and the amount of base added is known, then the ordinate can be expressed as pH or as the amount (α) of the substance titrated. The heat liberated is:

$$Q_x = (Q_s C_x/C_{HCl})\,(1 - \Delta H_i/\Delta H_f)\,(\alpha) \tag{7}$$

where α is the mole fraction titrated which according to the Henderson-Hasselbalch expression for the mass law is:

$$\alpha = N/(1 + 10^{(pK' - pH)}) \tag{8}$$

Equation 7 now expresses the observed heat as a function of the concentration of the sample, its heat of ionization, its ionization constant and the hydrogen ion concentration. Currently, it is possible to convert the time axis of the thermogram by using an independently obtained potentiometric curve as illustrated in Figure 2.

Results of these analytical procedures are shown in Table 1. The pK' values are those found for 32 μmoles of the model compounds and they are estimated to be accurate to $\pm$ 0.03 units. The accuracy of the ΔH_i values are estimated to be about $\pm$ 250 cals/mole for all the ionizations except for formic acid and the carboxyl group of glycylglycine.

When the carboxylic acids are titrated, there is a significant amount of the acid already ionized to form free hydrogen ions. This is illustrated in Figure 3. The free hydrogen ions have ΔH_i 0 and the ΔH_i of formic acid (388 cals/mole) is in error by the amount of hydrogen ion present. The approximate heat of ionization for formic may be obtained as shown in Figure 4. From the curves, 27.2 μmoles formic acid produces a millivolt rise of 0.80. An equivalent amount of HCl would produce a rise of 0.78. Using equation 3, ΔH_i of formic is - 188 cals/mole. This correction has not been applied to the mixtures because their titration begins at pH 4 where there is little (10^{-4} M) free hydrogen ion. At this pH, moreover, most of the carboxyl groups have already been ionized and only 15% of the carboxyl groups of glycylglycine and 35% of the formic acid are protonated. Under these conditions, the heats of ionization of the carboxylic

Table 1: IONIC PARAMETERS OF MODEL COMPOUNDS AT 20°C

COMPOUND	pK'		ΔH_i (cal/mole)			
	EQUATION 8[1]	LITERATURE[2]	EQUATION 3	EQUATION 4	EQUATION 7	LITERATURE[2]
Glycylglycine	3.23	3.09 - 3.29	347[3]	367[3]	---[4]	32 to 1600
	8.38	8.23 - 8.46	10340	10258	10434	10600 to 11600
Formic	3.73	3.75 - 3.74	438[3]	388[3]	---[4]	-400 to 1363
Pyridine	5.34	5.17 - 5.30	4870	4970	4805	4280 to 5230
Imidazole	7.16	6.99 - 7.20	8626	8650	8579	7100 to 8790
β-Mercapto-ethanol	9.74	9.72	6157	6441	6102	6210
Phenol	10.02	9.88 - 10.0	5886	5764	5584	4800 to 5660
Methylamine	10.98	10.59 - 10.79	12384	12620	12206	12720 to 13500

[1] Obtained by fitting equation 8 to the potentiometric titration of 32 μmoles of the compound.
[2] Compiled from references (6,8,9,10,11) using the data reported at low ionic strength (<.1) and at 18-25°C.
[3] These values are not corrected for the free acidity.
[4] A precise evaluation of these values will be reported in a later communication.

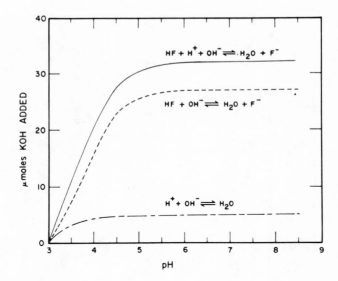

FIGURE 3: *Potentiometric titration curve of 32 µmoles of formic acid. The solid line is the total titration. The broken line is the curve for the free acidity. The dashed line is the actual titration of formic acid from pH 3.0.*

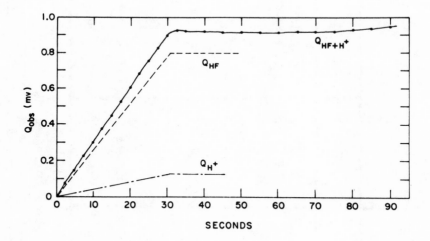

FIGURE 4: *Correction of thermal curve of formic acid for the free acidity at pH 3.0. The dashed line, Q_{HF}, represents the corrected thermal curve for the unionized formic acid.*

acid residues are not considered to be reliable.

An example of the analytical procedure is shown in Figures 5 and 6. The potentiometric curve for a compound is analyzed for the pK' and the concentration, N, of the compound. These values are then used to calculate the potentiometric curve. Agreement of the calculated curve and experimental titration is within 0.02 groups at any pH. For simple models, there is never any real discrepancy between the values found by fitting and the actual values. These values and the known heat of ionization are used to calculate the thermal curve (solid line, Fig. 6). The agreement of calculated and experimental values is shown.

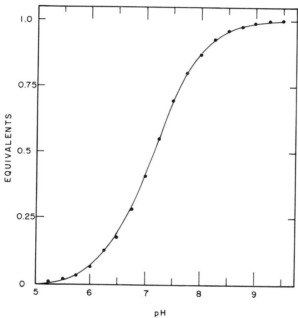

FIGURE 5: *Potentiometric curve for 32 μmoles imidazole. The solid line is that calculated for the ionization of an acid with pK' 7.16 in the concentration 32.32 μmoles. These values were derived from the curve fitting procedure using equation 8.*

It was of interest to see how small changes in any parameter might influence the calculated curve. This is shown for the thermal titration of pyridine (Fig. 7). When pK' was increased by 0.1 units, the thermal curve shifts along the pH axis (maximum deflection at pH 5.34 is 33 μvolts), and the calculated and experimental curves are convergent only at extremes of the curves. Increasing ΔH_i by 500 cals/mole causes a divergence of 33 μvolts. Increasing N by 10% causes a deviation of 60 μvolts. It appears that a thermal curve is

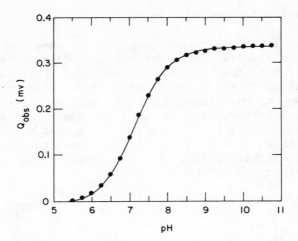

FIGURE 6:   Thermogram for 32 µmoles imidazole. The solid line is
the thermal titration calculated for an acid with pK' 7.16, N of 1.01
and ΔH_i 8574 cals/mole. The experimental points found for the titration of three separate samples of imidazole fall within the closed
circles.

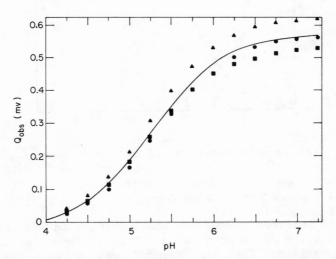

FIGURE 7:   The effect of changing parameter values on a thermal
titration. The solid line is that calculated for an acid with pK'
5.34, N of 1.0 and ΔH_i 5000 which are the values found for pyridine.
The triangles are the curve calculated for an increase in N to 1.1.
The circles are the curve calculated for a change in pK' to 5.44.
The squares represent a change in ΔH_i to 5500.

sensitive to small changes in these three parameters.

Calculated and experimental curves for all the compounds tested, except formic acid, are shown in Fig. 8. Only the carboxyl group of glycylglycine shows a deviation in excess of 5 μvolts for the reasons previously given.

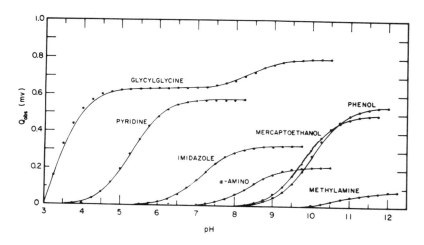

FIGURE 8:   Thermograms of model compounds.

Heats of Ionization for Mixtures

For the mixture consisting of 16 μmoles of pyridine (pK' 5.34) and 16 μmoles of imidazole (pK' 7.16) (Fig. 9), it is possible to determine each heat of ionization by slope analysis as previously described (6) because there is little overlap in the ionizations. It is not possible to calculate a thermal curve for timed addition of base because the two ionizations occur at differing hydrogen ion concentrations. When the thermal curve is presented as a function of pH (Fig. 10) then it is possible to calculate the thermal curve from:

$$Q_{obs} = (Q_s C_x \Delta H_f / C_{HCl}) \{ (N_1 (\Delta H_f - \Delta H_1))/(1 + 10^{(pK_1' - pH)}) +$$
$$(N_2 (\Delta H_f - \Delta H_2))/(1 + 10^{(pK_2' - pH)}) + \ldots$$
$$(N_8 (\Delta H_f - \Delta H_8))/(1 + 10^{(pK_8' - pH)}) \} - B \quad (9)$$

where C_x is the concentration of any component of the mixture (this is equivalent to protein concentration if the lowest concentration is selected) and N_1 to N_8 represents groups ionizing with pK_1' to pK_8' and with heats of ionization ΔH_1 to ΔH_8. The equation expresses the

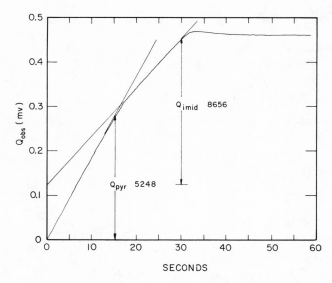

FIGURE 9: Thermogram for pyridine (16 μmoles) and imidazole (16 μmoles) illustrating derivation of ΔH_i by slope analysis (eq. 5).

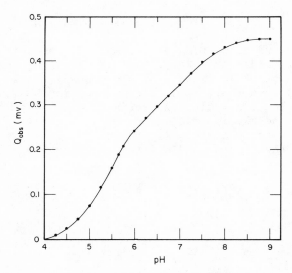

FIGURE 10: Thermogram of pyridine and imidazole as a function of pH. The solid line is calculated for the constants given in Table 1, using equation 9. The circles are the experimental points.

sum of all the heats from complete protonation to salt formation. Since the experiment need not begin at a sufficiently low pH to insure complete protonation, it is necessary to substract this amount of heat, B, which would have evolved prior to the pH at the beginning of the experiment.

Using the heats of ionization shown in Table 1 and the pK's and N's evaluated by the computer assisted curve fitting to the potentiometric titration, the calculated curve and the experimental points are coincident (Fig. 10). Coincidence is accepted as finding the experimental points to be within 10 μvolts of the calculated values. Under these conditions the variation in pK' can be less than 0.1 units, the N may vary less than 10% and ΔH_i must be within 500 cals/mole.

Curves for a number of equimolar mixtures were calculated. All of the calculated and experimental curves agreed within the limits specified. Some of these curves are shown in Figure 11. When the equimolar mixture of eight different ionizing groups were varied to include an additional equivalent of pyridine or imidazole, the curves were also of excellent concurrence (Fig. 12).

Evaluation of pK', N and ΔH_i by Curve Fitting

The excellence of the concurrence of the calculated curves with the experimental curves when the appropriate constants are used would lend support to the suggestion that pK' and N values calculated for a protein by the curve fitting procedure are of some substance.

We have attempted to evaluate pK' and N using the thermal curve alone. A number of sets of parameters may be obtained all of which represent the thermal curve equally well. However, only one set of parameters also represents the potentiometric curve. Obviously, a proper set of constraints for the potentiometric and thermal curves would arrive at a similar solution for both sets of data. It would also appear that potentiometric curve serves as a constraint on the evaluation of the thermal constants in equation 9 and the thermal titration serves as a constraint on the number of solutions for the potentiometric data. The curve fitting program was therefore utilized to simultaneously fit both sets of data with the parameters pK', N and ΔH_i. For all the curves analyzed, only one set of solutions was returned and these are shown in Table 2. Because the values are so nearly those actually used, it seems that it is possible to determine the ionization constants, the number of groups ionizing in any set and the heats of ionization of these groups without recourse to ancillary information. Should this analysis be equally applicable to proteins, it will be possible to define the equilibrium constants and thermodynamic constants for the ionic

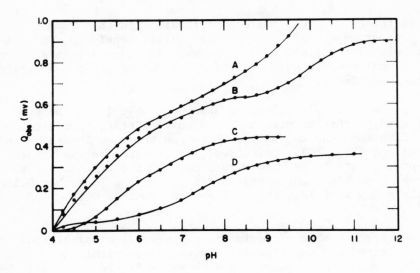

FIGURE 11:   Thermograms of equimolar mixtures.   Curve A is the thermogram for mixture EQ3.   Curve B is for a mixture containing formic,
pyridine, imidazole and phenol.   Curve C is for pyridine and imidazole.   Curve D is for glycylglycine and imidazole.

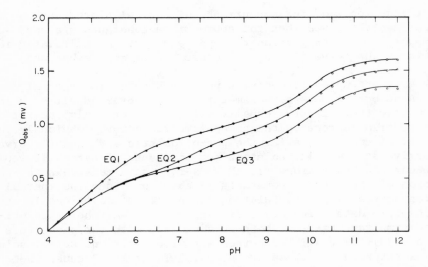

FIGURE 12:   Thermograms for mixtures.   EQ3 contains glycylglycine,
formic, pyridine, imidazole, β-mercaptoethanol, phenol and methylamine each in 0.004 M concentration.   EQ2 contains twice the concentration of imidazole and EQ1 contains twice the concentration of
pyridine.

Table 2: PARAMETER VALUES FROM SIMULTANEOUS CURVE FITTING*
OF POTENTIOMETRIC AND THERMAL TITRATIONS OF MIXTURES**

COMPOUND	EQ3			EQ2			EQ1		
	pK'	N	ΔHi	pK'	N	ΔHi	pK'	N	ΔHi
Glycylglycine									
pK_1'	3.09	0.99	141	3.17	0.99	-115	2.97	1.00	-298
pK_2'	8.38	1.02	10474	8.43	0.98	10452	8.38	0.98	10589
Formic	3.79	0.99	-345	3.76	0.99	-453	3.83	1.00	-588
Pyridine	5.35	0.99	4747	5.39	1.00	4709	5.33	2.00	5051
Imidazole	7.19	1.01	8652	7.18	1.98	8698	7.18	1.02	8802
β-Mercaptoethanol	9.73	1.02	} 12990	9.75	0.98	} 12980	9.67	1.01	} 12720
Phenol	10.02	1.01		10.07	0.98		10.09	1.03	
Methylamine	11.04	1.03	12447	11.20	0.98	12278	11.03	0.98	12711

*All the values were obtained without constraints.

**Titrations of EQ3 were conducted on mixtures containing 4 μmoles/ml of each component at pH 4.0.
EQ2 contained 8 μmoles/ml imidazole. EQ1 contained 8 μmoles/ml pyridine.

residues. This would, according to Edsall (12), permit the identification of all these groups in the protein. In another paper (13),
mathematical procedures have been developed for the calculation of
the ΔH_i values of the ionizable groups of proteins with the knowledge of the pK' and N values and the information contained in a
single thermal titration of the protein.

This process cannot be utilized to separate two sets of groups
with similar ionization constants. For β-mercaptoethanol (pK' 9.74)
and phenol (pK' 10.02), the ionizations overlap to such an extent
that the curve fitting program returns a number of values for the
two heats of ionization, the sums of which are correct (Fig. 13).
With this single exception, it now appears to be possible to obtain
the number of groups with their ionization constants and the heats
of ionization on the basis of the potentiometric and thermal curves
alone. The process is quite sensitive to small variations in each
of these parameters and requires a precise set of data for the dissociation curves. The thermal curves, in our hands, have proved to
be reasonably precise and of good quality.

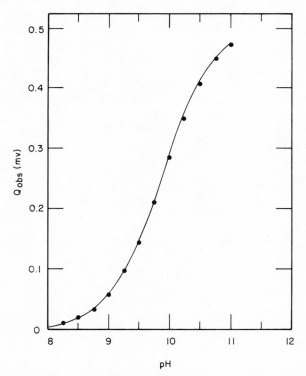

FIGURE 13: Thermogram for β-*mercaptoethanol and phenol. The line
is calculated for* ΔH_i *values of 6200 and 5800, respectively. The
points are calculated for* ΔH_i *values of 8000 and 4000.*

Eventually, it is hoped that the thermal and potentiometric procedures may be combined. Currently, we are working on this aspect which should offer a great saving in time and materials needed as well as providing an additional improvement in the precision.

SUMMARY

Mixtures of model compounds containing ionizing residues representing those commonly found in proteins have been titrated and the resulting potentiometric curves have been analyzed for the ionization constants (pK') and the concentrations of each group (N) by means of a computer assisted curve fitting procedure. Such an analysis allows a number of solutions and the correct constants are obtained when constraints are introduced. For many proteins, however, the proper constraints may not be known. It is possible to determine which set of ionic parameters is correct by means of thermal titrimetry. The experimental heat curve (thermogram) of the protein is obtained. Using the ionic parameters which have been evaluated from the potentiometric curve and the known heats of ionization, a thermogram is calculated. The calculated and experimental curves are coincident only where the proper ionic parameters are used. Alternately, computer assisted solution of thermogram for pK' and N yield values which may be compared to those found for the potentiometric curve. Only those values which are coincident can be used to adequately explain both sets of experimental data. Thus, utilizing the two separate experimental approaches, potentiometric and thermal titrations, it is possible to obtain the proper pK', N, and ΔH_i for the mixtures which have been studied. On this basis, it would appear to be possible to obtain the ionic parameters of proteins on the basis of a single thermal titration.

ACKNOWLEDGEMENT

This work was supported, in part, by research grant HE 14083 from the National Heart and Lung Institute and by research grant GM 10902 from the Institute of General Medical Sciences, National Institutes of Health, U. S. Public Health Service and by research grant GB 33473 from the National Science Foundation.

REFERENCES

1. R. K. Cannan, Chem. Rev., 30, 395 (1942).
2. J. Wyman, J. Biol. Chem., 127, 1 (1939).
3. J. Jordan and T. G. Alleman, Anal. Chem., 29, 9 (1957): H. J. V. Tyrrell and A. E. Beezer, "Thermometric Titrimetry", Chap-

man and Hall, London, 1969; L. S. Bark and S. M. Bark, "Thermometric Titrimetry", Pergamon, New York, 1969; and M. A. Marini and C. J. Martin, Methods in Enzymology, 27, 590 (1973).

4. M. A. Marini, C. J. Martin, R. L. Berger and L. Forlani, Submitted to Biopolymers. The Computer program is available upon request.

5. M. A. Marini and C. Wunsch, Biochemistry, 2, 1454 (1964).

6. M. A. Marini, R. L. Berger, D. P. Lam and C. J. Martin, Anal. Biochem., 43, 199 (1971).

7. J. J. Christensen, G. L. Kimball, H. D. Johnson and R. M. Izatt, Thermochim. Acta, 4, 171 (1972).

8. D. D. Perrin, "Dissociation Constants of Organic Bases in Aqueous Solution", Butterworths, London, 1965.

9. C. Kortum, W. Vogel and K. Andrussow, "Dissociation Constants of Organic Acids in Aqueous Solution", Butterworths, London, 1961.

10. R. M. Izatt and J. J. Christensen, in "Handbook of Biochemistry", (H.A. Sober, ed.), J-49, Chemical Rubber Co., Cleveland, 1968.

11. E. S. Domalski, data from the master file of the NBS Chemical Thermodynamics Data Center, personnal communication.

12. E. J. Cohn and J. T. Edsall, in "Proteins, Amino Acids and Peptides as Ions", Reinhold Pub. Corp., New York, 1943.

13. C. J. Martin, B. R. Sreenathan and M. A. Marini, These Proceedings.

CALORIMETRIC INVESTIGATION OF CHYMOTRYPSIN IONIZATION REACTIONS

Charles J. Martin and Bangalore R. Sreenathan
Department of Biochemistry
The Chicago Medical School
University of Health Sciences
Chicago, Illinois 60612

Mario A. Marini
Department of Biochemistry
Northwestern University Medical and Dental Schools
Chicago, Illinois 60611

On the basis of a series of investigations concerned with the effect of formaldehyde on chymotrypsin-catalyzed reactions, the cumulative evidence has led to the interpretation that formaldehyde is an active site label and forms an N-hydroxymethyl derivative with the imidazole moiety of the histidine residue, presumably His-57, involved in the rate-determining step for the catalysis of acylated aromatic amino acid esters. The consequence of this is that chymotrypsin becomes a less efficient catalyst (about 8-fold) although all active sites remain functional. The evidence leading to this conclusion has been presented in several publications (1-7), and encompasses experimental approaches involving potentiometric titration, kinetic, and spectroscopic studies.

In addition to the mechanistic involvement of the His-57 residue in chymotrypsin-catalyzed hydrolyses, presumably as a partner in a coupled "charge-relay" system that also involves Asp-102 and Ser-195 (8), evidence for the requirement of another group has also been presented (9-14). This group, with a pK' about 9, is believed to be the α-amino group of Ile-16 in its protonated state (15-19) and to exist in ion-pair bond with the carboxylate anion of Asp-194 (20,21). Functionally, this grouping has been postulated to regulate the concentration of the requisite active site conformation necessary for the formation of enzyme-substrate complexes. This concept has been questioned, however, from analysis of data obtained from potentiometric titration studies (3,5,22) and the results of chemical modification studies (23,24). The α-amino group of Ile-16 in δ-chymotrypsin

can be blocked with either an amidine group (23) or with a succin-
ate group (24) with full retention of activity.

Recently, we have explored the potentials of thermal titrimetry as an additional probe in the continuance of our investigations
in chymotrypsin-catalyzed reactions. Thus far, we have been concerned with the development of procedures for the determination of
the heats of ionization of weak protic acids, singly or in admixture, and of the dissociable groups in proteins. Over and above
the general applicability of such investigations, the determination of the heat of ionization of the α-amino group of the Ile-16
residue in chymotrypsin could bear on the postulated existence of
its participation in ion-pair with the carboxyl group of Asp-194.
In a previous publication (25), the heats of ionization of the
commonly occurring amino acids have been determined. Treatment
of calorimetric data obtained from the thermal titration of a
mixture of two amino acids with overlapping ionizations has also
been presented (26). The results of this paper extend these
studies and demonstrate that heats of ionization of the ionizable
group-sets of proteins can be calculated from a single thermal
titration provided that the pK's and the number of groups in each
group-set are known. In a related paper (27), a computer-assisted
curve-fitting procedure for the pK' values, the concentration of
groups, and the heats of ionization that adequately fit both the
potentiometric & thermal titration data of a mixture of ionizable
groups is described.

EXPERIMENTAL PROCEDURES

The chymotrypsinogen-A preparations used were obtained from
either the Armour Pharmaceutical Co. or the Worthington Biochemical
Corp. Amino acids and other chemicals were of reagent grade
quality. Fully acetylated chymotrypsinogen-A (Ac-CTG) and acetylated-δ-chymotrypsin (Ac-δ-CT) were prepared as previously described
(23). For use in calorimetric studies, both preparations were
purified by passage through Sephadex G-50 at pH 6.5, dialyzed
exhaustively and lyophilized.

Potentiometric titrations, from which pK' values for ionizable
groups or group-sets were derived, were done by published procedures (5,22,28). Thermal titrimetry was employed for all calorimetric measurements using a prototype instrument kindly supplied
by the American Instrument Co. A complete description of the
apparatus and of its operation has been published elsewhere (26).
The design of the reaction cell permits of rapid thermal equilibration of both the reaction solution and of titrant. Prior to titrant addition by means of a constant speed syringe pump, the reac-

tion solution can be effectively shielded from the external environment over a short time interval. Changes in the temperature of the reaction solution are sensed by a computer-matched twin thermistor probe (resistance, 17 kohms; response ca 25 msecs) with the output signal fed through a Keithley Microvoltmeter, model 155 and displayed on a model E1101S Esterline-Angus chart recorder. The constant temperature bath was regulated with a Tronac model 1040 controller. Once thermal equilibration of the reaction solution was achieved, the temperature variation without titrant addition showed an amplitude of 0.002° within a sinusoidal period of about 25 min.

Standardization of instrument response was routinely checked by the neutralization of 32 µmoles HCl in a 4.00 ml reaction volume by titration with M/1 KOH. In this reaction, the heat of mixing is negligible and requires 0.032 mls titrant (32 µmoles) for complete neutralization of the HCl. With titrant added at a constant rate (1.333 µmoles/sec) and since the fraction of H^+ neutralized is proportional to the equivalents of base added, the heat generated is proportional to the base added throughout the course of the reaction. A representative standard curve is shown in Fig. 1. The formation of 32 µmoles HOH generates a recorder response equal to 0.82 mV as measured by the height of the curve between the extrapolated low and high temperature baselines at the midpoint of the titration. Using a value of -13.5 ± 0.05 kcal/mole (29) for the heat of formation of HOH (ΔH_F), this yields a calibration constant (C_S) of 0.527 cal/mV and equals a temperature change of 0.132°/mV. From experiment to experiment, the determination of C_S fell within the range of 0.494 to 0.527 cal/mV. The accuracy of the calibration was confirmed by determining the heat of protonation of an analytical preparation of TRIS by thermal titration downscale with M/1 HCl. At 20° the value of -11.4 kcal/mole was obtained in good agreement with literature values of 11.35 to 11.39 kcal/mole (30) for the heat of ionization at 25°.

The mV response from the titration of varying concentrations of HCl in a constant volume is linear full-scale from 0 to 1.0 mV. A constant mV response is elicited from the titration of differing reaction volumes of the same HCl concentration. These results show that the observed mV changes adequately reflect temperature changes of the reaction solution and that for the short time duration wherein measurements are made, there are no appreciable heat losses from the reaction system.

All measurements have been made at the initial temperature of 20° and all reaction solutions (4.00 ml) contained 0.15 M KCl. Without exception, all thermal titrations were done by the addition of base to an initially acid solution (pH 4 to 5).

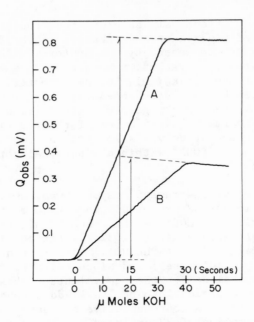

Fig. 1. The thermal titration of 32 μmoles HCl (curve A) or 40 μmoles imidazole (curve B) in 4.0 ml 0.15 M KCl.

RESULTS AND DISCUSSION

In the thermal titration of a weak protic such as imidazole, the heat liberated as a consequence of the ionization reaction is less than that seen in the formation of an equimolar amount of HOH from the HCl + KOH reaction by the amount of energy required to ionize the proton from imidazole (Fig. 1). Although such reactions are generally endothermic, the net reaction is exothermic. The thermogram exhibits only a single slope and the time span is equal to the μmoles base required for the complete titration of the acid present (40 μmoles). This serves as a control check on the analytical concentration of the sample. From such data, the heat of ionization, ΔH_i, of any acid HA can be calculated from

$$\Delta H_i = \Delta H_F - (VC_sQ/V_sN) \tag{1}$$

where V_s equals the reaction volume of the standard used for the determination of C_s and Q the observed heat (Q_{obs}) for the complete titration of N Moles HA in volume V. The value of Q is obtained from the height of the curve at the midpoint of the titration between the extrapolated high and low temperature baselines. The midpoint of the titration occurs at the pK' of the acid.

In the titration of a mixture of two acids with pK' values that are sufficiently separated so that their ionizations do not overlap appreciably and with different heats of ionization, the thermogram will consist of two distinct slopes joined by a small transitional curve and each slope will span a segment on the abscissa (μmoles KOH) equal to their respective concentrations in μmoles. The acid with the lower pK' will be the first acid neutralized in an upscale titration. From such data, evaluation of the respective slope heights permits determination of the ΔH_i value of each acid. Examples of this case have been given elsewhere (25).

In contrast, when the ionizations of two acids in a mixture occur within overlapping pH regions, any point on the thermogram will contain contributions from both dissociation reactions. When this occurs, attempts at the determination of the heats of ionization by simple slope height analysis will be in error with the magnitude dependent on the difference between the two pK's and the ΔH_i values (26). In the special case where the ionization constants are equal, only a single slope will be obtained and Q_{obs} will represent the average of the two heats of ionization.

For the general case of a mixture of $\underline{n}$ weak protic acids HA_1, HA_2, ... HA_j at concentrations N_1, N_2, ... N_j (in μmoles) with the negative logarithm of the ionization constants expressed as pK_1', pK_2', ... pK_j', the value of Q_{obs} at any point, P_i, on the thermogram (corrected for baselines) will be given by

$$P_i = a_{i1}N_1Q_1 + a_{i2}N_2Q_2 + \ldots a_{ij}N_jQ_j \tag{2}$$

where a_{i1}, a_{i2}, ... a_{ij} represents the ith fraction $(0 \leq i \leq 1)$ of the acids HA_1, HA_2, ... HA_j in the anionic form and Q_1, Q_2, ... Q_j are equal to the change in mV/μmole generated by their complete ionization. When all acids of the mixture are completely titrated, i.e., when i = 1, the total heat (Q_T) in mV is given by

$$Q_T = N_1Q_1 + N_2Q_2 + \ldots N_jQ_j. \tag{3}$$

At any stage of the titration, the abscissa coordinate, equal to the μmoles KOH added which in turn is equal to the group equivalents titrated (in µmoles), is given by

$$\text{Group Equivalents} = a_{i1}N + a_{i2}N_2 + \ldots a_{ij}N_j \qquad (4)$$

For any mixture of $\underline{n}$ acids then, the values of Q_1, Q_2, $\ldots$ Q_j can be calculated from the data of a single thermal titration using the coordinates of n-1 points on the thermogram and the coordinate values for the complete titration. For each point chosen, the fraction of each acid ionized must be known and this can be calculated from the pK' values using the Henderson-Hasselbalch equation. If we define

$$\Delta pK'_{12} = pK'_1 - pK'_2 \qquad (5)$$

$$\Delta pK'_{23} = pK'_2 - pK'_3 \qquad (6)$$

$$\vdots \qquad \vdots$$

$$\Delta pK'_{(j-1)j} = pK'_{(j-1)} - pK'_j \qquad (7)$$

with $pK'_j > pK'_{(j-1)} > \ldots pK'_1$, then, for any selected value of the fraction of the acid HA_1 ionized, say a_{i1}, the extent of the protonic dissociation of all other acids in the mixture relative to that of HA_1, i.e., at the same pH, can be calculated from

$$a_{i2} = \frac{a_{i1}(10^{\Delta pK'_{12}})}{1 + a_{i1}(10^{\Delta pK'_{12}} -1)} \qquad (8)$$

$$a_{i3} = \frac{a_{i2}(10^{\Delta pK'_{23}})}{1 + a_{i2}(10^{\Delta pK'_{23}} -1)} \qquad (9)$$

$$\vdots \qquad \vdots$$

$$a_{ij} = \frac{a_{i(j-1)}(10^{\Delta pK'_{(j-1)j}})}{1 + a_{i(j-1)}(10^{\Delta pK'_{(j-1)j}} -1)} \qquad (10)$$

Successive calculations using additional selected values for the fractional extent of ionization of the acid HA_1 yields sets of values for the fraction ionized of other acids in the mixture. Using any set, the group equivalents (in µmoles) titrated of all acids in the mixture can be equated to the µmoles KOH added at any stage of the titration by means of eq. 4. The values of Q_1, Q_2, $\ldots Q_j$ can then be calculated by solution of $\underline{n}$ simultaneous equations of the form

$$P_1 = a_{11}N_1Q_1 + a_{12}N_2Q_2 + \ldots a_{1j}N_jQ_j \tag{11}$$

$$P_2 = a_{21}N_1Q_1 + a_{22}N_2Q_2 + \ldots a_{2j}N_jQ_j \tag{12}$$

$$\vdots \qquad\qquad \vdots$$

$$P_{(n-1)} = a_{(n-1)1}N_1Q_1 + a_{(n-1)2}N_2Q_2 + \ldots a_{(n-1)j}N_jQ_j \tag{13}$$

and

$$Q_T = N_1Q_1 + N_2Q_2 + \ldots N_jQ_j \tag{14}$$

Once Q_1, Q_2, $\ldots Q_j$ are known, their corresponding ΔH_i values can be calculated from eq. 1 using 10^{-6} as the value of N.

A Monroe model 1860 calculator with computer program 1300A (method of Crout) has been used for the solution of the simultaneous equations. In practice, each thermogram is sampled for 5 to 7 sets of the necessary number of P points and the average of each collection of calculated Q values used for the calculation of the heats of ionization. If the number of different acids in a mixture are no more than three, Q values can be calculated by data substitution into equations for Q_1, Q_2, and Q_3 (vide infra).

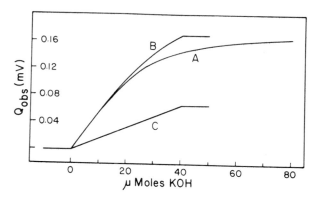

Fig. 2. Curve A. Thermogram for titration of 20 μmoles lysine corrected for low temperature baseline. Curve B. Thermogram of lysine (Curve A) corrected for the base uptake of the water solvent. Curve C. Thermal titration of 40 μmoles N^α-acetyl-lysine corrected for low temperature baseline and base uptake of the water solvent.

However, when the number of acids exceeds three, this type of calculation becomes progressively more tedious as n increases.

Thermal titration of lysine. This amino acid has two ionizable groups: an α-amino group with pK' 9.25 and ε-amino group with
pK' 10.42. The analysis of the thermogram obtained with this compound is complicated by the fact that the complete titration of
the ε-amino group extends into a pH region wherein the uptake of
base due to the titration of water becomes significant. Thus, in
the thermal titration of 20 μmoles lysine (40 μmoles of titratable
groups), an abrupt change in the slope of the thermogram, which
would be indicative of stoichiometric completion of the reaction,
is not observed (Curve A, Fig. 2). Rather, the increase in mV
(temperature) continues with the addition of KOH beyond this point
and reflects the continued heat change by the as yet incomplete
neutralization of the ε-amino group. This elongation of the
thermogram due to base uptake in the titration of water has been
labeled the "thermogram stretch" effect (26). Its correction is
important in the thermal titration of proteins and can be corrected
for by measurement of the pH of the reaction solution at various
additions of KOH and subtracting the μmoles of KOH required to titrate the same volume of solvent to the same pH. When such corrections are applied, Curve B of Fig. 2 is the result. Since the
pK's of the two groups are only about one unit apart, their overlapping ionization precludes the presence of two well-defined
slope regions. In this case, the Q values can be calculated from

$$Q_1 = \frac{P_1 - a_{12}Q_T}{N_1 (a_{11}-a_{12})} \tag{15}$$

$$Q_2 = \frac{Q_T - N_1Q_1}{N_2} \tag{16}$$

The data yields 10.47 kcal/mole for the heat of ionization of the
α-amino group and 12.23 kcal/mole for the heat of ionization of
the ε-amino group. In confirmation of the latter value, the thermal titration of 40 μmoles of N^{α}-acetyllysine (one titratable
group), after correction for the "stretch effect" (cf. Curve C,
Fig. 2), gives the value of 12.64 kcal/mole for the heat of ionization of its ε-amino group. The agreement between the latter two
values would indicate that the method of data treatment applied to
the solution of ΔH_i values for two groups with overlapping ionizations is valid.

Thermal titration of a three group model mixture. The thermogram for a mixture of 20 μmoles imidazole (pK'$_1$, 7.23), 10 μmoles
TRIS buffer (pK'$_2$, 8.21), and 30 μmoles acetyl-L-tyrosine ethyl
ester (pK'$_3$, 9.93) is shown in Fig. 3. This mixture represents

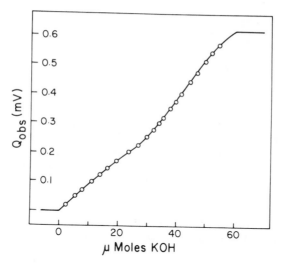

Fig. 3. Thermal titration of a mixture of 20 μmoles imidazole, 10 μmoles TRIS, and 30 μmoles acetyl-L-tyrosine ethyl ester.
The curve has been corrected for high and low temperature baselines. The circles represent the theoretical thermogram of the
mixture calculated from heats of ionization data extracted from
the experimental curve (cf. text).

three of the ionizable groups present in Ac-δ-CT; imidazoyl, amino,
and phenolic hydroxyl. The curve has been corrected for the slope
of the upper and lower temperature baselines. It will be noticed
that the slope of the curve becomes steeper after about 30 μmoles
of KOH have been added. To a degree dependent on its concentration, this will always be seen when a group such as a phenolic hydroxyl has a higher pK' but a lower ΔH_i than the group preceding
its titration. The Q values, and hence the heats of ionization,
were obtained from

$$Q_1 = \frac{(a_{22}-a_{23})P_1-(a_{12}-a_{13})P_2 + [a_{23}(a_{23}-a_{13})-a_{13}(a_{22}-a_{23})]Q_T}{N_1\ [a_{11}-a_{13})(a_{22}-a_{23})-(a_{12}-a_{13})(a_{21}-a_{23})]} \quad (17)$$

$$Q_2 = \frac{P_2 - (a_{21} - a_{23})N_1Q_1 - a_{23}Q_T}{N_2 \; (a_{22} - a_{23})} \tag{18}$$

$$Q_3 = \frac{Q_T - (N_1Q_1 + N_2Q_2)}{N_3} \tag{19}$$

The values obtained for the heats of ionization of each component of the mixture are shown in Table I. The agreement with the ΔH_i values obtained by the thermal titration of each compound singly is essentially exact. The value for the heat of ionization of TRIS is about 300 cal. less than obtained previously (26) by measurement of the protonation reaction.

Using the Q values calculated from the thermogram of the mixture, the theoretical thermogram was calculated (cf. Fig. 3) using

$$Q_{obs} = a_{i1}N_1Q_1 + a_{i2}N_2Q_2 + a_{i3}N_3Q_3 \tag{20}$$

and

$$\mu Moles \; KOH = a_{i1}N_1 + a_{i2}N_2 + a_{i3}N_3 \tag{21}$$

TABLE I

DETERMINATION OF THE HEATS OF IONIZATION OF
IMIDAZOLE, TRIS, AND ACETYL-L-TYROSINE ETHYL
ESTER IN ADMIXTURE

Compound in Mixture	Concentration (µMoles/4.0 ml)	ΔH_i (kcal/mole) Calc.	Found[a]
Imidazole	20	8.53	8.53
TRIS	10	11.08	11.06
ATEE	30	6.08	6.05

a - Calculated from the results of the thermal titration
of each compound singly.

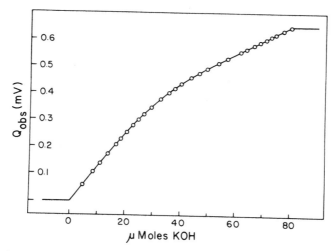

Fig. 4.   Thermal titration of a mixture of 20 μmoles imidazole, 20 μmoles histidine, and 20 μmoles TRIS corrected for high
and low temperature baselines.   The theoretical curve (o) was
calculated as described in the text.

TABLE II

DETERMINATION OF THE HEATS OF IONIZATION OF
THE IONIZABLE GROUPS IN AN EQUIMOLAR MIXTURE
OF IMIDAZOLE, HISTIDINE, AND TRIS

Group in Mixture	Concentration (μMoles/4.0 ml)	ΔH_i (kcal/mole) Calc.	Found[a]
Imidazoyl (pK_1')	20	6.64	6.47
Imidazole (pK_2')	20	8.89	8.53
TRIS (pK_3')	20	11.45	11.06
α–Amino (pK_4')	20	10.87	10.66

a – Calculated from the results of the thermal titration of each
    compound singly.

to calculate the ordinate and abscissa coordinates of the curve using successive sets of values representing the distribution of the ionized state of each component of the mixture at any point on the abscissa.

Thermal titration of a four group model mixture. It was of interest to determine the degree of success that could be achieved by attempting to calculate the heats of ionization of the groups of a mixture wherein the pK's of each group were about one pH unit apart from each other. For this purpose a mixture composed of 20 μmoles histidine (pK$_1'$, 6.23 for the imidazoyl ring; pK$_4'$, 9.38 for the α-amino group), 20 μmoles imidazole (pK$_2'$, 7.23), and 20 μmoles TRIS (pK$_3'$, 8.21) in a 4.0 ml volume was thermally titrated (Fig. 4). The calculated heats of ionization from the Q values obtained with the computer program are given in Table II. The agreement between calculated and found values is considered acceptable and the calculated ΔH_i values were used to generate the theoretical thermogram (circles, Fig. 4) using extensions of eq. 20 and 21.

Thermal titration of proteins. Although a gross oversimplification, a protein can be considered a complex mixture of weak

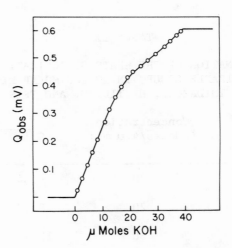

Fig. 5. Thermogram for the titration of 4.15 μmoles acetylated-chymotrypsinogen at the initial pH of 5.0 corrected for baselines (solid line). The circles represent the theoretical curve calculated from the parameters given in Table III.

protic acids. Except under special circumstances, the pK' of a single ionizable group in a protein cannot be determined although reasonable estimates of group set pK's can be obtained from the analysis of potentiometric titration data (3). Titration results, in conjunction with or supplemented by the results of amino acid analysis, spectrophotometric determination of the sulfhydryls and of the phenolic hydroxyls, etc. will also yield the number of ioni- zable groups in each group set, i.e., the number of carboxyls, imidazoyls, α- and ε-aminos, phenolic hydroxyls, and sulfhydryls. When such information is known for any one protein, it should be possible to calculate the heats of ionization of the dissociable groups of each group set by thermal titration of the protein.

We have chosen to treat two simpler cases first since their analysis provides a more direct path to the determination of the ΔH_i of the α-amino group of Ile-16. Thus, treatment of chymotryp- sinogen-A with acetic anhydride results in the acetylation of its single α-amino group (Cys-1) and the ε-amino group of all 14 of the lysine residues (23). Three of the four tyrosine hydroxyl groups are also acetylated (31). The two imidazoyl groups are also ace- tylated but deacetylate relatively rapidly. The tyrosine hydroxyl groups can be deacetylated by exposure to hydroxylamine or by a brief exposure to pH 11-11.5. In practice, deacetylation of the tyrosine hydroxyls was found to be necessary so that reproducible thermograms suitable for analysis could be obtained. Otherwise, the thermogram will not show a relatively abrupt change in slope at the completion of the titration due to the base-catalyzed time- dependent release of acetate. After the above treatments, Ac-CTG has only carboxyls, imidazoyls, and tyrosine hydroxyls as the ion- izable group sets.

The thermogram obtained for the titration of 4.15 μmoles Ac-CTG at the initial pH of 5.0 is shown in Fig. 5. Below this pH, Ac-CTG precipitates. Of the 15 carboxyl groups present per molecule, only 3.0 group equivalents should titrate above pH 5 on the basis of po- tentiometric titration data for chymotrypsinogen (3). Along with the 2 imidazoyl and the 3 tyrosine hydroxyl groups also present, a total of 8.0 ionizable group equivalents per μmole protein should be available for titration. Since the amount of protein ($\underline{m}$) used was 4.15 μmoles, the μmoles of each group set titrated should equal the product of $\underline{m}$ and $\underline{f}$ where $\underline{f}$ (in μmoles/μmole) is the frequency number of the groups in each group set. The thermal titration of 4.15 μmoles of Ac-CTG should then require only 33.2 μmoles KOH. However, the complete titration consumed 39.0 μmoles KOH and thus required that the value of $\underline{f}$ for the carboxyl group set be raised from 3.0 to 4.4 μmoles/μmole protein to arrive at 39.0 μmoles of titratable groups. This adjustment, i.e., for only the carboxyl group set, was also dictated by the initial slope of the thermogram; the heat change was too great to be accommodated by the other groups present. It is most unlikely, however, that the increase in the

TABLE III

HEATS OF IONIZATION FOR IONIZABLE GROUP SETS IN ACETYLATED-
CHYMOTRYPSINOGEN AND ACETYLATED-δ-CHYMOTRYPSIN

Protein	Group Set	Assigned pK'	f (μmoles/ μmole)	N (μmoles)	ΔH_i (kcal/mole)
Ac-CTG	Carboxyls	4.35	4.4	18.26	0.75
	Imidazoyls	7.0	2.0	8.30	10.53
	Tyr-hydroxyls	10.2	3.0	12.45	8.57
Ac-δ-CT	Carboxyls	4.35	3.16	11.724	0.44
	Imidazoyl	6.6[a]	1.0	3.71	9.28
	Imidazoyl	7.0	1.0	3.71	9.28
	α-Amino	7.8	1.0	3.71	10.60
	Tyr-hydroxyls	10.2	3.0	11.13	8.43

a – From kinetic measurements, the pK' of the group controlling
the rate-limiting step in the catalysis of ester substrates.

carboxyl group concentration comes from the protein. Rather, it is
probably due to the presence of acetate as an impurity since the
Ac-CTG preparation was passed through a Sephadex column in 0.005M
acetate buffer prior to dialysis and use. With this revision,
values for the heats of ionization of the group sets were calculated
from the data of the experimental curve, the N values, and the assigned pK' values (Table III). The ΔH_i values so obtained were
then used to generate the theoretical curve as shown in Fig. 5.

The μmoles KOH required to titrate 3.71 μmoles Ac-δ-CT at
the initial pH of 5 (Fig. 6) can be adequately described by the f
numbers given in Table III and agree with both potentiometric titration data and the known changes that accompany the trypsin-cata-
lyzed conversion of Ac-CTG to Ac-δ-CT, i.e., the appearance of two
new end-groups, the α-amino group of Ile-16 and the carboxyl group
of Leu-13. The values of ΔH_i calculated for the various group sets
are given in Table III and were used to calculate the theoretical
thermogram for Ac-δ-CT as shown in Fig. 6.

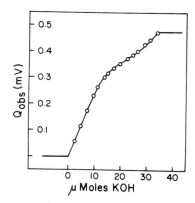

Fig. 6. Thermogram for the titration of 3.71 μmoles acetyla-ted-δ-chymotrypsin at the initial pH of 5.0 corrected for baselines (solid line). The circles represent the theoretical curve calcu-lated from the parameters given in Table III.

For both proteins, the calculated ΔH_i values for the various group sets give an acceptable fit to the experimentally obtained thermograms. Whether the solutions are unique is open to some question since the calculation of ΔH_i is dependent also on the pK' assignments for the group sets and the N values. When these are reasonably certain, the values calculated for the heats of ioniza-tion of the ionizable groups are probably close to being correct.

Comparison of the ΔH_i values calculated for the imidazoyls and the tyrosine hydroxyls of both Ac-CTG and Ac-δ-CT with those of model compounds (cf. Tables I and II) reveal that they are somewhat higher. It would seem reasonable that this need not be labeled a discrepancy. It more likely represents the inadequacy of low mo-lecular weight "model" compounds in solution to reflect their same ionization behavior when incorporated into the three dimensional polymer structure of a protein. For the single α-amino group (Ile-16) of Ac-δ-CT, however, the ΔH_i value of 10.6 kcal/mole is in excellent agreement with that found for the amino group of a number of different amino acids including that for isoleucine (10.60 kcal/mole) (25). If this amino group in its protonated state were to exist in ion-pair bond with a carboxylate anion, one would expect that its heat of ionization would be higher than normal.

SUMMARY

Using the technique of thermal titrimetry applied to the measurement of the heat change occurring in a solution as a result of a change in pH, procedures have been developed for the calculation of heats of ionization of a mixture of weak protic acids with overlapping ionization behavior. The method requires a knowledge of the pK' and concentration of each ionizable group (or group set) and when this is satisfied, the heats of ionization of the group sets of dissociable groups in proteins can also be determined from a single thermogram.

ACKNOWLEDGEMENTS

This investigation was supported in part by a research grant from the National Science Foundation (GB-33473) and by research grants from the Institute of General Medical Sciences (GM-10902) and the National Heart and Lung Institute (HE-14803), National Institutes of Health.

REFERENCES

1. Martin, C.J. and Marini, M.A., J. Biol. Chem., 242, 5736 (1967).
2. Marini, M.A. and Martin, C.J., Eur. J. Biochem., 19, 153 (1971).
3. Marini, M.A. and Martin, C.J., Eur. J. Biochem., 19, 162 (1971).
4. Martin, C.J., Oza, N.B. and Marini, M.A., Eur. J. Biochem., 20, 276 (1971).
5. Marini, M.A. and Martin, C.J., Anal. Biochem., 26, 231 (1968).
6. Dunlop, P., Marini, M.A. and Martin, C.J., Biochim. Biophys. Acta, 243, 320 (1971).
7. Martin, C.J., Oza, N.B. and Marini, M.A., Can. J. Biochem., 50, 1114 (1972).
8. Blow, D.M., Birktoft, J.J. and Hartley, B.S., Nature, 221, 337 (1969).
9. Bender, M.L., Clement, G.E., Kezdy, F.J. and Heck, H. d'A., J. Am. Chem. Soc., 86, 3680 (1964).
10. Bender, M.L., Gibian, M.J. and Whelan, D.J., Proc. Natl. Acad. Sci. U.S., 56, 833 (1966).
11. Himoe, A. and Hess, G.P., Biochem. Biophys. Res. Commun., 23, 234 (1966).

12. Keizer, J. and Bernhard, S.A., Biochemistry, 5, 4172 (1966).
13. Glick, D.M., Biochemistry, 7, 3391 (1968).
14. Wedler, F.C. and Bender, M.L., J. Am. Chem. Soc., 91, 3894 (1969).
15. Oppenheimer, H.L., Labouesse, B. and Hess, G.P., J. Biol. Chem., 241, 2720 (1966).
16. Ghelis, C., Labouesse, J. and Labouesse, B., Biochem. Biophys. Res. Commun., 29, 101 (1967).
17. Karibian, D., Laurent, C., Labouesse, J. and Labouesse, B., Eur. J. Biochem., 5, 260 (1968).
18. Ghelis, C., Garel, J.R. and Labouesse, J., Biochemistry, 9, 3902 (1970).
19. Dixon, J.W. and Hofmann, T., Can. J. Biochem., 48, 671 (1970).
20. Matthews, B.W., Sigler, P.B., Henderson, R. and Blow, D.M., Nature, 214, 652 (1967).
21. Sigler, P.B., Blow, D.M., Matthews, B.W. and Henderson, R., J. Mol. Biol., 35, 143 (1968).
22. Marini, M.A. and Martin, C.J., Biochim. Biophys. Acta, 168, 73 (1968).
23. Agarwal, S.P., Martin, C.J., Blair, T.T. and Marini, M.A., Biochem. Biophys. Res. Commun., 43, 510 (1971).
24. Blair, T.T., Marini, M.A., Agarwal, S.P. and Martin, C.J., FEBS Lett., 14, 86 (1971).
25. Marini, M.A., Berger, R.L., Lam, D.P. and Martin, C.J., Anal. Biochem., 43, 188 (1971).
26. Marini, M.A. and Martin, C.J., in Methods in Enzymology, C.H.W. Hirs and S.N. Timasheff, Eds., Vol. 27, Part D, p. 590, Academic Press, New York (1973).
27. Marini, M.A., Martin, C.J., Berger, R.L. and Forlani, L., these Proceedings.
28. Marini, M.A. and Wunsch, C., Biochemistry, 2, 1454 (1963).
29. Everett, D.H. and Wynne-Jones, W.F.K., Trans. Faraday Soc., 35, 1380 (1939); Harned, H.S. and Owens, B.B., Chem. Rev., 25, 31 (1939).
30. Bates, R.G. and Hetzer, H.B., J. Phys. Chem., 65, 667 (1961); Christensen, J.J., Wrathall, D.P., and Izatt, R.M., Anal. Chem., 40, 175 (1968).
31. Martin, C.J., Oza, N.B. and Marini, M.A., unpublished results.

DSC STUDY OF THE CONFORMATIONAL TRANSITION OF POLY-γ-BENZYL-L-GLUTAMATE IN THE SYSTEM: 1,3-DICHLOROTETRAFLUOROACETONE-WATER

J. Simon[a], G. E. Gajnos[b], and F. E. Karasz

Polymer Science and Engineering, University of

Massachusetts, Amherst, Massachusetts 01002

INTRODUCTION

The helix-coil transition in homopolypeptides is of interest not only because of its obvious relevance to _in vivo_ processes in biological systems but also because it serves as a useful model for co-operative transitions in general, with the attractive simplification of a quasi-one dimensionality, and because of the opportunities it affords for the study of hydrogen bonding and other non-covalent interactions in polymers under reasonably well-defined conditions. For these, and other reasons, a substantial body of experimental and theoretical observations in this area has been accumulated in the past decade and a half, (1,2): A large fraction of the experimental research in model homopoly-α-amino acids has concentrated on non-ionizable polymers soluble in organic solvent systems. These offer a simplicity in the pre-sumed molecular mechanism of transition, which may be conceived simply in terms of competitive intra-molecular (peptide-peptide) and inter-molecular (peptide-solvent) interactions. In the ab-sence of the latter, the order-disorder transition would still occur because of the gain in entropy afforded by conversion to the latter phase, but for most known polypeptides the transition would take place at an inconveniently high temperature. The in-troduction of a solvent capable of interacting with the peptide in

---

a. Permanent Address: Institute for General and Analytical Chemistry, Technical University, Budapest, Hungary.
b. Permanent Address: Department of Chemistry, Western New England College, Springfield, Massachusetts 01119

effect lowers the transition temperature to an experimentally accessible range. In this case, it is now the overall entropy of the system (polymer and solvent) that has to be considered and it is not unusual to find that the gain in entropy of the polymer in the transition is more than offset by a concommitant decrease in the entropy of the bound solvent. The consequence of this is the occurrence of the seemingly anomalous "inverse" (coil-to-helix) thermal transition that has often been observed, (1).

In many polypeptides the interactions favoring the ordered conformation (we shall be dealing here exclusively with the α-helix) are relatively strong and highly interacting solvents are required to obtain a transition around ambient temperatures. Frequently halogenated carboxylic acids, typically trifluoracetic (TFA) or dichloroacetic (DCA) acids, are employed; it is now generally agreed that the strong hydrogen bonding capability of these plays an important, if not predominant role, in their effectiveness. Some crude measure of the stability of a given polypeptide may be gained by determining the concentration of the interacting species in a mixture of the latter with an inert solvent that is necessary to induce the transition.

Solvent systems, other than the strong acids mentioned above, capable of disrupting the α-helix in organic soluble polypeptides have been studied on a limited scale. Amongst these (3,4) are halogenated ketones and their hydrate derivatives, and halogenated alcohols. Recently we have studied the conformational transition of a fairly stable polypeptide, poly-γ-benzyl-L-glutamate (PBG), as a function of temperature in the solvent system, 1,3-dichloro-tetrafluoroacetone (DCTFA)-water, (5). PBG is soluble in compositions containing between about 4 and 11 weight % water and within this range an inverse thermal coil-to-helix transition was observed. The transition temperature, T_c, (defined here as the temperature at which the fractional helical content of the polymer is one-half) was found to be a function of the solvent composition, as expected, and reached a maximum at 8.3 wt % water content. This composition corresponds precisely to that of an equimolar mixture of DCTFA and water ie. to the composition of the stable gem-diol, 1,3-dichloro-1,1,3,3-tetrafluorpropan-2,2-diol, formed by the addition of one water molecule to the carbonyl group in the ketone. Non-equimolar mixtures thus contain excess DCTFA or water with respect to the diol and the presence of either of the former decreases the activity of the interacting species and hence lowers the transition temperature. Quantitative analysis of the data in these terms corroborated this interpretation but also showed that water could not be regarded as a simple diluent, but to some extent competed for adsorption sites with the gem-diol, again a not unexpected result.

Recently the theory of the solvent effect in conformational transitions has been developed to permit the calculation of the individual contributions of the intra- and inter-molecular interaction energies to the overall process, (6). Such a determination is of interest for several reasons. Firstly, it provides a much more refined approach to an assessment of the intrinsic stability of the α-helix for a given polypeptide. Secondly, it provides a measure of the peptide-solvent interaction and of the fraction of peptide sites bound to a solvent molecule as a function of temperature and/or solvent composition. Thirdly, it permits the calculation of a phase diagram delineating the domains of stability for the ordered and disordered conformations over the entire temperature-solvent composition plane.

To use this theory it is necessary to have in addition to transition temperature-solvent composition data, corresponding transition enthalpies. As is well known such enthalpies for the helix-coil transition of a typical polypeptide in dilute solution are small (of the order of 10^{-2} cals/g. of sample) and correspondingly difficult to measure. Techniques that have been employed include heat capacity determinations in precision adiabatic calorimeters, measurements of heats of mixing or of solution in micro-calorimeters and the use of a variety of custom-built or commercial differential scanning calorimeters, (7 - 9). The last class of measurements in principle offers advantages with regard to sample size and data acquisition times. We have shown that it is possible to use one of the most sensitive of the commercial DSC's, the Perkin-Elmer DSC-2, for transition enthalpy measurements though even here a modification to permit a three-fold increase in usable sample size, and careful attention to maximization of the signal/noise ratio are necessary, (8).

In this work, we present results of a DSC determination of the transition enthalpies of PBG in DCTFA-water mixtures, and an analysis of these results in terms of the treatment cited above.

EXPERIMENTAL

Materials

The halogenated ketone, DCTFA, (Hynes Chemical Research Corporation) was distilled at 760 mm (b. p. 45.3°C) under dry nitrogen and stored in a desiccator at 0°C. The PBG (Pilot Chemicals) had a viscosity average molecular weight of 500,000 and was used as received. The water was doubly distilled.

Method

DCTFA-water mixtures of the required compositions were carefully prepared under dry nitrogen and used to obtain 1.0 - 1.2%
(w/v) solutions of PBG. As previously reported, complete solute
dissolution took place in solvent mixtures containing between 3.8
and 10.8 wt % water. Samples of from 40 to 60 mg of these solutions were hermetically sealed in special aluminum containers and
used in the DSC measurements. Precautions were taken to exclude
air or moisture during the sealing process and to check the
efficiency of the seals by careful weighing. All measurements in
the Perkin-Elmer DSC-2 were carried out at heating rates of 10°/
min and at full-scale sensitivities of either 0.5 or 0.2 mcal/sec.
Sub-ambient measurements were performed using the manufacturer's
refrigerator accessory.

RESULTS

DSC traces for the conformational transition of PBG were
obtained over the entire range of usable solvent compositions, 3.8
to 10.8 wt % water. The area under the observed "melting" peaks
yielded the transition enthalpy ΔH_{cal} while the peak maximum
corresponded to T_c. Half-widths of the transitions ranged from
7 to 15°C.

The observed values of T_c were in excellent agreement with
those obtained earlier (5) from optical rotation measurements,
(Fig. 1). With increasing water content in the DCTFA-water mixtures T_c increased to a maximum of 47°C at a water content of 8.3
wt %. At greater water contents T_c decreased as shown. ΔH_{cal}
behaved in an opposite manner, decreasing from a maximum of about
1100 cals/residue mole in a solvent containing 3.8 wt % water to
a minimum close to zero at 8.3 wt % water (the DSC peak at this
solvent composition was too small to yield a meaningful result)
and increasing again at higher water contents.

The enthalpy results are qualitatively consistent with the
explanation for the effect of solvent on the transition that has
already been advanced on the basis of T_c measurements alone and
was discussed in the Introduction. The observed transition
enthalpy is the sum of intra-polypeptide and polypeptide-solvent
contributions. The size of the latter, however, depends also on
the fraction of backbone sites bound to the solvent molecules at
the transition mid-point. This fraction increases with decreasing temperature. At the maximum in T_c, 47°C, the two contributions (which are of opposite sign) virtually balance and the observed transition enthalpy is thus close to zero. As T_c is
lowered by the presence of a diluent, either DCTFA or water, the

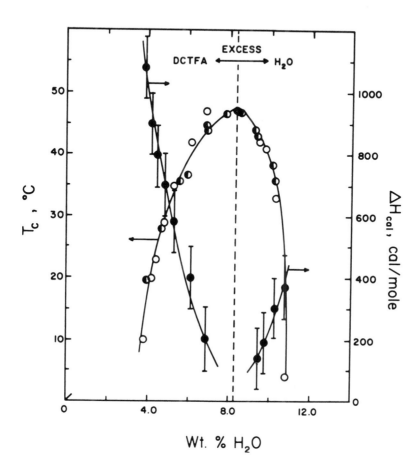

Figure 1. Transition temperatures, T_c, and enthalpies, ΔH_{cal}, of PBG as a function of solvent composition in DCTFA–water mixtures. The vertical dashed line corresponds to the composition of 1,3-dichloro-1,1,3,3-tetrafluoropropan-2,2-diol. "Excess" DCTFA or H_2O refers to compositions with respect to the gem-diol. Half-filled circles: ref. (5); open and filled circles: present results.

peptide-solvent contribution increases and a corresponding increase in ΔH_{cal} is observed. Similar results, with an analogous explanation, have been reported for the PBG-dichloroacetic acid (DCA) 1,2-dichloroethane (DCE) system, (6).

DISCUSSION

We assume a two-step mechanism for the order-disorder transition in which segments in the random-coil conformation (C) are transformed into the helical state (H)

$$....\text{HHCCCC}....\overset{\rightarrow}{\leftarrow}....\text{HHHCCC}....$$

$$(1)$$

while there is a simultaneous interaction of a fraction of the segments in state C with active solvent molecules A

$$...(CA)_m C_{n-m} \overset{\rightarrow}{\leftarrow}....(CA)_{m-1} C_{n-m+1}... +A$$

$$(2)$$

The data presented above may be used to determine the separate contributions to steps 1 and 2. It is not necessary to specify the mechanism any further in this analysis (for example we ignore the potential difference between amino and carbonyl bonding sites on the polypeptide) but we do take into consideration the fact, established earlier, that the diluents DCTFA and water are not equivalent. The latter cannot be regarded as an inert diluent; it competes with the gem-diol for binding sites on the polypeptide. The analysis of the results on the "excess water" side of the T_C maximum (Fig. 1) therefore requires a more complicated analysis, not feasible with the restricted range of data available. Thus the discussion below is restricted to results for solvent compositions containing less than 8.3% water.

The mechanism postulated in eqns. (1) and (2) yields the following expression for the equilibrium constant s, describing the addition of a single peptide unit to an existing helical sequence at temperature T in the presence of a mole fraction x_A of binding (active) solvent with an activity coefficient of unity,

$$- RT \ln s = \Delta H_1 - T\Delta S_1 + RT \ln \{1 + x_A \exp [\Delta H_2 - T\Delta S_2 / RT]\}$$

$$(3)$$

At equilibrium, $s = 1$, $T = T_c$, and eqn. (3) may then be regarded as an expression for T_c as a function of x_A in terms of the four parameters, ΔH_1, ΔH_2, ΔS_1 and ΔS_2 which represent the enthalpies and entropies of the two equilibria in eqns. (1) and (2). A primary objective of the analysis of experimental results is to obtain these parameters from observed phase boundaries, but it develops that unless they are available over a very wide range of solvent compositions and transition temperatures, it is necessary to have ΔH_{cal} measurements in addition.

It can easily be shown that in this model the transition enthalpy is given by (6, 10)

$$\Delta H_{cal} = \Delta H_1 + F_c \Delta H_2$$

$$(4)$$

where F_c is the fraction of occupied (solvent-bound) sites at the transition mid-point, and which may in turn be written as

$$F_c = \frac{x_A}{x_A + K_2}$$

$$(5)$$

where K_2 is the equilibrium constant for the binding step, eqn. (2). Finally, we note that the slope of the phase boundaries is itself related to an enthalpy which we have previously denoted ΔH_{comp}; the latter is related to the ΔH_1, ΔH_2 etc., through

$$\Delta H_{comp} = \Delta H_1 F_c^{-1} + \Delta H_2$$

$$(6)$$

Thus the analysis of the present data led us to first calculate F_c along the phase boundary by using the expression $F_c = \Delta H_{cal}/\Delta H_{comp}$ [obtained by combining eqns. (4) and (6)] calculating K_2 from eqn. (5) and then using the temperature dependency of this

equilibrium constant to obtain ΔH_2 and ΔS_2. ΔH_1 and ΔS_1 may then
be calculated in several ways; the former, for example, from eqn.
(4) or (6), the latter from a modification of eqn. (3). Alterna-
tively ΔH_1 and ΔH_2 may be found by plotting ΔH_{cal} or ΔH_{comp} vs.
F_c and F_c^{-1} respectively [eqns. (4) and (6)] and proceeding via
eqn. (5), etc. As an obvious test of the results one may use the
calculated ΔH_1, ΔH_2, etc. to compute T_c vs. x_A [eqn. (3)] and ΔH_{cal}
as a function of x_A and/or T_c [eqn (4)] and compare these with
the experimental data.

When applying these methods to the present data it was not
possible to find a single set of parameters which fitted both the
experimental T_c and the ΔH_{cal} results within the estimated limits
of accuracy ($\pm 1 - 2°C$ and $\pm 50 - 100$ cals/mole) over the entire
solvent composition range studied (0.4 to 1.0 mole fraction). This
is in contrast to our findings in the PBG-DCA-DCE system in which
very close agreement for both T_c and ΔH_{cal} was readily obtained.
This may reflect an aberration of the mechanism proposed, and also
the fact that agreement was sought over a much broader range of
solvent composition.

However, a very satisfactory fit of the transition temperature
could be obtained (Fig. 2) with, moreover, values for the "intrinsic" polypeptide parameters, ΔH_1 and ΔS_1 (-850 cal/mole and -1.56
e.u.), which were in excellent agreement with those obtained earlier
in the DCA-DCE solvent system, (-845 cal/mole and -1.55 e.u. respectively), (6). The corresponding parameters for the peptide-solvent
interaction were $\Delta H_2 = 5725$ cal/mole, $\Delta S_2 = 18.4$ e.u. Both of these
are almost twice as large as the values obtained in the PBG-DCA case.
The higher enthalpy may be related to the fact that each solvent
molecule has available two hydroxyl groups for bonding to the backbone and that these may be more advantageously placed for strong
hydrogen bonds to be formed in comparison to DCA. The higher
entropy ΔS_2 may simply be attributed to the greater decrease of
translational freedom upon binding as a consequence of the higher
molecular weight of the gem-diol.

The consequences of the increased ΔH_2 and ΔS_2 in terms of the
phase boundaries may be seen in Fig. 2. Whereas with DCA there is
no temperature range in which the PBG helix may exist in the pure
acid, for the undiluted gem-diol there exists a domain, over 200°C
wide, within which the helix is the stable conformation. This is
consistent with the observation that the optical rotation of PBG
in the undiluted gem-diol remains essentially constant up to at
least 90°C, (5). A large fraction of the phase diagram for both
solvent systems is, of course, experimentally inaccessible because
of solute degradation, insolubility, etc.

The crossing of the gem-diol and DCA phase boundaries at about

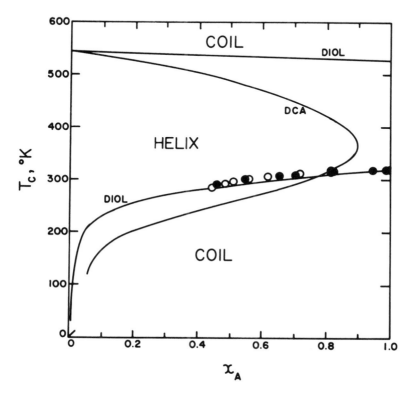

Figure 2.   Phase diagrams for PBG in gem–diol–DCTFA solvent and
            DCA–DCE mixtures; x_A indicates mole fraction of first
            component in each case. Filled circles: ref. (5);
            open circles: present results; DCA phase boundary:
            ref. (6).

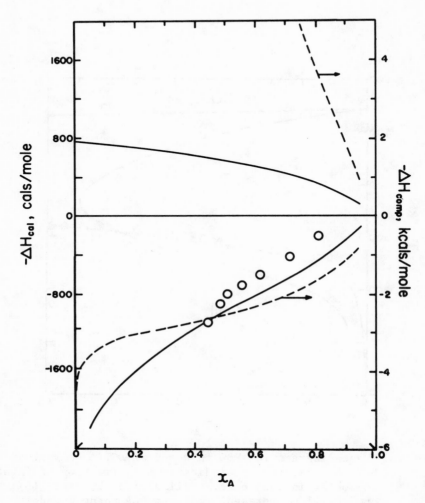

Figure 3.   Calculated calorimetric (full line) and composition
             (dashed line) transition enthalpies for PBG in gem-
             diol-DCTFA mixtures as function of mole fraction, x_A,
             of gem-diol. Open circles: present calorimetric
             measurements.

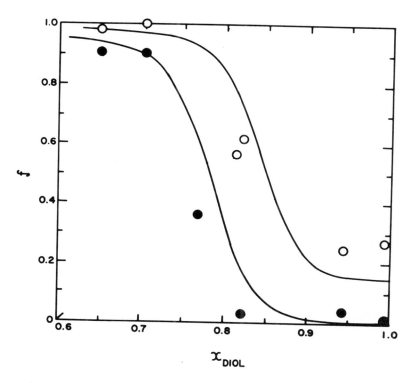

Figure 4. Fractional helical contents, f, of PBG in gem-diol-DCTFA mixtures as function of mole fraction diol. Full lines: calculated 40°C (left) and 45°C (right) isotherms; closed and open circles are corresponding experimental observations from ref. (5).

30°C also illustrates the point that assessing the "strength" of active solvents by noting, without further qualifications, the relative amounts required to induce conformational transitions in a given polypeptide helix can be somewhat misleading. In the present case the gem-diol would be regarded as the "stronger" solvent below 30°c; the reverse would be noted at higher temperatures.

The best fit to the transition enthalpies was not as satisfactory (Fig. 3). Although the decrease of ΔH_{cal} with increasing x_A could be reasonably well reproduced, the calculated heats were on the average some 100 cals/mole higher. Furthermore the ΔH_1, ΔH_2, etc., parameters necessary to obtain this fit differed by an average of 15% from the values cited above: ΔH_1 = -776 cal/mole, ΔS_1 = -1.94 e.u., ΔH_2 = 4550 cal/mole, and ΔS_2 = 16.6 e.u. These differences, while important in terms of the calculated phase boundary, are probably not outside the experimental error, however. Fig. 3 also shows the calculated composition enthalpies.

Finally we used the present results to predict transition widths and related quantities as a function of solvent composition. From the values of ΔH_{cal} given here and the van't Hoff heats ΔH_{VH} calculated from optical rotation data in ref. (5), we estimate the Zimm-Bragg co-operativity parameter σ (11) to average 2×10^{-4} in the temperature range under diccussion. Using this value and eqn. (3) we have calculated helical content contours as a function of solvent composition for the experimentally observed section of the phase diagram. [To do this, Applequist's relation for s as a function of f, the fractional helical content, and σ was used, (12).] Such a composition thus permits a presentation of f as a function of either temperature or solvent composition at any desired point on the phase boundary. In Figure 4 we show two isotherms of f vs. x_{diol}, the mole fraction of gem-diol, calculated at 40°C and 45°C together with experimental data from ref. (5).

Reasonable agreement is obtained not only in the overall shape of the isotherms but also in showing that at 45°C it is not possible to obtain complete conversion from helix to coil.

ACKNOWLEDGEMENT

This work was supported by NSF Grant GB 33484.

REFERENCES

1. POLY-α-AMINO ACIDS, G. D. Fasman, Ed., M. Dekker, N.Y., N.Y., 1967.

2. N. Lotan, A. Berger and E. Katchalski, Ann. Rev. Biochem. 41, 869 (1972).

3. R. Longworth, Nature (London), <u>203</u>, 295 (1964).

4. D. Balasubramanian and R. S. Roche, Polym. Prepr., Amer. Chem. Soc. Div. Polym. Chem., <u>11</u>, 127, 132 (1970).

5. G. E. Gajnos and F. E. Karasz, J. Phys. Chem. <u>76</u>, 3464 (1972).

6. F. E. Karasz and G. E. Gajnos, J. Phys. Chem. <u>77</u>, 1139 (1973).

7. BIOCHEMICAL MICROCALORIMETRY, H. D. Brown, Ed., Academic Press, N.Y., N.Y., 1969.

8. R. P. McKnight and F. E. Karasz, Therm. Acta <u>5</u>, 339 (1973).

9. V. S. Ananthanarayanan, G. Davenport, E. R. Stimson and H. A. Scheraga, Macromolecules <u>6</u>, 559 (1973).

10. M. Bixon and S. Lifson, Biopolymers <u>4</u>, 815 (1966).

11. B. H. Zimm and J. K. Bragg, J. Chem. Phys. <u>31</u>, 526 (1959).

12. J. Applequist, J. Chem. Phys. <u>38</u>, 934 (1963).

A THERMOMETRIC INVESTIGATION OF THE REACTION BETWEEN PROTEINS AND 12-PHOSPHOTUNGSTIC ACID

P. W. Carr, E. B. Smith, S. R. Betso and R. H. Callicott

Department of Chemistry
University of Georgia
Athens, Georgia 30602

INTRODUCTION

Proteins in acid solution may be precipitated by the addition of a wide variety of anions including perchlorate, trifluoroacetate and tungstate(1). We have recently developed (2) a thermometric titration procedure for total serum protein based on their quantitative precipitation by 12-phosphotungstic acid ($H_3PW_{12}O_{40}$). End point precision was~0.5% when the total protein concentration was in the range of 1-10 g/l. Outside this range the precision deteriorates and the stoichiometry is not constant. At low protein concentration the reaction kinetics are too slow thereby inducing a positive error(3); while at high protein concentration the titration curve becomes quite non-linear due to rather dramatic changes in the sample viscosity and concomitant changes in the rate of heat generation due to stirring (2).

The thermometric titration curves were complicated by the existence of two breaks or end points. The ratio of the two breaks is governed primarily by the concentration of added anions (vide infra). Scatchard (4,5) has shown that bovine albumin binds many anions and at low concentration has three distinct classes of binding sites which contain a total of~27 sites which bind anions. At high anion concentration it appears that up to 70 moles of chloride per mole of protein can be bound (4). The purpose of the present work was to elucidate the nature of the reaction between proteins and PTA.

EXPERIMENTAL

Due to the rather small temperature changes encountered, a phase lock amplifier of the type described previously (6) was used to measure the temperature changes. All titrations were carried out in a 30 ml Dewar of the type developed by Izatt, Christensen and Hansen (7). The rather viscous titrant solutions were added via a Radiometer buret (Model ABU-11) or with a syringe buret in conjunction with a linear position transducer (6). The titrant solution can not be standardized by weight due to its very nebulous degree of hydration. We have found that PTA reacts quantitatively with cesium and that the titrant can be standardized thermometrically via the reaction given below:

$$3Cs^+ + PTA^{3-} \leftrightarrows Cs_3PTA(\downarrow) \tag{1}$$

In all of the titrations reported here the protein was in the concentration range of 2-5 g/l and the titrant was added at a rate of 1.26μ equivalent/sec. Photometric measurements of protein concentration were carried out on a Cary 14 recording spectrophotometer. Amperometric measurements were carried out with a Heath polarographic unit (Model EUA-19-2).

RESULTS AND DISCUSSION

A set of thermometric titration curves of bovine albumin with PTA at acidities ranging from pH 2 to 5 are shown in Figure 1. Most titrations reported here were carried out at pH 1. Considerable end point curvature is evident even at pH 2, but end point location is quite accurate and precise at pH $\leqslant$ 2 principally due to the fact that "linear" titration methods such as conductometric, photometric and thermometric may be extrapolated from regions where the reaction is complete (8). The results for a number of proteins, all of which yield titration curves qualitatively similar to those of Figure 1 are summarized in Table I. The data are given as equivalents of reactive unit (see Reaction 1) per 100,000 g of protein to avoid ambiguity due to molecular weight.

In general the data correlate quite well with the sum of basic amino acids in the protein. The data for fibrinogen is out of line; however, the material used was only 85% clottable. The discrepancy may be due to the very high molecular weight (350,000) and the inaccessibility of some of the groups. Since the titrations were generally carried out in very acid media (0.1 $\underline{M}$ hydrochloric acid), virtually all basic nitrogen atoms are protonated and the data should be comparable to the maximum cationic charge as obtained from acid-base titrations. This is seen to be the case.

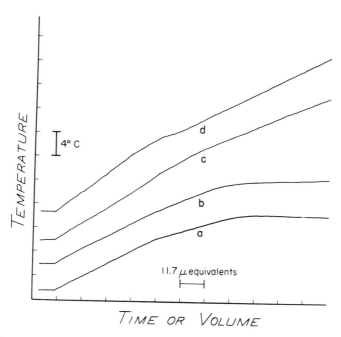

Figure 1. Thermometric titration of bovine albumin with 12-PTA. All protein concentrations 2.0 g/l. Solution conditions are: a. pH 2.0; b. pH 3.0; c. pH 4.0 and d. pH 5.0.

TABLE I – SUMMARY OF TITRATION END POINTS

Protein	Equivalents of PTA/10⁵g	Sum of Basic Amino Acids/10⁵g	Maximum Cationic Change/10⁵g	Ratio of Breaks[a]
Bovine Albumin	145.2	150.5	147.7	0.49
β-Lactoglobulin	114.2	104.3	112.6	0.53
Ovalbumin	92.0	91.1	91.1	0.55
Human γ Globulin	103.3	99.2	83.7 [b]	0.58
Bovine Fibrinogen	91.7[c]	124.9	n.a	0.63

a. Defined as the distance from the first to second break divided by the distance to the second end point.

b. For rabbit γ globulin

c. Material used was only 85% clottable.

Based on the above data the overall stoichiometry to the second end point is dictated by the net charge on the protein. In very acid media all carboxylic acids are protonated; therefore, the reaction is:

$$HnP^{n+} + \frac{n}{3} PTA^{3-} \rightleftharpoons HnP(PTA)_{\frac{n}{3}} \quad (\downarrow) \tag{2}$$

where n is the sum of the basic amino acid residues. It would be extremely useful if the titration could be used to measure the net charge at higher pH where some protons are dissociated. Unfortunately the titration curve becomes quite ill defined at pH> 3 (see Figure 1). Two breaks are no longer evident, and the post titration region is distinctly exothermic. This is probably due to a rearrangement or decomposition of the PTA. Schwartzenbach (9) has noted that the titration of PTA with base is time dependent, and we have noted that when the titrant is adjusted to pH> 3 it no longer precipitates cesium. It is possible, however, that some other factor is responsible for the distortion of the titration curve.

Initially the first break was assigned to differential reactivity of the three groups of basic amino acids (arginine, lysine, histidine) in much the same fashion as the thermometric acid-base titration of proteins (10); nevertheless, no consistent grouping can explain the ratio of breaks for the various proteins listed (see Table I). Indeed the titration of both poly-l-histidine and poly-l-arginine both show two breaks. The ratio of breaks as a function of various solution parameters is given in Table II. Clearly the anion concentration and not the hydrogen ion concentration (for pH≫ 2) is the most important factor.

Several experiments were carried out to determine the nature of the reaction at the first break. First the titrant was added in increasing quantity to several samples of protein, the precipitate was removed immediately by centrifugation and the protein content of the supernatant determined spectro-photometrically. The protein concentration decreased linearly with added PTA and virtually all protein was removed at the first break. Second the PTA concentration during a continuous titration was monitored amperometri-cally. No increase was found until the region of the second break was attained. Last the precipitate formed upon adding PTA beyond the first end point was removed and resuspended in fresh hydrochloric acid. The precipitate was titratable with PTA but showed only one end point.

Since the precipitate phase must be electrically neutral the sum of all positive charges must equal the sum of all negative charges. The equivalents of PTA added is considerably less than that required by electroneutrality;

TABLE II — SUMMARY OF END POINT DATA AS A FUNCTION OF ACID AND CHLORIDE
CONCENTRATION[a]

Solution	Second End Point	Ratio of Breaks[c]	$\bar{\nu}_{Cl^-}$
0.1 M HCl	145.2	0.49	46.2
1.0 M HCl	146.1	0.74	71.9
0.1 M NaCl[b]	145.8	0.54	52.5
0.3 M NaCl[b]	146.4	0.62	60.1
0.5 M NaCl[b]	144.1	0.64	62.1
0.9 M NaCl[b]	145.4	0.72	69.9

a. The protein is bovine serum albumin at a concentration of 2.0 g/l. Titration rate is 1.26 μ equivalents
of PTA/sec.

b. These solutions also contain 0.1 M̱ hydrochloric acid.

c. Defined in Table I.

thus we believe that the remaining negative charge is supplied by a second bound anion e. g. chloride. The sequence of reactions in $\underline{0.1}$ $\underline{M}$ hydrochloric acid is therefore assumed to be:

$$P + nH^+ + yCl^- \quad HnPCl_y^{(n-y)+} \tag{3}$$

(prior to titration)

$$HnPCl_y^{(n-y)+} + \frac{n-y}{3} PTA^{3-} \leftrightarrows HnP(Cl)_y (PTA)_{\frac{n-y}{3}} (\downarrow) \tag{4}$$

(up to first end point)

$$HnP(Cl)_y (PTA)_{\frac{n-y}{3}} + \frac{y}{3} PTA^{3-} \leftrightarrows HnP(PTA)_{\frac{n}{3}} (\downarrow) + yCl^- \tag{5}$$

(up to second end point)

The ratio of breaks should provide an estimate of the anion binding to the protein. Based upon a molecular weight of 65,000 for bovine albumin there are 97 reactive units per molecule at the second end point. The extent of chloride binding per molecule ($\bar{y}_{Cl^-}$) may be calculated as:

$$\bar{y}_{Cl^-} = R \cdot 97 \tag{6}$$

where R is the ratio of breaks defined as the distance between the two breaks divided by the length to the second break. Table II indicates that the extent of binding increases significantly with the chloride concentration but is independent of acid concentration at pH < 1. In order to correlate the data with Scatchard's equation (4,5) for multiple ion binding it was necessary to divide the bound chloride into two classes. The first class contains ~ 27 saturated sites; these correspond to the sum of Scatchard's first three classes. The second contains a total of 65-70 unsaturated sites. Thus it appears that under conditions of high acidity and anion concentration virtually all positively charged sites may bind chloride.

The thermometric titration curves indicate that the first stage of reaction is more exothermic than the second. This may be rationalized by assuming that chloride binding to proteins is exothermic and that the heat of binding PTA to a given site is independent of whether the neighboring sites are vacant or occupied by chloride. More extensive studies of the reaction with emphasis on the effect of varying the binding anion (bromide, nitrate, trifluoroacetate) and precise measurement of the reaction enthalpies are in progress.

ACKNOWLEDGEMENT

This work was supported by a grant (GM 17913) from the National Institutes of Health.

REFERENCES

1. R. J. Henry, Clinical Chemistry: Principles and Techniques, Harper and Row, New York, p. 163 (1964).
2. E. B. Smith and P. W. Carr, Anal. Chem., 45, 1688 (1973).
3. P. W. Carr and J. Jordan, Anal. Chem., 45, 634 (1973).
4. G. Scatchard, Y. V. Wu and A. L. Shen, J. Amer. Chem. Soc., 81, 6104 (1959).
5. G. Scatchard and W. T. Yap, J. Amer. Chem. Soc., 86, 3434 (1964).
6. E. B. Smith, C. S. Barnes and P. W. Carr, Anal. Chem., 44, 1663 (1972).
7. J. J. Christensen, R. M. Izatt and L. D. Hansen, Rev. Sci. Instrum., 36, 779 (1965).
8. D. Rosenthall, G. L. Jones and R. Megargle, Anal. Chim. Acta., 53, 141 (1971).
9. G. Schwartzenback, G. Geier and J. S. Sittler, Helv. Chim. Acta., 45, 2601 (1962).
10. N. D. Jesperson and J. Jordan, Anal. Letters, 3, 323 (1970).

DIFFERENTIAL SCANNING CALORIMETRY STUDIES ON DNA GELS

Horst W. Hoyer and Susan Nevin

Hunter College, Department of Chemistry

695 Park Avenue, New York, New York 10021

Introduction

The present report is concerned with a study of the enthalphy and kinetics of the helix-coil transition of DNA by differential scanning calorimetry (DSC). Although several investigators have reported calorimetric studies on DNA[1-4], the DSC experiment provides kinetic as well as thermal information about the system. We have applied the method of Borchardt and Daniels[5] to obtain information concerning the order and activation energy of the thermal denaturation of calf thymus DNA over a pH range of 5.4 to 8.2.

Experimental

All of the studies herein reported were made with the differential scanning calorimeter cell of the duPont 900 differential thermal analyser. This DSC cell has been characterized by Baxter[6] and need not be described here. Checks were made periodically to monitor the rate of temperature rise and to assure calibration of the instrument for enthalphy changes and for transition temperatures. Most of the thermograms in this study were recorded at a heating rate of $22^{\circ}C$ per minute and at maximum sensitivity of $0.01^{\circ}C$ per inch.

The DNA used was the sodium salt of calf thymus DNA, Sigma batch number 101C-9520 with a reported average molecular weight of 1,350,000. Samples were weighted directly into phosphate buffer containing 0.75 moles of $PO_4^{\equiv}$ per liter, giving concentra-

465

tions of approximately 1 milligram of DNA per 15 microliters.
Solutions were prepared directly in small aluminum cups supplied
by duPont which were hermitically sealed with a special die press.
In most cases the seal held under the pressures generated by
heating the solutions to 120°C, corresponding to an internal pre-
sure of approximately 2 atmospheres. Similar aluminum cups filled
with an equal volume of buffer solution served as references.
Leakage or rupture was readily detected by a sudden and substantial shift of the baseline.

Results

In the concentrations of about 4 to 6 weight percent of DNA
which were used in these studies the solutions are viscous gels.
The onset of the transition of the endotherm in these systems
occurs at approximately 95°C, consistently higher than that reported for the DNA helix-coil transition in more dilute solutions,
and is not completed until the temperature reaches about 110°C.
While no significant change of onset temperature was observed over
the pH range 5.4 to 8.2, there was a small variation of enthalphy
for the transition which is tabulated in Table I. The maximum of
9.4 Kcal per mole base pair in the enthalphy change for the reaction occurs at a pH of 7.0 in phosphate buffer containing 0.75
moles phosphate per liter solution.

Table 1
ENTHALPHY OF HELIX-COIL TRANSITION IN DNA GELS

pH (Phosphate buffer)	Kcalories per mole base pair (618 g per base pair)
5.4	7.7 $\pm$ 0.3
5.8	7.7 $\pm$ 0.4
6.2	8.5 $\pm$ 0.4
6.6	8.6 $\pm$ 0.2
7.0	9.4 $\pm$ 0.2
7.4	8.7 $\pm$ 0.4
7.8	8.5 $\pm$ 0.5
8.2	7.6 $\pm$ 0.4

Because the differential scanning calorimeter measures the rate of change of the enthalphy of a reaction as a function of temperature and time, it permits calculation of kinetic data for the reaction being studied. Borchardt and Daniels[5] derived the equation for the rate constant k of the reaction

1. $$k = \frac{(\frac{A\ V}{n_o})^{x-1} \frac{dH}{dt}}{(A - a)^x}$$

where A is the total area of the thermogram curve, a is the area swept out at some arbitrary time (or temperature) of the reaction, V the volume of the solution, n_o the intial number of moles of solute, x the order of the reaction and dH/dt the rate at which heat is supplied to the sample.

A plot of ln k as a function of the reciprocal of the absolute temperature should give a straight line with slope equal to the negative of the activation energy divided by the gas constant, provided that the correct value of the order of reaction was chosen in the calculation of the rate constant. In practice a computer was used to compute all values of a by numerical integration from values of dH/dt measured as the ordinate of the thermogram. These values were then used by the computer to calculate values of k' for different assumed values of x where k' is

2. $$k' = \frac{\frac{dH}{dt}}{(A - a)^x}$$

Figures 1, 2 and 3 show such plots of log k' vs 1/T for values of x = 0, 1, 2 and 3 for one representative experiment at pH = 7.0 in phosphate buffer. The best straight line, from approximately 10% reaction to 90% conversion, is obtained for an assumed second order plot. Because of the dependence of k and k' upon (A-a), errors in determining A and a become more significant as a approaches A and deviations would be expected at the higher conversions. The initial slope, when (A-a) is large should be less susceptible to error and therefore is significant as is the linear region which corresponds to the central 80% of the conversion. It has been suggested[7] that the initial stage of the helix-coil transition of DNA differs from the unwinding of the two strands and our experiments would seem to support this view.

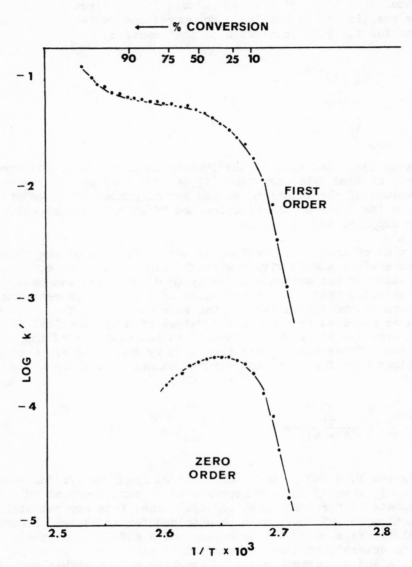

Figure 1. Zero and first order kinetic plots of log k' from equation 2 versus reciprocal of absolute temperature for 6.50% DNA in 0.75M phosphate buffer at pH 7.0.

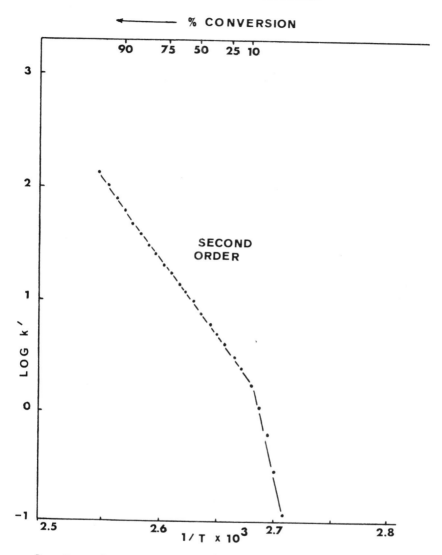

Figure 2. Second order kinetic plots of log k' from equation 2 versus reciprocal of absolute temperature for 6.50% DNA in 0.75M phosphate buffer at pH 7.0. To convert k' to k multiply by 0.12.

Values of the second order activation energies for the major portion of the reaction apparently do not change significantly over the pH range from 5.4 to 8.2, the average being 69 Kcal/+3. Corresponding values for the initial 10% of the reaction vary more widely, averaging 187±14 but this variation is undoubtedly

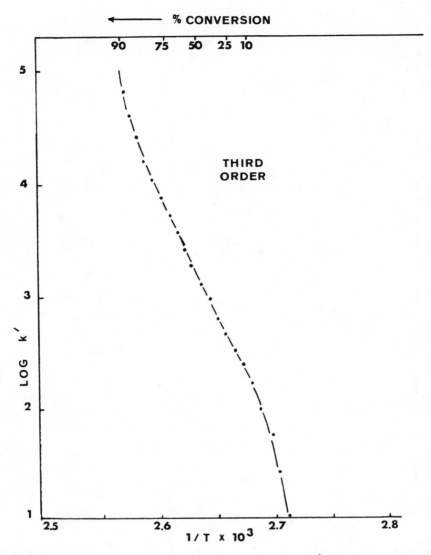

Figure 3. Third order kinetic plots of log k' from equation 2 versus reciprocal of absolute temperature for 6.50% DNA in 0.75M phosphate buffer at pH 7.0.

due to the small number of points from which the slope must be determined. Table 2 lists these activation energies for solutions of different pH. Because of the limited number of points during the initial 10% of the reaction, it is impossible to distinguish between first or second order in this region and both sets of

Table 2
ACTIVATION ENERGIES FOR HELIX-COIL TRANSITION IN DNA GELS
(Kcalories per mole kinetic unit)

pH	Second Order E_a		First Order E_a
	Initial 10%	Main Portion	Initial 10%
5.4	177	68	167
5.8	175	68	185
6.2	169	66	167
6.8	190	74	181
7.0	174	65	241
7.4	197	72	206
7.8	192	71	243
8.2	220	71	217
Average	187 ± 14	69 ± 3	201 ± 25

figures are therefore included. Note however that the average activation energies for these two sets are within experimental error of each other.

Discussion

Our enthalphy values for the thermal denaturation of DNA are in reasonable agreement with others reported in recent years. Privalov and coworkers[2] studied T_2 phage DNA and found the pH dependence of the enthalphy to be bell shaped but essentially flat in the pH range 5 to 8, with a value of 9.65 Kcalories per mole base pair at pH equal to 7.0. Barber[3] studied four different types of DNA in alkaline solution. After correction for heat absorbed in deprotonation the enthalphy for the helix-coil transition was found to vary from 6.2 to 7.6 Kcalories per mole base pair. Sturtevant and coworkers reported values of from 7.8 to 8.3 Kcalories for four different samples at pH 6.0 and at 25°C. Shiaro and Sturtevant[4] found a value of 7.2 Kcalories per mole base pair at pH 7.0 and at 77°C. The somewhat higher enthalphy values found in our DSC studies, as compared with those obtained by adiabatic calorimetry may be due to our need to work with more concentrated solutions. This however is compensated by the additional kinetic information available from the DSC experiment.

The observed second order behavior of the thermal denaturation of DNA is somewhat surprising. One expects the rate to be proportional to the concentration of the DNA but since helix separation involves the rotation of one strand about the other,

the rate limiting step must involved the hydrodynamic resistance
to rotation. Theoretical considerations[8] suggest that the fric-
tional resistance for one isolated molecule will be proportional
to the viscosity number η_{sp}/c of the DNA solution. The viscosity
is known to vary as a power series in the concentration of the
macromolecule according to

$$3. \qquad \frac{\eta_{sp}}{c} = [\eta] + k[\eta]^2 c$$

where $[\eta]$ is the intrinsic viscosity and k, the Huggins constant, has a value near 0.4 dl^2/g^2. In the concentration range
used in our studies the second term on the right becomes larger
than the first. The rate equation for the helix-coil transition
thus approximates a second order equation because of the contributions of DNA concentration and viscosity to the rate equation.
If this view is correct then at lower concentrations, where the
second term is significantly less than the intrinsic viscosity,
the rate equation should become first order.

Acknowledgement

We are grateful to Arthur Noguerola for assistance in working out the computer program.

References

1.   Bunville, L.G., Geiduschek, E.P., Rawitscher, M.A. and
      Sturtevant, J.M., Biopolymers, 3, 213 (1965).

2.   Privalov, P.L. Ptitsyn, O.B., Birshtein, T.N., Biopolymers,
      8, 559 (1969).

3.   Barber, R., Biochemica et Biophysica Acta, 238 60, (1971).

4.   Shaio, D.D.F. and Sturtevant, J.M., Biopolymers, 12, 1829
      (1973).

5.   Borchardt, H.J. and Daniels, F., J. Am. Chem. Soc., 79 41,
      (1957).

6.   Baxter, R.A., in "Thermal Analysis" Vol 1, Schwenker, R.F.
      and Garn, P.D., Academic Press (1969).

7.   Spatz, H. Ch., and Crothers, D.M., J. Mol. Biol., 42 191
      (1969).

8.   Longuet-Higgins, H.C., and Zimm, B.H., J. Mole. Biol., 2 1
      (1960).

APPLICATIONS OF GROUP ENTHALPIES OF TRANSFER

Richard Fuchs

Department of Chemistry, University of Houston

Houston, Texas   77004

Group enthalpies of transfer are a recent development in studies of solvent - solute interactions (1-3). By this calorimetric technique qualitative and quantitative studies of subtle intermolecular interactions may be made, which would be, at best, more difficult and less quantitative by other means. We shall cite some systems about which useful information has already been obtained, and some areas where group enthalpies of transfer might be applied in the future.

The enthalpies of solution (ΔHs) of a large number of aromatic compounds in methanol (MeOH) and in N,N-dimethylformamide (DMF) were recently reported (1,2), and additional unpublished values in benzene have been measured. From the ΔHs values in the different solvents enthalpies of transfer ($\Delta\Delta$Hs) were determined. The latter are a measure of the change in solvation (enthalpy) that a molecule undergoes upon transfer from one solvent to another. The important observation was made that, within experimental error, the values of $\Delta\Delta$Hs are additive functions of the substituent groups present, regardless of the relative positions of the substituents on the aromatic ring. The additivity principle makes possible comparison of the solvation of individual functional groups by various solvents.

ENTHALPIES OF TRANSFER

We will now make use of some data for the enthalpies of solution (Table I) to obtain enthalpies of transfer, and from the latter, group enthalpies of transfer will be derived.

Table I. Enthalpies of Solution of Some Aromatic Compounds in
          MeOH, DMF, and Benzene at 25°

	$\Delta Hs(MeOH)^{a,b}$	$\Delta Hs(DMF)^{a,b}$	$\Delta Hs(Benzene)^{a}$
Benzene	+0.36	+0.04	0.00
Toluene	+0.45	+0.16	+0.12
Mesitylene	+1.04	+0.83	+0.62
Fluorobenzene	+0.03	−0.35	+0.02
Chlorobenzene	+0.17	−0.25	+0.03
Bromobenzene	+0.30	−0.19	+0.05
Iodobenzene	+0.51	−0.40	+0.20
m-Diboromobenzene	+0.61	−0.10	+0.44
Aniline	−0.62	−2.68	+1.25
Anisole	+0.81	−0.07	+0.13
1,2-Dimethoxybenzene	+1.44	−0.10	+0.51
Nitrobenzene	+1.00	−0.19	+0.56
Ethyl benzoate	+1.21	+0.27	+0.05

[a] Calorimetric values ±0.05 kcal/mole for liquid samples.
[b] Data for eighty-two additional compounds are listed in Reference 2.

     The enthalpy of transfer of a solute from MeOH to DMF is de-
fined as follows: $\Delta\Delta Hs(M{\rightarrow}D) = \Delta Hs(D) - \Delta Hs(M)$, and has a value of
−0.32 kcal/mole for the solute benzene. Similarly, $\Delta\Delta Hs(M{\rightarrow}D)$ for
bromobenzene is −0.49 kcal/mole. If we assume that the phenyl group
of bromobenzene has a transfer value equal to that of benzene, and
the remainder of $\Delta\Delta Hs$ can be attributed to the bromine, a group
transfer value of −0.17 for bromine is obtained. If the group $\Delta\Delta Hs$
values are, in fact, additive, we should be able to estimate the
transfer value for m-dibromobenzene by summing the appropriate
group values: −0.32 + 2(−0.17) = −0.66. The experimental value is
−0.71, which is well within the probable combined error of the six
ΔHs values on which the comparison is based. In practice the group
transfer value for bromine is based on the best value derived from
twelve aromatic bromine compounds. Other group $\Delta\Delta Hs$ values, which
are similarly derived, are listed in Table II.

     An obvious use of group enthalpies of transfer is in the study
of group solvation by various types of solvents. For example, the
individual group $\Delta\Delta Hs(MeOH{\rightarrow}DMF)$ values have been discussed (2) in
terms of the greater dipole moment, hydrogen bond acceptor ability,
and solvent-solvent association forces of DMF (relative to MeOH),
and the resulting more exothermic dipole-dipole, dipole-induced
dipole, and hydrogen bonding interactions with polar, polarizable,
and hydrogen bond donating groups, respectively. Group enthalpies
of transfer from MeOH to benzene reflect comparable solvent-solute

Table II. Group Enthalpies of Transfer for Aromatic Substituents

Substituent	$\Delta\Delta H_S(M \to D)$[a]	$\Delta\Delta H_S(B \to M)$[a]
$-C_6H_5$	-0.3	+0.4
$-CH_3$	+0.03	0.0
$-CF_3$	+0.3	+1.0
$-OCH_3$	-0.6	+0.3
$-NH_2$	-1.8	-2.2
$-COCH_3$	-1.0	+0.5
$-NO_2$	-0.8	+0.1
$-CN$	-1.0	+0.4
$-F$	-0.1	-0.4
$-Cl$	-0.1	-0.2
$-Br$	-0.2	-0.1
$-I$	-0.6	-0.1
$-COOEt$	-0.6	+0.8
$-OH$	-1.7	-4.5
$-N(CH_3)_2$	-0.5	-
Naphthyl	-0.6	-
$-OC_6H_5$	-0.7	-
$-CH=CH_2$	-0.3	-

[a]Enthalpies of transfer in kilocalories per gram group weight.

dispersion interactions in both solvents; dipole-induced dipole
interactions of polar groups with benzene, which are comparable in
magnitude to the dipole-dipole interactions in methanol; but a substantially greater ability of methanol to accept a hydrogen bond.

APPLICATIONS OF GROUP ENTHALPIES OF TRANSFER

Additivity of group enthalpies of transfer implies that the
solvation of each group is substantially independent of the solvation of other groups within the same molecule. By contrast, what
is the significance of the several observed cases for which $\Delta\Delta Hs$
(MeOH→DMF) is not additive? In these situations one group is modifying the properties of a second group in such a way that the solvation of the molecule as a whole is not a simple additive function.
The first cases of this observed (2) were o-nitroaniline and o-nitro-
phenol, in which intramolecular hydrogen bonding is known to occur.

A second case of non-additivity appears in the enthalpies of
solution of a series of m- and p-substituted benzaldehydes in methanol and in DMF (4). These, together with the resulting enthalpies
of transfer, and the apparent $\Delta\Delta Hs$ values for the aldehyde group
are listed in Table III.

Benzaldehyde has a dipole moment (2.78D) only slightly less
than that of acetophenone (2.96D). Since the exothermic transfer of
the acetyl group (-1.0 kcal/g group wt) has been attributed (2)
mainly to the larger dipole-dipole interaction with DMF than with
methanol, an aldehyde group $\Delta\Delta Hs$ of perhaps -0.7 to -1.0 might be
anticipated. A value of -0.8 is, in fact, found (within experimental error) for the three benzaldehydes with strongly electron releasing substituents, p-N(CH$_3$)$_2$, p-OH, and p-OCH$_3$. Slightly less
negative values are calculated for the p-CH$_3$ and unsubstituted members, and increasingly positive values are found as the electron
withdrawing character of the substituents increases: p-Cl < m-Cl
< m-NO$_2$ < p-NO$_2$. Rather similar values of ΔHs (DMF) are observed
for the liquid aldehydes, whereas a large range of values is observed in methanol, with the values becoming less endothermic with
increasing electron withdrawal. It therefore appears that an exothermic reaction of the benzaldehydes occurs in methanol, such as
uncatalyzed hemiacetal formation, and the equilibrium is favored by
electron withdrawing groups.

The proton magnetic resonance spectrum of p-nitrobenzaldehyde
in methanol has peak areas in the formyl/methoxy proton ratio of
0.29/0.71 x 3, which indicates that 71% of the original aldehyde
exists in methanol solution as the hemiacetal. While $\Delta\Delta Hs$(MeOH→DMF)
for the aldehyde group is -0.8, the apparent group values (Table III)
differ from this by an amount directly dependent on the amount of
hemiacetal formation. p-Nitrobenzaldehyde differs from this by 3.6-
(-0.8) = 4.4 kcal, due to 71% conversion to hemiacetal. A 100% conversion would have led to a discrepency of 4.4/0.71 = 6.2 kcal.
From these considerations the following amounts of hemiacetal formation in methanol may be calculated: m-NO$_2$, 65%; m-Cl, 31.5%; p-Cl,
13%; H, 3.7%; and p-CH$_3$, 2.1%. The value for m-chlorobenzaldehyde

Table III. Enthalpies[a] of Solution and of Transfer of Substituted
             Benzaldehydes in Methanol and Dimethylformamide at 25°.

Substituent	$\Delta Hs(MeOH)$[b]	$\Delta Hs(DMF)$	$\Delta\Delta Hs(M{\to}D)$	$\Delta\Delta Hs(-CHO)$[c]
p-N(CH$_3$)$_2$[d]	6.3	4.6	-1.7	-0.9
p-OH[d]	4.0	1.2	-2.8	-0.8
p-OCH$_3$	1.53	-0.19	-1.72	-0.8
p-CH$_3$	0.88	-0.08	-0.96	-0.67
H	0.64	-0.25	-0.89	-0.57
p-Cl[d]	4.0	3.6	-0.4	0.0
m-Cl	-0.92	-0.17	0.75	1.15
m-NO$_2$[d]	2.1	4.2	2.1	3.2
p-NO$_2$[d]	2.2	4.7	2.5	3.6

[a]In kcal/mole. [b]Calorimetric values ±0.1 kcal/mole (solids), ±0.05
(liquids). Concentration range 10^{-4} to 10^{-3}M. [c]Apparent group
enthalpy of transfer for -CHO[= $\Delta\Delta Hs(M{\to}D)$ - calcd. $\Delta\Delta Hs$ for remainder of molecule]. [d]Solid compounds at 25°.

has been confirmed by nmr. The percent of hemiacetal formation by
the other aldehydes (p-OCH$_3$, 1.1%; p-OH, 0.7%; p-N(CH$_3$)$_2$, 0.06%) is
too small to affect, within experimental error, the enthalpies of
solution, but these values may be estimated using the Hammett equation.

SOME POSSIBLE FUTURE APPLICATIONS

The enthalpies of transfer alkane hydrocarbons from methanol
to dimethylformamide are quite accurately predictable [$\Delta\Delta Hs(CH_3)$ =
0.14; $\Delta\Delta Hs(CH_2$, CH, C) = 0.11]. In the compounds tetrabutyltin and
tetrapentyltin $\Delta\Delta Hs$ for the tin atom is -0.2. However, the corresponding value in tetramethyltin is +0.4 kcal. This might result
from the reversible coordination of methanol at the tin atom of the
less hindered methyl compound. Our preliminary nmr studies suggest
that tin-proton coupling constants are affected by the addition of
methanol to Me$_4$Sn, as would be expected for a change in the coordination number of tin.

The conformations of alkenes and alkyl halides in solution appear to influence ΔΔHs values. Our present hypothesis is that these molecules exist in folded form in MeOH and DMF solutions, and, depending on chain length, this results in more or less hindrance of the functional group to solvation. It may also be anticipated that ΔΔHs for a polar substituent will differ for that group in the axial and equatorial conformations on a ring.

The preceeding examples are intended only to suggest a few of the possible   applications of group enthalpies of transfer. Many other applications may well be developed in the future.

ACKNOWLEDGMENT

This research was supported by the Robert A. Welch Foundation (Grant E-136).

REFERENCES

1. R. Fuchs, L. L. Cole, and R. F. Rodewald, J. Amer. Chem. Soc., 94, 8645 (1972).
2. R. Fuchs and R. F. Rodewald, J. Amer. Chem. Soc., 95, 5897 (1973).
3. C. V. Krishnan and H. L. Friedman, J. Phys. Chem., 73, 1572 (1969); 75 3598 (1971). These papers deal mainly with the enthalpies of transfer of the methylene and hydroxyl groups of aliphatic molecules.
4. R. Fuchs, T. M. Young, and R. F. Rodewald, Can. J. Chem., In Press.

THERMODYNAMICS OF INTERMOLECULAR SELF-ASSOCIATION OF

HYDROGEN BONDING SOLUTES BY TITRATION CALORIMETRY

Earl M. Woolley and Noel S. Zaugg

Department of Chemistry, Brigham Young University

Provo, Utah 84602

INTRODUCTION

The hydrogen bond has been the object of extensive interest in the past few decades. In particular, the study of dilute solutions of hydrogen bonding solutes has made possible an increased understanding of the nature of the hydrogen bond. The tendency of many such solutes to self-associate intermolecularly has permitted systematic investigations of the effects of solvents and substituents on the strength of the self-association interaction, thereby contributing to the understanding of the theory of hydrogen bonding.

In their collection of pre-1958 literature on the hydrogen bond (ref. 1), Pimentel and McClellan list fifteen different experimental methods used in the study of self-association interactions in solution. There are disadvantages associated with each of these methods, however, which make many of the results have highly uncertain meaning. For example, in much of the distribution work there has been a failure to consider the effect of the second solvent (usually water) on the self-association equilibria. Infrared spectroscopic methods have been widely used, but there is a great variation in the numerical values of self-association equilibrium constants obtained by these methods and reported in the literature. Enthalpies of association, when reported, are usually derived from the van't Hoff equation. Any uncertainty in the values of the equilibrium constants thus greatly increases (ref. 2) the uncertainty in the derived value of the enthalpy.

Only in recent years has calorimetry been applied to the study of intermolecular self-association in dilute solution. The fact that calorimetry enables one to more directly determine enthalpies of interaction makes it a valuable tool in the investigation of hydrogen bonding interactions in solution. We will report on some of the recent applications of calorimetry to the study of inter-molecular self-association.

METHOD AND CALCULATIONS

Heats of dilution of hydrogen bonding solutes are more conveniently used in calculations if they are expressed as apparent relative molar enthalpies, ϕ_L. Values of ϕ_L are obtained from measured enthalpies of dilution, ΔH_i, as follows. A titrant solution containing F_t total moles of hydrogen bonding solute per liter of solution is "titrated" into the pure solvent in the calorimeter. Enthalpy data that correspond to the following processes are obtained conveniently by titration calorimetry.

$$F_t \longrightarrow F_1 \qquad\qquad \Delta H_1 \qquad\qquad\qquad (1a)$$
$$F_t \longrightarrow F_2 \qquad\qquad \Delta H_2 \qquad\qquad\qquad (1b)$$
$$\vdots$$
$$F_t \overset{\vdots}{\longrightarrow} F_i \qquad\qquad \Delta H_i \qquad\qquad\qquad (1i)$$

The enthalpy of dilution of titrant to infinite dilution is

$$F_t \longrightarrow 0 \qquad\qquad \Delta H_o = -\phi_{Lt} \qquad\qquad (2t)$$

Combination of eqns. (1) and (2) yields

$$F_1 \longrightarrow 0 \qquad\qquad -\phi_{L1} = \Delta H_o - \Delta H_1 \qquad (2a)$$
$$F_2 \longrightarrow 0 \qquad\qquad -\phi_{L2} = \Delta H_o - \Delta H_2 \qquad (2b)$$
$$\vdots$$
$$F_i \longrightarrow 0 \qquad\qquad -\phi_{Li} = \Delta H_o - \Delta H_i \qquad (2i)$$

The correct value of $-\phi_{Lt} = \Delta H_o$ can be found by extrapolation of a plot of ΔH_i versus F_i (or versus F_i^2, or F_i^3, etc.) to zero formal solute concentration. However, in some cases (eg, when the solute is self-associated to a large degree) the uncertainties in ΔH_i are large at small F_i, thereby making the extrapolation quite uncertain. In those cases where the stoichiometry of all major associated species is known, one can treat this extrapolated value of $\Delta H_o = -\phi_{Lt}$ as an independent variable. Once values of $-\phi_{Li}$ and F_i are available, the data can be evaluated in terms of three different reaction models which we will now describe in detail.

Monomer-Single Polymer Model

We may write the reaction for formation of a polymer, S_n, from n monomer solute molecules S, as

$$nS \rightleftharpoons S_n \qquad\qquad\qquad\qquad\qquad (3)$$

with associated molar enthalpy, ΔH_n^o, and equilibrium constant, K_n, defined as

$$K_n = [S_n] / [S]^n \qquad\qquad\qquad\qquad (4)$$

where brackets indicate molar concentrations and where activity coefficients are assumed to be unity (ref. 3).

For the dilution process of eqns. (2) we write

$$F_i \phi_{Li} = \Delta H_n^{\,o} [S_n]_i \tag{5}$$

Appropriate combination of eqn. (4) with the material balance equation

$$F_i = [S]_i + n[S_n]_i \tag{6}$$

leads one to numerical values of $[S_n]_i$ in each solution of formal concentration, F_i, if the numerical value of K_n is known. Eqn. (5) then leads to $\Delta H_n^{\,o}$ for reaction (3). Accordingly, one can iteratively choose various K_n values and calculate $\Delta H_n^{\,o}$ for each solution of formal solute concentration, F_i, until one obtains the value of $\Delta H_n^{\,o}$ for all concentration values, F_i, for the particular run. A similar approach based on differences in ϕ_L values at finite formal concentrations leads to the equation

$$\Delta H_n^{\,o} = (\phi_{Li} - \phi_{Lj}) / ([S_n]_i / F_i - [S_n]_j / F_j) \tag{7}$$

This approach obviously has the advantage of being independent of the extrapolated $\Delta H_o = -\phi_{Lt}$ value. Similarly to the method described for using eqn. (5), one uses eqn. (7) by choosing various K_n values to obtain $[S_n]$ values from eqns. (4) and (6). These $[S_n]$ values are then used in eqn. (7) to find $\Delta H_n^{\,o}$. One chooses K_n values until the same $\Delta H_n^{\,o}$ is obtained for all sets i and j.

A third approach is based on combination of eqns. (4),(5), and (6) to yield

$$\phi_L = (\Delta H_n^{\,o}/n) - (1/n)(-\Delta H_n^{\,o})^{(n-1)/n}(1/K_n)^{1/n}(-\phi_L/F^{n-1})^{1/n} \tag{8}$$

Eqn. (8) predicts that a plot of ϕ_L versus $(-\phi_L/F^{n-1})^{1/n}$ should be a straight line plot with intercept = $(\Delta H_n^{\,o}/n)$ and slope = $-(1/n)(-\Delta H_n^{\,o})(n-1)/n(1/K_n)^{1/n}$. It can be shown that low concentration data fall below a straight line drawn through the high concentration data when the chosen value of n is too big or when the extrapolated $\Delta H_o = -\phi_{Lt}$ is too large. Conversely, the low concentration data fall above a straight line drawn through the high concentration data when the chosen value of n it too small or when the extrapolated $\Delta H_o = -\phi_{Lt}$ is too small.

Monomer-Multiple Polymer Model

Solutes that intermolecularly associate to form hydrogen bonded species of more than one stoichiometry (in significant amounts) can be described by reaction (3) where n has more than one numerical value. Eqn. (4) still describes the concentration of each polymer in terms of monomer. However, eqns. (5) and (6) need further modification to give the following.

$$F_i \phi_{Li} = \sum_{n=2}^{\infty} \Delta H_n^{\,o} [S_n]_i \tag{9}$$

$$F_i = \sum_{n=1}^{\infty} n[S_n]_i \tag{10}$$

Combination of eqns. (2),(9), and (10) leads to the most consistent sets of K_n and ΔH_n^o for a given set of $\emptyset_{Li}$ and F_i data if one knows which n values to use in this analysis.

Monomer-Related Polymer Model

Let us consider a dimerization reaction
$$2S \rightleftharpoons S_2 \tag{11}$$
followed by a series of further stepwise associations
$$S_{n-1} + S \rightleftharpoons S_n \quad (n>2) \tag{12}$$
Corresponding equilibrium constant expressions are
$$K_2 = [S_2]/[S]^2 \tag{13}$$
$$K_s = [S_n]/[S_{n-1}][S] \quad (n>2) \tag{14}$$
Let us next specify that all K_s are equal (ref. 4). We may now relate eqns. (4),(13), and (14) to obtain
$$K_n = K_2 K_s^{n-2} \quad (n>2) \tag{15}$$
Combination of eqns. (4),(10), and (15), followed by algebraic manipulation and rearrangement gives
$$F_i = K_2[S]_i^2(2-K_s[S]_i)/(1-K_s[S]_i)^2 + [S]_i \tag{16}$$
Further, if we take all ΔH_s^o equal for reaction (12) we obtain for reaction (3)
$$\Delta H_n^o = \Delta H_2^o + (n-2) \Delta H_s^o \quad (n>2)$$
Appropriate substitution, manipulation, and rearrangement of eqns. (9) and (16) yields
$$\emptyset_{Li} = \frac{K_2[S]_i^2}{F_i(1-K_s[S]_i)}\left(\Delta H_2^o + \Delta H_s^o \frac{K_s[S]_i}{1-K_s[S]_i} \right) \tag{18}$$

Knowledge of K_2 and K_s values would enable us to evaluate $[S]_i$ values from eqn. (16). These $[S]_i$, K_2, and K_s values along with each $\emptyset_{Li}$ and F_i can be used with eqn. (18) to obtain least-squares values of ΔH_2^o and ΔH_s^o. One can then adjust K_2 and K_s values until the best fit of the data to eqn. (18) is obtained.

In a slight modification to the above approach, we take $\Delta H_2^o = \Delta H_s^o = \Delta H^o$ in eqn. (18) above to obtain
$$\Delta H^o = F_i \emptyset_{Li}(1-K_s[S]_i^2)/K_2[S]_i^2 \tag{19}$$
Combination of eqn. (16) with K_2 and K_s values leads to the value of ΔH^o for each F_i and $\emptyset_{Li}$. By iteratively choosing K_2 and K_s values to obtain the same ΔH^o value for all F_i and $\emptyset_{Li}$, one can arrive at the "best" K_2, K_s, and ΔH^o values.

RESULTS AND DISCUSSION

Results of the study of acetic acid, propionic acid, and chlorinated acetic and propionic acids in CCl_4 are summarized in Table I.

Table I. Thermodynamic Parameters for Hydrogen Bonding Solutes in CCl_4 Solvent at 25°C Based on the "Best Fit" of the Data to Various Models.

Solute	Polymers	K_n[a]	$-\Delta H_n^{\circ}$(kcal)	$-\Delta S_n^{\circ}$(cal/deg)[a]	Ref.
CH_3COOH	Dimers[b]	1400	8.4	13.8	5,6
$CH_2ClCOOH$	Dimers[b]	700	7.8	13.1	6
$CHCl_2COOH$	Dimers[b]	600	6.7	9.7	6
CCl_3COOH	Dimers[b]	150	5.6	8.7	6
CH_3CH_2COOH	Dimers[b]	1400	7.6	11.1	6
$CH_3CHClCOOH$	Dimers[b]	700	6.9	10.1	6
CH_2ClCH_2COOH	Dimers[b]	1200	7.5	11.1	6
C_6H_5OH	Trimers[b]	5.6	8.54	25.2	3,7
	Dimers[d]	1.0	3.10	10.5	
	Polymers[d]	3.3	3.10	8.0	
$o\text{-}CH_3C_6H_4OH$	Dimers[c]	0.7	3.4	12	7
	Trimers[c]	1.3	12.5	41	
	Dimers[d]	0.7	4.1	14	
	Polymers[d]	1.7	4.3	13	
$m\text{-}CH_3C_6H_4OH$	Dimers[c]	0.8	5.0	16	7
	Trimers[c]	5.0	13.6	42	
	Dimers[d]	1.3	4.7	15	
	Polymers[d]	4.0	4.3	12	
$p\text{-}CH_3C_6H_4OH$	Dimers[c]	0.35	5.5	20	7
	Trimers[c]	6.5	13.4	41	
	Dimers[d]	1.0	3.4	11	
	Polymers[d]	6.5	3.5	8	
CH_3CH_2OH	e	---	---	--	8,9

[a] Standard states based on molarities of solutes.

[b] Only one predominant hydrogen bonded species in solution when $0.001 \lesssim F_i \lesssim 0.5$ for carboxylic acids and $0.001 \lesssim F_i \lesssim 2.0$ for phenols.

[c] Based on monomer-multiple polymer model assuming only dimers and trimers predominate when $0.001 \lesssim F_i \lesssim 2.0$.

[d] Based on monomer-related polymer model when $0.001 \lesssim F_i \lesssim 2.0$.

[e] Data also taken at 10°C and at 45°C. Results of data analysis according to models given in this paper are not included in original papers.

The trends in K_2, ΔH_2^o, and ΔS_2^o for these solutes can be rationalized if we consider the hydrogen bond to be an electrostatic attraction between two molecules. The electronegative chlorine substituents withdraw charge from the carboxyl group causing the increase in ΔG^o and ΔH^o. The "unexpected" effect of the methyl group is under current investigation in our laboratory.

Phenol and the cresols also show some interesting trends in CCl_4 solvent. Phenol has been shown to follow the monomer-trimer model almost perfectly over the concentration range of 0.001 F to 2.0 F (ref. 3). By constraining the phenol data to the monomer-stepwise polymer model of eqns. (16) and (18) or (16) and (19), one can obtain "reasonable" values for the thermodynamic parameters. Comparison of the thermodynamic parameters for phenol and the three cresols also offers some interesting trends. O-cresol is clearly less associated than the other solutes. The monomer-dimer-trimer parameters permit one to calculate relative amounts of cyclic and linear dimers on the assumption that o-cresol is sterically hindered to such an extent that it forms only linear dimers which contribute -3.4 kcal/mole of dimer formed. On this

Table II. Thermodynamic Parameters for Hydrogen Bonding Solutes in Benzene Solvent at 25°C Based on "Best Fit"of the Data to Various Models. (Footnotes: same as Table I, except ^{f} $4 \times 10^{-5} < F_i < 4 \times 10^{-4}$)

Solute	Polymers	$K_n{}^a$	$-\Delta H_n^o$ (kcal)	$-\Delta S_n^o$ (cal/deg)^a	Ref.
CH_3COOH	Dimers^b	300	7.8	15	5,10
$CH_2ClCOOH$	Dimers^b	100	7.0	14	10
CCl_3COOH	Dimers^b	6.5	8.3	24	11
$CH_2BrCOOH$	Dimers^b	130	7.5	15	10
$CH_2(C_6H_5)COOH$	Dimers^b	350	8.8	18	10
$CH(C_6H_5)_2COOH$	Dimers^b	300	8.8	18	10
C_6H_5COOH	Dimers^b	400	7.3	13	12
$o\text{-}CH_3C_6H_4COOH$	Dimers^b	380	7.7	14	12
$m\text{-}CH_3C_6H_4COOH$	Dimers^b	170	6.8	13	12
▷-COOH	Dimers^b	650	7.5	12	13
◇-COOH	Dimers^b	350	7.7	14	13
⬠-COOH	Dimers^b	380	7.5	13	13
⬡-COOH	Dimers^b	450	8.0	15	13
⬡-COOH	Dimers^b	430	8.1	15	13
C_6H_5OH	Dimers^c	0.13	5.6	22.9	4
	Trimers^c	0.012	4.3	23	
CH_3CH_2OH	e	---	---	--	8,9
Cu(II)phthalo-cyanine	Dimers^{b,f}	1.3×10^4	10.0	15	14

basis, it can be shown that the ratios of cyclic dimer to linear dimer are 0, 0.9, and 1.7 for o-, m-, and p-cresols, respectively.

In Table II we give the results of several studies of various carboxylic acid solutes in benzene solvent. The results for the series of monohalogenated acetic acids follow the expected trend. The unusually large $-\Delta S_2^{\,o}$ values for the phenylacetic acids are a possible indication of a fixed trans configuration of the phenyl groups in the dimer. The unusually large $\Delta S_2^{\,o}$ and K_2 for cyclopropane carboxylic acid is a possible indication of enhanced electronic charge on the cyclopropane ring. Phenol in benzene, unlike phenol in CCl_4, exists primarily as dimers at concentrations less than about 1.0 F. The recently measured calorimetric values for the "ring-stacking" of copper (II) phthalocyanine in benzene illustrates the possible application of the technique to non-hydrogen bonding interactions. Comparison of the results for any given solute in Tables I and II indicates that benzene is less inert (more solvation) than CCl_4.

The results in Table III are for various solutes in toluene solvent. The cyclic carboxylic acids show no striking differences between toluene and benzene as solvents. However, there are some interesting differences for the benzoic and substituted benzoic acids in benzene and toluene. Benzoic acid in toluene might possibly form only one hydrogen bond per dimer, rather than the usual two, or the increased $\Delta H_2^{\,o}$ and $\Delta S_2^{\,o}$ could indicate preferential solvation of benzoic acid monomers by toluene.

Table III. Thermodynamic Parameters for Hydrogen Bonding Solutes in Toluene Solvent at $25^{\circ}C$ Based on the "Best Fit" of Data to Various Models.

Solute	Polymer[b]	$K_n{}^a$	$-\Delta H_n^{\,o}$(kcal)	$-\Delta S_n^{\,o}$(cal/deg)[a]	Ref.
▷-COOH	Dimer	950	7.8	12	13
◇-COOH	Dimer	480	7.8	14	13
▢-COOH	Dimer	380	7.6	14	13
⬡-COOH	Dimer	480	7.7	13	13
◯-COOH	Dimer	480	7.8	14	13
C_6H_5COOH	Dimer	200	4.8	6	12
$o\text{-}CH_3C_6H_4COOH$	Dimer	390	7.4	13	12
$m\text{-}CH_3C_6H_4COOH$	Dimer	490	7.2	12	12
CCl_3COOH	Dimer	3	7.4	23	11

^{a,b}All footnotes refer to the same footnotes as in Table I.

In Table IV we list the thermodynamic parameters for a number of self-associating solutes in various other solvents. It is interesting to note that cyclohexane is more "inert" than CCl_4, as evidenced by comparison of data for phenol in the two solvents. Rytting (ref. 16) has performed measurements on five n-alkanols in iso-octane solvent and interpreted the results in terms of the monomer-single polymer model of eqn. (8). These alcohols best fit the monomer-trimer or monomer-tetramer model, but a true interpretation should probably be in terms of monomer and more than one polymer.

Table IV. Thermodynamic Parameters for Hydrogen Bonding Solutes in Various Other Solvents at 25°C Based on "Best Fit" of Data to Various Models.

Solute-Solvent	Polymer	K_n[a]	$-\Delta H_n^o$ (kcal)	$-\Delta S_n^o$ (cal/deg)	Ref.
C_6H_5OH-cyclohexane	Dimer[c]	0.10	4.5	20	4
	Trimer[c]	21.5	11.0	31	
	Dimer[d]	1.8	3.68	11.2	
	Polymer[d]	7.6	3.68	8.3	
CH_3COOH-1,2-dichloroethane	Dimer[b]	3.0	8.0	25	15
CH_3CH_2OH-cyclopentane, -cyclohexane, -cycloheptane,-n-hexane, -carbon disulfide	e	---	--	--	8,9
CCl_3COOH-m-xylene	Dimer[b]	1.0	6.5	22	11
CCl_3COOH-mesitylene	Dimer[b]	0.6	5.3	19	11
$CH_3(CH_2)_2CH_2OH$-iso-octane	Trimer[b]	31	18.4	55	16
	Tetramer[b]	260	22.8	65	
$CH_3(CH_2)_3CH_2OH$-iso-octane	Trimer[b]	59	21.0	62	16
	Tetramer[b]	840	26.3	75	
$CH_3(CH_2)_4CH_2OH$-iso-octane	Trimer[b]	31	18.4	55	16
	Tetramer[b]	350	22.8	65	
$CH_3(CH_2)_5CH_2OH$-iso-octane	Trimer[b]	29	18.1	54	16
	Tetramer[b]	330	22.4	64	
$CH_3(CH_2)_6CH_2OH$-iso-octane	Trimer[b]	420	18.5	55	16
	Tetramer[b]	450	23.1	65	
6-methylpurine-water	Dimer[d]	8.6	6.0	16	17
	Polymer[d]	8.6	6.0	16	
Diketopiperazine-water	Dimer[d]	0.06	4.2	20	18
	Polymer[d]	0.06	4.2	20	
Urea-water	Dimer[d]	0.053	1.63	11.3	19,20
	Polymer[d]	0.106	1.63	9.9	

a,b,c,d,e All footnotes refer to the same footnotes as in Table I.

In conclusion, we note that the data for all carboxylic acids studied calorimetrically can be interpreted satisfactorily in terms of the monomer-dimer model. Only occasionally (eg, phenol in CCl_4) can the alcohols be interpreted in terms of a monomer-single polymer equilibrium. When it is apparent that the data cannot be interpreted in terms of a single equilibrium reaction, it is not always clear whether the various thermodynamic parameters can be assumed to be related or independent [ie, eqns. (9) and (10) or (16) and (18)]. The assumption that equilibrium constants and enthalpies are related is most reasonable when only linear polymers are formed. This assumption excludes the existence of smaller cyclic polymers.

ACKNOWLEDGEMENT

Acknowledgement is made to the donors of The Petroleum Research Fund, administered by the American Chemical Society, the Research Corporation, and the Brigham Young University Research Division for partial support of some of the research reported herein.

REFERENCES

1. G.C. Pimentel and A.L. McClellan, "The Hydrogen Bond," W.H. Freeman and Co., San Francisco-London, 1960.
2. E.J. King, "Acid-Base Equilibria," Permagon, New York-London, 1965.
3. E.M. Woolley, J.G. Travers, B.P. Erno, and L.G. Hepler, J. Phys. Chem., 75, 3591 (1971).
4. E.M. Woolley and L.G. Hepler, J. Phys. Chem., 76, 3058 (1972).
5. N.S. Zaugg, S.P. Steed, and E.M. Woolley, Thermochim. Acta, 3, 349 (1972).
6. N.S. Zaugg, L.E. Trejo, and E.M. Woolley, Thermochim. Acta, 6, 293 (1973).
7. E.M. Woolley and D.S. Rushforth, Can. J. Chem., 52 (1974), in press.
8. R.H. Stokes and C. Burfitt, J. Chem. Thermodynamics, 5, 623 (1973).
9. R.H. Stokes and K.N. Marsh, Third International Conference on Chemical Thermodynamics, 5/5, Vienna, Austria (1973).
10. N.S. Zaugg, A.J. Kelley, and E.M. Woolley, in preparation.
11. E. Calvet and C. Paoli, Compt. Rend., 257, 3376 (1963).
12. C.C. Panichajakul and E.M. Woolley, in preparation.
13. N.S. Zaugg, N.W. Petty, E.A. Ballard, and E.M. Woolley, in preparation.
14. R.C. Graham, G.H. Henderson, E.M. Eyring, and E.M. Woolley, J. Chem. Eng. Data, 18, 277 (1973); at 15°C and 35°C in preparation.

15. G. Olofsson and I. Wirbrant, Acta Chem. Scand., 25, 1408 (1971).
16. J.H. Rytting, personal communication.
17. P.R. Stoesser and S.J. Gill, J. Phys. Chem., 71, 564 (1967).
18. S.J. Gill and L. Noll, J. Phys. Chem., 76, 3065 (1972).
19. R.H. Stokes, Aust. J. Chem., 20, 2087 (1967).
20. G.C. Kresheck, J. Phys. Chem., 73, 2441 (1969).

RAPID QUANTITATIVE METHOD FOR BOUND WATER
DETERMINATION IN AQUEOUS SYSTEMS USING
DIFFERENTIAL SCANNING CALORIMETRY

Endel Karmas and C. C. Chen
Department of Food Science
Rutgers University
New Brunswick, NJ 08903

SUMMARY

The percent of "bound" water in an aqueous system
is obtained from the isothermal dehydration curve. The
"free" water peak in the sample curve is related empirically on a weight basis to the evaporation curve of
distilled water. The method is sensitive and precise.
The water binding of the sodium form of a soy protein
isolate at higher concentrations was larger than that
of sodium caseinate. Sodium tripolyphosphate imparted
higher water binding properties to foods than did
sodium chloride. It was further shown that there was
no correlation between water binding and water activity
as measured in food model systems.

INTRODUCTION

The functional and protective role of water in
biological as well as fabricated aqueous systems is
critical and presumably is related to water binding
(Karmas, 1973).

Raw foods are biological systems. Water in foods,
comprising some 60-95% of the total weight, is by far
the dominant component. The state of water and its

distribution in foods is of prime importance and it is
obvious that changes in water content and water distribution affect the texture, appearance, palatability,
and preservation of foods.

Other food components, particularly carbohydrates,
proteins, and minerals, adsorb and bind water by different mechanisms and in different quantities. For example,
the presence of salts, such as sodium chloride and/or
phosphates, increase the water binding of proteins
(Kuprianoff, 1958).

Work on bound water is complicated by the fact
that the term is not easily defined. Fennema (1970)
stated that the relative water binding associated with
biological materials changes as a continuum. At low
concentrations water is less mobile than when it is
present at high concentrations.

Any method devised to determine water binding
depends on the definition of bound water. Most literature defines bound water as the water that remains in
an unchanged form when the system is subjected to a
particular treatment, for example, water that is un-
freezable, water that does not combine with chemical
desiccants, or water that behaves differently from
pure water.

Methods for the determination of water binding
were reviewed by Karmas and DiMarco (1970). The literature revealed that different methods produced diver-
gent results. Wide deviation was found among results
on similar materials not only when different methods
were used, but also when the same method was used.

The most consistent results appear to be those
obtained by the freezing method developed by Riedel
(1961) which is based on calorimetric measurements.
However, the calorimetric freezing method can be performed more easily with differential scanning
calorimetry. Besides the need for a freezing attachment, a shortcoming in this method seems to be that
distortions may be introduced in the water binding
when liquid water is changed to a solid phase at an

empirically selected temperature.

It is, therefore, essential to have a rapid and convenient method for estimating water binding in aqueous systems at temperatures above the freezing point of water.

Karmas and DiMarco (1970) developed a method for the determination of relative water binding intensity, the so-called Water Binding Index (WBI). A differential Scanning Calorimeter (DSC-1B, The Perkin-Elmer Corp.) was used to measure the energy input necessary to liberate water from an aqueous system compared to the evaporation of an identical amount of distilled water.

The present method describes a further refinement of the WBI method which makes it possible to determine the relative quantity of bound water in aqueous systems with the DSC.

EXPERIMENTAL

Sodium soy protein isolate, lean beef muscle tissue, sodium chloride, and sodium tripolyphosphate were used as sample materials. As in the WBI method, samples containing approximately 2.0 mg of water were weighed accurately with a Cahn Electro-Balance. Each sample was dehydrated isothermally at 105°C in the DSC at 0.032 cal/sec sensitivity. A standard curve was prepared as described by Karmas and DiMarco (1970).

The energy spent to evaporate either the water from the samples or the distilled water reference is shown in the recorded endotherms in Fig. 1. The free water in both the reference and sample is shown as A in the figure and the bound water area is described by B. In other words, the bound water is retained by the sample beyond the free water area, A, and evaporated gradually tracing the curve bordering the bound water area, B. Only the areas under the free water peaks are integrated with a planimeter and used to calculate the percent of bound water.

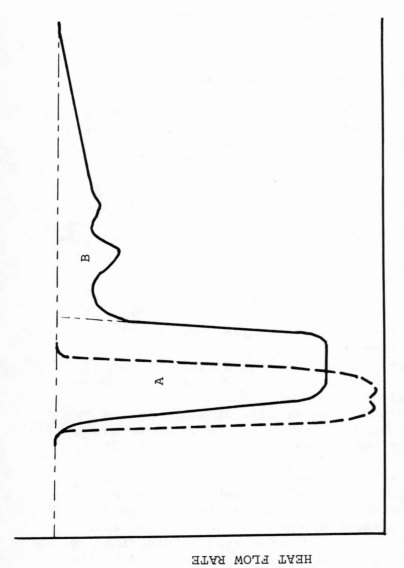

TIME

HEAT FLOW RATE

Fig. 1. Typical endotherms for evaporation of (1) distilled water (dashed line) and (2) water from a sample (solid line) iso-thermally at 105°C with the DSC. The baseline for both curves and the borderline between the free and bound water areas are indicated by irregular dashed lines.

A – free water areas; B – bound water area.

The determination of bound water by directly measuring the corresponding area is impractical. Very frequently the curve describing bound water never returns to the base line or the bound water area is too greatly elongated thus not allowing an accurate planimetric measurement.

The calculation of bound water is based on the assumption that free water in a sample behaves thermodynamically in a manner similar to pure or distilled water and that the heat of vaporization of free water from the sample is substantially equal to that of distilled water. The relative percent of bound water is calculated from the following formula:

$$B = 100 \left(1 - \frac{w}{W}\right)$$

where B = percent of bound water
w = weight (in mg) of distilled water to be evaporated which corresponds to the area of the free water peak of the sample (portion A under the solid line in Fig. 1); the value is obtained from a standard curve.
W = total sample weight (in mg); obtained with Cahn Electro-Balance.

RESULTS AND DISCUSSION

Table 1 gives the percent of bound water for some compounds, at various concentrations, which interact with water. Sodium chloride produced a large increase in water binding as its concentration approached the saturation level. The interaction of sodium chloride with water is ionic. The water binding of the sodium soy protein isolate was greater than that of caseinate at higher concentrations. Both compounds have complex water binding mechanisms comprising chemical interactions as well as physical entrapment of water. The precision of the method, as indicated by the standard deviation values, is analytically acceptable.

TABLE 1
PERCENT OF BOUND WATER IN SOME AQUEOUS SYSTEMS

Concentration of aqueous system	Sodium chloride	Sodium soy protein isolate[a]	Sodium caseinate
	Bound water (%)		
5%	1.3 ± 0.2[b]	nil	0.5 ± 0.1
10%	5.0 ± 0.4	1.1 ± 0.1	1.7 ± 0.1
15%	11.9 ± 1.7	13.8 ± 0.4	3.5 ± 0.2
20%	34.0 ± 1.6	26.4 ± 0.5	9.8 ± 0.3
25%	−	36.5 ± 0.5	19.4 ± 0.7

[a] Promine-D (manufactured by Central Soya, Chicago, Ill.)
[b] Standard deviation.

TABLE 2
PERCENT OF BOUND WATER IN BEEF COOKED IN
SOLUTIONS OF VARIOUS IONIC STRENGTH

Cooking solution	Moisture content of beef (%)	Bound water (%)
Water	66.3	43.3 ± 1.0
Sodium tripoly-phosphate, 1% aq.sol.	66.6	45.8 ± 0.6
Sodium tripoly-phosphate, 2%	67.1	47.5 ± 0.1
Sodium chloride, 2%	67.2	45.3 ± 0.8

TABLE 3

BOUND WATER VS. WATER ACTIVITY IN FOOD MODEL SYSTEMS

Sample	Bound water (%)	Water activity[a]
FMS-22-2[b]	46.34 ± 0.31	0.984 ± 0.002
FMS-20-4	45.86 ± 0.74	0.985 ± 0.003
FMS-18-6	41.34 ± 0.13	0.986 ± 0.001
FMS-16-8	39.13 ± 1.03	0.987 ± 0.002
FMS-14-10	40.87 ± 0.34	0.986 ± 0.003

[a] Determined with Electric Hygrometer, Model 15-3001 Hygrodynamics, Inc.

[b] FMS-x-y: composed of x% carboxymethyl cellulose, y% casein, 1% oil, and 75% water.

Table 2 demonstrates the remarkable sensitivity of the method when the moisture content of the treated samples, beef muscle tissue in this experiment, was kept virtually constant.

The data indicate that sodium tripolyphosphate is a better water binding agent than sodium chloride. The results were significant at $P=0.05$ level.

It is interesting to note that water binding, as determined by this method, does not correlate with water activity as is evident from the data presented in Table 3.

The correlation coefficient for these data is $r = -0.668$. The lack of correlation between water binding and water activity in meat proteins has been verified by other investigators (e.g., Vrchlabsky and Leistner, 1970) by more complicated methods. The data in Table 3 imply that carboxymethyl cellulose is a better water binding agent than casein.

These data represent only a few of the many possible applications of this method for determining the relative quantity of bound water. It must be emphasized that the method for bound water determination in aqueous systems is empirical. The method may be applied to

both aqueous solutions and aqueous mixtures. The principle may also be applied to other than aqueous systems
which probably would require different vaporization
temperatures.

Among other factors the sample size is an important
parameter. A standard curve constructed by evaporation
of various amounts of distilled water was not linear.
Therefore, it is recommended that the sample size be
kept as uniform as possible.

In conclusion it may be stated that this method
for determining water binding in various aqueous systems
is rapid, precise, sensitive, and empirically quantitative.

REFERENCES

Fennema, O. R. 1970. Nature and characteristics of
water in food and biological systems. Paper presented
at the symposium "Highlights in Food Science", March
23, East Lansing, Michigan.

Karmas, E. 1973. Water in biosystems. J. Food Sci.
38, 736.

Karmas, E., and DiMarco, G. R. 1970. Water Binding
Index of proteins as determined by differential microcalorimetry. In "Analytical Calorimetry", Vol. 2,
p. 135. R. S. Porter and J. F. Johnson, eds. Plenum
Press, New York.

Kuprianoff, J. 1958. "Bound water" in foods. In
"Fundamental Aspects of the Dehydration of Foodstuffs",
p. 14. (Society of Chemical Industry, London). The
Macmillan Co., New York.

Riedel, L. 1961. Zum Problem des gebundenen Wassers
in Fleisch. Kältetechnik 13, 122.

Vrchlabsky, J., and Leistner, L. 1970. Beziehung
zwischen Wasseraktivität und Wasserbindung von Rind-
und Schweinefleisch. Fleischwirtschaft 50, 967.

EFFECT OF DEHYDRATION ON THE SPECIFIC HEAT OF CHEESE WHEY

Elliott Berlin[+] and Phyllis G. Kliman[+]

Dairy Products Laboratory, Eastern Regional Research Center, Agricultural Research Service, United States Department of Agriculture, Philadelphia, Pennsylvania 19118[*]

ABSTRACT

Differential scanning calorimetry was used to determine the specific heat of cheddar cheese whey as a function of water content and thereby provide fundamental data useful for the further development of dried whey products. The specific heat of fluid cheddar cheese whey, which contains 7% solids, was $0.951 \pm .036$ cal/g/°C at 12°C. A linear relationship was maintained between specific heat and moisture content when dried whey solids were rehydrated to moisture levels between 3 and 93% H_2O. The apparent partial specific heat of the whey solids was 0.328 cal/g/°C and that of the water was 0.995 cal/g/°C, a value close to that of bulk water. An inflection, however, was noted in the relation between specific heat and water content at 50% H_2O when the specific heat data were obtained with concentrated whey samples prepared by evaporation of water from fluid whey. These data yielded apparent partial specific heat values for water of 0.966 cal/g/°C above 50% H_2O and 1.203 cal/g/°C below 50% H_2O. Apparently the water is in a more structured form in concentrated systems provided that the solids are initially fully hydrated. This conforms to the concept that a critical amount of water must be present in a proteinaceous system for the water to be held in a quasi-solid or "icelike" structure.

[+]Present address: Nutrition Institute, Agricultural Research Center, Beltsville, Maryland.

[*]Research conducted while Dairy Products Laboratory was located in Washington, D.C.

INTRODUCTION

Much of the nutritious solids in whey have been wasted in the past. However, recent legislation to reduce environmental pollution has discouraged this practice. Considerable quantities of whey are now being spray-dried for use as feed and food. This process, however, is relatively inefficient, at least partly because the various dissolved and suspended materials in liquid cheese whey bind substantial quantities of water. With the prospect of serious energy shortages it is becoming increasingly important for scientists to obtain thermodynamic data which should be useful to the developers of more efficient food processing techniques. In this study the relationship between water content and the specific heat of cheddar cheese whey was determined, both to define a portion of the energy requirements in whey processing and to provide an insight into the physical chemical state of water in concentrated whey.

We have reported (1) on the specific heat of β-lactoglobulin, the major whey protein, containing 20-250 mg sorbed water per gram protein. Whey is a more complex system containing carbohydrate and salt fractions which also bind substantial quantities of water (2). We have now extended our studies with the differential scanning calorimeter to measure the specific heat of cheddar cheese whey over a wide range of water contents from anhydrous whey solids to ordinary fluid whey, containing 93% water. We investigated the reversibility of the water binding process by measuring the changes in specific heat accompanying the dehydration of fluid whey and the rehydration of whey powder.

EXPERIMENTAL

<u>Materials</u>. A single lot of cheddar cheese whey powder, spray dried in the Dairy Products Laboratory Pilot Plant, was used throughout this study to insure identity of the solid components in all samples. Moisture levels in powder samples up to 0.25g H_2O/g whey solids were adjusted by exposing the powder to controlled humidity environments, maintained with appropriate saturated salt solutions (3). Samples at higher moisture levels were prepared by dispersing the whey powder in water to form preparations ranging from concentrated slurries to dilute solutions.

To study the properties of fluid whey, powder was reconstituted to 7% total solids content and held overnight to allow for completion of lactose mutarotation. Reconstitution of the dried whey was preferred over using separate batches of fluid whey that might vary slightly in composition. The reconstituted whey was concentrated with a rotary evaporator and/or freeze-dried to obtain whey samples at various stages of dehydration.

Methods. Specific heats were determined using the Perkin-Elmer Model DSC-1B[1] differential scanning calorimeter by a technique based on the methods of Wunderlich (4) and O'Neill (5). First, empty sealed aluminum capsules were heated in the sample and reference holders of the calorimeter from 0 to 25°C at a programmed rate of 5°C/min. to establish a baseline accounting for the asymmetry of the system as a function of temperature. The power output of the calorimeter was then calibrated by replacing the empty capsule on the sample holder with one containing a 0.03g sample of Al_2O_3 (standard sapphire plate supplied by the Perkin-Elmer Co.) and repeating the scan from 0 to 25°C at 5°C/min. Finally, this capsule was replaced with one containing a whey sample (5-35 mg), and the scan was repeated to determine the specific heat. Cooling below ambient temperature was accomplished by filling the low temperature cover of the DSC-1B with solid CO_2. More stable recorder tracings were obtained when solid CO_2 was used rather than liquid N_2, and there were less of the usual experimental nuisances resulting from moisture condensation on the instrument.

The sample pans used in these determinations were hermetically sealed with the Perkin-Elmer volatile-sample sealer accessory to prevent the condensation or evaporation of water vapor until after the completion of the specific heat determination. After the scan, the moisture content of each sample was determined by puncturing the sample container and drying to constant weight under vacuum at 65°C.

Experimental specific heats were calculated from the amplitudes of the recorder tracings using sapphire specific heat data of Ginnings and Furukawa (6). The sample and reference pans used were weighed and selected so that the specific heat of aluminum could be neglected. Apparent partial specific heats were determined using the equation of White and Benson (7), as employed by Bull and Breese (8) in the form:

$$(1 + W_1) \, C_p = \bar{C}_{P_2} + \bar{C}_{P_1} W_1$$

where W_1 is the weight of water per gram of whey solids, C_p is the experimental specific heat for whey solids plus water, $\bar{C}_{P_1}$ is the apparent partial specific heat of the water, and $\bar{C}_{P_2}$ is the partial specific heat of the whey solids.

Results and Discussion. The specific heat of the reconstituted fluid cheddar cheese whey was found to be $0.951 \pm .036$

---

[1]The mention of brand or firm names does not constitute an endorsement by the Department of Agriculture over others of a similar nature not mentioned.

cal/g/°C at 12°C. This value is the average of four separate determinations and the error indicated here and elsewhere in this
paper is the standard deviation. This value and the other specific
heat data presented in this paper are only the values computed for
12°C, the mid-point of the DSC scans; however, specific heats were
computed at 5° intervals between 7 and 22°C. No significant variation in specific heat with temperature was observable in this range
for any of the whey samples. The data at temperatures other than
12°C have therefore not been reported in detail.

It is not unexpected that the specific heat of fluid whey is
similar to that of water since fluid whey is approximately 93%
water. Accordingly, when water is removed from whey during concentration or drying, the specific heat value is diminished. Such
changes in specific heat during the course of dehydration are evident from the data in Figure 1 where the specific heats of whey
samples are presented graphically as a function of moisture content.

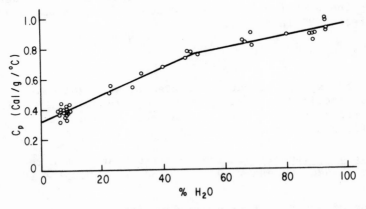

Figure 1. Graph of C_p, the measured specific heat of concentrated
and dehydrated whey at 12°C against percent water

Examination of these data indicates an inflection in the relation between specific heat and water content occurring at about
0.5g of water per gram of sample. Bull and Breese (8) reported a
similar inflection in the relation between specific heat and the
water content of ovalbumin crystals at 0.43g H_2O/g egg albumin.
In our studies (1) with crystalline ovalbumin and β-lactoglobulin
only linear relationships between specific heat and water content
were observed; however, those studies were limited to systems containing less than 25% water.

When dried whey was rewetted to various levels of rehydration,
a linear relation between heat capacity and water content was maintained over the entire range from zero to 93% water (Figure 2).
Least squares analysis of these data yielded the linear relation:

$$C_p = 0.312 + 0.007 \ (\% \ H_2O).$$

In this equation C_p represents the measured specific heat of the
whey solids-water samples, and the intercept on the C_p axis at

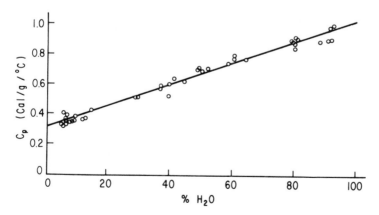

Figure 2. Graph of C_p, the measured specific heat of rehydrated
whey solids at 12°C against percent water.

zero water content is the extrapolated value for the specific heat
of anhydrous whey solids, i.e. 0.312 cal/g/°C. A similar value of
0.315 cal/g/°C was obtained from the data in Figure 1 for whey samples which were dehydrated to less than 50% moisture. Analysis of
the data of Figure 1 yielded the equations:

$C_p = 0.315 + 0.009$ (% H_2O)

for samples containing 0-50% H_2O and

$C_p = 0.562 + 0.004$ (% H_2O)

for the more dilute samples.

Comparison of these two sets of data for hydration and dehydration clearly suggest the presence of some fundamental physical
difference in these processes. Further insight into these phenomena may be obtained from the apparent partial specific heats of the
water and the whey solids in each case. Plots of $(1+W_1) C_p$ against
W_1 were made for: (1) the experimental points (Fig. 1) in the range
from zero moisture to the inflection point, (2) the experimental
points (Fig. 1) in the range from the inflection point to 93% moisture, and (3) all the data (Fig. 2) for rewetting the dried whey
solids. The results of these plots, analyzed by least squares, are
given in Table 1 with the variations expressed as the standard
deviations.

Table 1:   Apparent partial specific heat values for water, $\bar{C}_{p_1}$,
and whey solids, $\bar{C}_{p_2}$.

Sample and moisture range	$\bar{C}_{p_1}$ (cal/g/°C)	$\bar{C}_{p_2}$ (cal/g/°C)
Dehydrated whey, 0-50% H_2O	1.203 ± .028	0.316 ± .012
Concentrated whey, 50-93% H_2O	0.966 ± .020	0.467 ± .151
Rehydrated whey, 0-93% H_2O	0.995 ± .010	0.328 ± .046

It is notable that when whey is concentrated to less than 50%
moisture, there is an appreciable increase in the apparent partial
specific heat of the water associated with the whey solids over
that of water in the bulk. Bull and Breese (8) reported a value
of 1.247 ± .023 cal/g/°C for the apparent partial specific heat of
water associated with ovalbumin, and they attributed the increase
over the specific heat of bulk water to the large exothermic heat
of hydration of egg albumin. Heating a sorbate-sorbent combination
at constant pressure usually results in the desorption of gas or
vapor if the system is initially at equilibrium, thereby complicating heat capacity measurements of adsorbed films by unwanted

desorption effects which cause the coverage to be temperature dependent and the total heat capacity to increase above the isosteric value (9). In a previous publication (1) we demonstrated
that heat capacities of protein-water systems measured with the
differential scanning calorimeter using sealed containers are isosteric values. We concluded from the isosteric specific heat
values, $\bar{C}_{P_1}$, for water sorbed by ovalbumin (1.269 $\pm$ .103 cal/g/°C)
and β-lactoglobulin (0.947 $\pm$ .137 cal/g/°C) that the
sorbed water exists in some structured form involving multiple
hydrogen bonding. The excess heat capacity of liquid water over
that of ice or water vapor is sometimes called "structural" heat
capacity, and is attributed to the thermal breakdown of the associated structure present in the liquid (10). Water molecules
bound to isolated specific sites on the surface of a protein should
not exhibit such a structural heat capacity contribution, hence the
elevation of $\bar{C}_{P_1}$ over that of ice or water may be taken as evidence
for the   association of sorbed water into a structured hydration shell. The results with ovalbumin (1) and our present data
for concentrated and dehydrated whey ($<$ 50% H_2O) suggest an even
greater order of structuring of the associated water in these systems. In more dilute systems ($>$ 50% H_2O) it is conceivable that
so much water is present as ordinary water in addition to that associated with the whey solids that the apparent partial specific
heat approaches that of bulk water.

The similarity between $\bar{C}_{P_1}$ for water associated with whey
solids rehydrated to less   than 50% water (Table 1) and the
specific heat of bulk water demonstrated that no excess structure
is present unless the whey solids are more completely hydrated.
Changes in specific heat of biological materials during hydration
are not unusual. Chakrabarti and Johnson (11) measured the specific heat of tobacco-water complexes with the differential
scanning calorimeter and observed inflections in the specific heat-
moisture content relation above 40% H_2O. A transitional moisture
range was identified which they related to a change in water from
an adsorbed phase to a solution phase. Similarly, elevated heats
of desorption of water vapor from milk and whey powders at higher
moisture levels have been reported (12).

While studying water binding in proteinaceous systems (13) we
observed an increase in the heat of vaporization of sorbed water
once a critical amount of water (approximately 0.18g H_2O/g protein)
is sorbed by β-lactoglobulin, bovine serum albumin, bovine casein,
or calfskin collagen. It was concluded that at the higher moisture
levels the solid protein matrix had become swollen, and possibly
conformational changes occurred in the protein molecules permitting
more H_2O - surface contacts and ultimately the formation of a quasi-
solid or "icelike" structure. It is thus plausible that excess

structuring of associated water in whey only occurs at higher
moisture levels and is observable upon dehydration from such
dilute systems. Interactions of water with other components of
the whey solids should, however, not be neglected.

REFERENCES

(1) Berlin, E., P. G. Kliman, and M. J. Pallansch. Thermochim.
     Acta, 4, 11 (1972).

(2) Berlin, E., P. G. Kliman, B. A. Anderson, and M. J. Pallansch.
     J. Dairy Sci., 56, 984 (1973).

(3) Rockland, L. B. Anal. Chem., 32, 1375 (1960)

(4) Wunderlich, B. J. Phys. Chem., 69, 2078 (1965).

(5) O'Neill, M. J. Anal. Chem., 38, 1331 (1966).

(6) Ginnings, D. C. and G. T. Furukawa. J. Amer. Chem. Soc., 75,
     522 (1953).

(7) White, P. and G. C. Benson. J. Phys. Chem., 64, 599 (1960).

(8) Bull, H. B. and K. Breese. Arch. Biochem. Biophys., 128,
     497 (1968).

(9) Dash, J. G., R. E. Peierls, and G. A. Stewart. Phys. Rev. A,
     2, 932 (1970).

(10) Berendsen, H. J. C., in A. Cole (Ed.) "Theoretical and Experi-
     mental Biophysics," Marcel Dekker, New York, 1967, p. 26.

(11) Chakrabarti, S. M. and W. H. Johnson. 1971 Winter Meeting
     American Society of Agricultural Engineers, Chicago.

(12) Berlin, E., P. G. Kliman, and M. J. Pallansch. J. Dairy Sci.,
     54, 300 (1971).

(13) Berlin, E., P. G. Kliman, and M. J. Pallansch. J. Colloid
     Interface Sci., 34, 488 (1970).

SOLID STATE REACTION KINETICS IV:

THE ANALYSIS OF CHEMICAL REACTIONS BY MEANS OF THE WEIBULL FUNCTION

E.A. Dorko,* W. Bryant, and T.L. Regulinski

Departments of Aero-Mechanical and Electrical
Engineering, Air Force Institute of Technology
Wright-Patterson Air Force Base, Ohio 45433

There is a good deal of interest in the analysis of reactions which are initially heterogeneous, solid state reactions but which end up as homogeneous, liquid phase reactions.[1] This interest stems from the determination of the true melting point of a material which reacts during melting. Also, since organic systems are potentially useful as heat transfer and temperature sensing agents, it is of interest to have a mathematical model of their decomposition on melting.

Recent work on the development of models to take into account kinetic, melting, heat and mass transfer effects has been very seriously hampered by the difficulty of separating the reactions into their solid and liquid phases.[1a,1c] The present paper reports a simple, graphical algorithm for these reactions which separates the two reaction regimes and which allows the immediate determination of the characteristic parameters of each. The method utilizes the Weibull cumulative distribution function[2] to reduce the reaction history data.

The distribution function developed by Weibull has been a valuable analytical tool in many areas for the analysis of stochastic processes. The process under study in this report may be stated in stochastic notation[3] as eq 1.

$$P\{T \leq t\} = P\ \{\text{chemical reaction occurring in } (t_0, t)\,|\,\text{Temp}\} \quad (1)$$

Eq 1 describes a stochastic process in which the probability of chemical reaction occurring in the interval between the initial time (t_0) and any time (t) at some specified temperature (Temp) is given by the distribution function $P\{T \leq t\}$.

Quantifying eq 1 by the Weibull model yields eq 2 for the conditions shown.

$$F(t) = P\left\{ T \leq t \right\} = \begin{cases} 1 - \exp\left\{ -(t - \gamma)^{\beta}/\alpha \right\} & \begin{cases} t \geq 0 \\ t > \gamma \\ \alpha, \beta > 0 \end{cases} \\ 0 & t < 0 \end{cases} \tag{2}$$

where $F(t)$ is the cumulative probability of reaction occurring at a time equal to or less than t and α, β, and γ are the scale, shape, and location parameters respectively. If α and γ have fixed values and $\beta = 1$, then the Weibull function reduces to an exponential function characteristic of a first order mechanism.[4] In this case the rate constant is equal to $1/\alpha$. Many decompositions proceed by a first order reaction mechanism in the melt.[1]

If on the other hand $\beta > 1$, which is normally the case for the solid state portion of reaction, the Weibull function becomes identical to the Avrami - Erofeev equation[5],[6] shown as eq 3.

$$a = 1 - \exp\left(-kt\right)^{\beta} \tag{3}$$

where $F(t) = a$, the extent of reaction, and $k = 1/\alpha$. This function characterizes many solid phase reactions.

Weibull[2] and Kao[7] have shown that some populations do not follow a simple Weibull function; rather they consist of a mixture of two or more sub-populations each with its own set of parameters. Such a mixture of two sub-populations, $F_1(t)$, and $F_2(t)$, has a mixed Weibull function written as a linear combination of separate functions.

$$F(t) = pF_1(t) + (1-p) F_2(t) \qquad 0 < p < 1 \tag{4}$$

where p is an experimentally determined constant.

Several methods have been used to separate $F_1(t)$ from $F_2(t)$ in the mixed Weibull function. Kao[7] developed graph paper from which a determination of the parameters are made from a plot of $F(t)$ vs time. He also developed an algorithm for estimating the parameters of a mixed Weibull function. The presently reported algorithm is similar to the one originally prepared by Kao.[7]

Figure 1 shows a typical plot of data for the decomposition of benzoyl peroxide at 371°K (98°C). The data measurements were obtained on a differential scanning calorimeter operated in the isothermal mode. The experimental technique has been described previously.[1a],[1b],[8] The value of the reaction parameters are obtained by use of the following algorithm:

1. Plot the total reaction history (% reaction vs time).
2. Fit straight line segments F_1 and F_2 through the data points.
3. If the slope of F_2 is greater than the slope of F_1, draw a vertical line down from the intersection of F_2 and the upper abscissa.
4. If the slope of F_2 is less than the slope of F_1 draw a vertical line up from the intersection of F_1 and the lower abscissa.
5. Extend F_1 or F_2 to intersect with the vertical line and draw a horizontal line through the intersection.
6. The intersection of this horizontal line with the left ordinate gives P x 100.

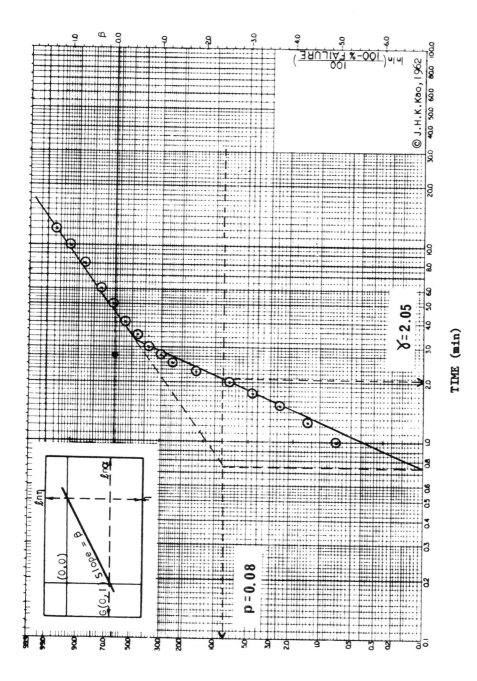

Figure 1. A "Weibull plot" of the reaction history of benzoyl peroxide at 371°K (98°C)

7. Draw a vertical line through the intersection of the horizontal line and F_1 or F_2.

8. A vertical line dropped from the intersection to the lower abscissa gives the value of γ.

9. Replot all data points from $t = t_0$ to $t = \gamma$ as $F_1(t)$ after dividing the value of each datum by p.

10. Let $\gamma = t_0$ and replot all data points from $t = \gamma$ as $F_2(t)$ after dividing the value of each datum by $(1 - p)$.

For the data of Figure 1, p was found to be 0.08. Completion of the algorithm gave $\gamma = 2.05$ min, $\alpha_1 = 11.0$, $\beta_1 = 2.3$, $\alpha_2 = 2.77$, and $\beta_2 = 1.0$. These results were verified by a computerized version of the Kolmogorov-Smirnoff test[9] which isolated the Weibull probability distribution function for the random variable (% decomposition).

The algorithm was applied to additional data for benzoyl peroxide and also to previously obtained data for cupferron tosylate.[8] The value of γ for those data corresponded precisely to the time at which the thermograms reached maximum deflection. In addition, data for malonic acid decomposition obtained by measuring pressure differences during the reaction by Hinshelwood[10] were reduced according to the algorithm. Representative results are shown in Table I for the liquid phase reaction. The solid state reaction, having been separated from the liquid phase, is shown to obey the model described mathematically by the Avrami-Erofeev equation for all the cases studied.

TABLE I
Values Obtained for the Kinetic Parameters of the Liquid Phase of Reactions which Occur During Melting.

Reactant	Temp °K	Rate Constant Present Work	Rate Constant Literature	Activation Energy (kcal/mol) Present Work	Activation Energy (kcal/mol) Literature
Benzoyl	371	6.13^a	----	69.0	----
Peroxide	361	0.46^a	----		
Cupferron	394	6.53^a	$8.89^{a,c}$	28.90	29.25^c
Tosylate	373	0.83^a	$1.40^{a,c}$		
Malonic	398.9	2.23^b	$2.04^{b,d}$	31.6	30.6^d
Acid	389.5	0.85^b	$0.81^{b,d}$		

^{a}sec^{-1} x 10^3
^{b}min^{-1} x 10^3
^{c}Ref. 8
^{d}Ref. 10

The present method of analysis produces results in substantial agreement with the earlier work where conditions are comparable. The analysis of benzoyl peroxide shows that the activation energy is substantially higher than the value (ca 30 kcal/mol) obtained for decomposition in solution[4] and under explosive conditions.[11] This difference may be explained by a change in the relative rates at which the reactions in the decomposition sequence are occurring. Specifically the slow step in the present case may be the breaking of the C-C bond in the benzoate radical formed by the initial 0-0 bond scissure.[12] The C-C bond strength is estimated to be 78 kcal/mol. Further work is being done to determine the cause of this marked difference.

REFERENCES

1. (a) E.A. Dorko, and R.W. Crossley, J. Phys. Chem., 76, 2253 1972;

    (b) R.N. Rogers, Thermochem. Acta 437 (1973);

    (c) Yu. Ya. Maksimov, Russian J. Phys. Chem., 41, 635 (1967);

    (d) F.I. Dubovitskii, G.B. Manelis, and A.G. Merzhanov, Dokl. Akad. Nauk SSSR, 121, 549 (1958);

    (e) G.B. Manelis and F.K. Dubovitskii, ibid., 124, 475 (1959).

2. W. Weibull, J. App. Mechanics, 18, 293 (1951).

3. A Papoulis, Probability, Random Variables, and Stochastic Processes, McGraw-Hill Book Company, New York, 1965.

4. A.A. Frost and R.G. Pearson, Kinetics and Mechanics, 2nd ed, John Wiley & Sons, Inc., New York, 1961, p 13.

5. (a) A.R. Allnott and P.W.M. Jacobs, Canadian J. of Chemistry, 46, 111 (1967);

    (b) M. Avrami, J. Chem. Phys. 7, 1103 (1939); 8, 212 (1940); 9, 177 (1941);

    (c) B.V. Erofeev, Comp. Rend. Acad. Sci. U.R.S.S., 52, 511 (1946).

6. For a review of this area see L.G. Harrison, "The Theory of Kinetics" Ch 5 in Comprehensive Chemical Kinetics, C.H. Bamford and C.F.H. Tipper, ed., vol 2, Elsevier Publishing Co., New York, 1969.

7. J.N.K. Kao, Technometrics, 1, 389 (1959).

8. E.A. Dorko, R.S. Hughes, and C.R. Downs, Anal. Chem., 42, 253 (1970).

9. F.J. Massey, American Statistical Association Journal, 46, 68 (1951).

10. C.N. Hinshelwood, J. Chem., Soc., 117, 157 (1920).

11. D.H. Fine and P. Gray, Combust. Flame, 11, 71 (1967).

12. W.H. Richardson and H.E. O'Neal, "The Unimolecular Decomposition and Isomerization of Oxygenated Organic Compounds (Other than Aldehydes and Ketones)," Ch 4 in Comprehensive Chemical Kinetics, C.H. Bamford and C.F.H. Tipper, ed., vol 5, Elsevier Publishing Co., New York, 1972, p 495.

PRELIMINARY RESULTS ON THE NATURE OF n IN EQUATION $(dx/dt)=(a-x)^n$

AS APPLIED TO THREE SOLID THERMAL DECOMPOSITION REACTIONS

D. M. Speros, Lighting Research Laboratory, General

Electric Co., Cleveland, Ohio 44112 and H. R. Werner, Lamp

Glass Dept., General Electric Co., Cleveland, Ohio 44143

INTRODUCTION

The relation

$$\frac{d\alpha}{dt} = k(1-\alpha)^n,\qquad(1)$$

a fundamental equation in homogeneous kinetics, has been widely applied to solid-state chemistry.[1] This application has been as widely contested because the physical significance of the equation in solid reactivity has not been elucidated.[2] This is particularly true for values of n other than 0, 1/2, 2/3 and 1[3], yet precise experimental results show that n frequently acquires values other than these.[4] Thus, the equation has been used mainly for purposes of analytical convenience.[3b]

It has been postulated on theoretical grounds[5] that for a reversible reaction the parameter n is not determined singularly by the overall geometry of the reacting interface as hitherto believed[1-4], but also by the interfacial epitaxy between reactant and product at the atomic level. If the latter is expressed by the parameter φ, characteristic of each product-reactant interface (see below), then

$$n = g \cdot \varphi \qquad(2)$$

where g acquires the values given above for various geometrical shapes of the interface.

The purpose of the phase of the work described in this paper

511

was to measure with the highest possible experimental accuracy values of n and g in two reversible and one irreversible solid thermal decompositions. This permitted the determination of the value of φ for each reaction by means of equation (2) and its comparison with the theoretically predicted values.

DETERMINATION OF g

1. Derivation of equation (1) and nature of g.

In order to express unambiguously the nature of g in terms of moles of reactant and product only, without introducing assumptions concerning the rate of interface advance, as is necessary for this work, equation (1) was derived for the case of phase-boundary controlled reactions. A reaction of this type has been defined[6] as a reaction in which the kinetic rate is determined by the available interfacial area, and specifically, as a reaction in which the rate is proportional to the surface area of unreacted material. An attempt to express the kinetics of such a reaction then involves, as a first step, the calculation of the surface area of the reactant.

The volume V of (a-x) moles of reactant is $\frac{M}{d}(a-x)$, regardless of the shape of the particle or sample of the (a-x) moles. M is the molecular weight of the reactant, d its density, a the number of moles of reactant at time $t = 0$ and x the number of moles of product at time $t = t$. Therefore, (a-x) is the number of moles of reactant at time t.

The area, A, is a function of shape. For example, for a spherical specimen

$$V = \frac{M}{d}(a-x) = \frac{4}{3}\pi r^3 \; ; \tag{3}$$

$$A = 4\pi r^2 \; . \tag{4}$$

Solving (3) for r and substituting in (4)

$$A = 4\pi \left[\frac{3}{4\pi}\frac{M}{d}(a-x)\right]^{2/3} \; , \tag{5}$$

and if d can be assumed constant,

$$A = \text{const}\,(a-x)^{2/3} \; .$$

Similarly, it can be shown that the two-thirds exponential dependence is also obtained for cubes, parallelopipeds, cylinders, etc., and with certain restrictions[7] even for powdered samples when the reaction does not involve the individual particles separately but follows a reaction front which progresses from the exterior surface of the sample to the interior (bulk effect).

For a cylinder or disk of radius r and h\ght h reacting so that h remains constant, the "contracting disk" relation is obtained: $V = (M/d)(a-x) = \pi r^2 \cdot const.$, $A = 2\pi r \cdot const.$, (because only the curved surface is involved in the reaction), $r = [(1/\pi)(M/d)(a-x)]^{\frac{1}{2}}$ and $A = const. (a-x)^{\frac{1}{2}}$ giving the one-half power dependence.

Zero power dependence results when A is constant as for example for a cylinder reacting along h only. Then r is constant and, therefore, $A = const.$ regardless of the change in (a-x).

Finally, for an infinitely thin plate, $V \propto A \propto (M/d)(a-x)$ resulting in first power dependence.

In general then, the area of a specimen can be given by

$$A = const. (a-x)^g \tag{6}$$

where the value of g is determined, as seen above, singularly by the overall geometrical shape of the interfacial area, e.g. the phase-boundary.

Of particular importance in this work are the values of g for the cases of a) a rhombohedron such as a single crystal of calcite and b) reaction proceeding simultaneously throughout the mass of the specimen.

The volume of a rhombohedron of sides a,b,c,and angles α, β,γ is:[8]

$$V = \frac{M}{d}(a-x) = a \cdot b \cdot c \cdot (1-\cos^2\alpha-\cos^2\beta-\cos^2\gamma+2\cos\alpha \cdot \cos\beta \cdot \cos\gamma)^{\frac{1}{2}}.$$

If during the reaction, the reaction interface proceeds always parallel to the original crystal faces, i.e. the original shape is maintained, we can set $\frac{a}{b} = \rho_1 = const.$ and $\frac{a}{c} = \rho_2 = const.$ and likewise $\alpha,\beta,\gamma = const.$ Hence, $V = (M/d)(a-x) = a^3 \cdot const.$ Since $A = 2[a \cdot b \cos \theta_1 + a \cdot c \cos \theta_2 + b \cdot c \cos \theta_3]$, where the angles θ are those between one side and the vertical to the adjacent side, then for constant angles $A = a^2 \cdot const.$ Since then $a \propto V^{1/3} = [(M/d)(a-x)]^{1/3}$, $A = const. (a-x)^{2/3}$, yielding again a two-thirds dependence.

It will be shown that for several of the cases studied in this work, the reaction, under the experimental conditions employed, proceeded similtaneously throughout the mass of the specimen in a completely random fashion. Study of specimens, reacted to various

extents, by means of photomicrography and associated methods showed that the probability of reaction occuring in any element of
volume of the specimen was approximately the same as in any other
element. Hence, in the limit, the reacting area was proportional
to the volume of remaining unreacted material or $V = (M/d)(a-x) \propto A$.
Therefore, g should be near unity and such a value was assigned
tentatively to g in these cases.

Returning now to equation (1): according to the above definition of phase-boundary controlled reactions, the rate is proportional to A as expressed by equation (6). If the rate is defined
as the rate of production of x, i.e. as dx/dt, as it must be for
experimental methods such as T.G.A. and D.S.C., then for phase-
boundary reactions

$$\frac{dx}{dt} = k_s (a-x)^g .\qquad (7)$$

By substituting $\alpha = \frac{x}{a} =$ (fraction reacted) in equation (7) we
obtain equation (1) where $k = k_s \cdot a^{g-1}$, if g were identical to n.

These equations appear to be the same as the fundamental
equation for homogeneous kinetics. However, referring to the
above derivation, as it stands to this stage, the differences
between the two equations become clear: a and x are not intensive
but extensive quantities, i.e. not concentrations such as moles/
unit volume but moles. The parameter g is not molecularity or
even reaction order but simply the geometrical parameter specified
above, and k_s in addition to being a proportionality constant between the rate and the area, contains the constants specified
above resulting from interfacial geometry.

2. Experimental measurement of g.

This understanding of the nature of g makes clear that, in
principle, g is an experimentally determinable quantity. Thus, it
would be expected that, in phase-boundary controlled reactions,
spherical specimens can be expected to follow the 2/3 dependence
as would rhombohedral calcite; brucite platelets or disks reacting
from the edges inward the 1/2 dependence and so on.

It was found that indeed g is experimentally determinable but
by no means predictable. It proved necessary to examine, by a
number of experimental means, not only each material (i.e. calcite,
brucite, powders, etc.) but each specimen of each material and at
many stages of reaction in order to determine g with any degree of
certainty. Furthermore, it was important to do this in conjunction
with the T.G.A. and D.S.C. runs, i.e. each specimen had to be
matched with the particular run, and, therefore, with the particu-

lar n to which it pertained.

The techniques and results in each case will be given below.

DETERMINATION OF n

According to equation (7), n can be determined in principle by measuring the differential quantity (dx/dt) and the integral quantity $(a-x)$. This can be done in homogeneous reactions either isothermally or with rising temperature T, particularly when T = const. t.

A voluminous literature exists [1d,9] on the controversy concerning the validity of the application of the isothermal or the non-isothermal approach to solid-state reactions. However, the fact remains that neither approach has yielded reproducible results. This is shown in tabulations listing published values of activation energy, ΔE, for two of the compounds of concern here, $CaCO_3$ and $Mg(OH)_2$, in any form and by either approach. Thorough literature searches by Ingraham and Marrier[10] on $CaCO_3$ and by Gordon and Kingery[11] on $Mg(OH)_2$ reveal that the ΔE values for $CaCO_3$ range from 35 to 230 kcal/mole, i.e. they differ by a factor of nearly seven. In the case of $Mg(OH)_2$, the values differ by a factor of over two, ranging from 19 to 43 kcal/mole. The values for n show greater relative disparity; predictably so because the determination of n is considerably more difficult than that of ΔE as indicated, for example, in Fig. 3. At the other extreme, i.e. that of pausity concerning the kinetic parameters of important compounds, ΔE and n for $CaHPO_4$ had not been measured previously.[12]

Consequently, it was decided to measure (dx/dt) and $(a-x)$ as directly as possible, i.e. to employ methods and apparatus yielding simultaneously both the differential and the integral quantities, such as thermogravimetry, under stringently controlled and reproducible conditions. Furthermore, because of the volume of the experimental work contemplated, non-isothermal calorimetry or D.S.C. was also employed in conjunction with the non-isothermal T.G.A.

Because equation (7) involves explicitly moles of reactant and product*, it affords a direct conversion of T.G.A. weight changes or D.S.C. heat changes to moles as follows:

---

* This, for example, avoids the confusion often introduced by the non-dimensional quantity α, equation (1), leading to use of inappropriate quantities to express initial, intermediate and final weights or heats. Likewise, it avoids further assumptions on the rate of interface motion.

For reactions of the type $Solid_1 \rightarrow Solid_2 + Gas$ in which the number of moles of $Solid_1$ decomposing results in as many moles of gas and $Solid_2$

$$a = \frac{W_o}{M_G} \; ; \; x = \frac{W_t}{M_G} \text{ and } \frac{dx}{dt} = \frac{1}{M_G} \cdot \frac{dW_t}{dt}$$

where M_G is the molecular weight of the evolved gas, W_o is the total weight lost and W_t the weight lost up to time t.

Substituting in (7) we obtain

$$\frac{dW}{dt} = k_s \, M^{1-g} (W_o - W_t)^g \; . \tag{8}$$

For D.S.C., for H_o being the total heat change, H_t the heat change up to time t, ΔH_T the enthalpy of the reaction at reference[13] temperature T, and by the Hess principle of additivity of heats, expressed[14] as $(dx/dt) = (1/\Delta H_T) \cdot (dH/dt)$, we obtain by substitution in (7)

$$\frac{dH}{dt} = k_s \, \Delta H_T^{\,1-g} (H_o - H_t)^g \; . \tag{9}$$

As stated in the introduction, the application of these equations to experimental results will yield the exponent n and not g as in derived equations (8) and (9). Thus, the next problem involves the evaluation of n from these equations as applied to the experimental results.

A voluminous literature also exists on the subject of the evaluation of n from empirical equations such as (1).[3b,9,15] In most cases, the evaluation methods proposed involve the integration of equation (1). Since, however, (dx/dt) was measured directly in this work it was preferred to apply to the heterogeneous reactions studied here, the method applied by Borchardt and Daniels[16] to homogeneous reactions studied non-isothermally by D.T.A. They simply solved an equation such as (9) for k and equated this to the Arrhenius equation. Then, through trial and error they adjusted the value of n until a straight line Arrhenius plot was obtained. For the T.G.A. equation then

$$k_s = \frac{\dfrac{dW}{dt}}{M^{1-g}(W_o - W_t)^g} = Ze^{\frac{\Delta E}{RT}} \; . \tag{10}$$

It should be noted that since the left hand side of equation (10) was _derived_ here for phase-boundary controlled reactions, three assumptions are involved in the application of equation (10) to heterogeneous systems:

1. A phase-boundary reaction is involved.

2. This reaction obeys an Arrhenius type relation.
3. g ≡ n.

Assumption 1 can be justified by applying the equation to reactions such as those chosen in this work, namely, the thermal decompositions of $CaCO_3$, $Mg(OH)_2$ and $CaPHO_4$ which have been characterized[4,12,17] as almost certainly phase-boundary controlled (however, see below).

The second assumption is considered to correspond to reality, in general, in the reactivity of solids.[18]

As explained, the reason for this work is to show that assumption 3 is not true and to utilize equations (10) and (2) to determine the difference between g and n.

THE NATURE OF φ

The equations derived above are based on the assumption that the rate is proportional to the physical extent of the reactant interfacial area. It is postulated[5] that this may not always be identical with the area available for reaction in a chemical sense. The theory leading to this conclusion will be given in detail elsewhere. However, an idea of the essence of the theory can be given by considering a specific example such as the thermal decomposition of $CaCO_3$ in which the interfacial CaO occupies a lesser area, Sp, than that of the interfacial $CaCO_3$, Sr, because of shrinkage. For a reversible reaction such as this it is indicated that the rate over the fraction of the reactant area covered by the interfacial product, $\frac{Sp}{Sr} = f$, can be less than the rate over the fraction of the reactant area, $1 - f = \varphi$ not covered by the interfacial product. This is the result of a higher probability of recombination between the interfacial CaO and the evolving CO_2 over the fraction of the area f than over the fraction of the area φ.

Thus, for reversible reactions involving product shrinkage such as these of the thermal decomposition of $CaCO_3$ and $Mg(OH)_2$, the value of φ can be expected to be larger than zero and less than unity and will be determined by the interfacial epitaxy. This epitaxy is known in the above cases primarily through electron diffraction studies.[11,19] These data permit the calculation of φ as being near 0.25 for $CaCO_3$ and near 0.75 for $Mg(OH)_2$ as will be documented in detail elsewhere.

For an irreversible reaction such as the thermal decomposition of $CaHPO_4$, there is no possibility of recombination between interfacial product and the evolving gas regardless of the epitaxy and, provided the reaction remains phase-boundary controlled, the rate of the reaction over the entire physical extent of the interface should be uniform. Hence in this case $\wp = 1$.

Consequently, a relevant experimental test of the theory is to demonstrate specific cases in which, regardless of the values of n and g, $\wp$ (equation 2) remains constant <u>and</u> equal to 0.25 for $CaCO_3$, 0.75 for $Mg(OH)_2$ and 1 for $CaHPO_4$. The first two substances were selected because the interfacial epitaxy is known permitting the calculation of $\wp$ in addition to fulfilling the requirements given above including reversibility[20], and the third because it is known to be irreversible.[12,21]

EXPERIMENTAL

1. Materials

All materials were studied as loosely packed powders, as compacts a) hand pressed at an estimated 10,000 - 13,000 p.s.i. b) pressed in an isostatic press at 51,000 p.s.i. and c) pressed in a hydraulic press to 175,000 p.s.i. All compacts were cylindrical with a diameter of 3 mm and height varying from a fraction of 1 mm to 3 mm. $CaCO_3$ and $Mg(OH)_2$ were also studied as single crystals obtained from Ward Scientific Co. The calcite (of Mexican origin) consisted of large (cm) clear crystals which were cleaved to the desired size. The brucite (from Quebec) consisted mostly of plate agglomerates but within these agglomerates it was possible to locate a few clear single crystals approximately 0.5 mm in thickness and several mm in the other dimensions.

We were unable to obtain monetite (i.e. single crystals of unhydrous $CaHPO_4$).

The $CaCO_3$ powder was A.C.S. reagent grade (Fisher Scientific Cat. No. C-65) and consisted of irregular rhombohedral particles up to 1-2 μ in the largest dimension. One of the $Mg(OH)_2$ powders studied was laboratory grade (Fisher Scientific Cat. No. M-42) and consisted almost entirely of transparent or transluscent spheres up to 30-40 μ in diameter (Fig. 1A). The second $Mg(OH)_2$ powder studied in this work was a high-purity material prepared by hydrolysis of magnesium methoxide. This powder consisted of irregular plates, a fraction of one micron in thickness (Fig. 1B).

The preparation of the two powders of $CaHPO_4$ employed, has already been described.[12] The first powder, of $(Ca/PO_4) = 1.02$, consisted of regular parallelopipeds up to 20-30 μ, while the second material, of $(Ca/PO_4) = 1.006$, consisted of irregular thin plates similar in appearance to those of $Mg(OH)_2$ in Fig. 1B.

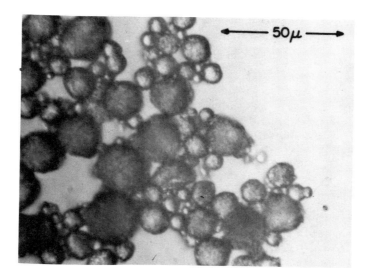

A

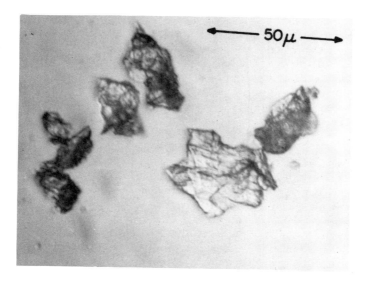

B

Fig. 1. Photomicrographs of particles of Mg(OH)$_2$ powders.
Transmitted light.
A: Powder from Fisher Scientific, Cat. No. M-42.
B. Powder from hydrolysis of Mg(OCH$_3$)$_2$.

Thus, it should be noted that in the cases of the hydroxide and the phosphate, two very different individual particle shapes were used which in usual past practice would have been automatically assigned g = 2/3 in one case and g = 1/2 in the other. That this would have been erroneous is demonstrated below.

2. Apparatus

The main instrument used was a Mettler Recording Vacuum Thermoanalyzer, instrument No. 91. In addition, two other recording thermobalances were used in other or duplicate experiments: an Ainsworth Semi-Micro RV-AV and a Dupont 950 Thermogravimetric Analyzer. The W_t curve was electronically and simultaneously differentiated in the Mettler instrument to give (dW_t/dt).

The differential scanning calorimetry work utilized the instrument already described[14a] capable of accuracy of $\pm$ 1% at temperatures in excess of the terminal decomposition temperature of over 800°C for $CaCO_3$, hence its use was mandatory for this purpose.

The sample size used in all studies varied from a few mg to over 100 mg.

In order to ascertain that the rate of the thermal decompositions was not limited or determined by the rate of heat input[22], the rate of temperature increase was varied above and below the rate utilized in the experiments. This study was extensive and utilized primarily the Ainsworth and D.S.C. instruments. For example it was found that, in the case of $CaHPO_4$, varying the heating rate from less than 0.5 to over 5°C/min. did not result in significant change of n, ΔE and Z (Eq. 10). Similar results have already been reported for the case of $CaCO_3$.[2b] The details of this work pertain to the question of validity of non-isothermal measurements and will be reported elsewhere. As a result of this preliminary exploratory work, all experiments performed with the Mettler instrument involved a heating rate of 1°C/min.; all other instruments involved rates of up to 2.5°C/min. with the exception of the Dupont thermobalance where the heating rate was varied up to 10°C/min. whenever the instrument was used not for kinetic but specimen preparation studies.

All experiments were performed in an atmosphere of flowing dry argon or nitrogen at a rate of 17 liters /hour for the Mettler, and up to 2 liters/min. in the D.S.C.

In all experiments the samples were contained in platimum cylindrical containers varying in diameter from 2 mm to 8 mm (the latter used in the Mettler instrument), the sample occupying various heights depending on the quantity used.

Microscopic studies and photomicrography employed a Leitz Ortholux Research microscope involving both transmitted and re-flected light, and the x-ray work employed standard G.E. equipment.

3. Treatment of Data

Figure 2 shows a photograph of a typical thermogram obtained by means of the Mettler instrument, showing W_t, (dW_t/dt) and $T, ^\circ C = t$, min. curves. Figure 3 shows a plot of equation (10) for several values of n obtained by means of the data in Figure 2. The D.S.C. thermograms are very similar[2b,13,14a]   but involve only the (dH/dt) and $T = const.$ t curves; the values of H_t are obtained by measuring the areas up to time t under the (dH/dt) curves. The treatment of D.S.C. data results in plots such as in Figure 3. Figure 6 is a plot obtained from D.S.C. data.

As in past work[2b,12,14a]   , plots such as those of Figure 3 were obtained by means of "manual" calculations and plotting from data such as in Figure 2. This procedure was employed for most of the work presented here. However, for a number of experiments the thermograms were read by a digitizer, the taped data from which was fed to a computer along with an appropriate program for the calculation of k in equation (10) and the results plotted automatically by means of a plotter attached to the computer output. Figure 3 is a photograph of such plots obtained in this fashion.

RESULTS

1. Values of n and their Precision

As already established[2b,12,14a]   , it is seen in Figure 3 that with precise experimental data, plots of equation (10) can distinguish between values of n as near each other as .05 or 1/20. Such precision is mandatory since experimental values of n as near each other as 2/3, (0.667) and 3/4, (0.75), must be dis-tinguished in the case of $Mg(OH)_2$ and values of 1/4, (0.25) and 1/6, (0.167), must be distinguished in the case of $CaCO_3$.

The form of equation (10) is such that the value of k should tend to zero at the beginning of the reaction and, since (dW_t/dt) decreases faster than the denominator, also at the end of the reaction. This is observed. However, in all cases, the plot

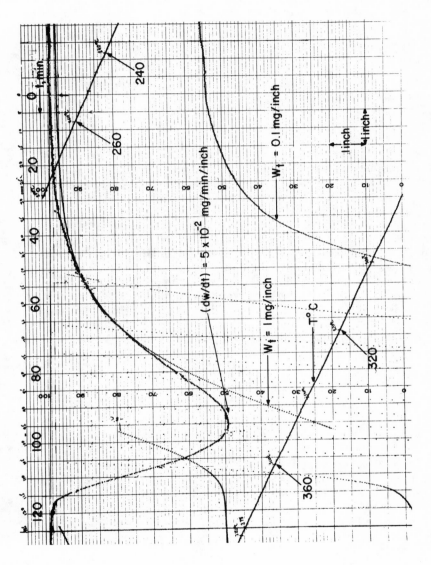

Fig. 2.   Photograph of T.G.A. thermogram for a Mg(OH)$_2$ compact, showing (dW/dt), W_t and T°C = 1·t min. traces.

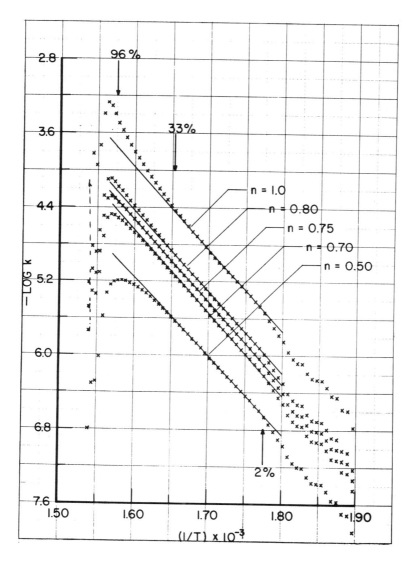

Fig. 3. Photograph of plot of equation 10 of data from Fig. 2 by means of digitizer-computer-plotter procedure (see text). Mg(OH)$_2$ compact. Plot is linear from 2% to 96% of reaction for n = 0.75. Points deviate from linearity, for other values of n, above approx. 33% of reaction.

departs from linearity for the first few percent of the reaction, and in most cases also for the last few percent of the reaction (Fig. 3). It was ascertained that these departures are real and cannot be accounted for on mathematical (Eq. 10) or instrumental grounds such as buoyancy corrections, inaccuracy in the measurement of small initial and final dW/dt values, uncertainty in the determination of t = 0 or t = t final, etc. The most certain cause for the initial departure seems to be desorption of adsorbed gasses, etc. surface effects; the most certain cause for the departure at the end appears to be the fact that the reaction then tends to become diffusion limited (see below).

Except for these departures, equation (10) seems to apply, giving plots very nearly linear <u>for the proper value of n</u> for, usually, about 90% of the reaction. This is quite an unusually large range for most solid-state kinetic studies.[4]

For inappropriate values of n, the plot becomes curved primarily at approximately, the last 30-50% of the reaction. This is to be expected because of the nature of the denominator in equation (10). However, this behavior illustrates the danger of considering lesser parts of the reaction while attempting to determine kinetic constants.

Close scrutiny of Figure 3 reveals that, even for the proper value of n, the experimental curve has a slight S shape. In some instances this is quite pronounced as in the case of Figure 5. We believe that this may be due to nucleation effects, i.e. the reactions are not <u>purely</u> phase-boundary controlled. This is presently under study. However, it is important that, for the purposes of this part of the work, even with the interference of the effects causing the S curvature, the value of n can be determined unambiguously even on Figure 5.

In some instances, such as measurements involving dense $CaHPO_4$ compacts, scattering of points to a somewhat larger extent than that seen on Figure 5 was obtained. Because of the accuracy of the data and its treatment, as seen above, we believe that this scattering is real, and previous work[12] strongly suggests that it may be due to the introduction of amorphous content into the crystal structure by the mechanical means of pressing. However, also in most of these cases, it was possible to determine n with little uncertainty. Figure 6 shows a plot from data obtained employing a loose powder sample of the Ca/PO_4 = 1.006 in the D.S.C.

The values of n thus obtained in each case are shown in Table I.

TABLE I

Compound	Specimen Form	$n_{exptl.}$	$g_{exptl.}$	φ from $n = g \cdot \varphi$
$CaCO_3$	Loose Powders: Rhombo-hedral particles $\sim 1~\mu$	0.25	1	0.25
	Compacts: a) hand pressed b) 51×10^3 p.s.i. c) 175×10^3 p.s.i.	0.25	1	0.25
	Single crystals of Calcite	0.167	2/3	0.25
$Mg(OH)_2$	Loose Powder I: Spherical particles $\sim 35~\mu$	0.75	1	0.75
	Loose Powder II: Irregular thin plates	0.75	1	0.75
	Compacts: a) hand pressed b) 51×10^3 p.s.i. c) 175×10^3 p.s.i.	0.75	1	0.75
	Crystal agglomerates or single crystals of Brucite	0.75	1	0.75
$CaHPO_4$	Loose Powder I: Para-lellopipeds $\sim 30~\mu$	1	1	1
	Loose Powder II: Irregular thin plates	1	1	1
	Compacts: a) hand pressed b) 51×10^3 p.s.i.	1	1	1
	Compact: Pressed to 175×10^3 p.s.i.	0.667	2/3	1

2. Values of g and their Precision

In most cases it was possible to see the demarkation between the reacted and unreacted portions of the samples even under small magnification (Figs. 4A and 4B) simply by the differences in the physical appearance. This was supplemented and a degree of identification provided, by covering the interface with a liquid of the same diffractive index as either product or reactant. This was especially important whenever it was necessary to distinguish between finely subdivided product, and microcracks in the reactant. The interfaces between CaO and $CaCO_3$ were further "developed" by means of an alcohol solution of methyl red.[10] Whenever it was necessary to further ascertain the identity of the product, x-ray techniques were employed.[12]

Brucite in any form (i.e. crystal agglomerate or single crystal) reacted simultaneously throughout its mass, and proof of this required most of the effort in this part of the work. In a number of experiments brucite pieces were reacted to various extent, from near absence of reaction to near completion of reaction. The specimen shown in Figure 4 was reacted to 19%. After reaction, the specimens were cleaved along the basal planes in thin sections thus exposing the interior of the crystal, as shown in Figure 4B, which was then compared to the original exterior surface (Fig. 4A). The situation depicted in Figure 4 was invariably obtained: while the exterior surface showed considerable <u>randomly</u> distributed areas free of MgO, the interior surfaces showed considerable <u>randomly</u> distributed areas of product. Focusing below the surface revealed that this was happening in depth. The presence of MgO on the surface of one of these cleaved specimens was confirmed by x-ray patterns showing the MgO lines.

As a result, the value of unity was assigned to g in these cases, Table I.

In contrast to brucite, calcite reacted in a variety of ways which may be grouped in four categories:

a. "Normal": At least during the early (low T) part of the decomposition the layer of CaO was porous, appearing white and opaque and maintaining the original rhombohedral shape of the calcite crystal. Scraping this layer with a fine spatula revealed the calcite crystal in the interior of the specimens. This remnant of the original crystal maintained its clarity and sharp edges, and measurements showed that the proportions of the lengths of its edges was the same as originally. Figure 5, lower, is the plot for such a crystal which was allowed to react to 34%. In a case such as this, g is clearly

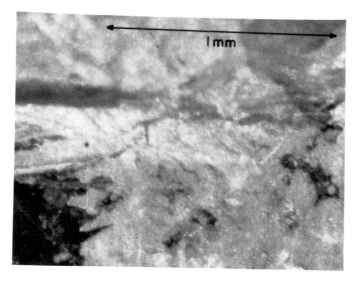

A

B

Fig. 4. Photomicrographs of brucite surfaces after 19% reaction.
Vertical dark field illumination. Dark areas: unreacted
$Mg(OH)_2$. Light areas: MgO. A: Natural (exterior)
surface. B: Cleaved (interior) surface.

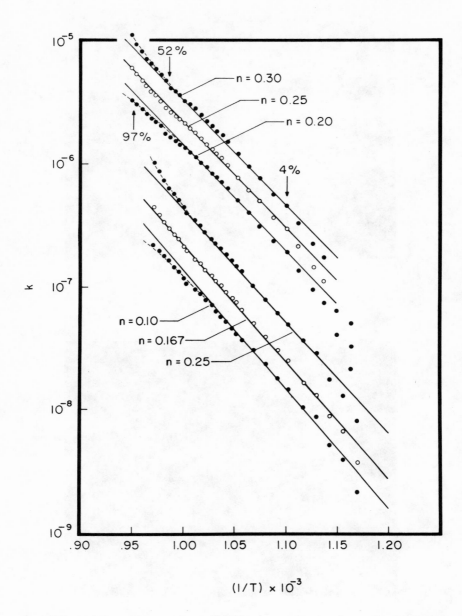

Fig. 5. Plot of equation (10) using T.G.A. data for $CaCO_3$.
Upper group: loose powder, n=0.25. Lower group:
calcite crystal allowed to react to 34% of theoretical completion; n=0.167.

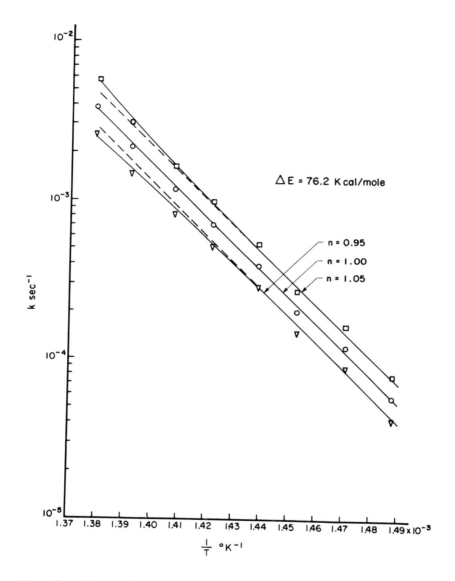

Fig. 6. Plot of equation (9) using D.S.C. data for $CaHPO_4$; n = 1.0.

equal to 2/3 (Table I).

b. Diffusion limited. Especially during the later or high
    temperature stages of the reaction the CaO often appeared
    to densify into a hard, nearly transluscent layer.   In
    these cases, the thermogram showed that the reaction
    stopped (although the temperature was allowed to rise at
    the same rate) considerably before the theoretical com-
    pletion of the reaction was attained. Subsequent exam-
    ination of the specimen revealed unreacted $CaCO_3$ in the
    interior. The plots of k vs 1/T in these cases were
    similar to that described in case d below.

c. Fractionation. In several instances the calcite crystals
    fractionated 1) with sufficient force to eject fragments
    as heavy as 37 mg out of the container, 2) slowly and
    continuously throughout the reaction. In the former case,
    the thermogram showed sudden breaks; in the latter case,
    the kinetic character of the thermogram was altered
    yielding values of n which seemed to increase with
    fractionation (i.e. mass reaction) toward unity.

d. Combination of the last two cases: In one instance,
    for example, the reaction ceased after 74.6% completion
    when the temperature reached the approximate value of
    840°C. The sample showed the characteristic transluscent
    appearance. The CaO layer was dissolved by repeated
    (50-60) rinsings in distilled water until a neutral
    reaction was obtained. The residue of unreacted $CaCO_3$
    was then found to exist in the state of fine particles.
    The plot of k vs 1/T for this experiment showed the
    curve to rise normally until approximately 700°C then
    bend toward the horizontal axis and finally further bend
    toward zero.

Obviously, only the "normal" case (a) is relevant to this
phase of the work. However, cases (b-d) are also important not
only kinetically, but also because they tend to suggest some of
the reasons for the great disparity of results in past studies of
calcite.

As far as could be determined with the means reported above,
all powders reacted simultaneously throughout their mass. Additional evidence that this was so is that powders of the same
substance but greatly different particle shape, as in the two
cases of $Mg(OH)_2$ and $CaHPO_4$ mentioned above, gave the same value
of n. This was also found to be the case for most of the compacts
regardless of the pressure at which they had been formed. Therefore,
in these cases, there was no other alternative but to assign to g

the value of unity.

To date, two instances were found in which two of the compacts pressed to 175,000 p.s.i. showed a definite reaction front proceeding from the outer surface inward. The first occurance involved CaHPO4, the reaction front proceeding in a fairly regular manner three-dimensionally toward the interior. Clearly, this case involved a value of g = 2/3 and is reported in Table I. The k vs 1/T plot showed some scatter; however the value of n determined from this plot seems to be clearly very near the value of 2/3.

The second case involved a CaCO3 compact, however the reaction front was irregular tending to form a hellipsoid with the long axis along the height of the compact. The nearest g then would be 1/2. Significantly, the value of n from the concomittant plot was closer to 0.125 than .167, i.e. approximately 1/2 x 0.25 (Eq. 2). Because of the ill-defined reaction front, however, this case is not included in Table I.

Except as indicated above, the basic causes for the variations in the behavior of individual samples of calcite and the compacts is not understood. Apparently they involve complex phenomena pertaining to crystal flaws, such as dislocations or microcracks, etc., and non-uniformities in pressure distribution during pressing.

However, this is not a direct concern here provided a value of g of high degree of certainty could be obtained. This seems to have been the case.

The values of g obtained in each case are shown in Table I.

DISCUSSION

Table I shows that:

1) φ is constant for each substance. Equation (2) is obeyed.

2) The value of φ, as experimentally obtained in each case, agrees to within narrow limits of error with the theoretically predicted value.

The conclusion that can be drawn from the first result is that regardless of the basic reason (theory), the parameter n in empirical equation (1) is not determined solely by the overall geometry but is a product of a geometrical factor, g - as shown by the derivation of equation (7) - and another factor, φ.

The second result suggests that the nature of φ corresponds
to the concept of non-uniformity of the reaction rate over the
interface, because the value of φ obtained experimentally corresponds to the value of $1-\dfrac{Sp}{Sr}$ calculated by implementing this
concept.

Because of the limited number of only three thermal decompositions measured to date, coincidence, although unlikely, is
not excluded. Hence, these results must be considered preliminary.

ACKNOWLEDGEMENTS

Indebtedness is expressed to J. Cooper for x-ray analyses,
to D. C. Henderson for a sample of magnesium methoxide, to J. H.
Ingold and R. J. Petti for mathematical assistance, to M. Jaffe
for guidance in photomicrographic techniques, to T. King for
general assistance, to M. Maier for the computer program, and
to numerous colleagues for discussions, but especially to D. C.
Fries for collaboration on interfacial crystallography.

REFERENCES

1. For example:

   a) H. B. Jonassen and A. Weissberger, "Technique of Inorganic Chemistry". Vol I, Interscience Publishers, New York, N. Y., 1963, p. 247.

   b) W. W. Wendlandt, "Thermal Methods of Analysis". Interscience Publishers, New York, N. Y., 1964, p. 174.

   c) P. D. Garn, "Thermoanalytical Methods of Investigation". Academic Press, New York, N. Y., 1965, p. 196.

   d) J. D. Hancock and J. H. Sharp, J. Amer. Ceramic Soc. 55, 74, (1972).

2. For example:

   a) Reference 1c, p. 220.

   b) D. M. Speros and R. L. Woodhouse, J. Phys. Chem. 72, 2849, (1968).

   c) W. Gomes, Nature, 192, 865, (1961).

   d) K. Hauffe, "Reactionen in und and Festen Stoffen", Springer-Verlag, Berlin, 1955, p. 639.

3. For example:

   a) S. F. Hulbert, J. Brit. Ceramic Soc., 6, p. 19, (1969).

   b) J. H. Sharp, G. W. Brindley and B. N. N. Achar, J. Amer. Ceramic Soc. 49, p. 380, (1966).

4. For example: D. A. Young, "Decomposition of Solids", Pergamon Press, New York, N. Y., 1966, p. 64.

5. D. M. Speros, Internal G.E. Report, March 1972; to be published.

6. K. J. Laidler, "Chemical Kinetics", McGraw-Hill, Inc., New York, N. Y., 1965, pp. 316-318.

7. P. Barret, R. Hartoulari and R. Perret, Comptes Rendues, 248, 2987, (1959).

8. G. H. Stout and L. H. Jensen, "X-ray Structure Determination". McMillan Co., New York, N.Y., 1968.

9. See, for example, the entire issue of J. Thermal. Analysis, 5, pp. 179-354, (1973), Symposium on "The Estimation of Kinetic Parameters, etc." held in Budapest, Hungary, July 6-8, 1972.

10. T. R. Ingraham and P. Marrier, Can. J. Chem. Eng. 41, 170, (1963).

11. R. S. Gordon and W. D. Kingery, J. Amer. Ceramic Soc., 50, 13, (1967).

12. D. M. Speros, R. L. Hickok and J. R. Cooper, in "Reactivity of Solids". J. W. Mitchell, et. al. Edit., John Wiley & Sons, 1969, p. 247.

13. D. M. Speros, in "Thermal Analysis". Vol. 2, Academic Press, New York, N. Y., p. 1191 (1969).

14. Expressed by a) D. M. Speros and R. L. Woodhouse, i) J. Phys. Chem. 67, 2164, (1963); ii) Nature, 197, 1261, (1963), iii) References 2b and 12; and later by b) J. M. Thomas and T. A. Clarke, J. Chem. Soc. (A) p. 457, (1968).

15. For example:

    a) P. S. Nolan and H. E. Lemay, Jr., Thermochim. Acta. 6, 179 (1973).

    b) M. D. Judd and A. C. Norris, J. Thermal Anal. 5, 179, (1973) and references therein.

16. H. J. Borchardt and F. Daniels, J. Amer. Chem. Soc., 79, 41, (1957).

17. K. H. Stern and E. L. Weise, Nat. Stand. Ref. Data Ser., #30 National Bureau of Standards, 1969.

18. For example:

    a) Polanyi and Wigner, Z. Physik. Chem. A, 139, 439, (1928).

    b) R. D. Shannon, Trans. Farad. Soc., 60, 1902, (1964).

19. For $Mg(OH)_2$

  a) Reference 11.

  b) J. F. Goodman, Proc. Roy. Soc. (A) 247, 346, (1958).

  c) P. J. Anderson and R. F. Horlock, Trans. Farad. Soc.
      58, 1993, (1962).

  For $CaCO_3$

  a) J. M. Thomas and G. D. Renshaw, J. Chem. Soc., (London)
      A, p. 2058, (1967).

  b) Reference 12, p. 216.

20. a) For $CaCO_3$ see, for example, E. P. Hyatt, I. B. Cutler
         and M. E. Wadsworth, J. Amer. Ceram. Soc., 41, 70, (1958).

  b) For $Mg(OH)_2$ see, for example, R. F. Horlock, P. L. Morgan
      and P. J. Anderson, Trans. Farad. Soc., 59, (1963).

21. J. G. Rabatin, R. H. Gale, and A. E. Newkirk, J. Phys. Chem.,
     64, 491, (1960).

22. For example, see K. Sveum and A. L. Draper, "The Kinetics of
     Endothermic Solid State Decompositions". Thesis, Texas Tech
     University, 1970.

KINETICS OF AN ANHYDRIDE-EPOXY POLYMERIZATION AS DETERMINED BY DIFFERENTIAL SCANNING CALORIMETRY

Paul Peyser and W. D. Bascom

Naval Research Laboratory

Washington, D.C. 20375

The use of differential scanning calorimetry (DSC) to study the thermal behavior of polymers has developed to the point where it is now used to establish the kinetics of polymer phase transformations, including polymerization reactions. We will describe the use of DSC to determine the polymerization kinetics of the anhydride-epoxy system, bisphenol A diglycidyl ether (DGEBA)-"nadic" methyl anhydride (NMA). This polymerization reaction, catalyzed by benzyldimethyl amine (BDMA) is generally viewed as a very complex curing process but the data here were fitted to a two stage model utilizing a single reaction order of $n=2$ and (only) two activation energies. Namely, in dynamic scans, after the reaction had proceeded with an activation of about 15 kcal/mole for about 12% completion, the rate increased until completion but with a higher activation energy of about 38 kcal/mole. Such an increase in rate was verified qualitatively by an isothermal run. Other workers studying similar systems have reported the lower activation energy. However, scrutiny of their data indicates a similar increase in rate. We will describe four commonly used analytical methods available for deriving kinetic results from DSC data and show how each method is affected by an increase in activation energy. Also we will show that some simple approaches to treating DSC polymerization data can lead to erroneous conclusions.

EXPERIMENTAL

Materials: The epoxy polymer of bisphenol-A diglycidyl ether (DGEBA, Dow Chemical DER-332, epoxy equiv. 175) was cured with "nadic" methyl anhydride ([2.2.1] heptane-2,3-dicarboxylic anhydride). The polymerization was catalyzed with benzyldimethylamine (BDMA, Eastman Kodak reagent grade). All materials were used as received. The NMA and BDMA were premixed and stored at 25°C for no more than 3-4 days. The solution invariably turned a dark brown within 24 hours but with no evident changes thereafter. The DGEBA was vacuum degassed and then thoroughly mixed with the NMA-BDMA to give 87.5 parts by weight of NMA and 1.5 parts by weight of BDMA per hundred parts of the epoxy monomer. The third column of Table 1 lists the time between mixing and the start of a DSC scan.

Differential Scanning Calorimetry: The DSC measures the rate of the exothermic or endothermic change of a material as a function of time and temperature. These changes may occur as the sample is heated or cooled at some constant rate or as the sample undergoes a spontaneous change while being held at a constant temperature. The enthalpic change is measured with respect to a reference sample - usually an empty sample pan - that is being given the same thermal treatment. The calorimeter used here was Perkin Elmer DSC-2. The sample and reference temperatures were calibrated using melting point standards and the calorimeter output was calibrated using a standard indium sample. Weighed amounts of reactant mixture was scanned over the temperature interval of 310° K to 500° K in aluminum pans with loose fitting aluminum covers. The sample and reference compartments were purged continuously with dried N_2 at a flow rate of 15 ml/min. The analog recorder output is indicated schematically in Figure 1.

TABLE 1

TOTAL HEATS OF REACTION AND GLASS TRANSITION TEMPERATURES

Batch No.	Heating Rate °/min	Elapsed Time (Hours)	Weight mg.	Total Heat of Reaction cal/gm	Tg °K
1	5	6	41.052	46.8	< 310
2	5	1.3	35.639	38.3	< 310
1	2.5	3	43.665	54.4	~ 330
2	1.25	3	57.580	53.4	363
2	0.62	7	63.855	63.2	399

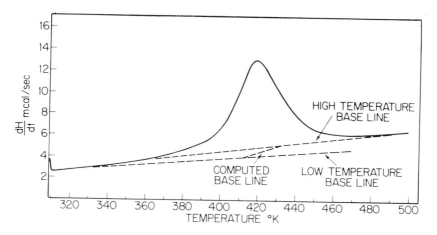

Figure 1 - Typical DSC Scan (5°/min) showing calculated base line.

Data were obtained for scanning rates of 10, 5, 2.5, 1.25 and 0.62 degrees/min. At 10° K/min gaseous products (which did not have an amine order) escaped from the sample and condensed on the underside of the cooling block cover of the enclosing chamber and to a lesser extent on the underside of the platinum sample cover. (Figure 2) At the end of this fast heating rate scan the sample was quite soft indicating incomplete reaction.

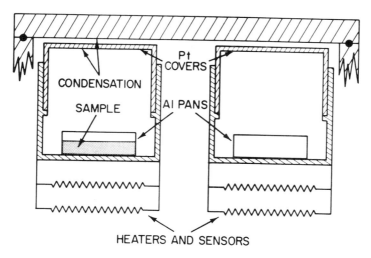

Figure 2 - Areas of condensation in the reference and sample compartments of the DSC.

At the lower heating rates there was far less vaporization, the weight loss was < 0.1% and the base lines before and after the exotherm were nearly coincident (see Figure 1 and the next section).

The data obtained from the analog output and used in the various computations were the change in heat output with time, dH/dt, and the temperature, T, taken at 1° intervals (or in one case at 2.5° K intervals). After the initial run the samples were rescanned at 10° K/min in order to obtain the glass transition temperature (T_g) (taken to be the temperature at which the change in heat capacity is one-half its maximum value).

Data Analysis: The basic assumption made in treating calorimetric data to obtain reaction kinetics is that the heat of reaction evolved at any time is proportional to the number of moles of reactant consumed. The Arrhenius rate equation then becomes (1)

$$\frac{1}{H_T} \frac{dH}{dt} = Ae^{-E_a/RT} \quad f(\frac{H}{H_t}) = Ae^{-E_a/RT} \left(\frac{H_T - H}{H_T}\right)^n \tag{1}$$

H_T = total heat of reaction (cal/gm) ∼ area under DSC curve.

H = heat of reaction at a given time and temperature ∼ partial area under curve.

E_a = Arrhenius activation energy (Kcal/mole).

A = Arrhenius equation frequency factor. For a second order reaction it includes the initial concentration of reactant (sec^{-1}).

n = order of reaction

R = gas constant

T = temperature in ° K

$f(\frac{H}{H_T})$ = a function of $\frac{H}{H_T}$ which is assumed here to be $\left(\frac{H_T - H}{H_T}\right)^n$

In this section brief descriptions are given of the four more commonly used methods of data analysis based on equation 1. A more detailed discussion can be found in reference (2). The treatments are for reactions having a single activation energy and reaction order but later in the DISCUSSION section it will be necessary to consider the analysis for an abrupt change in the kinetic parameters during the course of the reaction.

Method A: The simplest treatment is when $\underline{n}$ is known or assumed so that in logarithmic form eq. 1 becomes

$$\ln \frac{dH}{dt}\frac{1}{H_T} - \underline{n} \ln \left(\frac{H_T - H}{H_T}\right) = -\frac{E_a}{R}\frac{1}{T} + \ln A \equiv \ln k \qquad (2)$$

A plot of $\ln \frac{dH}{H_T dt} - \underline{n} \left(\frac{H_T - H}{H_T}\right)$ vs $\frac{1}{T}$ should be a straight line with slope $-\frac{E_a}{R}$ and intercept $\ln A$. Note that the heating rate does not appear in the above equation so that plots of data from different scans should lie on the same line.

Method B: The temperature at the maximum of a DSC curve shifts to lower temperatures the lower the heating rate. If we differentiate equation 1 with respect to $\underline{t}$ and set the value equal to zero we obtain at the maximum point

$$\frac{E_a}{n} = \left(\frac{RT^2}{H_r \beta}\right)\frac{dH}{dt} \qquad (3)$$

where $H_r = H_T - H$ and $\beta = \frac{dT}{dt}$, so that $\frac{E_a}{n}$ can be calculated directly.

Method C: As can be seen from equation 2, by comparing data from different heating rates at a constant value of fraction reacted we can plot $\ln \frac{dH}{H_T dt}$ vs $\frac{1}{T}$ to obtain a straight line whose slope equals $\frac{-E_a}{R}$, i.e.,

$$\left(\frac{\Delta \ln \frac{dH}{H_T dt}}{\Delta \frac{1}{T}}\right)_{\frac{H}{H_T}} = \frac{-E_a}{R} \qquad (4a)$$

For various heating rates the maximum temperature occurs, to a first approximation, at a constant value of $\frac{H}{H_T}$ and assuming that it does we can combine equation 3 with equation 4a. Since E_a, $\underline{n}$ and $\frac{H_r}{H_T}$ can be assumed to be independent of the change in maximum temperature with heating rate, we obtain

$$\left(\frac{\Delta \ln \frac{\beta}{T^2}}{\Delta \frac{1}{T}}\right)_{max} = \frac{-E_a}{R} \qquad (4b)$$

Equation 4b is equivalent to equation 4a at the maximum. Equation
4b is quite simple to use since only the temperature of the maximum
and the heating rate need be determined. Moreover equation 4b
determines E_a even if A, the frequency factor changed between runs.

Method D: By suitable differentiation and algebraic manipulation
of equation 1 the Freeman-Carroll equation is obtained [2], one form
of which is

$$\frac{\Delta \ln \frac{dH}{dt}}{\Delta \ln H_r} = \frac{\frac{E_a}{R} \Delta \frac{1}{T}}{\Delta \ln H_r} + n \tag{5}$$

A plot of the left side of equation 5 vs $\frac{\Delta \frac{1}{T}}{\Delta \ln H_r}$ will be a straight
line of slope $\frac{-E_a}{R}$ and intercept $\underline{n}$, independent of heating rate.

Various combinations of these equations can be used to compute
the kinetic parameters, E_a, $\underline{n}$ and A. If all four methods are used,
then multiple values of the parameters are obtained and serve to
check on the self-consistency of the arguments and the computations.

Base-Line and Data Computations: The DSC instruments can be
adjusted so that with both pans empty, the base line scan is a
straight line. These adjustments were easily made for the temperature range and instrument sensitivity (10-5 mCal/sec/10 mV) used
here. With material in the sample pan, the base line will shift
in proportion to the specific heat and weight of material, but
ideally will still be straight. Then the extension of the straight
line at the lower temperature before a reaction occurs coincides
with the straight line after the reaction, and the area under the
curve so drawn, is proportional to the heat of reaction. (see Figure
1).

Usually, when the sample has reacted, the two lines do not
coincide because the specific heat of the material changes during
the scan. The corrections for this problem have been discussed
in the literature[3,4]. In our experiments, two additional effects
may have contributed to the base line shifts.

(1) The escape of gaseous materials (previously described) is
an endothermic process compared to the exothermic curing reaction.
To the extent that the endothermic curve is non-linear, the base
line is affected and also the magnitude of the measured heat of
reaction.

(2) The condensation of gaseous materials on the underside of
the platinum cover (see Figure 2 of the sample compartment) will
change the "heat radiation" characteristics of the sample compartment and hence the base line.

Various methods can be used to correct or account for a change in base line. The approach adopted $(3,4)$ here was to assume the change in base line is proportional to the fraction reacted. A computer programmed, iterative computation was performed in which a an arbitrary base line was assumed initially and used to calculate the fraction reacted, i.e., H_r. Assuming proportionality, a new base line was then calculated. This procedure was repeated ten times, although no further change could be detected with 6-place accuracy after three iterations. The straight line constants of the initial and final base lines were obtained from a least square fit of at least ten data points. Simpson's Rule was used to integrate the area under the curve. No attempt was made to correct for variations of sample temperature due to different heating rates or for the effect of non-zero thermal resistance. These corrections are small (e.g. $\pm 0.2°$ K) compared to the experimental errors.

Using the corrected base line, the program computed and printed the variables of equation 2,

$$\ln \frac{dH}{dt} \frac{1}{H_T} - \underline{n} \ln \frac{H_T - H}{H_T} \quad \text{(for } \underline{n} = 2\text{) and } \frac{1}{T} \text{ and the}$$

variables of equation 5,

$$\frac{\Delta \ln \frac{dH}{dt}}{\Delta \ln H_r} \quad \text{and} \quad \frac{\Delta \frac{1}{T}}{\Delta \ln H_r} .$$

Also printed were T, H_T, $\frac{H_r}{H_T}$ and $\frac{dH}{dt}$.

RESULTS

The total heats of reaction and the glass transition temperatures (T_g) are listed in Table 1 along with the experimental conditions of each DSC scan. There was no clear trend in the heat of reaction with heating rate although the T_g values were higher for the slower heating rates. Holding the samples at 500° K overnight, i.e., post curing, increased the T_g to about 420° K in all cases.

Figure 3 is a plot of equation 2 with $\underline{n} = 2$ for the 5°/min. experiment. There was a distinct discontinuity in the plot such that the data could be fit by two straight lines. Similar breaks occurred in the data from the other experiments and in each case, two lines could be drawn as shown in Figure 4. The straight line fit was excellent except for the 0.62°/min. experiment in the lower-temperature region. Table 2 lists the values of the activation energies obtained from the slopes of the lines and their intercepts. The higher temperature activation energy is denoted as E_{a2} and the lower as E_{a1}. The goodness of the straight line fit is evidenced

by the least square regression coefficient. Note in Figure 4 that
the intercepts of the data in the low and high temperature regions
occurs at progressively lower temperatures the lower the heating
rate but from Table 2 it is evident that the fraction reacted at the
intercepts are all approximately the same ($\sim$ 0.12).

The Freeman-Carroll plots (equation 5) of the data also indicated
two regions of linearity. A typical plot is shown in Figure 5 for

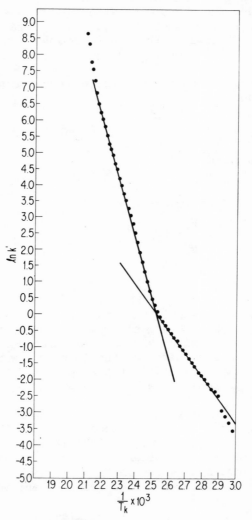

Figure 3 - Plot of equation 2 for 5°/min DSC scan. (ln k' = ln k
+ 7.33, ordinate shifted for plotting purposes).

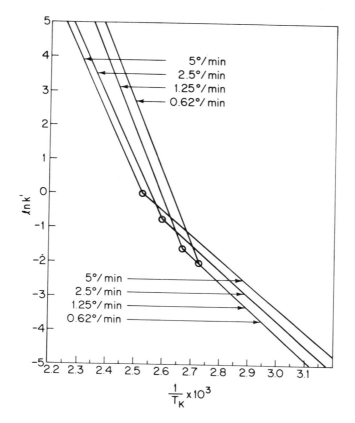

Figure 4 - Least square fit of equation 2. Each heating rate was fit by two straight lines. (ln k' = ln k + 7.33, ordinate shifted for plotting purposes).

the high temperature region and the derived values of E_a and n are given in Table 3. Note in Table 3 that the least square regression coefficients for these plots indicate a poorer fit than the plots using equation 2 (Table 2), especially in the lower temperature region. However, the activation energies, E_{a1} and E_{a2} obtained by the two methods are quite comparable. The Freeman-Carroll intercepts for the high temperatures data gave an n value very close to two, but the intercepts for the low temperature data (which is insensitive to variations in n since ln H_r is close to zero) were between $n = 1$ and $n = 2$.

TABLE 2

LEAST SQUARE FIT OF EQUATION 2

Heating Rate °/min	Number of Points	Range of Fraction Reacted	Least Square Regression Coefficient	E_a kcal/mole	ln A (sec^{-1})	Fraction Reacted at Intersection of Straight Lines
5	17	0.009 - 0.107	0.999	14.0	10.5	0.14
	27	0.152 - 0.987	0.999	36.7	39.4	
5	42	0.006 - 0.101	0.998	15.1	11.9	0.12
	40	0.119 - 0.832	1.000	36.9	39.8	
2.5	43	0.004 - 0.090	0.996	14.8	11.4	0.12
	45	0.110 - 0.998	0.999	35.2	38.0	
1.25	19	0.014 - 0.063	0.994	14.9	11.1	0.09
	52	0.142 - 0.987	1.000	41.0	46.1	
0.62	31	0.137 - 0.856	0.998	39.9	45.5	0.10

Averages: E_{a1} = 14.7, E_{a2} = 37.9

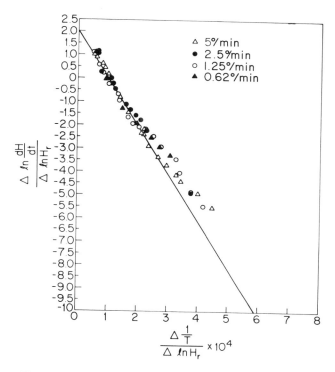

Figure 5 - Freeman-Carroll plot-high temperature region (E_{a2}).

The values of E_a obtained by the maximum term method, equation 3, are listed in Table 4. It was assumed that $\underline{n} = 2$ and as described in the DISCUSSION section the activation energy obtained using this method corresponds to E_{a2}. On the other hand, a plot of equation 4b (Table 5) for different heating rates had a slope corresponding to E_{a1}.

A typical isothermal run at $410°$ K is shown in Figure 6. Note that the very beginning of the reaction is lost due to the speed of the reaction and the time (about 30 sec) necessary for the instrument to adjust itself. Note also that the rate increases with time to a maximum at the very beginning of the reaction.

TABLE 3

FREEMAN-CARROLL LEAST SQUARE FIT

Heating Rate °/min	Number of Points	Range of Fraction Reacted	Least Square Regression Coefficient	E_a kcal mole	n
5	11	0.022 - 0.094	0.997	13.9	1.5
	17	0.261 - 0.949	0.996	42.3	2.2
5	36	0.241 - 0.920	0.991	40.3	2.1
2.5	19	0.024 - 0.073	0.735	13.2	1.2
	30	0.291 - 0.890	0.987	40.7	2.3
1.25	29	0.271 - 0.926	0.978	41.1	2.0
0.62	25	0.261 - 0.883	0.962	38.3	1.9

Averages: $E_{a1} = 13.6$, $E_{a2} = 40.5$, $n_1 = 1.3$, $n_2 = 2.1$.

TABLE 4

ACTIVATION ENERGY BY MAXIMUM TERM METHOD

Heating Rate °/min	$T_{K max}$	$\frac{E}{n}$ kcal mole	$E \frac{kcal}{mole}$ (n=2)	Fraction Reacted At Maximum
5	420.5	20.7	41.4	0.546
5	417	19.0	38	0.511
2.5	408	18.3	36.6	0.507
1.25	397	21.5	43	0.522
0.62	387	21.7	43.4	0.540

TABLE 5

LEAST SQUARE FIT OF EQUATION 4b

Heating Rate °/min (β)	$\dfrac{1}{T}$ x 10^3 K max	$\ln\left(\dfrac{\beta}{T^2}\right)_{max}$	Calculated from Least Square Fit $\ln\left(\dfrac{\beta}{T^2}\right)$
5	2.38	-10.47	-10.38
5	2.40	-10.46	-10.58
2.5	2.45	-11.11	-11.08
1.25	2.52	-11.74	-11.77
0.62	2.58	-12.39	-12.37

Least square fit equation:

$$\ln\left(\frac{\beta}{T^2}\right) = -9952.7\frac{1}{T} + 13.3094$$

Regression Coefficient = -0.996

$$E = 19.4 \text{ kcal/mole}$$

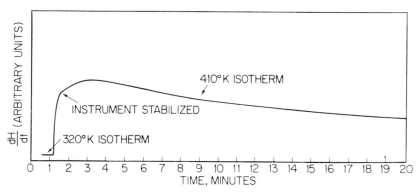

Figure 6 - An isothermal run. Note increase in reaction rate at beginning of reaction.

DISCUSSION

The four methods used to analyze our dynamic DSC data do not at first glance give consistent results. Method A and the Freeman-Carroll method D demonstrate that two regions exist with two activation energies E_{a1} and E_{a2}. However, the maximum term Method B measured E_{a2} and Method C measured E_{a1}.

Fava (5) using the DSC studied the polymerization kinetics of a system similar to ours (hexahydrophthalic anhydride, HHPA, was used instead of NMA). He plotted equation 4a at the maximum points and obtained 17.8 kcal/mole (E_{a1}), very close to our result using Method C (eq. 4b). Tanaka and Kakiuchi[6,7] studied isothermally the kinetics of the polymerization of higher molecular weight bisphenol-A diglycidyl ether (Epon 828 and Epon 1001) with NMA or HHPA in the presence of various catalysts including BDMA. The extent of reaction was determined by direct titration of the epoxy and the anhydride groups. They found, as we did, that the reaction was second order and also that the catalyst affected the value of A, the frequency factor (1st order with respect to amine whose concentration remains constant throughout the reaction). However, the activation energies they measured, even up to 60% reaction, was that of E_{a1}. On this point we do not agree.

The data of both Fava and of Tanaka and Kakiuchi both show that a region exists at the very beginning of the reaction where the reaction rate is increasing with time. For example, their isothermal plots of fraction reacted vs time of reaction are S shaped (similar to what one would obtain for an autocatalytic reaction). They, however, chose to ignore this fact. Our isothermal runs, if measured at a low enough temperature so that the beginning of the reaction is not missed, similarly demonstrate such a change in rate. Hence we believe that the increase in rate at about 12% fraction reacted, as demonstrated by Method A and D is meaningful and we will proceed to explain our results on this basis.

Method A: While Method A gives consistent results for E_{a1} and E_{a2} among runs at different heating rates, the curves do not lie upon each other either in the high (E_{a2}) or low (E_{a1}) temperature regions. The non-normalization at the low temperature region can be explained by one or both of the following experimental problems.

1. It is difficult to mix the three components of the reaction mixture in a reproducible way. The reactivity of the mixture is particularly sensitive to the concentration of the amine catalyst. Moreover, it was obvious that the amine and anhydride reacted in view of the color change that occurred upon mixing. Consequently the effect of concentration of amine at the start of a reaction

was not completely controlled and, since Tanaka and Kakiuchi found the frequency factor is affected by the amine concentration, we cannot expect a coincidence of the curves at the lower reaction temperature.

2. As previously mentioned, due to varporization of material, an endothermic reaction was super-imposed upon the exothermic one we tried to measure. The decrease in H_T with increasing heating rate and the increase in ln A with heating rate is consistent with this explanation.

If we were to normalize the curves in the low temperature region and ignore the slight curvature in the breaks of the straight lines and the slight variation in the value of the fraction reacted at the intersection of the two straight lines among the runs, then we can schematically present the data of Method A by Figure 7. Since the break in the curves occur for the same value of fraction reacted, it is easy to understand why the temperature at the break is greater the greater the heating rate. Simply stated, the lower the heating rate the longer the time of reaction and therefore the same fraction reacted can be achieved at a lower temperature.

However, even past the transition region at the same temperature the rate constant will depend on heating rate (refer to point A, B and C of Figure 7). Explanation of such a variability of results with heating rate has been the subject of controversy in the literature. Prime,[8] for example suggests that the effects of heating rate on E_a and ln A for the polymerization of an epoxy system are due to the fraction reacted as measured dynamically being a function of time and temperature. This explanation is not considered valid[9] because the time-temperature dependence is already controlled by the fixed heating rate.

A more reasonably explanation of a variable ln k is to be found in the dynamic nature of the DSC experiment. The isothermal condition, i.e. zero heating rate, is a standard against which the dynamic condition must be compared. In Figure 7 this standard is most closely approximated by the slow heating rate. Beginning at the point marked Start, the data proceeds to T_3, changes slope and continues along line T_3C. For the fast heating rate, β_1, the curve will follow the straight line until temperature T_1 is reached. To agree with our standard we would expect the rate to suddenly jump to the value of point C. Such a discontinuity is obviously impossible. In actual practice, especially in a viscous epoxy mixture, the reactants cannot diffuse together that quickly. The reaction cannot keep up with itself and is in effect retarded. Evidently for the system we studied this retardation can be expressed by multiplying the frequency factor by a constant.

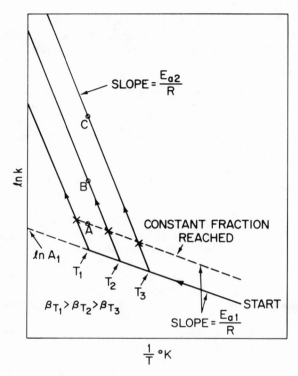

Figure 7 – Schematic representation of Figure 4 after suitable
normalization. Dotted line is a plot of equation 4a in high temperature (E_{a2}) region.

Methods B, C and D: What will occur if data from different
heating rates are compared at the same fraction reacted (Method
C) for temperatures past the break in the curves? Data of this
type are indicated in Figure 7 by the intercepts of the dotted
line with the $\frac{E_{a2}}{R}$ curves. The slope of this line is $\frac{E_{a1}}{R}$ and
so will yield a lower temperature activation energy even though
the reaction has proceeded into the higher temperature region.

The maximum term Method (B) resulted in a value of E_{a2} because,
at the maximum, the large bulk of the reaction occurred under
conditions where E_{a2} was prevalent. For the same reason A, (the
frequency factor) can be considered to be constant if measurement
is made far enough into the reaction so that the contribution of
the initial part of the reaction is small. Therefore equation 4b
which is derived upon the assumption of the constancy of A gave
a reasonable fit to the data.

On the other hand, the Freeman-Carroll Method (D) is based upon the differences between adjacent data points and is therefore a measure of the activation energy in the vicinity of those data points. Consequently, this method is sensitive to changes in the kinetics.

Glass Transition Temperatures: Others have demonstrated[10] that the epoxy polymerization reaction slows down considerably at the end of the reaction and hence the need for post curing to maximize the crosslinking. The DSC is not sensitive enough to detect this last part of the reaction. The kinetics measured therefore should be considered on this basis. The glass transition temperature is greatly dependent on this final crosslinking. The lower the heating rate in scans up to 500° K, the longer the sample is at the higher temperatures. In other words, the lower heating rate samples were given a greater amount of post curing which explains their higher glass transition temperatures.

Mechanism: The chemical mechanism involved in the complex polymerization reaction cannot be discerned from kinetic results alone. However, we believe that in a search for mechanisms the following observations would be pertinent:

(1) The amine and anhydride obviously interacted in the premix as evidenced by the color change.

(2) After about 12% reaction both the reaction rate and activation energy increased.

(3) The magnitude of E_a (38 kcal/mole), the dependence of k on heating rate in the high temperature region and the evident increase in sample viscosity during the course of a scan all suggest that the reaction is diffusion controlled after only 12% reaction.

The autocatalytic behavior exhibited at the beginning of the reaction may be due to the release of catalyst by the anhydride as it reacts with the epoxy. On the other hand the increase in viscosity that occurs as the reaction proceeds tends to slow down the reaction. Increasing the temperature helps alleviate this sluggishness by decreasing the viscosity with the net result of a greater temperature dependence of the rate constant (increased activation energy). Indeed it is surprising that it has not previously been suggested (except at the very end of the reaction) that an epoxy polymerization can be diffusion controlled.

SUMMARY

DSC is a powerful and relatively quick method which can be used to measure the kinetics of polymerization reactions. The data obtained is enormous and many methods are available to analyze this data. For complex reactions it is possible to miss changes in mechanism unless the analytical method examines data over the entire DSC scan. This is facilitated by computer programming of the data treatment. The epoxy anhydride polymerization studied here appears to be diffusion controlled over nearly 90% of the reaction.

REFERENCES

1. H. J. Borchardt and F. Daniels, J. Am. Chem. Soc. <u>79</u>, 41 (1957).

2. Joseph H. Flynn and Leo A. Wall, J. Research Natl. Bur. Standards 70A, 487 (1966).

3. William P. Brennan, Bernard Miller and John C. Whitwell, Ind. Eng. Chem. Fundam. <u>8</u>, 314 (1969).

4. H. M. Heuvel and K. C. J. B. Lind, Anal. Chem. <u>42</u>, 1044 (1970).

5. R. A. Fava, Polymer (London) <u>9</u>, 137 (1968).

6. Yoshio Tanaka and Hiroshi Kakiuchi, J. Appl. Poly. Sci. <u>7</u>, 1063 (1963).

7. Ibid, J. Appl. Poly. Sci. <u>2</u>, 3405 (1964).

8. R. Bruce Prime, "Analytical Calorimetry," Vol. 2 (Eds. R. S. Porter and J. F. Johnson) Plenum Press, New York, 1970, p. 201.

9. V. M. Gorbatchev and V. A. Logvinenko, J. Ther. Anal. <u>4</u>, 475 (1972).

10. K. Horie, H. Hiura, M. Sawada, I. Mita, and H. Kambe, J. Poly. Sci. A-1, <u>8</u>, 1357 (1970).

CHARACTERIZATION OF ENERGETIC MIXED

CRYSTALS BY MEANS OF TGA AND DSC CALORIMETRY

Scott I. Morrow

Propellants Division
Feltman Research Laboratory
Picatinny Arsenal
Dover, N.J. 07801

ABSTRACT

Closed cup DSC techniques along with TGA were used to compare the thermal decomposition behavior of AP-KP mixed crystals with that of physical mixtures of similar composition as well as pure components. The chief problem in comparing ΔH data so obtained with that established by other means, such as closed bomb calorimetry, is that the decomposition pathways may not be identical under the two sets of conditions. Nevertheless, for purposes of the indicated comparisons DSC was judged to be useful. It is necessary to keep in mind, though, certain limitations and relevant problems in order not to draw unwarranted conclusions.

INTRODUCTION

In a previous paper[1], we reported on the determination of structural homogeneity in isomorphous NH_4ClO_4(AP) - $KClO_4$(KP) mixed crystals by means of X-ray diffraction, DTA, and TGA procedures. Unfortunately these methods did not give any quantitative or semi-quantitative insight into the energetic properties of the materials in question. DSC, on the other hand, affords a means for the quantitative determination of ΔH values. It has been a matter of interest to characterize the thermal decomposition behavior of the AP-KP mixed crystals by a method which, through being simple and perhaps not as rigorously precise as calorimetry, might be more suitable than, TGA, for instance. Although DSC can give accurate ΔH information in more ideal cases, measurement of heats of decomposition or explosion of highly reactive, unstable materials can be a particularly difficult appli-

ćation. Nevertheless, we sought to apply DSC techniques to the
characterization of the thermal decomposition of pure AP and its
mixed crystals with the isomorph, KP.

EXPERIMENTAL

Crystal Synthesis

This has been described previously[1].

Thermal Analysis

A Du Pont 900 Differential Thermal Analyzer equipped for
closed-cup differential scanning calorimetry (DSC) work and with
a 950 Thermogravrimetric Analyzer (TGA) accessory unit was used in
this work. The equation applicable to use of the DSC mode of operation is as follows:

$$\Delta H \text{ (m.cal./mg.)} = E \left(\frac{A \Delta Ts Ts}{Ma} \right) \qquad (1)$$

$$
\begin{aligned}
\text{Where } E &= \text{Calibration Coefficient, m. cal/}^\circ\text{C-min.} \\
A &= \text{Peak area, sq. in.} \\
\Delta Ts &= \text{Y-axis sensitivity setting, }^\circ\text{C/in.} \\
Ts &= \text{X-axis sensitivity setting, }^\circ\text{C/in.} \\
M &= \text{Sample mass, mg.} \\
a &= \text{Heating rate, }^\circ\text{C/min.}
\end{aligned}
$$

The curve representing values of E for this equation at different
temperatures was obtained by measuring the areas of the endotherms
generated upon melting pure tin and zinc reference standards in
closed cups on the DSC stage. These cups were not gas tight.

Pure ammonium perchlorate and its physical and chemical mixtures with potassium perchlorate were then analyzed by the DSC
closed cup method. For reference purposes pure ammonium nitrate
(AN) was analyzed similarly, as was recrystallized ammonium dichromate. Most of the values for ΔH were obtained by averaging
the results of several determinations. Sample weights and losses
in weight thereof upon heating in the DSC cell were established by
means of weighings with a semi-microbalance. Since values for E
in equation 1 varied with temperature, it was necessary to obtain
ΔH for a given exotherm by means of a summation process carried
out according to the following equation:

$$\Delta H = \left(\frac{\Delta Ts Ts}{Ma} \right) \sum_{t=t_1}^{t_2} E_t A_t \qquad (2)$$

TABLE 1

THERMAL DECOMPOSITION BEHAVIOR OF NH_4ClO_4
AND MIXED CRYSTALS AND PHYSICAL MIXTURES WITH $KClO_4$

Mixed Crystal (MC), or Physical Mixture (PM)	$NH ClO_4$ content, % by weight	Loss in weight on heating, % (to nearest 5%)	Temperature range of exotherm °C	ΔH cal/gm
M.C.	25.87	30	343-447	142
P.M.	25.1	20	343-447	155
M.C.	37.35	30	311-438	257
M.C.	51.23	50	449-447	395
P.M.	49.9	70	334-449	371
M.C.	74.20	85	295-449	464
P.M.	75.03	74	221-447	466
M.C.	83.76	80	319-449	549
P.M.	83.7	90	329-435	380
NH_4ClO_4	100	100	329-443	563

Values of Δ H calculated from the DSC exotherms are given in
Table 1. Typical DSC thermograms for physical and chemical mixtures are shown in Figures 1 and 2. A graphical representation of
the variation of Δ H as a function of ammonium perchlorate content
of the mixture is given in Figure 3. It was not possible to measure exothermal behavior at temperatures higher than 460°C with this
DSC equipment.

DISCUSSION

The DSC equipment used in this investigation has a maximum
temperature capability of 600°C. On the other hand the TGA accessory enabled us to reach temperatures in excess of 700°C. Pure
potassium perchlorate itself is not completely decomposed at 600°C.
We were able to show by means of TGA, Fig. 4, that in AP-KP mixed
crystals upon heating, the AP component leaves the crystal first.
Curve 2, Figure 4, shows that the mixed crystal undergoes decomposition above 300°C in a manner similar to pure AP, curve 3. Then
at about 600°C the residue undergoes decomposition in a manner resembling but not necessarily identical to that of pure KP, curve 1.
Thus the mixed crystal behaves more or less as the two pure components when heated to temperatures sufficient to cause thermal decomposition. Of course, as shown in our previous work[1], the mixed
crystal is indeed entirely different from a physical mixture with
the same relative amounts of the two components. Because of the
600°C temperature limit of the DSC we were not able to completely
characterize the thermal decomposition behavior of the mixed crystals. We concerned ourselves, therefore, only with the behavior
of the AP component inasmuch as this fell within the temperature
range of our observational capabilities. Also it afforded a means
for estimating the heat of decomposition attributable to AP in the
mixed crystal. The Δ H's due to the AP component of mixed crystals
of varied compositions are compared to values of corresponding
physical mixtures in Fig. 3. This figure is obtained from the data
in Table 1. It is evident that there is no significant differences
in Δ H values between the mixed crystals and physical mixtures of
analogous composition.

In Table 1 losses in weight of the crystals are shown. Most
of the weight losses as might be expected, are fairly consistent
with the AP content of the mixed crystals or physical mixtures.
Figure 2 shows some of the curves obtained with AP-KP mixed crystals. Similar curves for AP and physical mixtures are shown in
Figure 1. Figure 5 compares the behavior of a 1:1 by weight mixed
crystal, a similar physical mixture, and pure AP. This reveals that
the mixed crystal, curve 3, undergoes decomposition at a somewhat
higher temperature than either the analagous physical mixture,
curve 1, or pure AP, curve. By implication KP from a thermal stand-

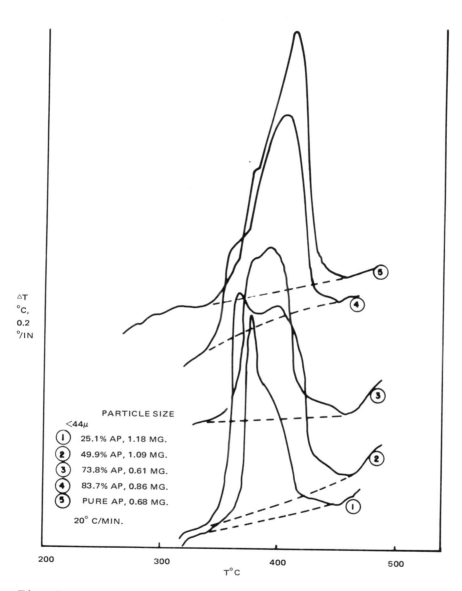

Fig. 1. Decomposition exotherms of AP, AP and KP physical mixtures by DSC calorimetry.

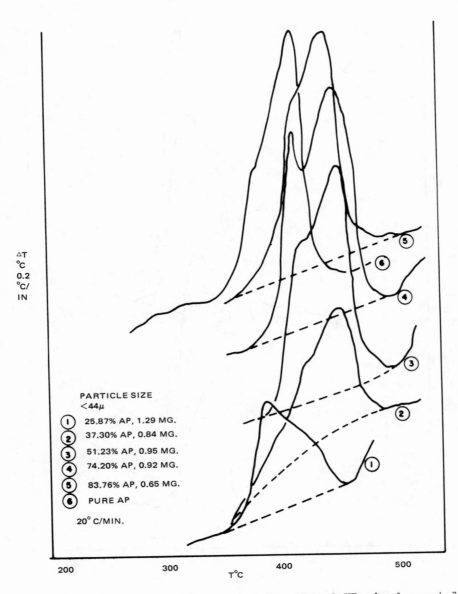

ΔT
°C
0.2
°C/
IN

PARTICLE SIZE
<44μ

① 25.87% AP, 1.29 MG.
② 37.30% AP, 0.84 MG.
③ 51.23% AP, 0.95 MG.
④ 74.20% AP, 0.92 MG.
⑤ 83.76% AP, 0.65 MG.
⑥ PURE AP

20° C/MIN.

200   300   400   500

T°C

Fig. 2.   Decomposition exotherms of AP, AP and KP mixed crystals
          by DSC calorimetry.

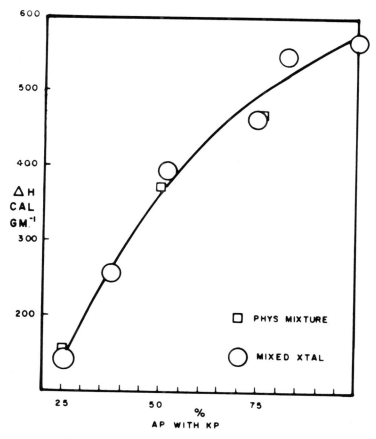

Fig. 3. Variation in Δ H in physical and chemical mixtures (mixed crystals) as a function of composition for AP-KP materials.

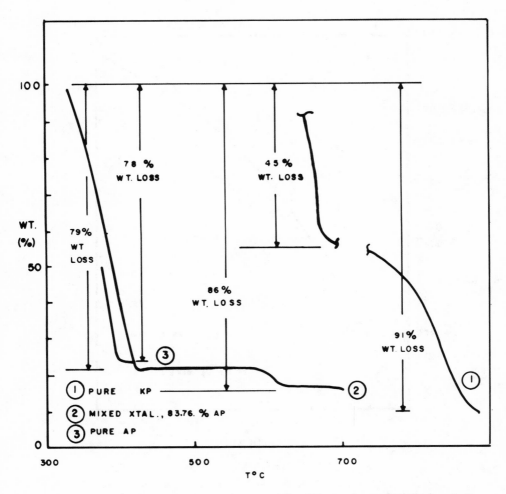

Fig. 4.   TGA thermogram of pure AP, KP, and a mixed crystal
          rich in AP.

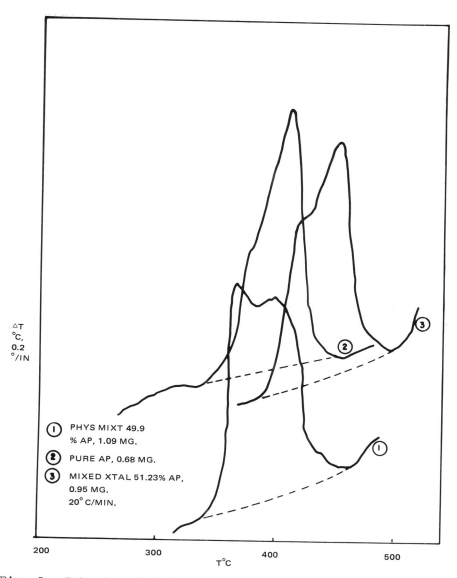

Fig. 5.   Behavior of 1:1 mixed crystal, physical mixture, and
          pure AP by DSC calorimetry.

point could be said to have a stabilizing effect upon the AP crystal.

Figure 6 illustrates the thermal behavior of pure AN. We obtained a value for Δ H of 345 cal/gm for the illustrated exothermal decomposition. The heat of detonation of AN has been reported by others[2] to be 346.3 cal/gm. The accompanying reaction is represented by the equation:

$$NH_4NO_3 = N_2 + 2H_2O + 0.5O_2$$

In spite of this apparent excellent agreement of our value for Δ H for AN with that previously published, a great deal of caution is necessary in making such quantitative comparisons with thermal data obtained by other means. It is possible that the respective chemical processes involved may not be identical. Gases can escape from the closed DSC cup used in our work. We did not find in our experience that any of the cups were visibly ruptured. It is worth noting that the Mettler Instrument Company offers a system for hermetically sealing DTA cups by means of cold welding. In closed bomb calorimetry gases are retained and can thereby undergo secondary reactions with one another. In the case of a highly energetic material such as AP the decomposition of which is complex and potentially sensitive to the conditions of the experiment, comparison of DSC/ΔH values with those observed by other methods is particularly hazardous. Our value of 569 cal/gm differs greatly from that of Waesche[4], who reported a value of 260 cal/gm. Inasmuch as this is even less than our value for AN, which is less energetic than AP, his result seems to be surprisingly low. If he did not use a closed cup technique, it is understandable why a low value was obtained. Our experience with open cup techniques was unsatisfactory, and for this reason we undertook the closed cup work. We should mention too that it has been shown by means of closed bomb calorimetry in this Laboratory that the heat of decomposition of ammonium perchlorate exceeds considerably our 569 cal/gm value.

We encountered more significant variation in Δ H values for repetitive experiments with AP than with the other materials. We thought this might be due in part to the relatively high temperature involved, compared to AN, for instance. Also there was a possibility of interaction of corrosive gases with the aluminum DSC cup. Unfortunately a more chemically resistant cup material was not available for our work. The DuPont Compant can now supply passivated aluminum cup components, which might be better for this type of work.

Because of this discrepancy in AP/ΔH values from ours and other work, we made further calibration efforts. Since it is a similar material for which accurate closed bomb thermal data is available, we investigated the thermal decomposition of ammonium

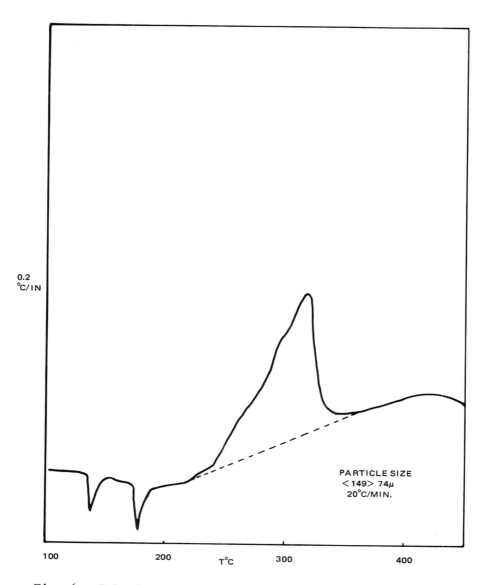

Fig. 6. Behavior of pure AN in closed cup DSC experiment.

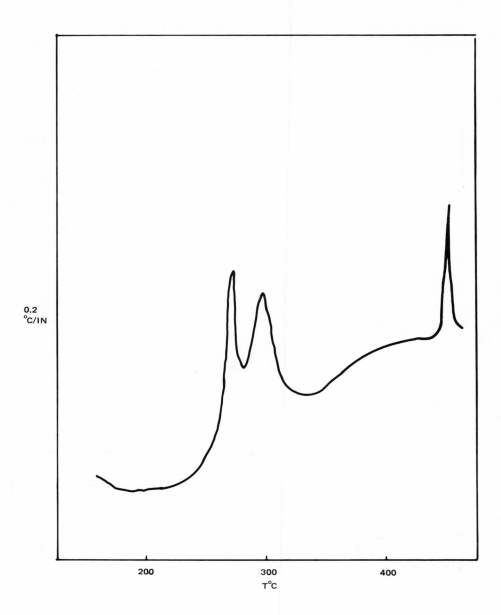

Fig. 7. Decomposition of pure ammonium dichromate by closed cup
         DSC calorimetry.

dichromate. The heat of reaction (decomposition) corresponding to the following reaction has been determined.

$$(NH_4)_2Cr_2O_7(S) = Cr_2O_3(S) + N_2(g) + 4H_2O(l)$$

Also a secondary reaction is known to occur:

$$1/2 \ N_2(g) + 3/2 \ H_2O(l) = NH_3(aq) + 3/4 \ O_2(g)$$

When we examined recrystallized material in the DSC, two separate modes of decomposition were detected, Fig. 7; one at about 270 and the other at $430°C$. Consequently, we were doubtful that our material was following the same decomposition pathway as that in closed bomb experiments. Thus it was not possible to compare ΔH values derivable from Fig. 7 to those obtained in closed bomb experiments. This result further served to underscore some of the problems involved in comparing DSC and other thermal data.

In spite of these problems in rigorously establishing absolute values for ΔH for a given chemical equation representing the decomposition reaction of an energetic material by DSC, we feel that the representations, Table 1 and Figure 3 are meaningful and help to characterize the AP-KP mixed crystal materials. The method proved to be useful in comparing the behavior or mixed crystals vs physical mixtures of corresponding compositions.

We came to the conclusion that when certain limitations are kept in mind, the DSC closed cup technique can be a highly useful adjunct to other methods of characterization of energetic mixed crystals.

REFERENCES

1. Morrow, S.I. and Bracuti, A., Thermochim. Acta. 1 (1970) 317-324.

2. Anon,. "Military Explosives" U.S. Army TM 9-1910 and U.S. Air Force to 11A-1-34, April 1955, U.S. Government Printing Office, Washington, D.C. pp. 119-23.

3. Keenan, A.G. and Siegmund, R.F., Quart Rev. 23, 430 (1969).

4. Waesche, R.H.W., "Research Investigation of the Decomposition
    of Composite Solid Propellants, Final Report," No. H910476-36
    United Aircraft Research Laboratories, East Hartford, Conn.

5. Neugebauer, C.A. and Margrave, J.L., $\underline{J}$. $\underline{Phys}.\underline{Chem}$. 61, 1429
    (1957).

HEAT OF FUSION OF CRYSTALLINE POLYPROPYLENE BY VOLUME DILATOMETRY AND DIFFERENTIAL SCANNING CALORIMETRY

JAMES A. CURRIE, E. M. PETRUSKA, AND R. W. TUNG

VILLANOVA UNIVERSITY

VILLANOVA, PENNSYLVANIA 19085

INTRODUCTION

Heat of fusion measurements on stereoregular forms of polypropylene have been the subject of a great number of investigations. The techniques most commonly used have been volume dilatometry, specific heat, DTA, TGA, DSC, empirical calculations, and copolymer studies. Furthermore, the published values for the heat of fusion of isotactic polypropylene using these methods have ranged all the way from 15.5 to 62 cal./g. Thus, an accurate value of ΔH_u for the hypothetical 100% crystalline polypropylene remains a subject of some uncertainty. The results of this work present information on a simultaneous study of identical samples using two of the techniques, namely, volume dilatometry and differential scanning calorimetry. The melting points of the pure homopolymer and binary mixtures of it with high purity transdecalin have been carefully determined in a series of dilatometers using mercury as the confining fluid. At the same time, a DSC technique has been employed to obtain the melting point depression as a function of the volume fraction of diluent. The Flory theory for melting point depression is applied to both data and results for ΔH_u thus obtained are compared to other estimates based on crystallinity calculated from density measurements and the actual heat of melting for the semi-crystalline samples. The aim is to demonstrate the ease and reliability with which DSC can be applied to measurements formerly requiring enormous effort and great expense of time.

EXPERIMENTAL

Materials

The isotactic polypropylene samples were obtained through the courtesy of Amoco Chemicals. The $\bar{M}_n$ in trans-decalin at 110° C was found to be 1.09×10^5 using a membrane osmometer. The diluent for these studies was trans-decalin having a 99+% isomeric purity as determined by gas chromatography. The polymer/diluent mixture samples are referred to as AI, AII, etc. and the pure homopolymer as A.

Dilatometry Technique

Isotactic polypropylene pellets were hot pressed between aluminum foils at 232°C with 7.25×10^7 newton/m^2 (10,500 psi) pressure and then annealed for 45 minutes at 135°C. Samples weighing about 3 grams and 1.25 mm thick were prepared. The sample density was then determined by hydrostatic weighing. The dilatometers were constructed from 66 cm lengths of 1.75 mm precision bore capillary tubing which was calibrated according to the method of Bekkedahl[1]. Two grams of homopolymer sample were inserted as .3 to .6 cm strips to fill the 12 mm bulb portion and were sealed off with a hollow glass sphere at the top of the bulb for protection from overheating while attaching the capillary tube. The polymer/diluent samples were prepared in a different manner using the trans-decalin as follows. The polymer pellets were quenched in liquid nitrogen and ground through a 40 mesh screen on a Wiley mill. After insertion and weighing, the trans-decalin was added to the ground polymer in the bulb through a special tube which prevented liquid from adhering to the capillary wall. A second weighing was used to determine v_1, the volume fraction of diluent. The bulb assembly was quenched in liquid nitrogen and evacuated to 10^{-6} mm Hg. with alternate thawing and freezing for a period of 12 hours. After sealing the ampoules in vacuo, the assembly was heat soaked for 6 days in a 180°C bath accompanied by periodic agitation. The 1.75 mm bore tubing was attached and evacuation to 10^{-6} mm Hg was repeated with alternate freeze-thaw cycles for 12 hours. Finally, the dilatometers were filled with mercury using a method described by Mandelkern[2]. The completed dilatometers were heated in a bath containing Dow Silicone 550 fluid heated by 500 watt knife blade immersion heaters controlled by a F. H. Mumberg thermoregulator and a saturable reactor controller. This system was capable of maintaining better than $\pm$ 0.1°C bath temperature. A typical heating schedule used for the volume-temperature measurements is shown in Table 1.

At each temperature level below the sample melting point, the mercury column in the dilatometer was allowed to continue to rise for 12 hours or more in achieving the equilibrium condition. After the melting point had been reached, the equilibrium level occurred in one hour or less. The equilibrium melting points were determined by plotting the polymer specific volume or mercury level versus temperature as shown in Figure 1. In the case of the pure homopolymer, the melting point was also determined by plotting the polymer-diluent volume-temperature data and extrapolating to $v_1=0$.

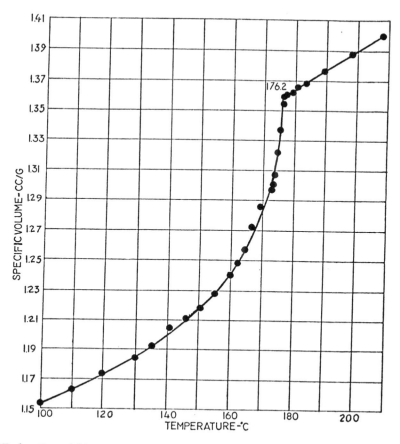

FIGURE 1. Specific Volume of Isotactic Polypropylene   (Sample A)

DSC Technique

In this phase of the experiments a standard DSC-1B instrument was employed and carefully calibrated for differential temperature, average temperature, and transition energy. The polymer sample was in the form of a finely divided powder and the diluent was the same as above. The volatile sample pans were partially filled with diluent by a microsyringe and weighed on a microbalance. Varying amounts of

TABLE 1

HEATING SCHEDULE FOR VOLUME—TEMPERATURE MEASUREMENTS

PURE POLYPROPYLENE

TEMPERATURE RANGE (°C)	HEATING RATE (°C/MIN.)	TEMPERATURE INCREMENTS (°C)	EQUILIBRIUM HOLDING TIME (HR.)
100–130	1.0	10	1
130–160	0.33	5	12–16
160–173	0.33	2.5	12–24
173–180	0.02	0.5	12–24
180–220	0.07	2	1
POLYMER/DILUENT MIXTURES			
138–163	0.01	0.5	12–16
163–184	0.02	0.8	12–16

the polymer powder were added to the pans and the pans sealed in the standard fixture. The sample pans were heated to 20°C above the melting point of the polymer and held for five minutes to permit complete fusion and mixing of the polymer and diluent. After the heat soaking period the samples were cooled to room temperature at .625°C/min. to promote an optimum crystallization process. Thus prepared, the samples were ready for subsequent temperature scanning to determine their melting behavior. Melting scans were performed at 5°C/min.

RESULTS AND DISCUSSION

Dilatometry

In order to pinpoint accurately the melting temperatures, two procedures were used; first, a series of drift curves for the mercury levels were plotted and it was noted that the time to reach equilibrium dropped drastically, from about 12 hours to 1 hour when the melting point was passed, and secondly, by large scale plotting of the volume-temperature data and extrapolation of the solid and liquid regions to the point where they intersected which was taken as the melting point. The two methods led to virtually identical temperatures and the average was taken as the best value for the equilibrium melting temperature. Table 2 shows the volume-temperature results for the pure homopolymer and five polymer/diluent mixtures.

TABLE 2

VOLUME-TEMPERATURE RESULTS FOR POLYPROPYLENE & POLYPROPYLENE/
DILUENT MIXTURES BY DILATOMETRY

SAMPLE #	VOLUME FRACTION TRANS-DECALIN (v_1)	EQUILIBRIUM MELTING TEMPERATURE ($^{\circ}C\pm$ 0.1)
A	0.0	176.2 (176.0)[*]
A I	0.106	169.6, 169.1
A II	0.212	163.3, 162.9
A III	0.306	157.1, 155.1
A IV	0.397	153.2, 151.8
A V	0.500	146.6, 144.2

[*] By extrapolation of volume-temperature data for polymer/diluent
mixtures to v_1 = 0.

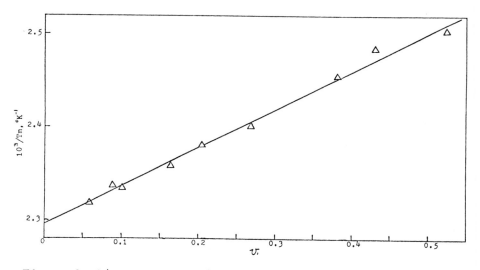

Figure 2. $1/T_m$ versus v_1 (volume fraction of diluent) by DSC

The volume fraction of diluent, v_1, is calculated from the density of the polymer in the liquid state at the depressed melting and the density of the diluent at the same temperature. The densities for trans-decalin are interpolated values from Seyer and Davenport[3]. The heat of fusion, ΔH_u, and the polymer/diluent interaction parameter, χ_1, are calculated from the intercept and slope respectively when a plot of $\left. \dfrac{1}{T_m} - \dfrac{1}{T_m^o} \middle/ v_1 \right.$ versus v_1 is made according to Flory's[4] equation:

$$\frac{1}{T_m} - \frac{1}{T_m^o} = \frac{R}{\Delta H_u} \frac{V_u}{V_1} \left(v_1 - \chi_1 v_1^2 \right) \tag{1}$$

where T_m^o = fusion temperature of polymer/diluent mixture of given volume fraction (^oK),

$\quad$ R $\quad$ = gas constant (1.987 cal/mole ^oK),
$\quad$ V_u $\quad$ = molar volume of polymer structural unit at T_m^o (cm^3),
$\quad$ V_1 $\quad$ = molar volume of diluent at T_m^o (cm^3)
$\quad$ v_1 $\quad$ = volume fraction of diluent at T_m,
$\quad$ χ_1 $\quad$ = BV_1/RT_m = polymer/diluent interaction parameter,
$\quad$ B $\quad$ = Van Laar heat of mixing parameter,
$\quad$ ΔH_u = latent heat of fusion per mole of polymer structural unit (cal/mole).

The entropy of fusion is then found by the thermodynamic relation as the ratio of the heat of fusion to the pure polymer melting temperature.

$$\Delta S_u = \frac{\Delta H_u}{T_m^o} \tag{2}$$

The pure polymer melting point is also determined by plotting diluent volume-temperature data and extrapolation to $v_1=0$. The crystallinity is calculated by (a) density ratios, using a crystalline density of 0.936 g/cc and an amorphous density of 0.855 g/cc (Frank[5]), and (b) ratio of sample crystalline-amorphous enthalpy difference (31.8 cal/g) to the dilatometric heat of fusion.

DSC

The determination of the melting temperature is not as direct nor nearly as absolute when scanning the melting range of pure polymers or polymer/diluents in the DSC when compared to the dilatometry experiments. There have been many methods employed to obtain significant melting points by other workers and the one judged most feasible for this experiment is the interpretation of the peak maximum as the melting point. Peak temperatures were subsequently corrected for instrument response using the leading edge for pure Indium. The interpretation of melting temperature becomes one at which the transition is proceeding at the maximum rate. All melting curves were analyzed in this manner and the data for melting point depression were determined accordingly. This results in a lower temperature level being recorded for the DSC experiments yet does not seriously affect the validity of the results so long as the procedure is handled properly. Table 3

TABLE 3

MELTING TEMPERATURE RESULTS BY DSC FOR POLYPROPYLENE AND POLYMER?
DILUENT MIXTURES

Sample No.	Volume Fraction Trans-Decalin,(v_1)	$10^3/T_m$, $(^\circ K^{-1})$
A	0.0	2.294
A I	0.089	2.339
A II	0.204	2.380
A III	0.268	2.411
A IV	0.381	2.455
A V	0.431	2.484
A VI	0.524	2.516
A VII	0.551	2.543
A VIII	0.575	2.547
A IX	0.611	2.569

TABLE 4

FUSION PROPERTIES OF POLYPROPYLENE BY TWO METHODS

	Dilatometry	DSC
T_m^o (measured $^\circ C$)	176.2	165.7[a]
T_m^o (extrapolated $^\circ C$)	176.0	164.2[b]
ΔH_u (from Flory eqn. cal/g)	44.0$\pm$ 2.0	43.8$\pm$ 2.0
ΔH_f (measured heat of melting cal/g)	31.8 (DSC)	31.8$\pm$ 1.0
ρ (measured density g/cm^3)	0.912	0.912
X% (by density ratio)	72.2	72.2
X% (by $\Delta H_f/\Delta H_u$)	72.8	72.6
ΔS_u (cal/g$^\circ K$)	0.10	0.10
X_1 (interaction parameter)	-0.44	-0.49

[a] Peak maximum, not maximum melting temperature

[b] From peak maxima of polymer/diluent samples

lists the results for the pure homopolymer and nine polymer/diluent mixtures.

The melting temperature data from Table 3 is plotted in Figure 2. The remaining procedure involving the Flory equation and deriving the thermodynamic properties was the same as for the dilatometry experiments and leads to the following comparison of results for the two methods.

Discussion

Many results for the fusion properties of crystalline polypropylene have been published in the literature. Because of the wide variation in results, a study was made using volume dilatometry and very slow heating rates to reach equilibrium conditions. At the same time, the fusion properties were measured by a more efficient method in a differential scanning calorimeter. The main feature of the results using these widely differing techniques is the agreement in the values for ΔH_u. These are encouraging because the DSC method is capable of enormous time saving in addition to yielding results closely related to the equilibrium methods of dilatometry and calorimetry. The magnitude for ΔH_u found here corroborates the work of several others who have recently studied the fusion of polypropylene as well. Knox[6] measured ΔH_u by a similar DSC technique and reported a range of 46-50 cal/g. Krigbaum and Uematsu[7] determined values of 174°C for the melting point and 47.7 cal/gm for ΔH_u for polypropylene crystallized by slow cooling from the melt over a 5 hour period. By the specific heat method, Wilski[8] obtained 40.0 cal/gm for ΔH_u, and Passaglia and Kevorkian[9] reported 45.0 gal/gm. All of these values are significantly less than the value of 62.0 cal/g obtained by the diluent method of Danusso, et al[10]. Thus the recent preponderance of results by the three methods dilatometry, DSC, and conventional calorimetry, lead to the conclusion that the ΔH_u for isotactic polypropylene is fixed in the region of 44-48 cal/g. Sample and experimental variations remain as the reason for the $\pm$ 2 cal/g range for reported values of ΔH_u.

CONCLUSION

The heat of fusion of crystalline polypropylene was determined by volume dilatometry using the melting point depression technique and also by differential scanning calorimetry. Pure polypropylene and polypropylene-trans-decalin mixtures were heated slowly in capillary dilatometers to yield volume-temperature data and equilibrium melting points. Equivalent polymer samples were heated in the calorimeter of the DSC to yield the melting point depression and the energy absorbed during the fusion process. Polypropylene with $\overline{M}_n$ = 1.090 x 10^5 had a melting point of 176.2°C. The average heat of fusion by dilatometry was 44.0 cal/g; and by calorimetry, 43.8 cal/g. The entropy of fusion was 0.10 cal/g°K by both methods of measurement. It is noteworthy that the heat of fusion properties of a crystalline polymer can be easily and accurately measured by means of a differential scanning calorimeter. The tedious dilatometric method is not necessary for most uses of the fusion properties and can be supplanted by the DSC method. The DSC method offers a time saving technique for obtaining these valuable properties.

REFERENCES

1. Bekkedahl, N., J. Res. Nat. Bur. Std., 42, 154 (1949)

2. Mandelkern, L., Polymer, 5 (12), 637 (1964)

3. Seyer, W. F., and Davenport, C. H., Journal of American Chemical Society, 63, 2425 (1941)

4. Flory, P. J., J. Chem. Phys., 15, 684 (1947)

5. Frank, H. P., "Polypropylene", 1st Ed., Gordon and Breach Science Pub., New York, N. Y., 1958, p. 51

6. Knox, J. R. in "Analytical Calorimetry", 1st Ed., Porter, R. S. and Johnson, J. F., Ed., Plenum Press, New York, N. Y., 1968, pp. 9-14.

7. Krigbaum, W. R. and Uematsu, I., J. Polym. Sci., Part A-3 (2), 767 (1965)

8. Wilski, H., Kunststoffe, 50, 335 (1960)

9. Passaglia, E. and Kervorkian, H. K., J. Appl. Phys., 34, 90 (1963)

10. Danusso, F., Marglio, G., and Flores, P., Atti. Accad. Nazl. Lincei. Prnd., 25, 420 (1958)

Self-Seeded PE Crystals;

Melting and Morphology

IAN R. HARRISON AND GARY L. STUTZMAN

THE PENNSYLVANIA STATE UNIVERSITY

UNIVERSITY PARK, PA. 16802

Introduction

Using a novel DTA technique, it has been reported that polyethylene (PE) single crystals exhibit narrow multiple-peaked thermograms[1]. The half-width of the individual peaks is usually less than 1.5°C. These peaks widths are comparable to those obtained using pure non-polymeric organics. Previous work suggested that discrete transformation temperatures exist for the {100} and {110} fold sectors[2]. The work that led to this suggestion was performed on lamellae produced by isothermal crystallization from a dilute (.1%) solution of PE in xylene. The solution was heated to the boiling point of xylene prior to being quenched to a particular crystallization temperature. Several questions evolved from this latter paper. The first of these concerns the homogeneity of lamellae produced in this way. Since not all lamellae began to crystallize at the same time, a distribution of sizes appears. Crystals often possessed overgrowths and/or defects. One might also question both the heating rate effects reported and the justification for resolving multiple-peaked thermograms into discrete components. It is the object of this paper to attempt to answer these questions.

In this present study lamellae are used which have been prepared by a self-seeding technique. The technique utilizes nuclei in the solution to initiate crystallization of lamellae at approximately the same time. In this way it is possible to produce a sample of PE lamellae with a homogeneous size distribution. In addition the extent of overgrowths and/or defects on the crystals can be somewhat controlled. These crystallites were then used for experiments which lead into a study of heating-rate

579

effects and peak resolving. However the original hypothesis of
discrete transformation temperatures for the {100} and {110} fold
sectors remains the main concern of the paper.

Experimental

The technique of self-seeding crystals was described in
detail in a paper by Blundell and Keller[3]. For completeness, we
shall list the salient points here. Self-seeding nucleation of
crystals was accomplished by preparing a sample suspension of the
polymer in the appropriate solvent. In this study the polymers
used were Marlex 6001[4] (M_W = 110 x 10^3 to 250 x 10^3), Marlex
55035[4] (M_W = 250 x 10^3 to 1.5 x 10^6), and AC 1222[5] (M_W = 1.5 x
10^6). The last two were used only at the highest crystallization
temperature (T_c). These polymers are all reported to be linear.
A dilute solution (.1%) was prepared by heating the polymer under
reflux in boiling xylene until it was totally dissolved. Nitrogen
was purged through the solution to provide an inert atmosphere.
After total dissolution the sample was allowed to cool to room
temperature and crystallization took place during the cooling
process. Part of this suspension was then slowly heated (< 10°C/
min) to 103°C and immediately placed in a constant temperature
bath at the appropriate T_c. After crystallization for several
days the crystals were washed three times by decantation with
fresh xylene at T_c. This washing reduces the concentration of
impurities, mostly in the form of low molecular weight PE, to
less than one thousandth of its original value.

After three washes, excess mother liquor was decanted and the
residual crystal suspension was divided into three samples:
a. A small sample was taken for transmission electron microscopy
(TEM). This sample was simply pipetted from the crystallization
vessel and placed on 200 mesh copper grids supporting a carbon
film. The xylene was allowed to evaporate at room temperature.
b. Most of the sample was exchanged to silicone oil for differential thermal analysis (DTA). Exchange was achieved by adding
silicone oil to the sample in xylene and removing most of the
xylene by vacuum using an aspirator and then a mechanical pump.
Residual xylene was removed by application of a high vacuum for
several hours. Some of this sample was back exchanged to xylene
by washing the oil exchanged sample many times with xylene, using
centrifugation, to remove the silicone oil. c. A sample was
exchanged to acetone and dried down under high vacuum. This is
the "solid" sample.

DTA was performed on oil exchanged samples from each of the
crystallization temperatures. All scans were made using 2mm glass
sample tubes and a DuPont 990 TA and Cell Base with a Standard
Cell. Several heating rates were used for all samples but most of

the data reported here was recorded at 20°C/min. Maximum
sensitivity (.05°C/in) was used on the ΔT axis unless otherwise
noted. All scans were at least duplicated and often repeated
several times.

In order to minimize effects such as variations between
thermocouples and other fluctuations in procedure, it was desirable to have a reference position on each chart. To accomplish
this, a scan of phenacetin* suspended in silicone oil was run on
the same chart as the PE samples. The same thermocouple was used
for both the PE sample and the phenacetin. Phenacetin exhibits a
single sharp endotherm near that of PE. The extrapolated onset
of this peak was taken to be 134.5°C; the melting point of pure
phenacetin.

In some experiments two samples having identical thermal
histories were required. This was achieved by placing material
in both the reference well and the sample well of the DTA heating
block. The DTA was then used as a preparative instrument to
condition the samples. Using this procedure, one sample was
removed after conditioning and washed free of silicone oil for
use in TEM. The second sample was reheated through the melting
point and changes in peak position and shape as a result of conditioning were recorded. Identical results were obtained regardless of which well was used for DTA or TEM. The samples were
often interchanged for duplicate runs.

One conditioning procedure used was to heat a sample rapidly
to 100°C using the isothermal program. The sample was then
heated at a programmed rate of 10°C/min to the final desired
temperature from which it was quickly cooled by forced air at
room temperature. This was termed a "quench" experiment.

Results

The effect of T_c on both position and shape of the endothermic
peaks is shown in figure 1. The scans were all made at a heating
rate of 20°C/min. The multiple peaks will be referred to as
first, second, and third in order of increasing temperature of
their endothermic maxima. As T_c increases several changes occur:
Three sharp peaks can be seen in the 85°C and the 87°C crystals;
whereas only two peaks plus a shoulder appear in the 90°C and
only one peak and a shoulder in the 93°C crystals. At the highest
T_c, 95°C, only one peak can be seen. The ratio of the first peak
to the second peak apparently increases as T_c increases. The

---

*Obtained from a test substance kit by Reichert (Wien) for their
Kofler Hotbench.

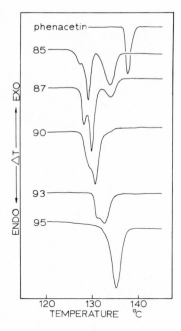

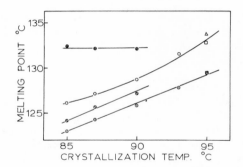

Figure 2. Graphical representation of the data in figure 1. The peak positions have all been normalized to phenacetin.
⊙ - onset, ◐ - 1st peak, ○ - 2nd peak, ❸ - 3rd peak, □ - Marlex 55035, ▲ - AC 1222.

Figure 1. DTA scans of PE single crystals grown isothermally at a series of crystallization temperatures. The number to the left of each scan refers to T_c in °C. All scans were made at a heating rate of 20°C/min and a sensitivity of .05°C/min. Unless otherwise noted this and all subsequent DTA scans are made on exchanged to oil crystals.

apparent onset of melting increases as T_c increases and similar changes are seen for the first and second peak position. The third peak remains essentially unchanged in position for T_c from 85°C to 90°C but becomes smaller as T_c increases and finally disappears at the higher crystallization temperatures. These results are presented graphically in figure 2 and have all been normalized to a phenacetin peak; that is, the onsets of the phenacetin peaks run with each sample were all taken as 134.5°C and the PE peak temperatures were adjusted correspondingly.

Micrographs of crystals from the T_c series are shown in figure 3. Crystals were generally 4 to 5μ in their longest dimension. It should be noted that there was no observable difference in crystal morphology between crystals taken directly from xylene and the same crystal preparation after back exchange to xylene via silicone oil. Micrographs show an increase in the {100} sector angle with an increase in T_c. By directly measuring the area of the {110} and {100} sectors a graph of the ratio {110} area/{100} area as a function of T_c can be obtained. This ratio

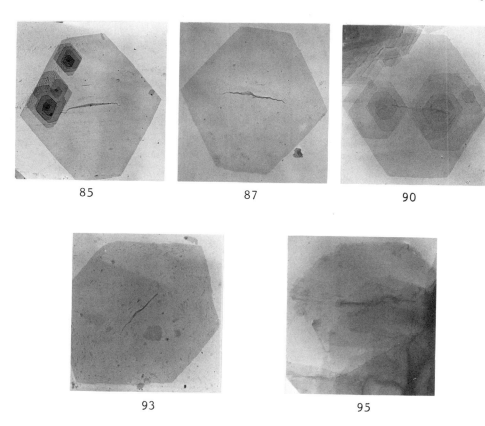

85   87   90

93   95

Figure 3. Micrographs of PE single crystals. The T_c in °C is indicated below each photograph. All crystals are unshadowed and about 4 to 5μ in their longest dimension.

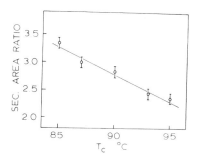

Figure 4. Sector area ratio (area {110} sector)/(area {100} sector) plotted against T_c.

has been denoted the sector area ratio (SAR) and is shown in
figure 4. A linear increase in SAR is observed as a function of
T_c. The increase in ratio of the first peak to second on the DTA
scans has already been noted (figure 1).

In order to determine if peak area ratios as recorded by DTA
were comparable with sector area ratios by TEM it was desirable to
resolve the complex thermograms into their component peaks. It
was assumed that if the height of the second peak was kept approximately constant, its shape and half-width would remain unchanged
throughout the T_c series. Using a T_c = 85°C crystal preparation
as a standard, the first and second peaks were resolved. This
second peak could then be readily superimposed on the thermograms
of other crystal preparations. By subtracting out the area of the
standard second peak, the area, shape, and position of the first

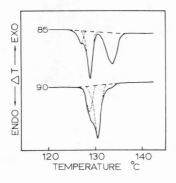

Figure 5. DTA scans of two
crystal preparations are
shown as an example of the
curve resolving methods
described in the text. The
number to the left of each
scan indicates the T_c in °C
of the crystal preparation.

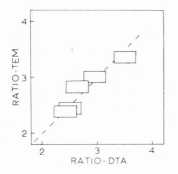

Figure 6. A graph of peak
area ratio obtained by
resolution of DTA curves
vrs SAR by TEM. The broken
line has been drawn at 45°.

peak of the other T_c preparations were determined. In some
cases it was necessary to change the baseline and size of the 85°C
resolved second peak. However, in all cases the half-width and
shape were kept the same. An example of this type of resolution
is shown in figure 5. A scan of the T_c = 85°C preparation is
shown and a scan of T_c = 90°C has been resolved. The procedure
for resolving other scans in the T_c series was identical. If a
one to one relationship exists between the SAR by TEM and the peak
area ratio by DTA then a graph of area by DTA versus area by TEM
should show a series of points falling on a 45° line. The results
are plotted in figure 6, the line is drawn at 45°.

In the introduction it was pointed out that as a result of
data presented in a previous paper[2], some questions were raised
regarding heating rate effects on the crystals. The results of
various heating rates on a T_c = 87°C preparation are shown in
figure 7. Note that heating rate changes both position and shape
of the peaks. In an effort to understand this result the effects
of heating rate on phenacetin were studied. Figure 8 shows the

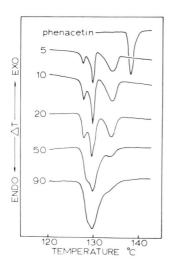

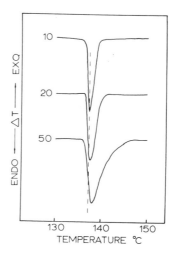

Figure 7. Thermograms of
a T_c = 87°C preparation for
a variety of heating rates.
Numbers to the left of a
scan refer to heating rate
in °C/min.

Figure 8. DTA scans of
phenacetin made at the
heating rate in °C/min,
shown to the left of each
scan. Heating rate
affected both shape and
position of the peak.

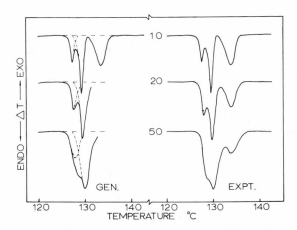

Figure 9. DTA scans of a T_c = 87°C crystal preparation, made at the heating rates indicated, are labeled 'EXPT'. To the left of the experimental curves are two generated curves GEN. The 20 and 50 GEN curves are generated using the resolved 10 curve.

scans for heating rates of 10, 20 and 50°C/min applied to phenacetin. Half width of the peak increased as heating rate increased. In addition, the position of the peak maximum moved to a slightly higher temperature. It should be possible to use the effects observed on phenacetin to predict heating rate effects in PE. This has been attempted in figure 9 using crystals grown at 87°C. The curve at 10°C/min is taken as a starting point and the first two peaks resolved. By permitting the two resolved peaks to change proportionally to the changes observed in phenacetin as a function of heating rate, one can generate curves for 20 and 50°C/min. Both generated and experimental curves are shown for comparison.

In an effort to further identify the origin of the first and second peak maxima a series of quenching experiments were undertaken. These consisted of heating samples to various points in the thermogram and then rapidly quenching in the hope of retaining whatever structure was present at that point. By using the experimental method previously outlined, two samples were given identical thermal histories. They were then examined using DTA and TEM. These experiments were performed on a T_c = 87°C crystal preparation. DTA scans were all run at 10°C/min. Four quench points were selected: a. at the onset, b. midway between the onset and the first endothermic maximum, c. at the first endothermic maximum, d. at the minimum between the first and second endothermic maxima. Figure 10 shows a DTA curve of an unquenched T_c = 87°C crystal preparation and a phenacetin "standard". The

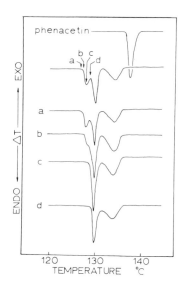

Figure 10. DTA scans of a
T_c = 87°C preparation samples
of which had been previously
quenched at different temperatures. The quench
temperatures are labeled "a"
through "d" and the resulting
DTA curves lettered accordingly.
All scans were made at 20°C/
min.

four quench points are labeled and curves resulting from these
quench positions are shown below and labeled accordingly.
Quenching at point 'a' did not change either the positions of the
peaks or the ratio of the height of the first peak to the second
peak. When a sample was quenched at point 'b' the positions of
the peaks did not change but the first peak did become smaller
while the second peak remained about the same. Quenching at 'c'
showed a total loss of the first peak and a slight change in
shape of the second and third peaks. The position of the second
peak remained unchanged. When a sample was quenched at point 'd'
the second peak was shifted by about .5°C to a higher temperature.
The shape of the second peak was similar to the second peak in
'c'.

Samples given identical thermal histories to the ones used
for the DTA scans in figure 10 were prepared for TEM. The
micrographs in figure 11 show morphology changes that occurred
as a result of this treatment. The micrographs are labeled to
correspond to their respective thermograms. An unquenched
crystal back exchanged from silicone oil to xylene is shown for
comparison. In 'a' of figure 11 there was no apparent change in
crystal morphology. Crystal 'b' showed thickening along the edge
of the {100} sector. A slight bowing of this edge was also
apparent. Crystal 'c' showed increased bowing of the {100}
sector edge but little or no change along the edges of the {110}
sectors. Striations were seen in the interior of the crystal and

seemed to be confined to the {100} sectors. Crystals corresponding to preparation 'd' showed little or no detail and micrographs are not shown. In those few cases where structure was
apparent in 'd' crystals, striations appeared throughout the
crystal interior with 'scalloping' and thickening of all edges.
In some cases the crystals assumed the shape of a 'puckered'
disk.

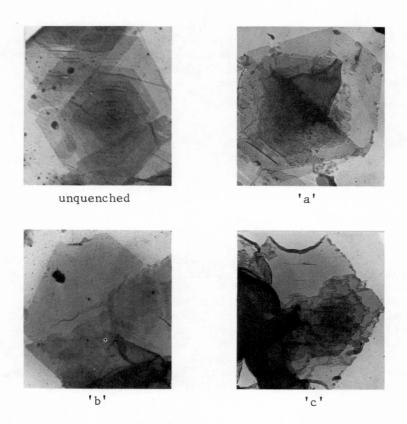

unquenched

'a'

'b'

'c'

Figure 11. TEM micrographs of quenched T_c = 87°C crystals.
The crystals were quenched at the positions indicated in
figure 10 and are labeled accordingly. An unquenched
crystal is shown for comparison.

     In figure 12 three micrographs are shown from a T_c = 85°C
preparation. The crystals were quenched at a temperature
corresponding to position 'c' previously mentioned. The micrograph labeled '1' is that of a monolayer with thickening and
extensive bowing of the {100} sector edge. The edges of the
{110} sectors showed only slight thickening and little or no

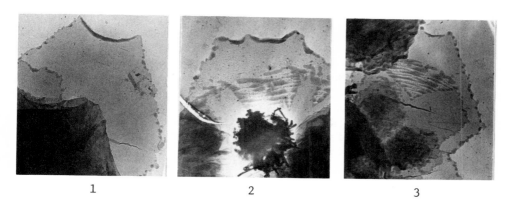

1   2   3

Figure 12.   TEM micrographs of a T_c = 85°C preparation
quenched at the maximum of the first peak. Morphological
changes such as bowing, thickening and striations are all
evident in the {100} sector.

bowing. In '2' the thickening and bowing is evident along with
some striations in the interior of the {100} sector. The
striations were only seen where an overgrowth appeared on the
crystal and within the {100} sector. The last micrograph, '3',
shows a morphology similar to '1' and '2' but the striations,
though largely confined to the {100} sector, are spreading to
the adjacent {110} sectors. Again, striations only appeared
where an overgrowth appeared on the original crystal.

Discussion

     Evidence will be examined for the existence of discrete
transformation temperatures for the two fold sectors in single
crystals of PE.   The examination will be concerned primarily
with DTA and TEM data presented in this paper. A discussion on
the thermodynamics related to melting phenomena will be reserved
for a later date after more detailed studies have been completed.

     Almost all of the previously reported thermal studies of PE
single crystals have been conducted on dried down crystal
preparations[6]. As reported earlier, exchanging PE crystals to
silicone oil without first drying them resulted in multiple
peaked thermograms. These thermograms have individual peaks
which are much narrower than has been reported for high molecular weight PE. Studies reported on low molecular weight samples,
also dried down, have shown somewhat broader multiple peaks.
In many cases these peaks showed a heating rate dependence which
has been interpreted as evidence of recrystallization or
annealing[7].

Heating rate studies in this paper were carried out on a pure, non-polymeric organic. They showed an increase in the width and a slight change in position of the endotherm as heating rate increased. The information gained from this study was applied to modify the first and second peaks of PE and hence generate observed heating rate effects without invoking annealing or reorganization processes. However the third peak can only be explained as the melting of material which has already undergone some transformation. This statement is based primarily on the heating rate dependence of the third peak which could not be explained using the effects noted for phenacetin.

TEM of crystals grown at a variety of crystallization temperatures showed an increase in the $\{100\}$ fold area as T_c increased (figure 3). In this paper it is described as SAR (figure 4). Other authors have used methods such as sector angle[8] (SA) and truncation ratio[8] (TR) in describing this phenomena. Application of all of the methods has been made to our preparations and the results reported in table 1 for comparison. These results are consistent with previously reported data on similar crystal preparations. Resolution of DTA scans for the T_c series was performed as described earlier. While large errors may be present in this type of measurement, a one to one relationship between peak area ratio by DTA and SAR was observed (figure 6).

Table 1

T_c (°C)	SAR	DTA	TR	SA(°)	ℓ (Å)*
85	3.4 ± .1	3.5 ± .2	.29	34	126 ± 6
87	3.0 ± .1	2.9 ± .2	.33	35	---
90	2.8 ± .1	2.6 ± .2	.37	40	145 ± 6
93	2.4 ± .1	2.5 ± .2	.56	59	---
95	2.4 ± .1	2.4 ± .2	.65	69	178 ± 15

*The data in this column are for the crystals cited in reference 7.

In the quench experiment it was shown that changes in peak position gave corresponding changes in crystal morphology. Preferential changes in the {100} sector such as bowed ends, thickening and striations are present as evidence of discrete transformation temperatures for the two sectors. Striations were only observed when overgrowths were present. The reason for this is not apparent but is perhaps related to interpenetration of mobile {100} sectors. The presence of bowing in isolated lamella with no overgrowths indicates that overgrowths are not a requirement for sector transformation.

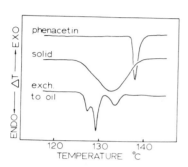

Figure 13.   Two thermograms of samples of a T_c = 87º C crystals are shown below a standard phenacetin peak. The samples were taken from the same crystal preperation.   All scans were made at 20ºC/min. The scans are labled according to the method used to prepare them for DTA.

Figure 13 is simply an illustration of the differences which can be obtained in DTA scans of the same material. Heating rate is the same for both the solid and exchanged to oil samples.   Both samples were obtained from the same crystal preperation.   The main differences are sample weight and of course the exchanged to oil sample has never been dried down. The solid sample weight is approximately 2.5 mg, the exchanged to oil sample contains about 0.1 mg of polymer.

Conclusions

1. Data presented in this work using a series of self-seeded crystals are consistent with preliminary studies made using heterogeneously crystallized polymers. Specifically the technique of exchanging to silicone oil, without prior drying, demonstrates the presence of overlapping peaks within an apparently single broad peak.

2. Heating rate effects observed in the thermograms of
    exchanged to oil polymers can readily be explained without
    recourse to reorganization, annealing or recrystallization
    phenomena; at least for the first two peaks.

3. Quenching experiments followed by DTA and TEM demonstrate
    preferential reorganization of the {100} sector at lower
    temperatures than the {110} sector. Changes observed in
    the thermograms can be related to morphological changes.
    In agreement with earlier work the first and second endo-
    therms are assigned as transformation temperatures of {100}
    and {110} fold sectors respectively.

Acknowledgement
      The authors would like to express their appreciation to the
National Science Foundation for their support of this work.

Refrences

1. Harrison I. R., J. Poly. Sci., A2 11, 991 (1973).
2. Harrison I. R., Polymer Reprints, 14, 502 (1973).
3. Blundell D. J. and Keller A., J. Macromol. Sci. - Phys.
       B2(2), 337 (1968).
4. Phillips Petroleum Company, Plastics Division, personal
       correspondence.
5. Allied Chemical Company, personal correspondence.
6. For example, see Mandlekern L., Prog. Poly. Sci., 2, 165 (1968)
7. For example, see Bair H. E., Huseby T. W. and Salovey R.,
       Analytical Calorimetry (1968).
8. Nagai H. and Kajikawa N., Polymer 9, 177 (1968).

THE ENTHALPY OF FUSION OF LOW MOLECULAR WEIGHT POLY-
ETHYLENE FRACTIONS CRYSTALLIZED FROM DILUTE SOLUTION

S. Go, F. Kloos and L. Mandelkern

Department of Chemistry and Institute of
Molecular Biophysics, Florida State University
Tallahassee, Florida 32306

INTRODUCTION

The thicknesses of linear polyethylene crystals
formed in dilute solution depend only on the isothermal
crystallization temperature and are independent of chain
length for molecular weights greater than about 20,000.
[1][2][3][4] Thermodynamic properties, such as the
density and enthalpy of fusion, are also independent of
chain length in this range.[1][4][5] These quantities
are only dependent on the crystallite thickness which is
determined by the crystallization temperature and solvent.
In an apparent exception to this generalization, it has
been reported that both the density and enthalpy of
fusion depend markedly on molecular weight.[6][7] However, subsequent reports have indicated that these experiments were in error due to sample contamination with
celite and grease.[8]

It has recently been found that for molecular
weights less than about 15,000, i.e., the order of a
thousand chain atoms, the crystallite thickness becomes
dependent on the chain length.[9] The question is then
raised as to whether the associated thermodynamic quantities are molecular weight dependent in this range.
The results of experiments designed to answer this question provide the substance of the present report. In
the lower molecular weight range fractionation occurs
quite easily during crystallization from dilute solution.
[9][10][11] It is therefore necessary that sharp fractions

593

be used in order to avoid spurious and misleading results. The details of the fractionation procedure used
here as well as the sample characterization have been
described elsewhere.[9] A summary of the weight and
number average molecular weights of the fractions, as
determined by gel permeation chromatography, are given
in Table I.[12] The procedures for determining the
density, enthalpy of fusion, and crystallite thickness
have been given previously.[1][2][5][13]

Table I

Characterization of Fractions by
Gel Permeation Chromatography[12]

Sample	M_W	M_N	M_W/M_N
(a)	1142	1055	1.08
(b)	1756	1586	1.10
(c)	4116	3769	1.09
(d)	4908	4405	1.11
(e)	6592	5986	1.10
(f)	9529	8619	1.10
(g)	11805	10675	1.10
(h)	15169	14299	1.06

RESULTS AND DISCUSSION

Density: - In presenting the experimental results it is
convenient to discuss the higher molecular weight samples
studied (c to h) separately from the two lowest ones.
The results obtained for these latter two fractions will
be discussed subsequently. Figure 1 represents a composite of the densities, measured at room temperature,
for samples (d) to (h) plotted against the reciprocal of
the crystallite thickness. Included, for comparative
purposes, are the data for much higher molecular weights
which were reported previously.[5] These latter data
are represented by the solid line which extrapolates to
the unit cell density of 1.00 when $1/\zeta = 0$.[5] The other
solid lines represent the smooth values of the measured
densities for the experimentally attainable crystallite
sizes for fractions (d) to (h). The dashed lines represent the extrapolation to the unit cell density in
each case. Except for sample (d) the density corresponding to the extrapolated extended chain length agrees
quite well with that of the unit cell. There is an inversion in the densities between samples (d) and (e).
The density differences are, however, quite small so that
the discrepancy, or inversion, may be more apparent than
real.

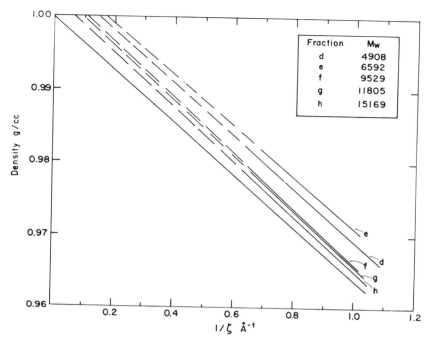

Fig. 1. Density against reciprocal of crystalline thickness in Angstroms. Complete solid line high molecular weight samples from reference (5).

The major portion of the density results can, within experimental error, be represented by a set of parallel straight lines. Thus, at a given value of $1/\zeta$ the density increases with decreasing molecular weight. The difference between the high molecular weight asymptotic value and the lowest molecular weight represented in Figure 1 is relatively small, being only about 0.0068 g/cm^3. The differences in the observed densities arise primarily because for these low molecular weights the extended chain lengths which correspond to the unit cell density can be resolved.

Enthalpy of Fusion: - The enthalpies of fusion for samples (c) to (h) are presented as a plot of ΔH against $1/\zeta$ in

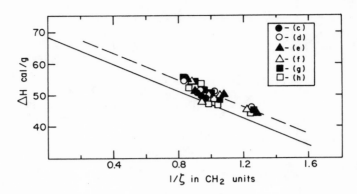

Fig. 2.   ΔH against reciprocal of crystallite thickness in CH$_2$ units.
Complete solid line high molecular weight samples from reference (5).

Figure 2.   The solid line in this figure represents the
previously reported results for high molecular weight
fractions.[5]   An examination of the individual experimental points in this figure indicates that the samples
studied here display no molecular weight dependence.
Within the experimental uncertainty of ±1 cal/g,[13] ΔH
depends only on ζ and not directly on the chain length.
However, the dashed line representing the results for the
low molecular weight samples is displaced upward about
3 cal/g from the high molecular weight asymptote.   When
the degree of crystallinity is taken into account,[1][5]

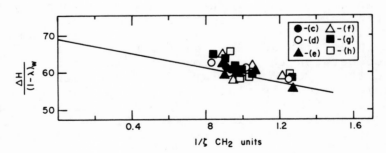

Fig. 3.   ΔH/(1-λ)$_w$ against reciprocal of crystallite thickness in
CH$_2$ units.   Solid line high molecular weight samples from reference
(5).

however, then as is shown in Figure 3 the enthalpy of fusion data for the low molecular weight samples and the high molecular weight asymptote delineate essentially the same straight line. For this plot ΔH is divided by the mass fraction crystallinity, $(1-\lambda)_w$.* Another way of

---

* The degree of crystallinity is calculated from the smoothed density values. The molecular weight dependence of the amorphous density is taken from the work of Fox and Loshaek.[14]

---

presenting the data,[15] as was suggested by Fischer and Hinrichsen, is given in Figure 4. Here $\Delta H/(1-\lambda)_w$

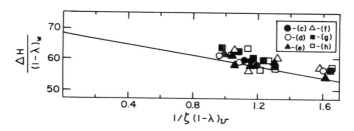

Fig. 4. $\Delta H/(1-\lambda)_w$ against $1/\zeta(1-\lambda)_v$. Solid line high molecular weight samples from reference (5).

is plotted against $1/\zeta(1-\lambda)_v$. The results are similar to those in Figure 3 in that no dependence on molecular weight is found.**

** This method of plotting examines ΔH in terms of the thickness of the crystalline portion rather than of the complete crystallite. This latter dimension also contains contributions from the amorphous overlayer.

In summary when the degree of crystallinity is not taken into account the lower molecular weights yield slightly higher values for ΔH, when compared at the same value of ζ, then is indicated by the high molecular weight asymptote. The magnitude of this difference is small and is nowhere near that which has been supposed.[6][7] As has already been noted, however, this latter

work can be discounted.[8] When the degree of crystallinity is taken into account, by either of the methods indicated, then as is illustrated in Figures 3 and 4 ΔH is
independent of molecular weight from very high molecular
weights to those as low as 3000 when the comparison is
made at the same crystallite thickness. For molecular
weights greater than 3000 the gross morphological structures, as revealed by electron microscopy, are very
similar.[16] Thus the essential invariance of the thermodynamic properties with molecular weight is not surprising and a systematic variation in the properties with
ζ would be expected.

When the molecular weight is less than 3000 new
elements of morphology can be observed. In particular
interfacial dislocation networks are displayed by bilayer
crystals for certain well defined crystallization temperatures.[16] It is therefore necessary to examine
separately the results for the two lowest molecular
weight samples: (a) M_W = 1142, M_N = 1045; and (b) M_W =
1756, M_N = 1586.

Very Low Molecular Weights: - The results for the two
lowest molecular weight samples studied are given in
Table II. Here T_C represents the crystallization temperature. The crystallite thicknesses, 1, are also
listed since in some cases their dimensions are comparable to those of the extended chain. For the other
samples studied here as well as for higher molecular
weight fractions this situation does not occur.

Table II
Properties of Very Low Molecular Weight Fractions

	Sample	$T_C °C$	ΔH cal/g	ρ g/cc	1 Å
(a)	M_W = 1142 M_N = 1055	45	54.8	0.9686	125
		55	54.5	.9725	124
		60	55.9	.9695	118
(b)	M_W = 1756 M_N = 1586	25	56.0	.9582	85
		45	52.4	.9636	95
		60	48.3	.9823	128
		70	52.4	.9823	153

For sample (a) the crystallite thicknesses are independent of the crystallization temperatures studied and are_* comparable in magnitude to the extended chain length.* Correspondingly, ΔH which is in the range of

*Strictly speaking the extended chain length for sample (a) should be 103 Å. The small discrepancy can be attributed to minor errors in the molecular weight determination by gel permeation chromatography.(15)

54.5 to 55.9 cal/g is, within experimental error, independent of the crystallization temperature. For molecular crystals a value of 64.8 cal/g is expected from the calculations of Flory and Vrij.(17) This value for the enthalpy of fusion is not obtained even for this very low molecular weight fraction. Molecular crystals would not, however, be expected to be formed for any real polymer system no matter how well fractionated.(17) The enthalpy of fusion values for sample (a) are plotted in Figure 5. For reference purposes the dashed and solid

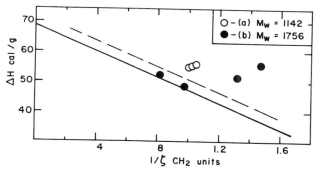

Fig. 5. ΔH against 1/ζ for two lowest molecular weight samples. Solid and dashed lines redrawn from Figure 2.

lines in this figure are reproduced from Figure 2. The results for sample (a) lie about 4-5 cal/g above the composite line for the higher molecular weight samples described previously. Similar results are obtained if

$\Delta H/1-\lambda$ is plotted against $1/\zeta$. All of these crystallites display interfacial dislocation networks.(16) This element of morphology has no major influence on the density. The slightly enhanced value of ΔH, as compared to higher molecular weight fractions of the same thickness, can be attributed to a reduced interfacial enthalpy associated with the extended crystallite size.

Sample (b), M_W = 1756; M_N = 1586, displays a very definite dependence of crystallite thickness on T_C. Interfacial dislocation networks are not, however, observed at all crystallization temperatures. For the highest crystallization temperature, T_C = 70°, the crystallite thickness is equal to the extended chain length of 157 Å and interfacial dislocation networks are found for this crystallization temperature. ΔH is 52.4 cal/g for this sample. This value is slightly less than the corresponding quantity for sample (a). It is approximately the same as is found for very high molecular fractions of comparable thickness.

Samples of this fraction crystallized at 60° and 45° are characterized by very poorly developed crystal habits as revealed by electron microscopy.(16) The crystallite thicknesses are reduced below the extended length and interfacial dislocation networks are not observed in these cases. The sample crystallized at 60° has properties which are similar to those expected for higher molecular weight species of comparable size. However, the sample crystallized at 45° has a much greater enthalpy of fusion and a lower density. This trend in the thermodynamic quantities is continued for the sample crystallized at 25°.

The crystal habit is normal for the sample crystallized at 25°. Interfacial dislocation networks are, however, observed in this case. This represents the only situation where such dislocations are observed and the crystallite thickness is not comparable to the extended chain length. In this case the ratio of crystallite thickness to extended chain length is about 0.5. However, slightly higher molecular weight fractions (not reported here), where a similar ratio in dimensions can be attained, do not display dislocation networks.(16) This special morphological situation is accompanied by very distinctive thermodynamic properties. For the very small crystallite thickness of 85 Å one observes a very high ΔH, 56 cal/g and a very low density, 0.9582 g/cc. The density in this particular case does not appear to

be a reflection of the degree of crystallinity of the
sample, as is usual for other solution formed crystals.
The high enthalpy of fusion associated with such a small
crystallite size requires a low interfacial enthalpy.
Consequently a unique interfacial structure must be
formed.

For the usual solution formed crystals there is
strong evidence that more than 90% of the end-groups
are located in the amorphous overlayer.[18][19] The
possibility exists, however, that because of the unique
dimensional relations that are found in this special
case and perhaps for crystallization at 45° as well,
an appreciable number of end-groups could enter the
crystallite lattice. If this were the case then the
low density could be explained by the expansion of the
lattice. A unique interfacial structure would develop
which would allow for the interaction and interpenetration of layers to form the observed interfacial dislocation networks.

Conclusion: - The thermodynamic properties of fractions
of linear polyethylene crystals formed in dilute solution
have now been studied over an extensive molecular weight
range. For molecular weights greater than 3000 the properties are determined solely by the crystallite thickness.
Molecular weights less than 3000 depart from the usual
pattern of results as has already been indicated from
morphological studies[16] and the relationship between
crystallite thickness and crystallization temperature[9]
and by dissolution temperature studies.[9] For this
molecular weight range interfacial dislocation networks
are formed when the crystallite thickness is comparable
to the extended chain length. Dislocation networks
formed under these conditions, although showing some
changes do not cause any major alteration in the observed thermodynamic quantities. When the structural
basis for this type of dislocation is recognized then
there is no reason to expect differences since an amorphous
layer exists.[20][21] The very special situation,
sample (b) crystallized at a temperature so that the
crystallite thickness is about half the extended chain
length, yields very different properties. Although this
case is clearly not of general applicability the unique
interfacial structure, could possibly be similar to the
one postulated by Sadler and Keller[22] for the formation
of all dislocation networks. It does lead to quite
different thermodynamic properties.

ACKNOWLEDGEMENT

This work was supported by the National Science Foundation under Grant GH33794.

References

(1) L. Mandelkern, A. L. Allou, Jr. and M. Gopalan, J. Phys. Chem. 72, 309 (1968).

(2) J. F. Jackson and L. Mandelkern, Macromolecules 1, 546 (1968).

(3) E. Ergöz and L. Mandelkern, J. Polymer Sci. 10B, 631 (1972).

(4) H. E. Bair and R. Salovey, J. Macrom. Sci.-Phys. B3, 3 (1969).

(5) R. K. Sharma and L. Mandelkern, Macromolecules 3, 758 (1970).

(6) F. Hamada, B. Wunderlich, T. Sumida, S. Hayaski, and A. Nakajima, J. Phy. Chem. 72, 178 (1968).

(7) B. Wunderlich, Macromolecular Physics, p. 405. Academic Press, 1973.

(8) A. Nakajima and F. Hameda, J. Pure and Appl. Chem. 31, 1 (1972).

(9) F. Kloos, S. Go and L. Mandelkern, to be published.

(10) D. M. Sadler, J. Polymer Sci. A-2 9, 779 (1971).

(11) D. M. Sadler and A. Keller, Kolloid-Z-Z Polymer 239, 641 (1970).

(12) We wish to thank Dr. G. W. Knight, Dow Chemical Company for graciously performing these analyses for us.

(13) J. F. Jackson and L. Mandelkern in Analytical Calorimetry, R. S. Porter and J. F. Johnsen, eds., p. 2. Plenum Publishing Company, New York, 1968.

(14) T. G. Fox and S. Loshaek, J. Poly. Sci. 15, 371 (1955).

(15) E. W. Fischer and G. Hinrichsen, Kolloid-Z-Z Polymer 213, 93 (1966).

(16) F. Kloos, S. Go and L. Mandelkern, J. Polymer Sci., Polymer Physics, 00 000 (1974).

(17) P. J. Flory and A. Vrij, J. Amer. Chem. Soc. 85, 3548 (1963).

(18) D. E. Witenhafer and J. L. Koenig, J. Polymer Sci. A-2 7, 1279 (1969).

(19) A. Keller and D. J. Priest, J. Macrom. Sci. 2B, 479 (1968).

(20) L. Mandelkern, J. Phys. Chem. 75, 3909 (1971).

(21) L. Mandelkern, Progress in Polymer Science, ed. A. D. Jenkins 2, 165 (1970), Pergamon Press.

(22) D. M. Sadler and A. Keller, Kolloid-Z-Z Polymer 242, 1081 (1970).

CRYSTALLIZATION OF POLYETHYLENE FROM

XYLENE SOLUTIONS UNDER HIGH PRESSURE

S. Miyata, T. Arikawa and K. Sakaoku

Faculty of Technology
Tokyo University of Agriculture and
Technology, Koganei, Tokyo, Japan

SYNOPSIS

Linear polyethylene has been isothermally crystallized from xylene solutions of various concentrations under high pressure up to 8,000 Kg/cm^2. The degree of supercooling ($\Delta T = 10^\circ C$) and the crystallization temperature were kept constant. The differential scanning calorimetry and the electron microscopy of the crystallized polyethylene revealed that the formation of extended chain crystals (ECC) were decreased according to the increase of the xylene concentration. The characteristic peak of the differential scanning calorimeter (D S C) for ECC disappeared at the vicinity of 70% of polyethylene in xylene solution. On the other hand the size of the crystals in the direction perpendicular to its lamellar thickness grew as extremely large as more than 100μ.

INTRODUCTION

It was shown by Wunderlich[1] et al. that linear polyethylene forms extended chain crystals when isothermally crystallized from melt under high pressure. Since then, a number of investigators have reported on the morphology and crystallization of the extended chain crystals. On the crystallization from the binary components under high pressure, however, there exists only one report by Wunderlich[2], in which he discussed the growth of polyethylene crystals from dilute toluene solution under high pressure. These crystals were not extended chain crystals but folded chain crystals just like those crystals at atmospheric pressure

under similar super coolings. At constant super cooling the
thickness of fold period of those crystals was held approximately
constant with pressure.

On the formation of the extended chain crystals, there have
been proposed two distinct mechanisms. One of which[3,4] is due to
the thickening of the folded chain crystals and the other[5] is the
direct formation of extended chain crystals from melt. In the
case of the latter, the extended chain crystals are formed from
intermolecular nucleus, while folded chain crystals are grown from
intramolecular nucleus. It is supposed that the addition of diluent to polyethylene to change the intermolecular interaction
plays a more important part on the formation of extended chain
crystals rather than that of folded chain crystals.

In this paper the formation of extended chain crystals from
xylene solutions of various concentrations under high pressure is
described. In addition, the question of crystallization mechanism
for the extended chain crystal is analyzed using the information
of the thermal analysis experiments and the electron microscopy
on the concentrated solution grown crystals under high pressure.

EXPERIMENTAL

The pressure apparatus used in this investigation, shown in
Figure 1, consists of three major components: a low pressure press,

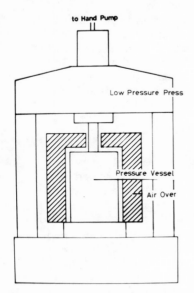

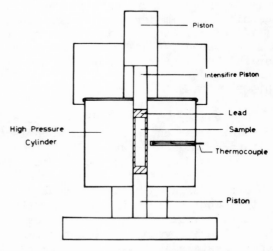

Fig. 1. High pressure
vessel and frame assem-
bly.

Fig. 2. Sample holder and
seal assembly.

an intensifier piston and a high pressure cylinder which holds the sample. The cylinder and the seal assembly are shown in Figure 2. The cylinder and the intensifier piston are machined from steel alloy and they are polished after heat treatment to yield a smooth, low friction surface. The sample holder is made of lead. Lead is in this case used to prevent solution from leaking. Pressure was determined by using the relationship between pressure and the electric resistance of a manganin wire. The calibration of the relationship was accomplished by observing the phase transition of benzene which was determined accurately by Bridgman[6]. In this manner the actual pressure in the sample holder was determined to an accuracy of ±10 atm. The volume changes in the sample holder were measured by detecting the vertical displacement of the intensifier piston to an accuracy of 0.001 mm in conjunction with a differential transformer. The high pressure cylinder was wound around with a nicrom wire to control its temperature and was submerged in a modified air oven. In order to measure the temperature in the sample holder, a chromel-alumel thermocouple was placed in tiny hole drilled at the side of the cylinder. It was found that the temperature was controllable within ±0.5°C up to 300°C.

Commercial linear polyethylene (Sholex F 6150) having the viscosity average molecular weight of 64,300 (Mw/Mn=4.12) and reagent grade xylene distilled and degassed were used in this investigation.

The isothermal crystallization under high pressure was carried out by the following procedures. The sample was heated at atmospheric pressure to the crystallization temperature T_c, chosen to yield ten degrees supercooling under a fixed crystallization pressure. The fixed T_c was determined after measuring melting temperature of the polyethylene-xylene diluent system with the differential transformer. After 1 hour at T_c to attain complete melting and temperature equilibrium of the sample, the pressure was rapidly raised to the fixed pressure and keeping at T_c, until no more volume change was detected by the differential transformer. Then the sample was cooled down to room temperature and was removed from the sample holder after releasing the pressure.

The melting behavior (a heating rate of 10°C/min) of the crystallized polyethylene and the structure of the fracture surface were examined with a differential calorimeter (Rigaku Co., 8001 CS) and a 120 KV electron microscope (Hitachi Co., 11 E) respectively.

RESULTS AND DISCUSSION

The obtained relationship between the pressure of melting and the weight percentages of folded chain crystals of polyethylene in polyethylene-xylene system (for various melting temperatures)

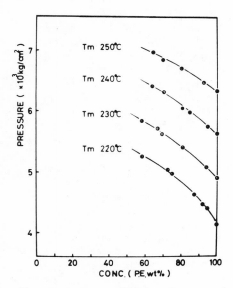

Fig. 3. The effect of concentration of xylene on melting pressure at constant melting temperature.

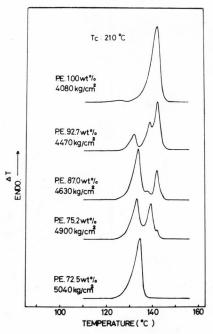

Fig. 4. DSC thermograms of polyethylene crystallized at 210°C from various xylene concentrations. The heating rate was 10° C/min.

is shown in Figure. 3. The crystallization pressures were determined to be 10 degrees below the melting temperatures using Figure 3. The isothermal crystallization curves for various xylene concentrations and for various crystallization pressures are shown in Figure 4. In general increase of pressure accelerates the rate of crystallization at the same temperature of supercooling. In this case, however, the addition of xylene to polyethylene retarded the rate of crystallization and prevailed against the effects of pressure.

The DSC tracing of the crystallized polyethylenes of various polyethylene weight percents at 210 and 230°C are shown in Figure

5 and Figure 6 respectively. In both cases, the heights of characteristic peaks at about 140°C, which indicate the melting of extended chain crystals, were decreased according to the increase of the xylene content in the polyethylene-xylene system, while those of folded chain crystals of polyethylene at 130°C were increased. The peak due to the melting of extended chain crystals of polyethylene disappeared when the xylene content increased nearly to 30% under any crystallization conditions.

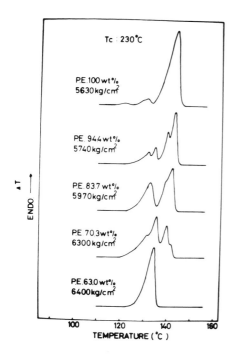

Fig. 5. DSC thermograms of polyethylene crystallized at 230°C from various xylene concentrations. The heating rate was 10°C/min.

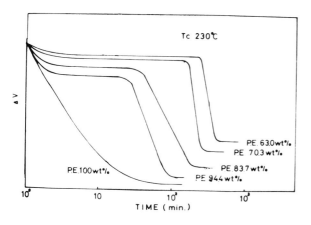

Fig. 6. Crystallization curves at 230°C under various pressures and xylene concentrations.

The change in morphology of polyethylene crystallized under
high pressure at 230°C on addition of xylene is displayed in
Figure 7-9. Figure 7 is the typical structure of the extended
chain crystals of polyethylene, which is similar to those obtained
elsewhere. As shown in Figure 7-9, when the xylene content of
the system was increased, the formation of extended chain crystals
became suppressed and the amount of folded chain crystals increased.
The result of the change of each crystalline amount is in good
accordance with the results of thermal analysis.

Fig. 7. Electron micrograph of
a fracture surface of crystallized form of pure polyethylene
(5630 Kg.cm^2, 230°C).

Fig. 8. Electron micrograph of a fracture surface of crystallized
polyethylene (xylene 22.7 wt %, 5970 Kg/cm^2, 230°C).

Fig. 9. Electron micrograph
of a fracture surface of crystallized polyethylene$_2$(xylene
29.7 wt %, 6300 Kg.cm^2, 230°C

     These observations lead to the supposition that the addition
of xylene to polyethylene prevents the nucleation of extended
chain crystals. However once the nucleus was formed, it grew to
the giant crystal. If the thickening process of the lamellae of
folded chain crystals is essential for the formation of extended
chain crystals, it is difficult to explain that only a little
quantity of xylene prevents the formation of the extended chain
crystal. For the explanation of the result, it is necessary to
consider that the extended chain crystals grow only form the extended chain nucleus.

REFERENCES

1. B. Wunderlich and T. Arakawa, J. Polymer Sci. A-2, 3697 (1964).
2. B. Wunderlich, J. Polymer Sci., A-1, 1245 (1963).
3. B. Wunderlich and T. Davinson, J. Polymer Sci., A-2,2043 (1969).
4. D. V. Rees and D. C. Bassett, Nature 219, 368 (1968).
5. P. D. Calvert and D. R. Uhlmann., J. Polymer Sci., C-8, 165
    (1970).
6. P. W. Bridgman, Phys. Rev. 3, 126 (1914).

THERMAL PROPERTIES OF THE POLYLACTONE OF DIMETHYLKETENE,

A NEW MODEL POLYMER

Edward M. Barrall II, Duane E. Johnson and
Barbara L. Dawson
International Business Machines Corporation
Research Laboratory
Monterey and Cottle Roads, San Jose, California 95193

During the past three decades much of the increase in our knowledge of polymeric systems, their physics and thermodynamics, has been based on a relatively few model polymer materials. These are polymers which can be made to certain specifications, whose molecular weight is easy to determine and whose molecular weight distribution and chemical properties may be controlled within certain predetermined limits. For this reason the number of cases where polystyrene and polyformaldehyde (polymethyleneoxide) have been studied far outweighs their commerical importance. To the polymer physicist a new or previously unrecognized model polymer is always of interest.

In recent studies of polylactones at this laboratory the product of dimethylketene was examined by the usual physical methods: GPC, scanning calorimetry (DSC), microscopy and spectrophotometric methods. The polymer proved to be crystalline with an easily formed glass state, to have relatively narrow molecular weight distribution, to contain reactive end groups and to be relatively tractable in a wide variety of solvents. Upon further study the thermal decomposition of the polymer proved to be controllable and to involve primarily end groups. In general, the qualifications of this polymer fit very well those of a model system. The subject of this report is the gross physical properties and synthesis of the polymer from the dimer of dimethylketene, poly(3-hydroxy-2,2,4-trimethyl-3-pentenoic acid-β-lactone):

$$\text{(I)}$$

EXPERIMENTAL

Synthesis

The synthesis and structure of I have been given by several authors (1-3). The following technique was used for the preparation of I for this study. Into an oven dried glass pressure bottle 75 ml of tetrahydrofuran was distilled from calcium hydride under dry nitrogen. To this fresh solvent was added 2 ml of 1.9M n-butyllithium in n-hexane. A light straw color resulted. Twenty ml (~19 g) of 3-hydroxy-2,2,4-trimethyl-3-pentenoic acid-β-lactone was added to the above solution under nitrogen. A dark yellow solution resulted. This was held at ambient temperature (~23°C). Within three hours a fine precipitate began to form. Polymerization continued overnight. The product, I, was isolated from excess acetone with a yield of 60%. The elemental analysis calculated was carbon, 68.54%; hydrogen, 8.63%; oxygen, 22.83%. The polymer I analyzed as follows: carbon, 68.21%; hydrogen, 8.65%; and oxygen, 23.03%.

The 2,4-dinitrophenylhydrazone of I, used later in this study, was prepared by weighing in 1 g of I and 0.6 g of 2,4-dinitrophenylhydrazine into an Erlenmeyer flask. This mixture was partially dissolved in 50 ml of glacial acetic acid. The mixture was warmed on a steam bath and ~40 ml of chloroform was added to complete solution. The solution was filtered and gave a clear orange solution. The solution was warmed to reflux for two hours and then allowed to stand at ambient temperature for two days. By this time a yellow solution appeared. This was concentrated by gently warming under argon to leave a mixed precipitate. This was washed three times with glacial acetic acid to remove unreacted hydrazine material. The washed precipitate (II) was air dried.

Gel Phase Chromatography

The samples as 2 ml 0.1% solutions in chloroform were chromatographed through a bank of five columns (60 to 2000 Å) in a Water's gel permeation chromatograph. The mobile phase was chloroform at room temperature. The column bank was calibrated with Pressure Chemical's polystyrene standards. The chromatographic data was acquired by an IBM computer data system (System/7) the general outlines of which have been given elsewhere (4). The molecular weights were calculated assuming that the exclusion volume of I was the same as polystyrene. The difference in absolute molecular weights has been accounted for by a simple ratioing of the molecular weight of styrene repeating unit to the repeating ketene unit of I. Obviously, this is not an ideal situation, but the only alternative possible under the instrumental and chemical conditions imposed.

Differential Scanning Calorimetry

Heats and temperatures of transition as well as the glass transition temperature of both I and II were measured on a modified Perkin-Elmer DSC-1B. The temperature and heat axes were calibrated with semiconductor grade gallium, indium and tin. Polymer sample sizes were in the range of 2 to 3 mg and were heated at 10°C/min at sensitivity of 2 mcal/inch of chart. Data was acquired by an IBM System/7. The calibration technique, sample encapsulation (under nitrogen) and system modification are given in detail elsewhere (5-7).

Thermogravimetry

The thermogravimetry in this study was carried out using a duPont 990 thermal analysis module attached to a 951 thermo-gravimetric analyzer (TGA). Samples of about 7 mg were heated at 5°C/min in 20 cc/min flowing nitrogen. The sample pan was made of platinum. The weight loss was recorded both integrally and differentially at 1 mg/inch.

Microscopy

Thermal transition temperatures and the changes in the physical texture of I and II were followed with a Zeiss Photo Microscope III equipped with a Mettler FP52 hot stage and polarizing optics. Such an examination is by no means complete but is intended to furnish a beginning for future studies.

RESULTS

Microscopy

The structure of both I and II between crossed polarizers
was that of a moderately birefringent powder. On heating at
2°C/min from 160°C, polymer I exhibited a change in texture
starting at 174° and melted (viz. flowing and darkening) from
200.6° to 202.2°C. The final melting, as noted by vanishing of
birefringence, was very sharp. On cooling, the sample formed a
glass and did not crystallize between coverslips in two weeks at
room temperature. However, at 180°C crystallization was accomplished in five hours. This gave a finely grained birefringent
texture which vanished sharply at 202.1°C. There was no evidence
of spherulite structure or of long leaf-like plates encountered
in polypropylene and high pressure polyethylene, respectively.
The solid phase appeared much like that of nucleated polypropylene. The melt was subject to flow birefringence while the
top coverslip was displaced. Solids produced by the evaporation
of chloroform solutions of the polymer were amorphous and
non-birefringent.

Polymer II under the same heating program started melting at
194.8°C, birefringence sharply increased at 195.5° and rapidly
vanished at 203.2°C. The phase between 195.5° and 203.2°C had
a distinctively liquid crystalline (nematic) appearance. The
same difficulties were experienced in recrystallization of II
from the melt. A fine birefringent texture was obtained after
annealing the glass at 180°C for five hours. The liquid phase
above 203.2°C was very sensitive to shear and exhibited bright
flow birefringence around bubbles.

Thermogravimetry

The thermogravimetric (TG) trace of polymer I weight loss as
a function of temperature in dry nitrogen is shown in Figure 1.
A small weight loss (~1% of the sample) starts at 100°C and is
concluded at 150°C. This is probably due to sorbed water and
solvent. The loss vanished from samples dried for two days at
80°C at 10^{-4} Torr. The sample abruptly decomposes at 242°C to
lose 89.5% of its initial weight by 280°C. Above 280°C this
final residue pyrolizes completely in nitrogen by 325°C. The
general shape of the TG trace is characteristic of an
autocatalytic reaction.

The TG trace of the hydrazone capped polymer (compound II)
is also shown in Figure 1 and is substantially different from I.
The onset of weight loss starts at 285°C and is concluded in a
single step by 405°C to yield no residue. The rate is much

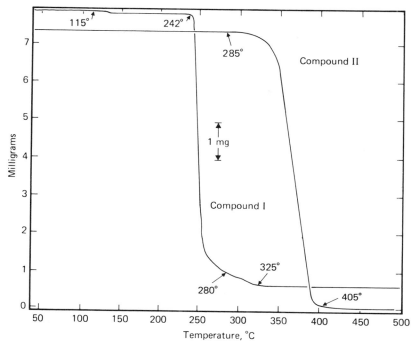

Figure 1. Thermogravimetric trace of Compounds I (7.46 mg) and II (7.410 mg) heated at 5°/min in nitrogen. (Compound I has been displaced upward by 0.6 mg for this illustration.)

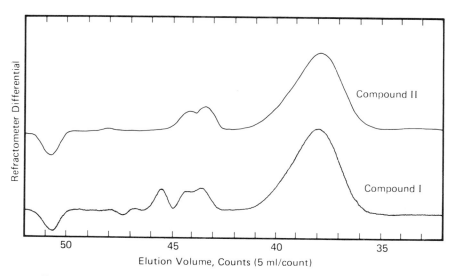

Figure 2. Gel Permeation Chromatogram of Compounds I and II.

slower. Such behavior is obviously indicative of the mode of decomposition of I. By elimination of the terminal ketone group (at two per molecule since other oxygens are carboxylic in nature) with 2,4-dinitrophenylhydrazine, a cyclic decomposition mechanism beginning at chain ends is made impossible. The only route open requires a much higher temperature process of carbon-oxygen or carbon-carbon cleavage probably by some random process. The hydrazone bond is fairly stable since many high molecular weight hydrazones may be distilled above 300°C.

The capping reaction to form II must have been reasonably complete since no trace of a two step weight loss is seen in Figure 1 for II. In a separate experiment the two polymers were mixed and a two step weight loss curve was obtained with the steps proportional to the ratio of I to II. Certainly the TG data are a good example of the role of terminal functionality on some types of polymer thermal degradation.

Molecular Weight Determination

Gel chromatography of I and II produced normal distributions with no particular skewing, see Figure 2. If we assume that the hydrodynamic radius of polystyrene is comparable to that of I and II and adjust for the molecular weight difference in the repeating unit, the data shown in Table I are obtained.

Table I

GPC Results on Uncapped and Capped Polymer

Compound	Molecular Weight			Polydispersity
	$\overline{M}_n$	$\overline{M}_w$	$\overline{M}_z$	$\overline{M}_w/\overline{M}_n$
I	17900	27400	37000	1.53
II	16900	31800	50100	1.88

The slight shift in the distribution for II may be due to some fractionation during the reaction combined with degradation on drying. Indeed, the derivative, II, exhibits less low molecular weight material than the parent polymer, I, note missing peak in the chromatogram of II near count 47.4. Although these materials do not enter into the molecular weight calculation, the absence of a trace component does indicate a cleaner sample.

Since it is possible to cap I with a chromophor group, 2,4-dinitrophenylhydrazine, it should be possible to determine the molecular weight spectrophotometrically Essentially, this is a determination of the number of chain ends and requires calibration with a phenylhydrazone derivative of known molecular weight. In addition it must be assumed that the optical extinction coefficient of the calibration compound is close to that of the capped form of I, i.e., II. The UV absorption spectrum of I is shown in Figure 3 along with that of II. Compound II has an intense absorption centering around 360 mμ. These spectra were obtained on a Cary Ultraviolet absorption spectrophotometer with chloroform as solvent.

For calibration purposes the 2,4-dinitrophenylhydrazone of 3-heptanone was prepared and purified. Scanning calorimetry indicated that the material was 99.6% pure. A stock solution containing 0.00354 g/10 ml was made up in spectral grade chloroform. This was diluted as follows: 1, 2, 3, 4 ml stock made to 100 ml. These dilutions were scanned and produced an analytically useful absorption at 365 mμ, see Figure 3. Peak height in terms of optical density was plotted as a function of moles of 2,4-dinitrophenyl hydrazone in the usual manner. This produced a linear plot which passed near the origin. The chloroform solution of II contained 0.05018 g of polymer and the peak height indicated from the calibration plot that there were 3.92×10^{-6} moles of hydrazone present. This is equivalent to twice the number of chains (two reactive groups per molecule). Thus, in 0.05018 g of polymer there are 1.96×10^{-6} moles of chain ends. This gives a molecular weight of 25600. This is a number average molecular weight. This figure is substantially different from the GPC determined number of 16900 to 17900, see Table I. This difference can be traced to one of two sources: (1) the extinction coefficient of the 2,4-dinitrophenylhydrazine moiety in II is significantly different from the same group in the 3-heptanone derivative; and (2) the hydrodynamic volume or radius of gyration of I and II is significantly smaller than polystyrene of equivalent weight. Item 1 could be answered quickly if a good nitrogen analysis could be obtained. However, II would contain only 0.438% nitrogen by calculation. There are specialized methods which certainly could be applied, but these are unavailable to the present authors. Item 2 is susceptible to attack by viscosity detector GPC and will be pursued (8). That technique does not require narrow fractions of the polymer or a knowledge of the Mark-Houwink constant of the polymer so long as a plot of η log M versus elution volume is known for narrow fractions of polystyrene.

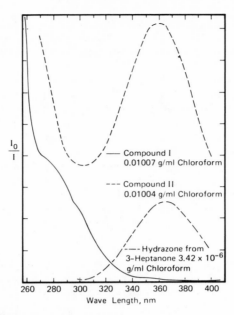

Figure 3. Ultraviolet Absorption Spectrum of Compounds I and II.

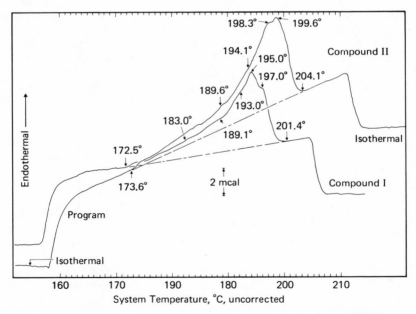

Figure 4. Differential scanning calorimetry traces of Compounds I (4.004 mg) and II (4.108 mg) heated at 2°/min.

Calorimetry

Differential scanning calorimetry clearly indicates the crystallinity seen under the microscope. Melting of the polymer I takes place over a broad range, from 172.5° to 201.4°. These temperatures are in good correlation with events noted in the microscope (the two instruments were cross calibrated with the melting of benzoic acid and anthracene as well as internally calibrated with more customary hot stage microscopy and DSC standards). The shape of the DSC curve in Figure 4 is irregular and appears to consist of a number of overlapping peaks. Certain features proved to be highly reproducible such as the shoulders at 183°, 189.1°, 193° and 197.0° and the onset, vertex (195.0°) and conclusion temperatures. These features are probably due to melting of lamellae of various sizes and not to some solid effect. Since the polymer does not recrystallize rapidly the series of meltings and recrystallizations encountered in polyethylene (9) or polypropylene (10) does not occur to a great extent.

The heat of fusion of polymer I is 9.45 cal/g. This number probably does not represent the maximum value obtainable on more careful precipitation or annealing. The value is certainly within the expected range for a polymer of this molecular weight and structure. Upon cooling at 5°/min in the DSC pan the polymer does not show any appreciable crystallinity on reheating. A glass transition of very large shift in heat capacity appears between 70° and 95°C. This second order transition also appears in the crystalline sample over the same range but 100 times smaller in magnitude. This indicates that a portion of the crystalline sample was amorphous. If the glass transitions were due to a configurational change possible in the crystalline phase then quenching should have little effect on its magnitude by analogy to the lower glass transition in polypropylene (11). This appears to be relatively insensitive to the sample crystallinity. Considering the polarity of the polymer, this transition is probably analogous to those of poly(ethylene terephthalate) and Nylon 66.

Capping of polymer I with 2,4-dinitrophenylhydrazine (II) appears to have little effect on the melting properties of the polymer. The thermogram of II is shown in Figure 4. The onset occurs at 173.6°, the maximum at 199.6° and ending at 204.1° with shoulders at 189.6, 194.1°, and 198.3°C. The heat of fusion is somewhat higher, 11.2 cal/g. All of these effects may be due to alterations in the solid state caused by recrystallization from the reaction mixture. However, chain end effects cannot be ruled out. The polymer II exhibits a glass transition over the same range as I and shows the same lack of recrystallization on cooling.

CONCLUSIONS

From the above data it is apparent that I is a reasonably high molecular weight polymer of relatively narrow polydispersity. In the solid state it may either exist in the crystalline or glassy state. By alteration of the end groups the decomposition temperature and probably the decomposition mechanism may be controlled. Using 2,4-dinitrophenylhydrazine modification of the end groups appears to be analytically complete. The number average molecular weight may be determined with reasonable accuracy by spectrophotometry (II). In general I appears to have many of the attributes required by a model polymer system.

LITERATURE CITED

1.   R. H. Hasek, R. D. Clark, E. U. Elam, and J. C. Martin, J. Org. Chem. 27, 60 (1962).

2.   H. Ohse and H. Cherdron, Makromol. Chem. 97, 139 (1966).

3.   H. Staudinger, F. Felix, P. Meyer, H. Hander, and E. Stirnemann, Helv. Chim. Acta 8, 322 (1925).

4.   A. R. Gregges, B. F. Dowden, E. M. Barrall II, and T. T. Horikawa, Separation Sci. 5, 415 (1970).

5.   E. M. Barrall II, Thermochim. Acta 5, 377 (1973).

6.   E. M. Barrall II and R. Diller, Thermochim. Acta 1, 509 (1970).

7.   E. M. Barrall II and B. Dawson, Modification of a DSC-1B Scanning Calorimeter, Thermochim. Acta, in press.

8.   A. C. Ouano, J. Polymer Sci., A-1, 10, 2169 (1972).

9.   A. P. Gray and K. Casey, Polymer Letters 2, 381 (1964).

10. T. W. Huseby and H. E. Bair, Polymer Letters 5, 265 (1967).

11. D. L. Beck, A. A. Hiltz, and J. R. Knox, SPE Trans. 279 (1963).

MELTING BEHAVIOR OF SOME OLIGOMERS OF HETEROCYCLIC POLYMERS

BY DIFFERENTIAL SCANNING CALORIMETRY

Hirotaro Kambe and Rikio Yokota

Institute of Space and Aeronautical Science

University of Tokyo, Komaba, Meguro-ku, Tokyo, Japan

ABSTRACT

The softening point of polybenzimidazole involving aromatic ether linkage in the main chain is not measurable below degradation of the polymer. The various oligobenzimidazoles and benzothiazole and benzoxazole derivatives with analogous structures were synthesized and their melting point Tm and heat of fusion ΔH were measured by differential scanning calorimetry. The entropy of fusion ΔS was calculated by $\Delta S = \Delta H/Tm$. The accuracy of the method was examined and the effect of imidazole ring on the melting behavior was investigated.

Hydrogen bonds owing to N-H bond of imidazole rings enhance ΔH markedly compared to oxazole and thiazole rings. It causes the higher melting point of imidazole derivatives. The values of ΔH and ΔS for benzimidazole dimer are smaller than for corresponding oxazole dimer and it suggests that a kind of structure is remaining even after melting of imidazole dimers. Along with the higher ΔH the oligobenzimidazole having diphenyl ether linkage shows a large ΔS due to molecular mobility by the internal rotation around ether linkages.

INTRODUCTION

The thermal stability of the polymer is evaluated by softening and degradation temperatures. Aromatic or heterocyclic polymers are generally known to show excellent characteristics in these properties. They do not soften and retain their rigidity up to high temperatures, also show higher decomposition temperatures, and do not exhibit any significant weight loss at higher temperatures. The glass transition temperature Tg and the melting point Tm are useful as respective measures of softening temperature for amorphous and semi-crystalline polymers. The Tm is expressed thermodynamically by the ratio of enthalpy ΔH and entropy ΔS of melting, i.e. $Tm = \Delta H / \Delta S$. The relation of Tm with ΔH or ΔS has been investigated for some polymers. However, for the aromatic polymers having rigid ring structures the degradation appears below the softening temperature; therefore, Tm for these polymers are difficult to be measured.

The polybenzimidazole having diphenyl ether linkages in the main chain is known to show a high thermal stability. In the present paper, the oligomers having the fundamental molecular structure of this polymer have been synthesized and melting behaviors were investigated by differential scanning calorimetry (DSC). The benzoxazole and -thiazole derivatives with analogous structures were compared with benzimidazoles. The effect of imidazole structure on melting behavior is discussed on the relation of Tm with ΔH and ΔS and compared with those of oligophenyls and oligoacenes.

EXPERIMENTAL

Materials

Benzimidazole, benzoxazole, and benzothiazole were commercial products. Derivatives of benzazoles were synthesized in polyphosphoric acid by the condensation of methyl benzoate with diaminobenzene, aminophenol, and aminothiophenol or diamino-, dihidrooxy-, and dihydrothiobenzidine, as follows:

where X' = NH_2, OH or SH, and X = NH, O or S.

Dimethoxycarbonyl phenylene oxide-based benzimidazole (DMOB) was synthesized as follows:

Samples were purified by recrystallization and/or sublimation.

Procedures

The measurements of Tm and ΔH were carried out with Perkin-Elmer DSC-I, at heating rates of 8 and 16 °C/min. The liquid cell was used for preventing evaporation of the sample. The several mg of powdered samples were measured in duplicate. The melting point Tm was determined at the top of melting peak. Temperature scale was calibrated in each range. ΔH was measured from the area of the peak, calibrated by melting peak of Sn at high temperature range and by that of Hg at low temperature range. ΔS was calculated from $\Delta S = \Delta H/Tm$.

Accuracy and Precision

The accuracy and precision of the calorimetry by DSC were examined by several standard materials. The precision of heat of fusion for metallic   tin   and mercury were within 3 %, and considered sufficiently good. Table I shows data for o-terphenyl sample supplied by ICTA Committee on Standardization. The data for a stable low-molecular compound gives ΔH and ΔS with an error below several %, and Tm with an accuracy of ± 0.5 °C in comparison to the

TABLE I

Melting Behavior of o-Terphenyl (ICTA sample).
Tm: 55 ± 0.1 °C (A. R. Ubbelohde)

Experiment	$\dfrac{Tm}{°C}$	$\dfrac{\Delta H}{\text{kcal. mol}^{-1}}$	$\dfrac{\Delta S}{\text{cal. mol}^{-1}\text{K}^{-1}}$
E-1-1	54.7	3.96	8.14
-2	55.0	3.88	7.98
-3	55.1	3.97	8.17
E-2-1	56.0	3.83	7.86
-2	55.7	3.94	8.07

TABLE II

The Effect of Purity for Anthracene Samples.

Purification	$\dfrac{\Delta H}{\text{kcal. mol}^{-1}}$	$\dfrac{\Delta S}{\text{cal. mol}^{-1}\text{K}^{-1}}$
Sublimation	5.3	11.0
Zone Melting	6.4	13.2

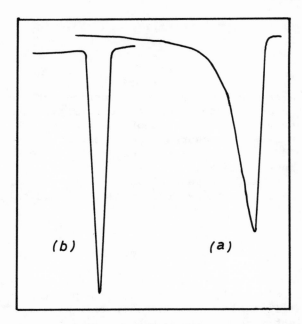

Figure 1   Melting Peak for Differently Purified Anthracene
           (a) Sublimed, (b) Zone-Melted

data by Ubbelohde[1].

     The effect of purity was examined for two samples of anthracene
purified by sublimation and by zone melting. They gave melting
curves shown in Figure 1 and the data in Table II. The melting
peak for zone-melted sample is very sharp, but that for the sublimed one is rather broad. The accurate peak area for the latter

TABLE III

The Comparison with Reference Data

Material	T_m °C	ΔH kcal. mol^{-1}	ΔS cal.mol^{-1}K^{-1}	Reference
Benzene	6	2.4	8.6	JCS,Chem.Tables[2]
	5.5	2.2	7.9	Present Data
Naphthalene	80.2	4.5	13	Westrum[3]
	80.1	4.1	11.7	Present Data
	80.1	4.5	12.7	Present Data
Anthracene	217	7.0	14.4	Westrum[3,4]
	216	6.21	12.7	Present Data
	216	6.49	13.3	Present Data

case is difficult to be measured, because of the difficulty of determining the base line. The precision of the melting temperature is with 0.5 °C.

The data obtained for typical compounds the present method are compared in Table III with the calorimetric data in reference. The present data gave a little lower values than those in reference.

Results and Discussion

The DSC curves obtained for samples are shown in Figure 2, and the data are summarized in Table IV. The values are reproducible within 10 %.

The Tm for benzimidazole (BI) is over 150 °C higher than those for benzoxazole (BO) and benzothiazole (BT). The ΔS's for them are about 10 cal.mol^{-1}K^{-1}, but ΔH for BI is much larger than for others. This is the reason that Tm for BI is much higher than for others. For the compounds introduced a phenyl group to these structures, PBI, PBO and PBT, ΔS's are not different from each other. However, ΔH for PBI is much larger than for the others, as for the basic compounds. This also causes the high Tm for PBI. With these two series, the effect of intermolecular interactions is significant, and particularly the hydrogen bonding by N-H in benzimidazole ring contributes characteristically to the high ΔH for BI derivatives.

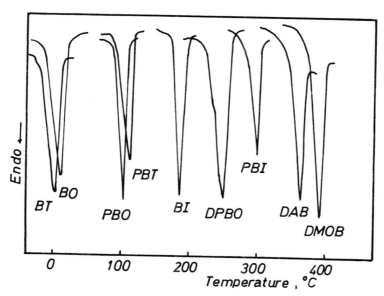

Figure 2   DSC curves for Benzazole Derivatives

TABLE IV

Melting Properties of Benzazole Derivatives

Symbol	Structure	M.w	$\frac{Tm}{(k)}$	$\frac{\Delta H}{\text{kcal.mol}^{-1}}$	$\frac{\Delta S}{\text{cal.mol}^{-1}K^{-1}}$
BI		118	443	4.59	10.4
BO		119	303	3.22	10.6
BT		135	273	2.78	10.2
PBI		194	568	5.57	9.89
PBO		195	375	4.22	11.2
PBT		211	386	4.48	11.6
DAB		386	620	7.23	11.7
DPBO		388	520	9.59	18.4
DMOB		402	653	13.1	19.9

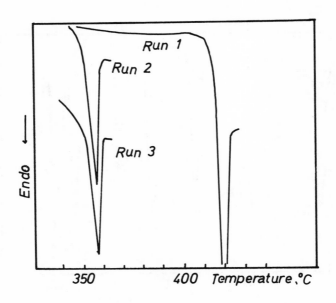

Figure 3   DSC curves for DAB.   Solid phase
transition of DAB at the first run[5]

DAB and DPBO in Table IV correspond to the dimers of PBI and PBO, respectively.

DAB showed a markedly high-temperature transition at the first scanning[5]. After the second run, it presented an ordinary reproducible peak, which was measured at the determination of the heat of melting. The data for this compound are much dispersed, and the precision for this material may involve the error over 30～40 %. Oxazole derivative DPBO shows larger ΔS, as for BO and PBO, due to its freer molecular motion. ΔH for DPBO is about twice as large for the monomer PBO. On the other hand, imidazole derivative DAB shows ΔH only 1.5 times as large for PBI and ΔS increases a little. The very high T_m for DAB is due to this small ΔS. It suggests in the molten state of this compound there remains some structure, owing to its restricted molecular motions.

Dimethoxycarbonyl phenylene oxide-based benzimidazole (DMOB) has a similar dimension as DAB, but it shows free rotation which makes the chain flexible. Therefore, it show large ΔS, while its ΔH is also large owing to the hydrogen bonding by N-H in imidazole rings. Then, DMOB shows higher T_m than that for DAB.

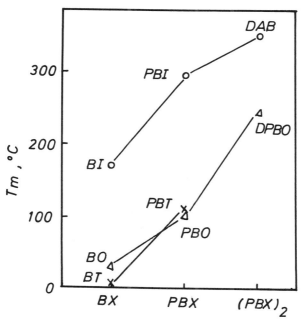

Figure 4   Melting Points of Benzazole Derivatives

In Fig. 4 are plotted T_m's for these compounds. The contribution of imidazole ring to the higher T_m is apparent in the figure. The behaviors of oxazole and thiazole rings are similar with each other.

REFERENCES

1) A. R. Ubbelohde, Proc. Roy. Soc. London, 288, 435 (1955)
2) Chemical Tables, Chem. Soc. Japan, Part 1, Table 7.47 (1966)
3) P. Goursot, H. L. Girdhar, and E. F. Westrum, Jr., J. Phys. Chem., 74, 2538 (1970)
4) W.-K. Wong and E. F. Westrum, Jr., J. Chem. Thermodyn., 3, 105 (1971)
5) H. Kambe, I. Mita, and R. Yokota, Thermal Analysis (Ed. H. Wiedemann), Birkhäuser Verlag, Basel, Vol. 3, p387 (1972)

THERMOGRAVIMETRIC ANALYSIS OF POLYMETHYLMETHACRYLATE AND POLYTETRAFLUOROETHYLENE

JAMES A. CURRIE AND N. PATHMANAND

VILLANOVA UNIVERSITY

VILLANOVA, PENNSYLVANIA 19085

Kinetics is of fundamental importance in the understanding of the mechanism of thermal decomposition of high polymers. Since most thermal decomposition reactions are associated with weight changes, thermogravimetry is being employed to a great degree in kinetic studies of polymer decomposition. Two approaches are commonly used to obtain the various kinetic parameters of these reactions. The static method or "isothermal thermogravimetry" produces precise weight loss curves for various single temperatures as a function of time. Thus the rate parameters are obtained directly. This method is inherently time consuming and the results are sometimes questionable due to experimental uncertainies arising from the initial thermal lag of the sample. More recently, "dynamic thermogravimetry" data has been used for kinetic studies. In these, a single weight loss curve obtained by means of programmed heating is used to provide information equivalent to an entire family of isothermal weight loss curves. The new approach is time saving, being able to cover a wide range of temperatures in a relatively short time. For this work the kinetics evaluation procedure by Freeman and Carroll[1] and its extension by Anderson and Freeman[2], is applied to dynamic thermogravimetric data in order to evaluate the rate parameters. Activation energy and order of reaction for the thermal decomposition of PMMA and PTFE are determined and discussed. The experiments were conducted in an inert atmosphere of nitrogen using several programmed temperature rates for heating. The proposed mechanisms for the decomposition reactions are used to interpret the experimental results and correlate with previous work on these materials.

THEORY

Kinetic Studies

Considerable attention has been recently directed toward the exploitation of thermogravimetric data for the determination of kinetic parameters. Many of the methods of kinetic analysis which have been proposed are based on the hypothesis that, from a single thermogravimetric trace, meaningful values may be obtained for parameters such as activation energy, pre-exponential factor, and reaction order. Thus, many using these methods make two assumptions, viz., these parameters are useful in characterizing a particular polymer degradation reaction and that the thermogram for each particular set of these parameters is unique.

The three most notable methods of kinetic analysis of thermogravimetric data are "integral" methods utilizing weight loss versus temperature data directly, "differential" methods utilizing the rate of weight loss, and "difference differential" methods involving difference in weight data. All of the above are based on the assumption of the reaction rate dependence on concentration being

$$\frac{dC}{dt} = - kC^n \qquad\qquad (1)$$

where C = concentration, mole fraction or amount of reactant,
         k = specific rate constant,
         n = order of reaction,
         t = time,
It is generally assumed that the Arrhenius equation is applicable: viz.,

$$k = A \exp (-E/RT) \qquad\qquad (2)$$

where A = frequency factor,
         E = energy of activation (kcal/mole),
         R = gas constant (1.987 cal/mole°K),
         T = absolute temperature (°K).

The "integral" methods have been applied by many others [3-7] and are not detailed here except to note that when applied to a single thermogravimetric trace a "best value" of activation energy and order of reaction are inevitably fitted to the data whether or not these parameters have any significance or even utility in the understanding of the mechanism. In addition, to obtain the activation energy, the order must be known or visa versa. "Differential" methods based on the rate of weight loss versus temperature data have been devised[8,9] which are much simpler in application, and in some cases, are able to circumvent difficulties found in many integral methods. However, they suffer from an inherent weakness, the magnification of experimental scatter often rendering their application to experimental data diffucult. Finally, the "difference differential" method [1,2] previously noted is becoming a widely accepted method for the kinetic analysis of thermogravimetric data. It has been applied to both the investigation of inorganic materials and high polymers. It is this method with which we are most concerned and details of its principle and use follow.

Consider a reaction in the solid state, where one of the products B is volatile, and all other substances being in the condensed state, viz.,

$$aA(s) = bB(g) + c\ C(s) \tag{3}$$

then, solving for k in equation (1) and substituting into equation (2) gives

$$A\ \exp(-E/RT) = \frac{(dC/dt)}{C^n} \tag{4}$$

The logarithmic form of equation (4) is differentiated with respect to, dC/dt, n, and T, resulting in the equation

$$\frac{E}{R}\frac{dT}{T^2} = d\ \ln\ (dC/dt) - N d\ln\ C. \tag{5}$$

Integrating the above relationship gives

$$-\frac{E}{R}\ \Delta(1/T) = \Delta\ln\ (-dC/dt) - n\Delta\ln C. \tag{6}$$

Dividing (5) and (6) by d lnC and ΔlnC, respectively, one obtains equation (7) and (8).

$$\frac{E}{RT^2}\frac{dT}{d\ln C} = \frac{d\ln\ (-dC/dt)}{d\ln C} - n \tag{7}$$

$$\frac{-E/R\ \Delta(1/T)}{\ln\ C} = \frac{\Delta\ln(-dC/dt)}{\ln C} - n \tag{8}$$

From equations (7) and (8), it is apparent that plotting of

$$\frac{dT}{T^2\ \log C}\ \text{versus}\ \frac{d\ \log(-dC/dt)}{d\ \log\ C}$$

and

$$\frac{\Delta(1/T)}{\log\ C}\ \text{versus}\ \frac{\Delta\ \log(-dC/dt)}{\Delta\ \log\ C}$$

should result in straight lines with slopes of $\pm$ E/2.3R and an intercept of -n.

We must consider three cases where C refers respectively to mole fraction of A, molar concentration of A, and amount of reactant A.

Case 1. Mole fraction of A;
$$C = x_a/M = N_A$$

where, x_a = number of moles of A at time t,

M = total number of moles in reaction mixture.

(a) Total Number of Moles is Constant During Reaction

Substituting for C in equation (4) results in the relationship

$$\ln k = \ln M^{n-1} + \ln(dx_a/dt) - n \ln x_a \qquad (9)$$

and

$$\frac{-E/R \; \Delta(1/T)}{\ln x_a} = -n + \frac{\Delta\ln(-d \; x_a/dt)}{\Delta\ln \; x_a} \qquad (10)$$

Equation (10) may also be written in differential form as equation (5).

(b) Total Number of Moles is not Constant

For this case,

$$\ln k = (n-2) \ln M - n\ln x_a + x_a(dM/dt) - M(dx_a/dt) \qquad (11)$$

with

$$\frac{E/RT^2 \; dT}{d(\ln M - \ln x_a)} = n + \frac{d\ln x_a(dM/dt) - M(d \; x_a/dt) - 2 \; d\ln M}{d(\ln M - \ln x_a)} \qquad (12)$$

and

$$\frac{E/R \; \Delta(1/T)}{\ln M - \ln x_a} = n + \frac{\Delta\ln x_a(dM/dt) - M(dx_a/dt) - 2 \; \Delta\ln M}{\Delta \; d(\ln M - \ln x_a)} \qquad (13)$$

Case 2. Molar Concentration of A;

$$C = x_a/V$$

where, V = volume of reaction mixture, the equations which result are identical to the case of mole fraction with the exception that V replaces M.

Case 3. Amount of Reactant A

$$C = x_a$$

For this case,

$$\ln k = -n\ln x_a + \ln(-dx_a/dt). \qquad (14)$$

The final equation is then identical to equation (10). The above relationships may be applied to simplify measurements of weight or volume changes by the appropriate substitutions for M and x_a. The following expression gives the relation between the decrease in the number of moles of reactant and the rate of weight loss:

$$\frac{-d \; x_a}{dt} = \frac{x_o}{w_c} \frac{dw}{dt} \qquad (15)$$

where, x_o = initial number of moles of A,

w_c = total weight loss at completion of the reaction,

dw/dt = rate of change of weight.

W_r, the weight remaining, is proportional to the amount of reactant and is defined as

$$W_r = w_c - w \qquad (16)$$

where w = weight loss up to any given time corresponding to the point at which (dw/dt) is taken.

Figure 1. illustrates the relationship between (dw/dt), w, w_c, and W_r. Combining equations (15) and (16), equation (17) is obtained

$$\Delta\log(dw/dt) = n\cdot\Delta\log W_r - (E/2.3R)\Delta(1/T) \qquad (17)$$

To evaluate the constants in equation (17), $\Delta\log(dw/dt)$ may be plotted against $\Delta\log W_r$ if $\Delta(1/T)$ is kept constant. The order of reaction, n, can be determined from the slope and the energy of activation, E, from the intercept at $\Delta\log W_r = 0$.

EXPERIMENTAL

Materials

The Teflon^R sample was a TFE-7A type obtained through the courtesy of the DuPont Co. The PMMA sample was obtained from Polymer Science, Inc. This sample was reported to have $\overline{M}_n \simeq 5\text{-}6 \times 10^5$. Both materials were in powder form. They were dried in vacuo at 50°C for three hours before use. Sample weights were ~3-3.5 mg for the thermobalance.

Thermogravimetry Technique

The thermogravimetric experiments were performed on the Perkin-Elmer TGS-1 thermobalance accessory equipped with a Cahn RG electrobalance. A Curie-point method of instrument temperature calibration and a time derivative computer which gives the rate of weight loss of the sample directly was also employed. Several programmed heating rates were used to determine the effect of heating rate on the weight loss behavior. The materials were examined in an inert atmosphere of nitrogen at one atmosphere pressure. The heating rates used were nominally 20, 10, 5, and 2.5°C/min. The TGS-1 furnace, capable of operating up to 1000°C, acts as both heater and temperature sensor. In the temperature sensing mode, it forms one side of a bridge circuit, the other side being driven by the output voltage from the program potentiometer in the DSC control unit. An error signal proportional to the temperature error is developed and fed into an amplifier. In the heating mode, the amplifier output is connected to the heater by means of a SCR that delivers 60 Hz power pulses designed to correct the temperature error.

The operation of the electrobalance is based on the nullbalance principle, which is generally accepted as being the most accurate and reliable method of measurement. When the sample weight changes, the balance beam tends to deflect momentarily. The flag moves with it, changing the light input to the phototube and thus the phototube current. This current is amplified in a two-stage servo amplifier and the amplified current is applied to a coil attached to the beam, which is in a magnetic field. The current in the coil acts like a d.c. motor, exerting a force on the beam to restore it to the original balance position. Thus, the change in e.m.f. is equal to the change in sample weight. Two modes of operation were simultaneously recorded on a two-pen Texas Instruments recorder. In the integral

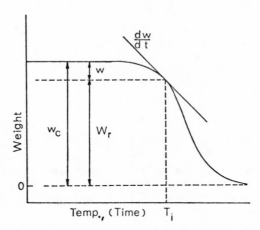

FIGURE 1. The Relationship between (dw/dt), W_c, W and W_r. Weight
          vs Temperature, (Time).

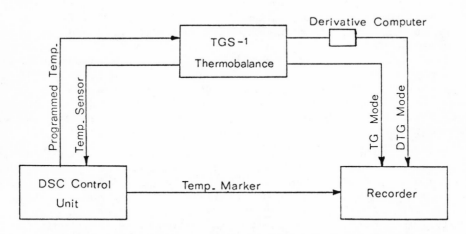

FIGURE 2. Block Diagram of the Thermobalance System.

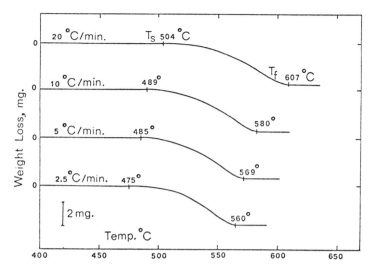

FIGURE 3.   Thermogravimetric Curves for the Thermal Decomposition
of Teflon in an Inert Atmosphere of Nitrogen at One
Atmosphere Pressure.

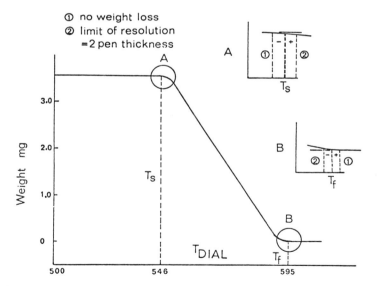

FIGURE 4. Method for Determining T_s and T_f.   Sample Teflon Heated
at 10°C/min.

or TG mode, the recorder traces out a record of the change in weight
of the sample. In the differential or DTG mode, the record shows the
rate of change of weight of the sample during a temperature scan.
The output signal of the Cahn RG electrobalance is in the 0-1 mv.
range. Figure 2 shows the block diagram of the thermobalance system.

RESULTS AND DISCUSSION

Teflon^R

Figure 3 shows the thermogravimetric curves for Teflon^R redrawn
from the actual recorder charts. These curves were obtained by averaging the values of weight change of two identical runs at each of
the respective heating rates (20, 10, 5, 2.5^{o}C/min). Shown on the
figure are the temperatures at which the decomposition reactions
started, T_s, and ended, T_f, for each heating rate. The method for
determining these temperatures is shown in Figure 4. The reaction
duration times where calculated from this data by taking the ratio
of the temperature difference (T_f-T_s) to the actual heating rate.
These results are listed in Table 1. The plots of T_s and T_f versus
actual heating rate are shown in Figures 5 and 6 respectively. Upon
linear extrapolation to zero heating rate, T_s (OHR)and T_f(OHR) were
found to be 474°C and 554°C respectively. The temperature difference
between T_f(OHR) and T_s(OHR) is 80°C for Teflon^R. The T_s(OHR) temperature indicates that in a hypothetical isothermally controlled
system, Teflon^R will start to decompose at 474°C. The interpretation of T_f(OHR) is less clear in that Teflon^R will undergo degradation at varying rates for any temperature above 474°C and the
significance of 554°C is restricted to its use in the dynamic experiments. It was found that T_s and T_f became higher with an increase in
the rate of heating, but the reaction temperature range or (T_f-T_s)
is relatively constant. The only significant difference in the
temperature range is at the 20°C/min. heating rate. The effect of
heating rate may be more clearly demonstrated by Figure 7 where the
derivative of the weight loss with respect to temperature is redrawn from the actual DTG curves. At the higher heating rates, the
maximum rate of the reaction, located at the peak in the curve, is
higher than at the lower heating rates. The maximum decomposition
rate is 0.29 mg/min at 553°C, 0.48 mg/min at 562°C, 0.88 mg/min at
572°C, and 1.36 mg/min at 596°C, for the heating rates of 2.5, 5,
10, and 20°C/min respectively.

The kinetics of the decomposition are evaluated by the method
discussed in the Theory section. The method involves the use of a
single weight loss curve obtained during the experiment where the
temperature is continuously increased. For precision in the results
two or more identical runs were always conducted on identical samples
for each heating rate.

Figure 8 shows the weight loss,W_r,and the rate of weight loss
(dw/dt) versus reciprocal absolute temperature for the thermal

TABLE 1

Temperature at which the Decomposition Reaction Started (T_s) °C and Ended (T_f) °C. Also Reaction Duration Time in min. for the Thermal Decomposition of Teflon at the Heating Rate of 20, 10, 5 and 2.5 °C/min. At One Atmosphere Pressure, Nitrogen Gas Flow Rate 40 cc/min.

Heating Rate (HR) °C/min		Temperature at which Decomposition Reaction Started (T_s) °C	Temperature at which Decomposition Reaction Ended (T_f) °C	ΔT ($T_{f\,°C} - T_s$)	Reaction Duration Time (min.)
Norminal	Actual				
2.5	2.3	475 ± 1.5	560 ± 1.0	85 ± 2.5	85/2.3 = 37.0
5.0	4.2	485 ± 2.0	569 ± 1.0	85 ± 3.0	85/4.2 = 20.2
10.0	8.5	489 ± 2.0	580 ± 2.0	91 ± 4.0	91/8.5 = 10.7
20.0	17.5	504 ± 2.0	607 ± 1.5	103 ± 3.5	103/17.5 = 5.9

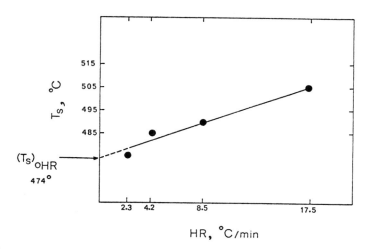

FIGURE 5. Temperature at which the Decomposition Reaction Started (T_s) °C vs Heating Rate °C/min. Sample Teflon in an Inert Atmosphere of Nitrogen at One Atmosphere Pressure.

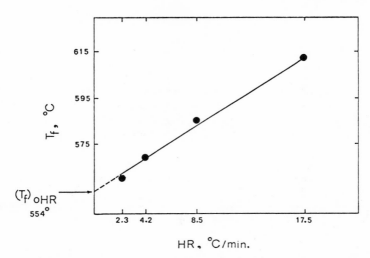

FIGURE 6. Temperature at which the Decomposition Reaction Ended (T_f) °C vs Heating Rate °C/min. Sample Teflon in an Inert Atmosphere of Nitrogen at One Atmosphere Pressure.

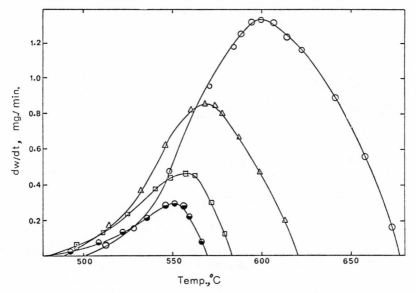

FIGURE 7. The Effect of Heating Rate on the Rate of the Thermal Decomposition of Teflon as a Function of Temperature 20°C/min=0, 10°C/min=△ 5°C/min= ▢ , 2.5°C/min = ◑.

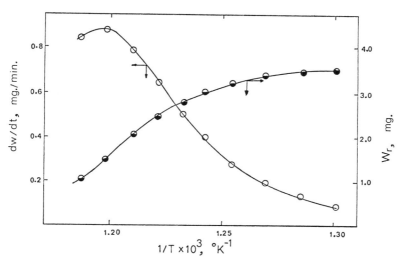

FIGURE 8. Thermal Decomposition of Teflon as a Function of 1/T.
Heating Rate 10 $^{\circ}$C/min. ● W_r, mg vs 1/T; O(dw/dt), mg/min. vs 1/T.

TABLE 2

Data Taken from Figure 8. for the Thermal Decomposition of Teflon
over a Constant Interval of (1/T).

	dw/dt mg/min	log(dw/dt)	W_r mg	log W_r
1.	0.0923	−1.035	3.480	0.542
2.	0.1347	−0.870	3.436	0.536
3.	0.1911	−0.720	3.352	0.525
4.	0.2700	−0.570	3.208	0.506
5.	0.3774	−0.423	3.040	0.483
6.	0.5115	−0.291	2.792	0.446
7.	0.6475	−0.189	2.440	0.387
8.	0.7873	−0.104	2.032	0.308
9.	0.8800	−0.056	1.560	0.193
10.	0.8490	−0.071	1.040	0.017

TABLE 3

Difference Table for Parameters Used to Evaluate the Rate Equation
for the Thermal Decomposition of Teflon.

Δ log	(dw/dt)	Δ log W_r
1-2	-9.165	0.006
2-3	-0.150	0.011
3-4	-0.150	0.019
4-5	-0.147	0.023
5-6	-0.132	0.037
6-7	-0.102	0.059
7-8	-0.085	0.079
8-9	-0.048	0.115
9-10	+0.015	0.176

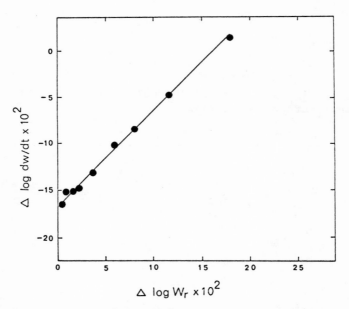

FIGURE 9. The Kinetics of the Thermal Decomposition of Teflon in an
Inert Atmosphere of Nitrogen at One Atmosphere Pressure. Heating
Rate 10 °C/min.

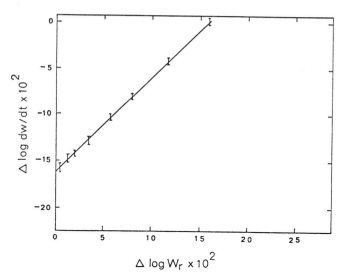

FIGURE 10. The Kinetics of the Thermal Decomposition of Teflon in an Inert Atmosphere of Nitrogen at One Atmosphere Pressure from the Average Value for Eight Samples at the Heating Rates of 2.5, 5, 10 and 20°C/min.

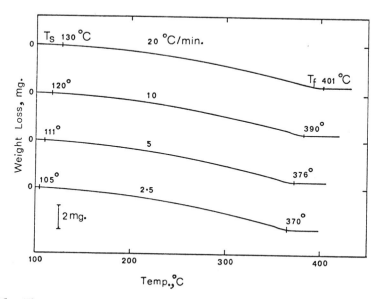

FIGURE 11. Thermogravimetric Curves for the Thermal Decomposition of PMMA in an Inert Atmosphere of Nitrogen at One Atmosphere Pressure.

decomposition reaction of Teflon[R] at 10°C/min. The data for these
curves were obtained from Figures 3 and 7. Using Figure 8 and equation (17), the kinetic parameters E and n are evaluated by taking the
values of W_r and dw/dt from Figure 8 over small constant intervals of
$(1/T)$, $\Delta(1/T) = 1.0 \times 10^{-5}$ °K^{-1}. These values of W_r and log (dw/dt)
are given in Table 2. The parameters used to evaluate E and n from
equation (17) were obtained by subtracting line 2 from line 1, line
3 from line 2, etc., as shown in Table 3. A plot of Δlog(dw/dt)
versus Δlog W_r and a best fit straight line is drawn. The activation
energy, E, is determined from the intercept at Δlog W_r=0. and the
order of reaction, n, is evaluated from the slope of the line. This
graph is shown in Figure 9. From Figure 9, we can make the necessary
calculations. The intercept = $-(E/2.3R) \cdot \Delta (1/T) = -16.5 \times 10^{-2}$,
where R = 1.987 cal/mole°K and $\Delta(1/T)= 1.0 \times 10^{-5}$ °K^{-1}. Thus,
E becomes 75.0 kcal/mole and n determined from the slope becomes
1.01. Figure 10 is a plot of Δlog(dw/dt) versus Δlog W_r obtained
from the average values of all the heating rates. From this graph
the average value of E is found to be 73.0 kcal/mole and n is 0.98.

PMMA

Figure 11 shows the thermogravimetric curves for PMMA. The
temperature at which the decomposition reactions started,T_s,and ended
T_f, as well as the reaction duration time for each heating rate are
listed in Table 4. The plots of T_s and T_f versus actual heating rate
are shown in Figures 12 and 13. Upon linear extrapolation of these
plots to zero heating rate, T_s(OHR) and T_f (OHR) were found to be
105°C and 369°C respectively. This indicates that in a hypothetical
isothermal system PMMA will start to decompose at 105°C. The temperature difference between T_f(OHR) and T_s(OHR) is 264°C. The temperature range at other heating rates were found to be 265°C, 270°C, and
271°C at 2.5, 5, 10, 20°C/min which is also relatively constant. The
effect of heating rate is also shown in Figure 14, where at the higher
heating rates, the maximum rate of the reaction is higher than at lower heating rates. The rate of weight loss and W_r plotted versus reciprocal absolute temperature are shown in Figure 15. The data for these
plots were obtained from Figures 11 and 14 for the thermal decomposition reaction of PMMA at 10°C/min. Using Figure 15 and equation (17),
the parameters E and n are evaluated by plotting Δlog(dw/dt) versus
Δlog W_r at the constant interval of 1/T, $\Delta(1/T) = 1.0 \times 10^{-5}$ °K^{-1}.
This graph is shown in Figure 16 for the 10°C/min heating rate. There
are two regions to this decomposition reaction. The initial region
begins with 0% weight loss and proceeds to approximately 45% of the
weight loss in the reaction. The activation energy and the order of
the reaction obtained from Figure 16 for the 0-45% weight loss region
are 20.6 kcal/mole and 0.06 respectively, and 47.0 and 0.2 respectively for the 45-100% region. Figure 17 shows the plot of Δlog(dw/dt)
versus Δlog W_r derived using the average values for all the heating
rates combined. In the initial region, the activation energy is 20.2
kcal/mole and n is 0.02. For the 45-100% reaction they are 51.0
kcal/mole and 0.15 respectively.

TABLE 4

Temperature at which the Decomposition Reaction Started (T_S) °C and Ended (T_f) °C. Also Reaction Duration Time in min. for the Thermal Decomposition of PMMA at the Heating Rates of 20, 10, 5 and 2.5 °C/min. at One Atmosphere Pressure, Nitrogen Gas Flow Rate 40 cc/min.

Heating Rate (HR) °C/min		Temperature at which Decomposition Reaction Started (T_s) °C	Temperature at which Decomposition Reaction Ended (T_f) °C	ΔT ($T_{f} - T_{s}$) °C	Reaction Duration Time (min.)
Norminal	Actual				
2.5	2.3	105 ± 1.5	370 ± 1.0	265 ± 2.5	265/2.3 = 105.0
5.0	4.2	111 ± 1.5	376 ± 1.5	265 ± 3.0	265/4.2 = 63.0
10.0	8.5	120 ± 2.0	390 ± 1.0	270 ± 3.0	270/8.5 = 31.8
20.0	17.5	130 ± 1.5	401 ± 1.0	271 ± 2.5	271/17.5 = 15.5

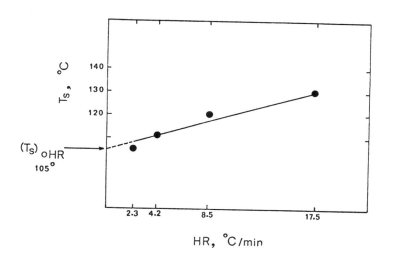

FIGURE 12. Temperature at which the Decomposition Reaction Started (T_S) °C vs Heating Rate °C/min. Sample PMMA in an Inert Atmosphere of Nitrogen at One Atmosphere Pressure.

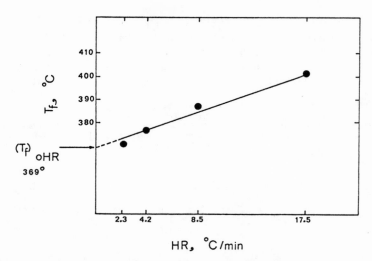

FIGURE 13. Temperature at which the Decomposition Reaction Ended
(T_f) $^\circ$C vs Heating Rate $^\circ$C/min. Sample PMMA in an Inert Atmosphere
of Nitrogen at One Atmosphere Pressure.

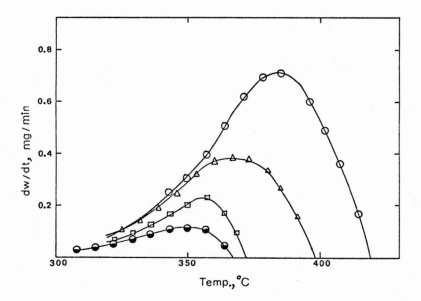

FIGURE 14. The Effect of Heating Rate on the Rate of the Thermal Decomposition of PMMA as a Function of Temperature 20°C/min = 0,
10°C /min = Δ 5°C/min = ☐ , 2.5°C/min = ◑.

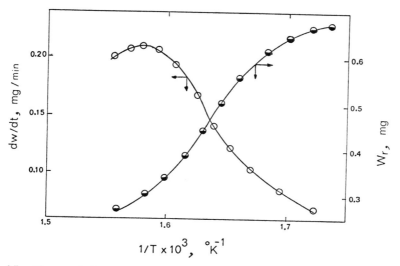

FIGURE 15. Thermal Decomposition of PMMA as a Function of 1/T.   Heating Rate 10°C/min; ◕ W_r, mg vs 1/T; O (dw/dt), mg/min vs 1/T.

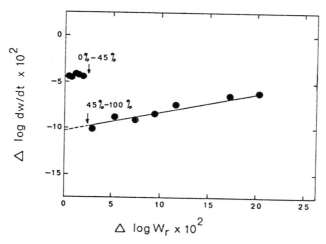

FIGURE 16. The Kinetics of the Thermal Decomposition of PMMA in an Inert Atmosphere of Nitrogen at One Atmosphere Pressure. Heating Rate 10°C/min.

Discussion

$Teflon^R$

The determination of $T_{s(OHR)}$ and $T_{f(OHR)}$ as $474^{\circ}C$ and $554^{\circ}C$ respectively in this work, which is interpreted as the temperature necessary for the decomposition to begin and presumably go to completion if the system were heated isothermally, agrees satisfactorily with the work of Reich et al[10] who conducted a similar experiment in vacuo and heated their sample isothermally at temperatures between $473^{\circ}C$ and $533^{\circ}C$ to study its decomposition. The kinetic analysis of our data yielding an activation energy of 73.0 kcal/mole and an order of 0.98 also agrees with their reporting of first order reaction kinetics with an activation energy of 74 kcal/mole when heating at the rates of $6^{\circ}C/min$ and $3^{\circ}C/min$ respectively. The effects of various gaseous atmospheres on the thermal decomposition of $Teflon^R$ has been reported by Wall and Michaelson[11] with the conclusion that the rate of decomposition is the same for nitrogen gas and vacuum. A mass spectrometry analysis of the gas evolved during the decomposition showed the volatile matter is virtually 100% nonomer leading then to the conclusion that little or no transfer reaction occurred. The mechanism of depolymerization was interpreted as unzipping of the chain to produce free monomer fragments with the initiation taking place at the terminal ends rather than randomly along the backbone. This is in line with the fact that the best estimate for the average C-C bond strength

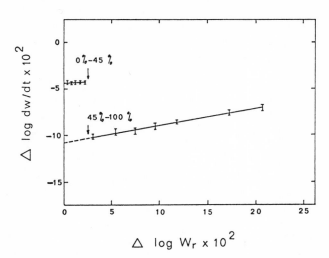

Figure 17. The Kinetics of the Thermal Decomposition of PMMA in an Inert Atmosphere of Nitrogen at One Atmosphere Pressure from the Average Value for Eight Samples at the Heating Rates of 2.5, 5, 10 and 20 $^{\circ}C/min$

in Teflon^R being about 110 kcal/mole[12], whereas the activation energy obtained from the decomposition reaction using hexafluoroethane was only 80.5 kcal/mole.

PMMA

Madorsky [13] found that the decomposition of PMMA contained two distinct regions. The first region started from approximately 5% to 40% of the decomposition reaction. The activation energy of this stage was 30 kcal/mole. The second region from 40% to 90% completion had an activation energy of 50 kcal/mole. His experiment was conducted isothermally at 325°C. Our dynamic thermogravimetric method results are in satisfactory agreement with this earlier study. In another study, Bywater[14] decomposed PMMA in vacuum and measured the rate of reaction by the increase in pressure in a closed system fitted with a Pirani gage. An activation energy of 22 kcal/mole and 48 kcal/mole were found for the two regions with zero order kinetics for both. He also reported the product evolving from the reaction as 100% monomer. Based on his findings, he interpreted that the initiation took place at the chain ends and that the propagation step was essentially the reverse of the normal polymerization propogation step. Our kinetic analysis is consistent with the results of this work as well.

CONCLUSION

The rate equation: $\Delta\log(dw/dt) = n \cdot \Delta\log W_r - (E/2.3R) \cdot \Delta(1/T)$ fits the thermogravimetric data obtained in the study of the thermal decomposition of both Teflon and PMMA excellently. Support for this statement is the fact that the activation energy and the order of the reaction as determined from the plot of $\Delta\log(dw/dt)$ versus $\Delta\log W_r$ are in very good agreement with many previous workers. We are led to conclude that the data obtained from the Perkin-Elmer TGS-1 is of high resolution and quite reliable for efficient studies of polymer decomposition reactions.

Summary

A nonisothermal method has been used to evaluate the activation energy and order of the reaction for the thermal decomposition of PMMA and PTFE. Since this method of kinetic analysis employs a continual increase in sample temperature, the uncertainties due to initial thermal lag associated with isothermal experiments are eliminated. Furthermore, the activation energy and the order of the reaction can be determined from a single experimental curve. The thermobalance was operated at one atmosphere with N_2 as the carrier gas. Heating rates were from 2.5 to 20°C/min. The decomposition rate was found to be a function of the heating rate used. With lower heating rates the decomposition began and ended at lower temperatures. The decomposition of PTFE was first order with an activation energy of

73 kcal/mole. With PMMA, two distinct regions of decomposition
were observed. The first 45% weight loss had an activation energy
of 20.2 kcal/mole and the remaining weight loss had an activation
energy of 51 kcal/mole. Zero order was found for both regions, the
method of analysis followed that of Freeman and Carroll[1] and Anderson and Freeman[2].

REFERENCES

1. Freeman E. S. and B. Carroll, J. Phys. Chem., 62, 394 (1958)

2. Anderson D. A. and E. S. Freeman, J. Polymer Sci., 54, 253 (1961)

3. Doyle C. D., J. Appl. Polymer Sci., 5, 285 (1961)

4. Coats, A. W. and J. P. Redfern, Polymer Letters, 3, 917 (1965)

5. Horowitz, H. H. and Metzger, Anal. Chem., 35, 1464 (1963)

6. Reich, L. and Levi, D. W., Makromol Chemie., 66, 102 (1963)

7. Flynn, J. H. and Wall, L. A., Polymer Letters, 4, 323 (1966)

8. Freidman, H. L., J. Polymer Sci., C, 6, 183 (1965)

9. Chatterjee, P. K., J. Polymer Sci., A, 3, 4253 (1965)

10. Reich, L., Lee, H. T. and Levi, D. W., Polymer Letters, 1,
    535 (1963)

11. Wall, L. A. and Michaelson, J. D., J. Res. N.B.S., 56, 27 (1956)

12. Duus, H. C., DuPont Co., presented at A.C.S., Sept., 1964, N. Y.

13. Madorsky, S. L., J. Polymer Sci., 11, 491 (1953)

14. Bywater, S., J. Phys. Chem., 57, 879 (1953)

APPLICATIONS OF QUANTITATIVE THERMAL ANALYSIS

TO MOLECULAR SIEVE ZEOLITES

W. H. Flank

Union Carbide Corporation
Saw Mill River Road at Route 100C
Tarrytown, New York   10591

INTRODUCTION

Molecular sieve zeolites are a class of stable mineral and synthetic crystalline inorganic compounds characterized by the presence of an open oxide framework structure. This open structure gives rise to a regular network of uniform pores of molecular dimensions that pervades the crystal. The open framework is generally an aluminosilicate one, in which the aluminum and silicon atoms are tetrahedrally coordinated to oxygen atoms in a continuous array. However, they form subunits of rings and cages that can be linked together in a variety of ways to provide, after suitable activation, channels and cavities of predetermined size regularly disposed throughout the crystal. The charge deficiency resulting from the tetrahedral coordination of the aluminum atoms is balanced by the presence of cations positioned by electrostatic forces at relatively fixed locations in the channels and cavities. These charge-balancing cations, however, are not an integral part of the oxide framework. They are, in fact, generally ion-exchangeable by other cations with varying degrees of ease depending on the nature of the cations involved, the conditions under which the exchange is carried out and the location of the various types of ion exchange sites in the structural network, among other things. In a sense, molecular sieve zeolites can be considered as "salts" of strong bases and weak acids, with many of the implications for chemical behavior that such a description implies. The importance of the nature of the cation or cations present derives from their effect on the properties of the molecular sieve zeolites. These properties have profound significance in adsorption, catalysis and ion exchange, and determine utility in such areas of application as petroleum and natural gas processing, bulk chemical processing and purification, moisture and trace gas removal to extreme degrees, and air and water pollution control. A thorough discussion of the structure, physical and chemical properties and chemical reactions of molecular sieve zeolites can be found in a recent monograph by Breck (1).

REVIEW OF EARLIER WORK

Although thermoanalytical techniques have been used in many earlier studies involving molecular sieve zeolites, the results were usually used for cursory characterization, qualitative comparison and, occasionally, semi-quantitative comparison or ranking. A typical example of such data can be found in the work of Barrer and Langley on ion-exchanged forms of natural chabazite (2) and synthetic phases structurally related to chabazite (3). Within the last decade or so, improved instrumentation has become widely available at the same time that the industrial application of molecular sieve zeolites has directed increasing attention to their properties, the ways of modifying these properties, and the increasingly subtle relationships between them. Quantitative correlations and innovative use of thermoanalytical data for diagnostic purposes began to appear in the literature. This can be illustrated by papers on the determination of the degree of NH_3 removal by thermal treatment of ammonium-exchanged type Y zeolite via measurement of exotherm intensity in an oxidizing atmosphere (4), and measurement and interpretation of isosteric heats of adsorption by gas chromatographic techniques of ethylene sorbed on various ion-exchanged forms of type X zeolite (5). A fundamental study of the adsorption of ethylene on ion-exchanged forms of type X zeolite (6) showed that the calorimetrically determined heats of adsorption, which varied widely with different exchanged cations, were primarily due to ethylene-cation interactions via the double bond rather than to ethylene-aluminosilicate interactions. These thermodynamic data were also found to correlate with the shift in the double bond infrared stretching frequency of the adsorbed ethylene, confirming the nature of the interaction. A study of high-temperature heats of adsorption on zeolites (7), using different kinds of adsorbates, indicated that cationic adsorption sites as well as Lewis and Bronsted acid sites can be present and suggested that catalytically active sites were a small fraction of the total. Another study, using isosteric heats of ammonia adsorption (8), suggested that the number of catalytic sites on a decationized type Y zeolite was, nonetheless, much larger than on a gel-type aluminosilicate.

Quantitative comparisons have been made of the thermal stability of a series of closely related zeolite materials. It has been shown, for example, that the temperature of framework

structure collapse in a series of partially ammonium ion-exchanged
type Y samples is a smooth function of the residual sodium content
(9). Following the elucidation of the effect of atmosphere in the
thermal treatment of NH4Y to change its thermal and hydrothermal
stability (10,11), the effects of temperature and partial pressure
of steam in the calcination atmosphere have been examined to better
define optimum conditions for stabilization of the structure (12,
13). The hydroxyl groups discussed in these studies are related to
aluminum atoms in the structure, which had earlier been related to
the silica to alumina ratio in aluminosilicate frameworks and were
correlated with changes in infrared band positions (14). The
behavior of catalytically significant hydroxyl groups in calcined
NH4Y has been further studied, and the temperature at which dehydroxylation occurs has been related to the degree of ammonium ion
exchange and the silica to alumina ratio in the synthesized
framework (15).

Thermogravimetric data have been used to elucidate mechanistic
questions regarding the behavior and reactions of molecular sieve
zeolites. In a study of the nature of rare earth ion-exchanged
type Y (16), it was shown that, at temperatures up to 350°C, one
molecule of water is associated with each rare earth cation.
Reversible removal of this water was found to persist even at
temperatures as high as 900°C, with no change in stoichiometry.
It could thus be postulated that the rare earth cation does not
undergo extensive hydrolysis, which would give rise to a different
water-to-cation stoichiometry, but that observed hydroxyl groups
arise from hydrogen ion exchange during the treatment of the zeolite
with the acidic rare-earth salt solution.

The influence of a number of different cations and degrees of
exchange on the thermal stability of type Y zeolite, as measured
by DTA, has been described and interpreted with the aid of IR and
ESR data (17). Three different types of stability behavior were
identified and related to specific interactions between the cation
and the zeolite framework. The relationship of thermal properties
of several type Y zeolites to their catalytic activity and the
conditions of activation has also been shown (18). A critical
review of the literature regarding hydrogen, decationized, super-
stable and aluminum-deficient forms of type Y zeolite and their
physical, chemical and catalytic properties has recently been
presented by Kerr (19).

RESULTS AND DISCUSSION

An amplification of some of the relationships obtaining in
the thermal treatment in flowing air of ammonium-exchanged type X
and Y zeolites, which can be related to some of the studies cited
earlier (9,12,13,15,19), is shown in Figs. 1 and 2 (20). The

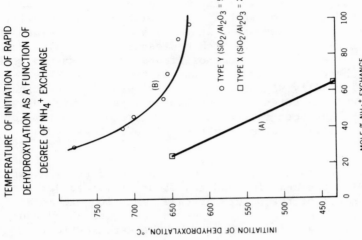

TEMPERATURE OF INITIATION OF RAPID
DEHYDROXYLATION AS A FUNCTION OF
DEGREE OF NH$_4^+$ EXCHANGE

Fig. 2. Data on type X and Y zeolites were obtained with a duPont Model 950 TG Analyzer, using a program rate of 18°C/min.

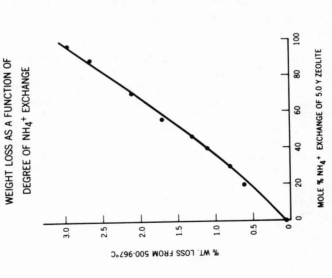

HIGH-TEMPERATURE DEHYDROXYLATION
WEIGHT LOSS AS A FUNCTION OF
DEGREE OF NH$_4^+$ EXCHANGE

Fig. 1. Type Y zeolite had a molar silica to alumina ratio of 5.0. Data were obtained with a duPont Model 950 TG Analyzer, using a program rate of 18°C/min.

weight loss due to dehydroxylation, after ammonium ion decomposition has been essentially completed, is seen in Fig. 1 to be a smooth function of the degree of ammonium ion exchange of the starting material. This correlation quantitatively confirms a suggestion made earlier (21) that increasing Na^+ removal from NaY zeolite (by an increasing extent of ammonium ion exchange) will result, after thermal activation, in increasing hydroxyl content. The probable effect of this hydroxyl content trend in increasing the paraffin to olefin ratios and C_3 to C_1 ratios in hexane cracking over a zeolite catalyst was pointed out in a discussion of the role of the proton in cracking reaction mechanisms. The lack of stoichiometric correspondence of hydroxyl loss with the alumina content of the zeolite indicates that the deammination and dehydroxylation processes cannot be completely separated from each other.

The correlation of degree of exchange with the temperature of initiation of rapid dehydroxylation, shown in Fig. 2, illustrates the sensitivity of the thermal stability of hydroxyl groups to the degree of ammonium exchange and to the molar silica to alumina ratio in the zeolite which is exchanged. It might be noted that the dehydroxylation "step" for an even more highly exchanged type X zeolite could not be readily detected because it is masked by ammonia evolution over the lower temperature range where it occurs (20).

Hydroxyl group thermal stability appears to be related to the energy required to effect the reaction $2 (-OH) \rightarrow H_2O + (-O-)$, and this depends on the proximity of hydroxyl groups to each other. In the ideal case, the structure contains one hydroxyl group for each aluminum atom which has had its original charge-balancing sodium cation exchanged for an ammonium ion which was then thermally decomposed. It is thus evident that both the silica to alumina ratio and the degree of ammonium ion exchange control the disposition of hydroxyl groups in the structure and, hence, their stability with respect to temperature. It must be remembered, however, that a Lewis acid site can be formed by dehydration of a protonic or Bronsted acid site, in a reaction of the type shown above. In the absence of thermally induced proton delocalization, it will remain relatively stable up to the temperature at which the structural framework collapses. The latter transition may easily be measured in relative terms, but it is difficult to measure absolutely because of the interaction of thermodynamic and kinetic effects.

Cumene cracking data have been reported showing that the pretreatment temperature at which catalytic activity begins to decrease is a function of the silica to alumina ratio in the zeolite (22). This can be explained in part by the inhibition of dehydroxylation due to the lower density of hydroxyl groups with increasing silica

to alumina ratio, and in part by the increasing thermal stability
of the crystal structure with increase in the silica to alumina
ratio (17,23).

Another method which has been used to confer thermal stability
in type X and Y zeolites used for catalytic reactions is ion
exchange by rare earth cations. The increase in stability has been
found to be distinctly non-linear with respect to the degree of ion
exchange. The relationship is shown in Fig. 3 (24), where the
structural collapse temperature as measured by DTA is plotted as a
function of the degree of exchange of La^{+3} ions for Na^+ in NaY
zeolite. The nature of the stabilizing effect achievable with rare
earth ions is markedly different from that achievable with carefully
treated ammonium ion-exchanged materials. The latter have been
shown to increase only slightly in stability at low degrees of
exchange, and much more sharply at higher degrees of exchange, to
a maximum of about 1050°C at very high exchange levels (9). In
contrast, a relatively low degree of La^{+3} exchange is seen to effect
a substantial enhancement in thermal stability, while higher degrees
of exchange effect decreasing increments of improvement and provide
a maximum of about 1020°C. A possible explanation for these
differing modes of behavior may lie in the differences in cation
positions in the structure that are assumed by these ions, and
their hydrolysis products, and the resultant differences in interaction with the aluminosilicate framework. The inductive effect of
the high charge density associated with rare earth cations would be
expected to be higher than for other species present.

The importance of thermal stability and the interest in high
temperature properties of molecular sieve zeolites is largely
related to their catalytic applications, especially the large-scale
use in petroleum cracking reactions. This factor has led to
emphasis on studying materials most likely to be useful in such
applications. The thermal behavior of molecular sieve zeolites
used primarily for adsorption has not been studied as extensively,
but thermal property relationships are quite important here too.
Adsorption and desorption of various gases as a function of
temperature has been studied in great detail to allow design of a
number of important adsorption processes, but this area is outside
the scope of the present discussion. It is, however, of interest
to examine the change in thermal stability of the type A structure,
which is extensively used in adsorption processes, as it relates
to the cation content of this material.

Two of the cations that are commonly introduced into the type A
structure by ion exchange to modify adsorption properties are
potassium and calcium. The effect of the degree of potassium ion
exchange on the temperature at which structural collapse occurs,
as measured by DTA, is shown by the middle of the three solid

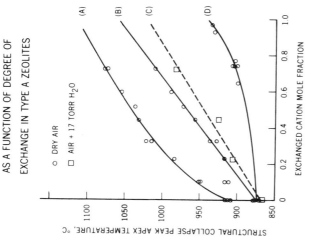

DTA THERMAL STABILITY
AS A FUNCTION OF DEGREE OF
EXCHANGE IN TYPE A ZEOLITES

○ DRY AIR
□ AIR + 17 TORR H₂O

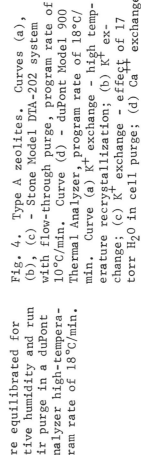

Fig. 4. Type A zeolites. Curves (a), (b), (c) – Stone Model DTA-202 system with flow-through purge, program rate of 10°C/min. Curve (d) – duPont Model 900 Thermal Analyzer, program rate of 18°C/min. Curve (a) K⁺ exchange – high temperature recrystallization; (b) K⁺ exchange; (c) K⁺ exchange – effect of 17 torr H₂O in cell purge; (d) Ca⁺⁺ exchange.

DEPENDENCE OF DTA THERMAL STABILITY
ON DEGREE OF La⁺³ EXCHANGE IN NaY

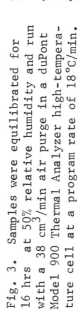

Fig. 3. Samples were equilibrated for 16 hrs at 50% relative humidity and run with a 38 cm³/min air purge in a duPont Model 900 Thermal Analyzer high-temperature cell at a program rate of 18°C/min.

curves in Fig. 4 (curve b), and is seen to be approximately linear. The lower solid curve (d) represents calcium ion exchange. The upper solid curve (a) shows the temperature at which recrystallization of the potassium-containing material takes place to form a thermodynamically stable non-zeolitic alkali aluminosilicate phase; this type of reaction is almost universal among molecular sieve zeolites but is sometimes difficult to detect. It illustrates the point, however, that under most sets of conditions, molecular sieve zeolites exist as thermodynamically metastable species. Thermal stability data, then, as represented by measurement of the temperature at which structural collapse occurs, may sometimes be misleading in that the degradation of the structure is a kinetic phenomenon beyond some threshold temperature, and is dependent on time-temperature-atmosphere relationships and the thermal history of the sample. Wide variations in results can easily be produced by changes in the conditions used or the technique employed. Various factors in the preparation of the sample for testing can also exert important effects (25).

It is thus clear that care must be taken in both obtaining and interpreting data, but relative comparisons can often be made if the relevant factors are recognized. One important factor which is difficult to assess, and which often leads to lack of agreement between laboratory data and actual performance, is the effect of water vapor in the atmosphere to which the material is exposed. The dashed line (curve c) in Fig. 4 illustrates the effect of adding a water partial pressure of 17 torr to the DTA cell purge gas on the thermal stability of several potassium-exchanged samples. Not only is the stability reduced by this small amount of water vapor, but the extent of the reduction in stability increases as the degree of exchange increases. The consequences of this type of behavior may be profound, and attention must obviously be paid to the distinction between thermal stability and hydrothermal stability. The difference between these types of stabilities in partially ammonium-exchanged type A materials has been recently described (26) and further illustrates this point.

Thermal properties can be related to other physicochemical characteristics of molecular sieve zeolites and thus provide some predictive utility. For example, the DTA structural collapse temperature of several stabilized type Y samples is correlated in Fig. 5 with a structure-sensitive symmetrical stretching frequency in the framework infrared region investigated earlier by Flanigen and co-workers (27). The structural collapse temperature can also be correlated with the unit cell parameter of the crystal, as seen in Fig. 6. Since a decrease in aluminum content leads to a decrease in some of the aluminosilicate framework bond lengths and an increase in bond force constants, resulting in higher infrared vibration frequencies (14,27), the data in Fig. 5 show that removal

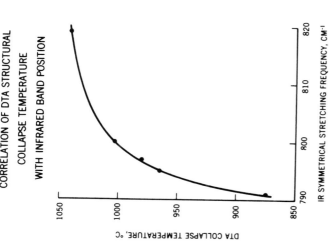

CORRELATION OF DTA STRUCTURAL
COLLAPSE TEMPERATURE
WITH INFRARED BAND POSITION

CORRELATION OF UNIT CELL SIZE
AND THERMAL STABILITY

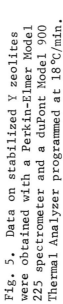

Fig. 5. Data on stabilized Y zeolites were obtained with a Perkin-Elmer Model 225 spectrometer and a duPont Model 900 Thermal Analyzer programmed at 18°C/min.

Fig. 6. Data on stabilized Y zeolites were obtained with a Norelco diffracto-meter with pulse-height analyzer and a duPont Model 900 Thermal Analyzer pro-grammed at 18°C/min. Average deviation for the DTA data is about 2.7 °C.

CORRELATION OF UNIT CELL SIZE
WITH MULLITIZATION TEMPERATURE

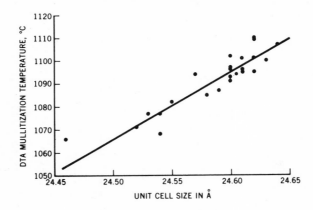

Fig. 7. Data on stabilized Y zeolites were
obtained with a duPont Model 900 Thermal
Analyzer high-temperature cell programmed
at 18°C/min.

of aluminum from the framework contributes to thermal stability.
This is also supported by the data in Fig. 6, where a decrease in
unit cell size, indicative of increasing silica to alumina ratio
in the structural framework, correlates with increasing thermal
stability.

It is interesting to note that this type of explanation
regarding the effects of changes in the silica to alumina ratio
of the structure does not apply once the zeolite structural framework has collapsed. The temperature of recrystallization of the
amorphous collapsed structure to a stable high-temperature phase
is shown in Fig. 7 to increase with increasing unit cell size, or
decreasing silica to alumina ratio. An explanation for this,
involving nucleation inhibition effects and the presence of glass-
type phases formed after breakdown of the zeolite structure, had
been advanced earlier (9), based on composition data. The high-
temperature crystalline phase has been identified as mullite, a
phase known to be formed from other aluminosilicates as well.

It is apparent from the preceding discussion that thermal
measurements of various kinds have been extensively used in
studying various aspects of molecular sieve zeolites, and that

they have contributed appreciably, when carefully interpreted, to an understanding of the material science and reaction mechanism relationships regarding this important class of materials. Thermal methods have been used also in quality control of zeolite products and raw materials (23). While their use is not limitless, or always preferred to other techniques, thermal methods possess the capacity for contributing to the answers for many of the questions outstanding in the relatively young and growing field of molecular sieve zeolites, as has been shown in some of the work presented here.

Acknowledgment - The cooperation of my colleagues in the Molecular Sieve Department of Union Carbide Corporation at Tarrytown, New York, in supplying some of the samples and data used in this work is gratefully appreciated. The experimental contributions of Ms. D. Kimak are also acknowledged.

REFERENCES

(1) D. W. Breck, "Zeolite Molecular Sieves", Wiley-Interscience, N.Y., 1974.
(2) R. M. Barrer and D. A. Langley, J. Chem. Soc., 3804 (1958).
(3) ibid, 3811 (1958).
(4) P. B. Venuto, E. L. Wu and J. Cattanach, Anal. Chem., 38:1266 (1966).
(5) H. W. Habgood, Can. J. Chem., 42:2340 (1964).
(6) J. L. Carter et al., J. Phys. Chem., 70:1126 (1966)
(7) H. C. Tuang, B. V. Romanovskii, K. V. Topchieva and L. I. Piguzova, Kinetics and Catalysis (transl.), 8:594 (1967).
(8) J. E. Benson, K. Ushiba and M. Boudart, J. Catal., 9:91 (1967).
(9) W. J. Ambs and W. H. Flank, J. Catal., 14:118 (1969).
(10) G. T. Kerr, J. Catal., 15:200 (1969).
(11) J. Cattanach, E. L. Wu and P. B. Venuto, J. Catal., 11:342 (1968).
(12) J. Scherzer and J. L. Bass, J. Catal., 28:101 (1973).
(13) J. W. Ward, J. Catal., 27:157 (1972).
(14) R. G. Milkey, Amer. Mineral., 45:990 (1960).
(15) A. P. Bolton and M. A. Lanewala, J. Catal., 18:154 (1970).
(16) A. P. Bolton, J. Catal., 22:9 (1971).
(17) H. Bremer, W. Morke, R. Schodel and F. Vogt, in "Molecular Sieves", Advan. Chem. Ser., Vol. 121, American Chemical Society, Washington, D.C., 1973, p. 249.
(18) D. A. Hickson and S. M. Csicsery, J. Catal., 10:27 (1968).
(19) G. T. Kerr, in "Molecular Sieves", Advan. Chem. Ser., Vol. 121, American Chemical Society, Washington, D.C., 1973, p. 219.
(20) J. J. Behen and H. F. Hillery, unpublished data.
(21) A. P. Bolton and R. L. Bujalski, J. Catal., 23:331 (1971).
(22) K. Tsutsumi and H. Takahashi, J. Catal., 24:1 (1972).

(23) W. H. Flank, unpublished work.

(24) H. F. Hillery and R. L. Bujalski, unpublished data.

(25) W. H. Flank, J. Thermal Anal., $\underline{3}$:73 (1971).

(26) G. H. Kuhl, J. Catal., $\underline{29}$:270 (1973).

(27) E. M. Flanigen, H. A. Szymanski and H. Khatami, in "Molecular Sieve Zeolites", Advan. Chem. Ser., Vol. 101, American Chemical Society, Washington, D. C., 1971, p. 201.

SOME RECENT RESEARCH ON THERMAL PROPERTIES OF MILK FAT SYSTEMS

J. W. Sherbon

Department of Food Science
Cornell University
Ithaca, New York   14850

In the first symposium on analytical calorimetry, there were
two papers on the thermal properties of milk fat (22, 32). At
that time, differential scanning calorimetry was just beginning
to be in wide spread use. In these papers, attention was focused
on the heat capacities and heats of fusion of the fat and on the
influence of thermal history of the fat upon melting phenomena.
Progress in relating thermal phenomena to fat composition was
just beginning. Since then, the thrust of research has been
towards refinement of techniques, behaviour of various fractions
of milk fat, and the influence of specific components upon the
overall thermogram.

Because of the wide melting range of milk fat, it is difficult to apply recent advances in interpretation of DSC thermograms.
The major difficulty lies in determination of the onset of the
melting transition, since so little of the milk fat melts at that
point. This leads to problems in location of the base line, such
as when following procedures described by Brennan, Miller, and
Whitwell (3) or by Heuvel and Lind (12). The latter method has
been successfully used for milk fat (24) except when the base
line appeared to be curved due to uncontrolled variations in
machine-sample heat transfer characteristics (17). At scan rates
of 8 C/min., corrections by the Heuvel and Lind technique to the
temperature of the major milk fat endotherm maximum amounted to
.5 C or less. An example is shown in Figures 1 and 2. More importantly, corrections to the heat of fusion reached nearly 10% of
the final value for intact milk fat. For samples of more
restricted melting range, heat of fusion corrections were closer
to 1%. In every case, the initial estimates of the heats of fusion
were high. One error that still has not been accounted for in

661

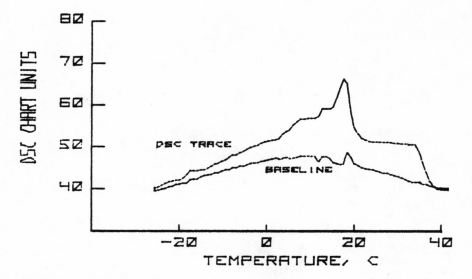

Figure 1.   A DSC thermogram of milk fat and the base line according to Heuvel and Lind (12).

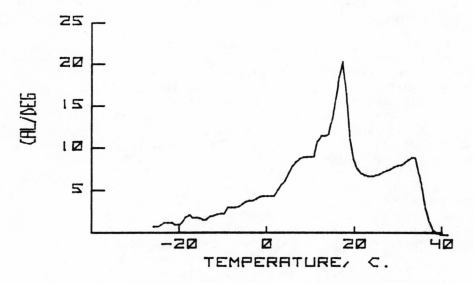

Figure 2.   Corrected plot of milk fat from Figure 1.

milk fat research is the well-known variation in heat of fusion
(10, 14) with melting point of the fat fraction.

There has been interest in relating seasonal changes in the
composition of milk fat to seasonal changes in melting characteristics. Antilla (1) found that Finnish winter fat was harder
because it contained more high melting triglycerides rich in
stearate esters. Norris, Gray, and Dolby (17) studied 17 butter-
fats collected over a production season in New Zealand, and
obtained significant positive correlations between proportions of
the fat liquified at most temperatures and concentration of short
chain plus unsaturated acids (either cis or total) constituting
the triglycerides. Typical results are shown in Figure 3. At
all temperatures below 30 C, the maximum percentage of liquid fat
occurred in the early spring while the minimum occurred in midsummer. The changes were most pronounced at 12 C. New Zealand
butter is softest in early spring and hardest in midsummer. One
traditional measure of butterfat hardness, the softening point
(temperature at which a standard ball penetrates the sample under
its own weight) correlated well with the temperature at which 94%
of the fat was melted. An interesting result of their research was
that the cessation of melting of all samples was relatively constant,
just below 40 C. Although the fats examined by Antilla (1) had not
been cooled sufficiently to obtain complete solidification,
(14, 17, 24) his thermograms are sufficiently similar to those
presented by Norris, et al (17), to confirm that the same factor,
namely fat composition, is operative in spite of the fact that
winter fat was the hardest in the one case and summer fat the
hardest in the other. Norris, et al, pointed out that the shoulder
at 12 C is more pronounced in the harder fat, and can be used as
a visual index of fat hardness.

Applications of calorimetry to butter, containing 16% aqueous
phase, instead of milk fat have been made. It is possible to
load a DSC without erasing thermal history of the sample, thereby
allowing study of the effect of variations in processing conditions (24). Shock cooling of the cream prior to churning, known
to produce hard butter, resulted in a fat having less liquid fat
at 22 C than slow cooling or crystallization initiation at 4 C
and tempering at 19 C, both producing softer butter. The differences in liquid fat content were more pronounced at 10 C. Differences between slow cooling and the 4/19 treatment were minor. The
serum of unsalted butter had a single melting endotherm at just
below 0 C, while serums from salted butters showed double melting,
as shown in Figure 4. The dependency of the double peaks on the
salt concentration was typical of solutions capable of forming
eutectic compounds; the lower peak occured at -22 C regardless of
salt concentration but its magnitude increased with increasing
salt concentration and both the magnitude and melting temperature

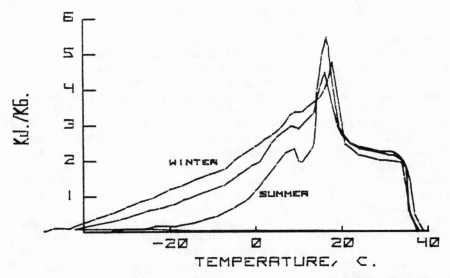

Figure 3. Melting thermograms of winter, spring and summer New Zealand milk fat. From Norris, et al (17).

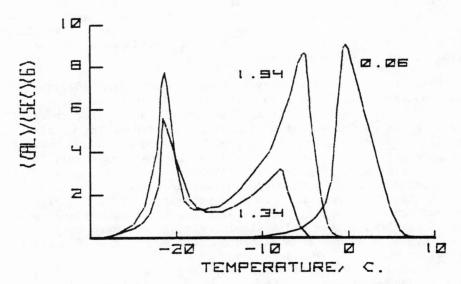

Figure 4. Thermograms of serums from butters of different salt contents (24). (Salt contents of butters shown)

of the upper peak decreased with increasing salt concentration.
There appeared to be no significant interaction between melting of
the aqueous and lipid phases.

Krautwurst (14) has published an extensive article on mixed
crystallization of triglycerides and milk fat. He examined by
DSC and GLC fractions prepared by progressive fractional crystallization and recrystallization, and was able to correlate composition with specific regions of a milk fat thermogram using
melting temperatures of the isolated fractions. The highest
melting region, 27 to 37 C, was attributed to fat fractions rich
in palmitate and stearate. When tested as isolates, these fractions exhibited three major types of DSC patterns. The highest
melting type of pattern was typical of a fat having a narrow
melting range and showing some combination of two polymorphs,
crystallization into the more stable form after melting of the
less stable form, or direct transformation of the unstable to
stable form. An intermediate type of pattern had sharp transitions
but three polymorphs. The third type had three polymorphs and
peak broadening indicating mixed crystal formation. The major
melting region of milk fat, from 5 to 27 C, was attributed to fat
fractions rich in short chain acids, but otherwise little difference from the higher melting group. DSC patterns of these isolates
were more complex, but polymorphic capability may be evident.
The melting region between -8 and 5 C was attributed to fractions
rich in C18 unsaturates. DSC curves of these isolates were simple
but with one shoulder on the low temperature side of the main
endotherm. No exotherms were seen. The lowest melting region,
below -5 C, was associated with fractions having higher contents
of C18 unsaturated acids than the previous melting regions. The
thermograms were essentially the same, aside from the lower
melting temperatures. X-ray diffraction was used to show the
probability of mixed crystallization when an isolate is "dissolved"
in the milk fat.

As mentioned above, Krautwurst (17) prepared a large number
of fractions of milk fat by double progressive crystallization
from acetone solutions. The melting points of the stable forms
ranged from 42 C to -40C and the heats of fusion decreased with
melting point from 42.5 cal/g to 10.4 cal/g. Antilla (1)
fractioned milk fat by both progressive fractional crystallization
from acetone and by molecular distillation and found that the
method of preparation did not have a major effect on the thermograms of fractions melting in the same ranges. Most of the
recent research has involved the high melting fraction as its
removal drastically lowers the melting point of the residual (7).
Vergelesov (29) found that the highest melting fraction crystallized
in the "β_L-2" form but would not transform to the β form. The
next lower melting fractions crystallized in a "β_L-2" form. The

high melting glyceride fraction (HMG) as originally defined (13)
can be further fractionated. Progressive crystallization into
four fractions and a residue from acetone results in fractions
melting from 56 C for the highest melting down to 27 C for the
residual and the heats of fusion of the most stable form decreased
from 41.0 to 17.9 cal/g with decreasing melting point (26). As
shown in Figure 5, two types of polymorphic transitions were seen,
unstable-supercooled liquid-stable and the direct unstable-stable.
Three transitions are marked by an endotherm followed by an exotherm
then the final melting endotherm, and by an exotherm followed by
the melting endotherm, respectively. The highest melting fraction
initially crystallized in the β_L form, the next in β_L', and the
next two lower melting fractions were in the β_L' form, according
to the infrared spectra (2). These fractions were predominately
tripalmitin, dipalmitylstearin, and distearylpalmitin (2). In
the four highest melting fractions, the majority of the palmitate
was found in the Sn-2 position of the glycerol while the majority
of the stearate and elaidate was found in the Sn-3 position (2).

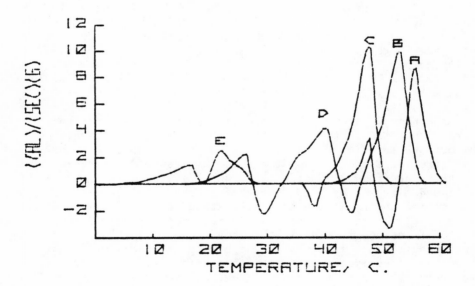

Figure 5. Thermograms of various fractions of high melting glyceride from milk fat (26).

Less work has been done on the low melting fractions. Aside
from those prepared by Krautwurst, referred to above, I will have
to rely upon some preliminary unpublished results from my own work.
Fractions melting in the 0 to -23 C range have increasing contents
of short chain and unsaturated C18 acids (especially C18:2 and
C18:3) as the melting point of the fraction decreases. The heats
of fusion of the stable forms decreased with melting point, as
stated by Krautwurst (17). Thin layer chromatography shows
increasing amounts of more polar lipids than triglycerides in the
lowest melting fractions. There is preliminary evidence showing
an increase in molecular weight and polyunsaturation of the lowest
melting fractions.

Fractionation of milk fat on a commercial scale has been done
using a process described by Fjaervoll (8). This fractionation
done in an aqueous suspension, essentially removes the high melting
material appearing as a shoulder or minor peak on the high melting
side of the main melting peak (18). This material is suspended
in lower melting fats, thus the yields of the two fractions are
approximately 1:1. Refractionation at 28 C of the solid fat did
not materially change the resultant solid phase, but refractionation
at 31-32 C resulted in a more saturated, higher melting product
(25). Liquid phases from these refractionations had thermograms
resembling intact milk fat. It appeared possible to crystallize
the highest melting products in an unstable form.

The behavior of milk fat in bulk is different from the same
fat as an emulsion in milk or cream. In 1938, Rishoi and Sharp
(20) reported a lag in the crystallization of the fat when cooled
in the form of cream. Maximum solidification occurred in 4 hours
at 0 to 10 C, but required months at 15 or 20 C. After the initial recrystallization, there was a slight decrease in extent of
solidification upon long term storage. It was this latter change
that was correlated with properties of the emulsion most closely
associated with interfacial tension of the globule. The extent
of fat solidification is known to affect the churning time of the
cream, the higher the solid fat content the longer the churning
time (4). When globular milk fat is cooled to 20 C, very little
of the fat solidifies but if it is first tempered for 2 hours at
10 C, the fat readily solidifies (30). Electron microscopy has
been used to show that crystallization of the high melting fractions starts at the membrane surface, forming a shell up to 50 Å
thick (5, 11). No other distinct layering can be seen. The
smaller the globule, the larger the decrease in solidification
when cooling fat to a given temperature (27). Addition of 5%
emulsifier (1 Span 40:2 Tween 40) to a milk fat fraction melting
at 56 C reduced the melting temperatures of both unstable and
stable forms of the fat by about a degree (23). These melting
points decreased about one more degree when the mixture was homo-
genized into water containing 0.5% sodium caseinate, probably an

effect of the dispersion rather than of the emulsifiers. Reduction
of the average globule size from 2.8 µ to 2.5 µ did not affect the
melting points of the fat, but did increase the proportion of fat
solidifying the unstable phase.

Thermograms of milk fat globule membrane preparations have
three major endotherms, at 15, 44, and an irreversible one at 80 C
(6). The first two were due to triglycerides in the preparation
but the 80 C transition may be associated with the release of a
copper containing lipoprotein from the membrane into the aqueous
phase. Two triglyceride fractions were isolated from the membrane
preparation. The acetone soluble triglycerides contained some
short chain acids and palmitic, stearic, and oleic acids in significant amounts and showed two melting endotherms, at 26 C and the
major transition at 38 C. The acetone insoluble triglycerides
contained primarily palmitic and stearic acids and only 7% of
C18:1 and showed evidence of unstable polymorph formation. The
stable melting point was 53 C. The latter preparation has many
points of similarity to preparations of the HMG referred to above.

Explanation of the behavior of emulsified milk fats cited
above can be based on results obtained using model systems. Bulk
fats crystallize under the influence of heterogeneous nucleation
and dissolved emulsifiers have only minor effects (28). If the
mixture is dispersed finely enough, 10^{10} globules/ml fat, nucleation
becomes homogeneous but the nature of the emulsifier becomes very
important, especially through interfacial tension effects. A
phase diagram of the system water/monocaprylin/tricaprylin shows
two mesomorphic phases, one neat soap phase, and a micellular
solution capable of dissolving up to 40% water (9). Milk and
cream fall into one of the mesmorphic phases. The presence of
liquid crystals involving the emulsifier apparently is not only
important to emulsion stability (9) but also to the crystallization
pattern of the emulsified fat (31). Thus, the lag in solidification
of emulsified milk fat can be attributed to the change from heterogeneous to homogeneous nucleation, and the liquid fat within the
globule probably is supercooled. The crystallization of the high
melting fractions on the globule surface is induced by the specific
structure of the membrane. The lack of crystal layering beyond
this is probably related to the poor heat transfer characteristics
of solidified fat. The increased incidence of unstable polymorphs
in emulsified fat can be related to both the higher degree of
supercooling and specific effects of the membrane.

As Rishoi and Sharp pointed out in their early paper (20),
crystallization of milk fat not only involves the melting points
of different fractions of the fat, but the solubility relationships
existing between the various constituent triglycerides. Progress
in this area will be slow, because of the complexity of milk fat,

but phase diagrams of triglyceride mixtures are becoming available (15) as are effects of chain length and unsaturation on crystallization of triglycerides (10, 16).

References

1. Antilla, V. 1966. Meijeritiet. Aikak. 27:1.
2. Barbano, D. M. 1973. M. S. Thesis, Cornell University.
3. Brennan, W. P., B. Miller, and J. C. Whitwell. 1969. Ind. and Eng. Chem. Fundamentals. 8:314.
4. Brunner, J. R. and E. L. Jack. 1950. J. Dairy Sci. 33:1950.
5. Bucheim, W. von. 1970. Milchiwissenschaft. 25:65.
6. Chandon, R. C., J. Cullen, B. D. Ladbrooke, and D. Chapman. 1971. J. Dairy Sci. 54:1744.
7. DeMan, J. M. 1961. J. Dairy Res. 28:117.
8. Fjaervoll, A. 1970. Dairy Ind. 35:502.
9. Friberg, S. and L. Mandell. 1970. J. Am. Oil Chem. Soc. 47:149.
10. Hagemann, J. W., W. H. Tallent, and K. E. Kolb. 1972. JAOCS. 49:118.
11. Henson, A. F., G. Holdsworth, and R. C. Chandon. 1971. J. Dairy Sci. 54:1752.
12. Heurel, H. M. and K. C. J. B. Lind. 1970. Anal. Chem. 42:1044.
13. Jenness, R. and L. S. Palmer. 1945. J. Dairy Sci. 28:653.
14. Krautwurst, J. 1970. Kieler Milchwirt. Forschiengs. 22:255.
15. Lutton, E. S. 1967. J. Am. Oil. Chem. Soc. 44:303.
16. Lutton, E. S. and A. J. Fehl. 1970. Lipids. 5:90.
17. Norris, G. E., I. K. Gray, and R. M. Dolby. 1973. J. Dairy Res. 40:311.
18. Norris, R., I. K. Gray, A. K. R. McDowell, and R. M. Dolby. 1971. J. Dairy Res. 38:179.
19. Pitas, R. E., J. Sampugna, and R. G. Jensen. 1967. J. Dairy Sci. 50:1332.
20. Rishoi, A. H. and P. F. Sharp. 1938. J. Dairy Sci. 21:399.
21. Sherbon, J. W. 1963. Ph.D. Thesis, Univ. Minn.
22. Sherbon, J. W. 1968. In: Analytical Calorimetry, ed. by Porter, R. S. and J. F. Johnson. Plenum Press, N.Y. p.173.
23. Sherbon, J. W. and R. M. Dolby. 1971. N.Z.J. Dairy Sci. and Techn. 6:118.
24. Sherbon, J. W. and R. M. Dolby. 1972. J. Dairy Res. 39:319.
25. Sherbon, J. W., R. M. Dolby, and R. W. Russell. 1972. J. Dairy Res. 39:325.
26. Sherbon, J. W., and R. M. Dolby. 1973. J. Dairy Sci. 56:52.
27. Van Berestyn, E.C.H. and P. Walstra. 1971. N.I.Z.O.-nieuws. 197.
28. Van den Tempel, M. 1968. Soc. Chem. Ind. Monographs. 32:22.

29. Vergelesov, V. M., M. F. Kurkov, E. A. Il'Chenko, and E. A. Sitchenko. 1970. XVIII Intl. Dairy Cong. 1E:210.

30. Wauschkuhn, P. and E. Knoop. 1970. Milchwissenschaft. 25:70.

31. Wilton and S. Friberg. 1971. J. Am. Oil Chem. Soc. 48:771.

32. Yoncoskie, Y. A. 1968. In: Analytical Calorimetry, ed. by Porter, R. S. and J. F. Johnson. Plenum Press, N.Y. p. 167.

THERMAL BEHAVIOR OF CHEMICAL FERTILIZERS

C. Giavarini

Institute of Applied and Industrial Chemistry

University of Rome, v.Eudossiana 18, 00184 Italy

The fertilizers industry is one of the major industries of the world, with a total output in 1970 of about 200 million tons of products and a growth rate of about 8,5% per year (1). Among the main directions of research to be taken into consideration in this field are a) the study of the thermal behavior of fertilizers and b) the research of a method for a sufficiently rapid and simple characterization of such products (2):
a) thermal reactions may occur, after manufacture or mixing, during drying, storage and handling (3,4,5);
b) the rapid determination of the chemical compound actually present at the various stages of production and in the finished products allows to act promptly in directing the process underway in the desired sense

In the present paper the thermal behavior of chemical fertilizers and the possibilities of differential scanning calorimetry for their analysis are examined. The chosen products are representative of some major classes of industrial chemical fertilizers (granular superphosphates and granular complex fertilizers with two or three primary plant nutrients). The experimental results are reported separately according to the class to which a product belongs.

671

EXPERIMENTAL

The tests were carried out using a Perkin-Elmer
scanning calorimeter model DSC-1, working in a dynamic
nitrogen atmosphere. Runs were generally performed within a temperature range of 25 to 300°C, with powdered
samples weighting about 10 mg. The heating rates were
8°C/min. Superphosphates C and D were prepared by mixing tricalcium phosphate with 60% w/w sulfuric acid
(P_2O_5/H_2SO_4 molar ratio was 0,5) at 20°C (C) and at
70°C (D). The other samples were of industrial origin.
Chemical analysis were made according to the methods
of European Economic Community (7).

RESULTS AND DISCUSSION

Superphosphates

As used in the fertilizer trade, the term superphosphate refers to the material wich results when either sulfuric acid or phosphoric acid, or a mixture of
both is caused to react with phosphate rock. Common u-
sage designates the product containing 16 to 21 per cent
phosphoric anydride (P_2O_5) as "normal" superphosphate,
and the product containing 43 to 48 per cent P_2O_5 as
"triple" superphosphate. P_2O_5 is normally in the form
of monocalcium phosphate, but can also be in the form
of dicalcium phosphate; the first compound is water
soluble, the second is soluble in a neutral ammonium
citrate solution. In the U.S.A. it is legal to consider as "available" the sum of that phosphate which is
soluble in water and in a neutral ammonium citrate solution (8); in the U.K. and in other European countries
on the other hand, a water solubility test is required.
From a practical standpoint however it is important to
know whether P_2O_5 is in the monocalcium or dicalcium
form.

Figure 1 shows the DSC curves of the four normal
superphosphates listed in table I. Such products show
generally three main endothermic peaks of which the
first, between about 80 and 100°C, (not present in

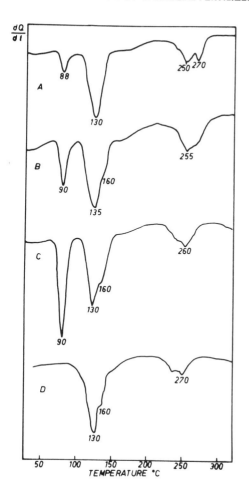

Fig.1. DSC curves of industrial normal superphosphates **A** and B, and of two laboratory prepared superphosphates (C,D). Sensitivity 2 mcal/sec for full-scale deflection.

Fig.2. DSC curves of two triple superphosphates (4 mcal/sec).

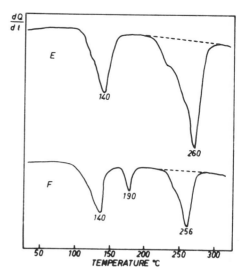

Table I

Chemical Analysis of Superphosphates (phosphate const.)

Constituent %	Sample					
	A	B	C	D	E	F
$Ca(H_2PO_4)_2$	31,1	32,7	43,9	41,5	63,0	39,1
$CaHPO_4$	1,9	2,3	–	–	2,8	30,7
Avail.P_2O_5	20,9	22,2	26,6	25,2	43,5	42,1
Declared P_2O_5	18-20	18-20	–	–	46	46

sample D) is due to the aridizing effect of calcium sulfate dihydrate; in fact when salts such as some chlorides, or monocalcium phosphate monohydrate, are added to gypsum, this compound undergoes a first dehydration at a temperature lower then that usually necessary to produce the hemihydrate form (9). The endotherm between 80 and 100°C therefore indicates the presence of calcium sulfate dihydrate and can be used for an approximate determination of the content of the dihydrate form; the aging of the samples results in a reduction of this peak, an this is an indication that a slow transformation of the dihydrate form into hemiydrate form takes place. The second main endotherm (first for sample D) is due both to the melting and dehydration of monocalcium phosphate monohydrate (10,11) and to the dehydration of calcium sulfate hemihydrate (shoulder at about 160°C). The third endotherm (second for D) between about 200 and 290°C is due to the formation of acid-pyrophosphate from monocalcium phosphate; the presence of gypsum contributes to lower and broaden this peak (11).

Samples C and D show the importance of the temperature during the production process; with the same concentration of H_2SO_4, lower temperatures favour the formation of $CaSO_4 \cdot 2H_2O$ (sample C); higher temperatures give mainly hemihydrate or anhidrite (sample D). From the exa-

mination of the DSC curves the presence of dicalcium
phosphate seems to be excluded; it was observed however
that the peak at about 190°C due to the dehydration of
$CaHPO_4 \cdot 2H_2O$ is reduced by the presence of monocalcium
phosphate and therefore small amounts of such salt are
not shown.

Figure 1 reports also the curves of two triple superphosphates (see table I): sample E consists mainly of
monocalcium phosphate monohydrate and shows the typical
curve of this compound; sample F instead contains appreciable quantities of dicalcium phosphate, responsible for
the peak between 170 and 200°C. In triple superphosphates
it is possible to calculate roughly the percentage of monocalcium phosphate by measuring the area of the whole
endotherm at about 260°C (fig.2 shows the interpolated
base lines) and subsequently comparing the corresponding
ΔH with the ΔH of the same peak of pure $Ca(H_2PO_4)_2 \cdot H_2O$.
The error is generally in the range 5 ÷ 20%, as seen on
many different samples.

Complex Fertilizers with two Primary Plant Nutrients

We may call "complex" those fertilizers which contain more than one of the three primary plant nutrients,
i.e. N, P and K, obtained by a chemical combination realized through a process of chemical reactions in which
other factors are present besides the simple addition
of water, or steam, for granulation (12,2). We shall consider first some complex fertilizers with two plant nutrients (see table II and figs. 3 and 4).

The DSC curve of sample G is similar to the curve
of ammonium nitrate (fig.3); the main differences appear
in the melting (155°C vs 171°C) and in the decomposition
temperature (lower for G); both these temperatures are
affected by the presence of other salts (5). The temperatures of the other phase changes, i.e. from rhombohedral
to monoclinic phase at 42°C, to tetragonal phase at about
80°C and to cubic phase at 125°C, are quite unaffected(13).
The endotherm at 80-85°C is generally more visible in compounds containing NH_4NO_3 rather then in ammonium nitrate

Table II

Fertilizers with two Plant Nutrients

Constituent %	Sample		
	G (24-24-0)	H (18-46-0)	I (20-20-0)
Water	0,2	1,1	0,8
Water soluble P$_2$O$_5$of which	25,3	41,6	6,4
from monos.phosphates	23,7		5,4
from diammonium phosphate	-		1,0
Water-citrate soluble P$_2$O$_5$	26,3	41,9	20,2
Insoluble P$_2$O$_5$	-	0,1	0,1
Ammonium N	14,7	17,8	11,0
Nitrate N	8,8	-	8,7
MgO	0,2		0,2
CaO	0,7		11,8
Water soluble SO$_4$	2,6		3,1
Total SO$_4$	3,0		3,6
SiO$_2$	0,2		1,5

Table III

Example of Quantitative Analysis of NH_4NO_3 from DSC data

Sample	mg	Peak at 125°C		$NH_4NO_3\%$ from DSC data	$NH_4NO_3\%$ from chemical analysis
		area mm^2	ΔH cal/g		
NH_4NO_3	9,94	309	7,2	100,0	100,0
G	9,96	160	3,8	52,8	50,3

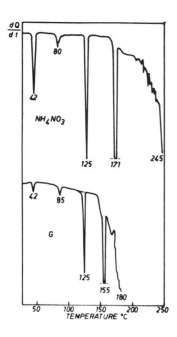

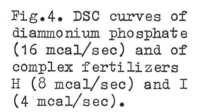

Fig.3. DSC curves of ammonium nitrate and of complex fertilizer G (4 mcal/sec).

Fig.4. DSC curves of diammonium phosphate (16 mcal/sec) and of complex fertilizers H (8 mcal/sec) and I (4 mcal/sec).

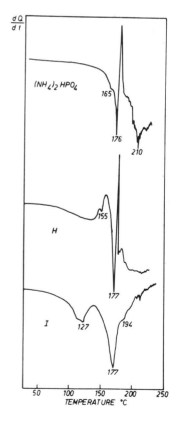

itself; the opposite occurs for the endotherm at 42°C, in-
dicanting that such phase changes are affected by the presence of other compounds. The peak at 125°C is the most
characteristic of NH_4NO_3 and may be used to proportion such
salt in the fertilizers where it is contained (naturally
when no other peak occurs in close sequence). Table III
shows an example of the measurement of the ammonium nitrate content in sample G. The error is generally in the range 2 + 8%. The chemical analysis shows that the main phosphate present in fertilizer G is monoammonium phosphate;
the presence of such salt, which decomposes at about 180°C
(as shown by DSC curves), may explain the differences noted in the final part of the curves relatives to G and to
pure NH_4NO_3 samples.

The thermal behavior of sample H is similar to that
of diammonium phosphate, which is therefore the main constituent (fig.4); the brosad effect before the peak at
177°C is typical of triammonium phosphate, wich can be present in sample H together with other minor components.

The main constituents of sample I are ammonium nitrate (about 50%) and dicalcium phosphate (table II); o-
ther minor constituents are monocalcium and diammonium
phosphate and ammonium sulfate. On heating, however, some ammonium phosphates must form (6):

$$2NH_4NO_3 + CaHPO_4 \longrightarrow Ca(NO_3)_2 + (NH_4)_2HPO_4 \qquad 1)$$

$$Ca(H_2PO_4)_2 + (NH_4)_2SO_4 \longrightarrow 2NH_4H_2PO_4 + CaSO_4 \qquad 2)$$

The curve I in fact is similar to the curve of the impure diammonium phosphate and the phase changes of ammonium
nitrate are not visible, except for the one which occurs
at 127°C (125°C for pure NH_4NO_3). The residual dicalcium
phosphate is probably responsible for the broadening of
the main endotherm at 177°C. The effect between 100 and
150°C may be partially due to the presence of some residual monocalcium phosphate.

Complex Fertilizers with three Primary Plant Nutrients

The chemical compositions of the studied NPK complex

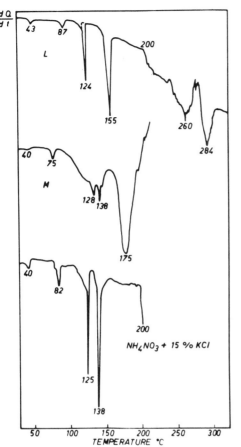

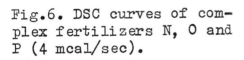

Fig.5. DSC curves of complex fertilizers L (4 mcal/sec), M (2 mcal/sec) and of an admixture of NH_4NO_3 and KCl (4 mcal/sec).

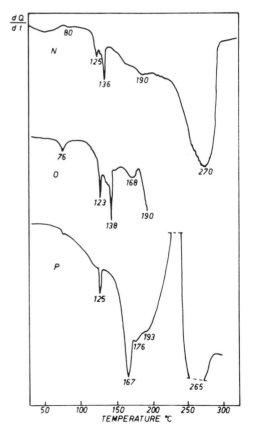

Fig.6. DSC curves of complex fertilizers N, O and P (4 mcal/sec).

Table IV

Fertilizers with three Plant Nutrients

Constituent %	Sample				
	L	M	N	O	P
	7.15.30	17.17.17	12.12.12	17.8.9	10.20.20
Water	1,2	0,7	3,0	1,5	1,6
Water soluble P_2O_5 of which	15,6	17,4	0,2	0,2	10,7
—from mono.phosphates	9,3	9,0	-	-	10,4
—from diammonium phosphate	6,3	8,4	0,2	0,2	-
Water—citrate soluble P_2O_5	16,4	18,4	11,4	7,8	20,2
Insoluble P_2O_5	0,1	0,1	1,6	0,1	0,2
Ammonium N	6,6	11,2	5,8	8,3	6,4
Nitrate N	3,3	5,8	5,8	8,3	4,2
K_2O	28,7	17,6	12,3	8,9	19,8
MgO	0,1	0,1	2,0	2,2	0,1
CaO	1,2	0,7	17,9	12,2	8,4
Water soluble SO_4	31,0	-	0,7	0,5	0,3
Total SO_4	32,0	0,9	1,2	0,7	1,8
Water soluble Cl	1,6	13,3	9,3	6,8	15,3
SiO_2	0,4	0,8	0,4	0,2	0,6

fertilizers are shown in table IV; figs. 5 and 6 show
their DSC curves. The thermal behavior of sample L is,
up to about 160°C, equal to that previously seen for
sample G. Sample L contains in fact 9,4% of ammonium
nitrate, as indicated by chemical analysis; the percentage calculated considering the extrapolated area of the
peak at 124°C, is 10,2%. The shoulder at the base of the
peak, which appears when potassium compounds, especially chloride, are present undoubtedly interferes with the
measure of the ΔH. In sample L potassium is mainly in
the form of K_2SO_4, but some KCl (up to 3,4%) is also
present and can react, on heating, with ammonium nitrate (5):

$$KCl + NH_4NO_3 \longrightarrow NH_4Cl + KNO_3 \qquad 3)$$

There is a tendency for a solid solution of the form
$NH_4NO_3 \cdot KNO_3$ to form and this fact can explain the form
of the peak at 124°C. Pure ammonium chloride is characterized by a wide and high endotherm at about 260-270°C
and can therefore be responsible for the peak at 260°C,
in sample L; such peak occurs when the decomposition of
NH_4NO_3 has already started (at 200°C).

Other constituents of sample L are monoammonium
phosphate (about 20%) and diammonium phosphate (11,7%);
ammonium phosphates are not clearly visible probably
because reactions can occur, in the presence of a molten phase, between such salts and potassium sulfate: the
resulting ammonium sulfate may be responsible for the effect at 284°C.

The curve M is, a part for minor thermal effects,
substantially analogous to the curve of sample I and
shows therefore the presence of $(NH_4)_2HPO_4$. Clearly visible are also the peaks of ammonium nitrate (at 40, 75,
128 and 138°C). Potassium chloride is indirectly shown
because, when present in appreciable quantities, it lowers considerably the melting point of NH_4NO_3 (wich occurs at 138°C). A curve showing this effect is reported
in fig. 5: it was obtained by heating ammonium nitrate
containing 15% of KCl.

The fertilizer N shows the presence of ammonium nitrate together with KCl; also clearly visible is the characteristic endotherm at 270°C, due to the decomposition of NH_4Cl formed through reaction 3). The presence of monocalcium phosphate is revealed by the endotherm at 190°C. Sample N contains an appreciable amount of moisture, indicated by the initial part of the DSC curve.

Sample O shows a thermal behavior similar to sample N and contains therefore ammonium nitrate (in major amounts) and KCl. The strong decomposition effect of NH_4NO_3 does not allow to detect in this case the peak of NH_4Cl; the amount of $CaHPO_4$ is smaller than in sample N and probably for such reason it cannot control the decomposition of NH_4NO_3 by diluiting it.

The chemical analysis of fertilizer P shows mainly the presence of KCl (31,4%), monocalcium phosphate (18,2%), NH_4NO_3 (24%) and monoammonium phosphate (about 17%). In curve P the characteristic phase change of NH_4NO_3 at 125°C can be observed, but not the melting peak, hidden in the main endotherm between 130 and 230°C; this endotherm is due mainly to the presence of ammonium phosphates, of which $(NH_4)_2HPO_4$ may be formed as indicated in reaction 1). The presence of appreciable quantities of gypsum (about 2,3% as $CaSO_4$) contributes to enlarge this peak. The residual monocalcium phosphate is probably responsible for the shoulder at 193°C. The endotherm at 265°C is due to NH_4Cl formed as previously mentioned (reaction 3).

CONCLUSION

Chemical fertilizers undergo by heating a series of interesting chemical reactions and phase changes, also at a relatively low temperature. By means of the scanning calorimetry it is possible to identify many chemical compounds present in fertilizers, in a simple and quick way. Semiquantitative analysis are possible and are useful both for characterisation of products and for routine control.

REFERENCES

1. United Nations: "Recent developments in the fertilizer industry", Report of the Second Interregional Fertilizer Symposium held at Kiev and New Delhi, 21 sempt. to 13 oct. 1971. New York 1972.

2. G. Scaramelli: "Present aspects and presumable developments in the technology of complex fertilizers", in G. Fauser: "Chemical Fertilizers", Pergamon Press--Tamburini Publ., New York-Milano, 133-160, 1968.

3. V. Sauchelli: "Chemistry and technology of fertilizers", Reinhold Publ., New York, 1960.

4. K.S. Barclay, J.M. Crewe, J.B. Dawson, K.F.J. Thatcher: "The cause of self-heating of NPK compound fertilizers containing nitrates". J. Appl. Chem. 15, 531-540, 1965.

5. K.S. Barclay: "Physical-chemical studies on decomposition reactions and the safe handling of ammonium nitrate-bearing fertilizers", in G. Fauser:"Chemical Fertilizers", 31-48, 1968.

6. W.A. Mitchell: "An investigation into the caking of granular fertilizers", J. Sci. Food Agriculture, 9, 455-456, 1954.

7. Commission des Communautés Européennes: "Méthodes pour l'analyse des engrais", 10.388/III 67-F. Rev. I 10 dec. 1968, Add.I 3 dec. 1969, Add.II 3 march 1970, Add.III 16 apr. 1970.

8. Assoc. Official Agr. Chemists: "Official Methods of Analysis ". 8th. Ed. Washington, 195, 1955.

9. M. Sekiya, Y. Sugiyama, S. Okamoto: "Effects of some inorganic salts on the dehydration of calcium sulfate dihydrate". Gypsum and Lime, 61, 263-271, 1962.

10. V.L. Hill, S.B. Hendricks, E.J. Fox, J.G. Cady :
     "Acid pyro- and metaphosphates produced by thermal
     decomposition of monocalcium phosphate", Ind. Eng.
     Chem., 39, 12, 1667-1672, 1947.

11) C. Giavarini, F. Pochetti: "Applicazione di tecni-
     che termoanalitiche allo studio dei superfosfati
     normali", Annali di Chimica, 61, 682-694, 1971.

12. I.S.M.A. (Intern. Super. Manuf. Ass.): Report LF/
     65/69 (Annex I), 1965.

13. P. Pascal: "Nouveau traité de chimie minérale",
     Tome X, 209, Masson ed., 1956.

POSITIONAL EFFECTS OF THE PHOSPHATE GROUP ON THERMAL

POLYMERIZATION OF ISOMERIC URIDINE PHOSPHATES

A. M. BRYAN and P. G. OLAFSSON

DEPARTMENT OF CHEMISTRY, STATE UNIVERSITY OF NEW YORK

AT ALBANY, ALBANY, NEW YORK 12222

Synopsis

A calorimetric study of the abiogenic synthesis of biopolymers utilizing high temperature and relatively anhydrous conditions (the volcanic-areas rationale) has led to an even greater understanding of the process by which life may have originated on this planet. Accordingly the energetics associated with the thermal polymerization of the isomers of crystalline uridine monophosphates (2'-UMP, 3'-UMP and 5'-UMP) have been determined by differential scanning calorimetry. The thermal spectra are characteristic of a particular isomer and indicate the absence of a common intermediate in the condensation reactions of 2'-UMP and 3'-UMP. Enthalpimetric values indicate a preference for the following order in polymerization:-

$$5'-UMP > 3'-UMP > 2'-UMP$$

When equimolar mixtures of each of these isomers with uridine are subjected to thermal condensation, the lower melting base provides a fusion mixture in which polymerization can be initiated at a lower temperature, extends over a wider temperature range and can proceed to a greater degree than is possible in the absence or uridine.

A temperature limit of 340-350°C exists for the formation of short chain nucleotides of 3'-UMP and 5'-UMP, as a result of thermal decomposition of residual monomer units, involving loss of uracil. The greater thermal stability of 2'-UMP is attributed to interaction between the phosphate group and the base.

INTRODUCTION

There has been remarkable progress in synthesizing the more
complex essential components of living organisms and of obtaining
ever greater understanding of the processes by which life may have
originated on this planet. The formation of organic molecules from
a primordial atmosphere under thermal conditions, as postulated by
Oparin (1) and Haldane (2) is strongly supported by experimental
investigation. Thus the synthesis of bases sugars (3) and nucleo-
sides(4) has been accomplished under simulated primitive earth conditions, while monophosphates have been obtained on heating crystalline nucleosides at 150° with phosphoric acid or its monobasic
salts(5,6). A typical chemical evolution experiment has shown that
cytidine phosphate can be condensed to oligonucleotides (average
number of monomers per chain = 5.6) at 65°C by the action of polyphosphoric acid(6). The resulting oligonucleotides were found to
contain the 3'-5'-phosphate linkages that are characteristic of
natural nucleic acids. Such oligonucleotides have also been obtained on heating uridine phosphate and uridine phosphate in the
presence of uridine(7).

Now that a plausible means has been found for promoting the
production of oligonucleotides under primitive Earth-like circumstances, two main questions concerning the nature of this polymerization remain to be answered:-

a) Why, despite the variety of conditions utilized, does
polymerization of nucleotides yield only oligomeric fragments
composed of five or less monomeric units.

b) What prebiotic environment is necessary to ensure a predominance of 3'-5' linkages in the biogenic and abiogenic
biopolymers.

Thus the main thrust of our research is to provide a better
understanding of the abiogenic synthesis of these oligonucleotides.
We have previously demonstrated that differential enthalpic analysis
can account for the formation of a number of oligonucleotides of
differing chain lengths, corresponding to products isolated on a
macroscopic scale following similar thermal treatment of the corresponding nucleotides(8). The present investigation extends our
study of the self interaction individual monomer units of crystalline monomeric uridine monophosphates (2', 3' and 5'-uridine mono-
phate) and 1.1 mixtures of each with the free nucleoside, uridine.
A significant feature of our study is that it permits us to determine the influence of the position of the phosphate group on the
relative stability of each isomer and its influence on the degree
of polymerization. The thermal conditions employed may be considered as analogous to those which might have contributed to a prebiotic synthesis.

EXPERIMENTAL

The samples of uridine (u), uridine-2'-phosphate (lithium salt) (2'-UMP), uridine-3'-phosphate (sodium salt) (3'-UMP) and uridine 5'-phosphate (sodium salt) (5'-UMP) were of the highest purity commercially available (Sigma Chemical Co.) The Thermograms were recorded on a Perkin Elmer DSC-1B differential scanning calorimeter which had been calibrated in terms of temperature and energy, following standard procedures. The samples utilized were in the 0.3 to 1.0 mg range and were scanned at 20°/min under a nitrogen atmosphere over the range of 50° to $420^\circ C$. A small pinhole was pierced in the cap of the sealed cup to prevent any undue baseline shift at higher temperatures resulting from build-up of excessive pressure within the sealed cup, following decomposition.

The following equation was utilized to calculate the enthalpy values for the endo and exothermic processes.

$$\Delta H \left(\frac{m.cal}{mg.}\right) = \frac{calorimeter}{sensitivity} \left(\frac{m.cal}{sec.\ in}\right) \times \frac{155}{\substack{chart\ speed \\ (cm./min.)}} \times \frac{area\ (in.)^2}{wt.\ (mg.)}$$

The minimum number of processes involved in each polymerization was established by curve analysis of the exothermic processes in the thermograms, on a Du Pont 310 Curve Resolver.(8)

Results and Discussion

The main endeavor in the study of chemical evolution has been to reconstruct the route by which components of the nucleic acid molecule could have been formed on primordial earth prior to the appearance of life. The synthesis of the bases, sugars[3] and nucleosides[4] has been accomplished under simulated primitive earth conditions, while the monophosphates of adenosine, cytidine, guanosine, uridine and thymidine were obtained on heating crystalline nucleosides at 150° with phosphoric acid and its monobasic salts[5,9].

In achieving the formation of nucleotides, phosphorylation of the nucleoside introduces two additional degrees of freedom to the molecule. Consequently, in the P-O-C_n linkage (n = 2',3' or 5' position of the sugar moiety), the orientations of the P-O bond and that of the phosphate group as a whole must be observed, relative to the tetrahedral bond of the C_n. Thus depending on the various angles that can be assumed in nucleotides, very different conformations may result. These may vary - from extended ones[10-12] where the phosphate is stretched away from the base[13], to compact ones with the phosphate, sugar and base crowded together - thereby effecting the stacking patterns and modifying the crystalline

structures. As a result, the three phosphate isomers would be expected to exhibit differences in behavior on thermal treatment.

Differential enthalpic analysis of each of the phosphate
isomers as well as the parent nucleoside, uridine, was carried out
under identical conditions, over the temperature range of 50°-420°C,
in order to provide comparable data. The thermal profile for uridine, showed a sharp fusion endotherm at 165°C and a second endotherm at 335°C associated with the melting of uracil, formed as a
result of thermolytic cleavage of the glycosidic bond[14]. The
thermograms of the corresponding monophosphate esters are complex
in comparison, and reveal the presence of exothermic as well as
endothermic processes.

The thermogram for the 2'-isomer (Fig. I) shows the commencement of crystal degradation at 252°C. The broad fusion endotherm

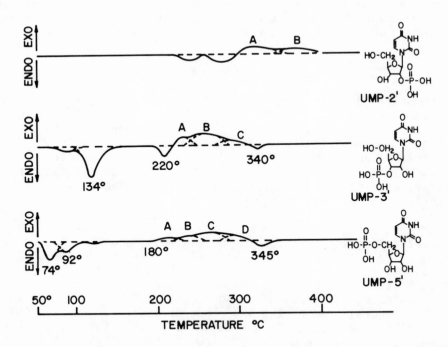

Fig. I Thermograms associated with thermal processes occurring
on heating of the individual monomeric isomers of uridine monophosphates (UMP-2', UMP-3' and UMP-5') from 50 to 420°C.

is indicative of the occurrence of simultaneous processes - fusion and condensation in the resulting melt to form a pyrophosphate which Pongs and Ts'o[15] regard, not as a side reaction but rather as an intermediate in the course of thermal polymerization of nucleotides. The thermolysis of diuridine-2'-pyrophosphate (I, R=uridine) at 310°C is followed by

I

polymerization associated with the broad exotherm from 333° to 420°C.

While the thermal spectra of 2'-UMP and 3'-UMP (Fig. 1) show similar general characteristics they do not reveal any segment of identical behavior throughout the thermal cycle. Consequently these isomeric phosphates do not pass through a common intermediate such as 2', 3'-cyclic monophosphate (II) on their route to polymerization.

II

Thermal treatment of 3'-UMP reveals a large endothermic peak with a maximum at 134°, indicative of dehydration.

The thermogram of 5'-UMP (Fig. I) shows two endotherms below 100°C associated with the removal of two moles of water held in the crystals[16] and differs markedly from those of 2'- and 3'-UMP in that 5'-UMP shows only an extremely weak endotherm above 100°C. This endotherm immediately precedes the strong exothermic process, indicating that fusion is immediately followed by the exothermic process of polymerization.

Polymerization involves phosphodiester bond formation[5,17,18] and consequently enthalpy values derived from ther thermograms for this process, provide estimates of the degree of internucleotide bonding in each case. The relative energies associated with the exothermic process for 2'-UMP, 3'-UMP and 5'-UMP, 51:87:113 respectively, indicate the much greater ability of the 5'-isomer to undergo polymerization. This compares favorably with the results of chemical polymerization of mononucleotides. Furthermore, initiation of polymerization occurs at a lower temperature for 5'-UMP than for 3'- or 2'-UMP (Fig. I).

An analysis of the exothermic portions of the thermograms by curve resolution indicates that a number of consecutive and sometimes simultaneous reactions take place. These data depict a minimum of two main components in 2'-UMP, whereas 3'- and 5'-UMP involve at least 3 and 5 components respectively, with a preponderance of a single structure in the mixed products (Table I).

A comparison of the above thermograms for 2'- and 3'-UMP with that of the 2'(3')-UMP mixture[8] shows the latter to be a composite representation of the individual nucleotides. Self condensation of 3'-UMP is preferred, suggesting the influence of stereochemical factors on the course of the reaction.

Components equivalent to those for the thermic reaction of 3'-UMP have been isolated by paper and ion exchange chromatography from the mixture obtained by heating uridylic acid on a mass scale[20].

It has been demonstrated experimentally that under suitable thermal or non-enzymatic conditions, the common types of linkages for polymer formation principally involve 2'-5' and 3'-5' phosphodiester bonds with the main products being dinucleotides and trinucleotides[21].

The thermographic data indicate that an increase in the number of components in a polymer may be associated with a positional effect of the phosphate group on the self interaction between reactant molecules, and shows clearly that while there is an inhibition or restriction of polymerization for 2'-UMP, where the P-group is closer to the base, there is increasing ease of self interaction as the phosphate group is moved to the 3'- and then

TABLE 1

Sample	Dehydration		Fusion and Decomposition		Polymerization			
	Tm °C	ΔH cal/g.	Tm °C	ΔH cal/g.	Temp. Range	ΔH cal/g. Total	Component	% Component
2'-UMP			270° 313°	16.2 24.0	333-455	50.9	A B	54% 46%
3'-UMP	72° 95° 134°	78.0	220° 342°	18.7	232-333	87.6	A B C	19% 62% 19%
5'-UMP	75° 93° 135°		345°		188-329	113.1	A B C D	12.7% 21.5% 48.3% 17.8%
2'-UMP+ U			167° 350°	17.7	240-336	40.4	A B	68% 32%
3'-UMP+ U	74° 94° 126°	44.4	165° 350°		208-340	149.7	A B C	9.3% 67.5% 23.5%
5'-UMP+ U	72° 94° 135°		166° 350°		180-338	149.9	A B C D	13.5% 16.5% 51.8% 18.2%

to the 5'-position. The distribution pattern in each case is represented by the formation of a product of maximum intensity and it
may be deduced that a preferential course of steric reactions is
preferred. Spacefilling models of these nucleotides provide a
better understanding of steric interactions of the phosphate group
as it is located at different sites on the ribose moiety. An inspection of such models indicates that 2'-UMP is more compact and
condenses with itself less readily than can 3'-UMP, whereas the
more extended phosphate group of the 5'-isomer is a more available
site for increased reaction. In each case it would seem that the
conformation for maximum interaction is complex and that the matrix
components for each molecule tend to assume a position at a certain
distance from each other with a high degree of directional alignment while orienting for maximum interaction.

The pyrocondensation of the nucleoside phosphate isomers with
the free nucleoside uridine were also investigated. (Fig. 2)

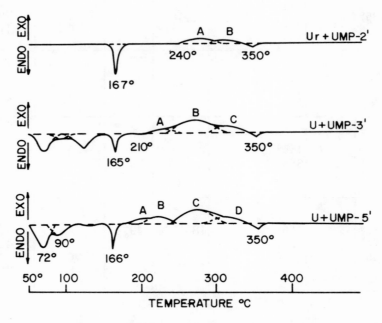

Fig. 2 Thermograms associated with thermal processes occurring on
heating of 1:1 mixtures of uridine with each of the uridine monophosphates over the temperature range of 50-420°C.

Enhanced yields are obtained if the free nucleoside is added to the monophosphate in equimolar proportions. Besides providing an extra reaction site, the addition of uridine yields a fusion mixture in which polymerization can be initiated at a lower temperature, extended over a greater temperature range and proceed to a greater degree of completion than is possible in the absence of the nucleoside. The yields of polymer are increased by more than 50% (147 cal/g) for both 3' and 5'-UMP (Table 1).

The 2'-isomer provides an exotherm of only 40 cal/g. This value may be somewhat low since bond forming in the mixture occurs in the same temperature range at which bond breaking occurs for 2'-UMP. A correction for the overall enthalpy still however, provides a value which is low in comparison to that for the other isomers. Consequently, the values obtained for the exothermic processes suggest that the 3',5'-diester linkages are more easily formed than the corresponding 2', 5' linkage[20,21].

The thermograms not only provide information regarding the reaction sequences and distribution of products but also indicate a characteristic temperature range of 340-350°C at which all polymerization reactions cease with the exception of the self-interaction of 2'-UMP. In the case of uridine, as previously discussed, thermolysis of the glycosidic bond results in the formation of uracil which melts in this region. The similar behavior of 3'- and 5'-UMP indicates that the presence of the phosphate group on these isomers does not prevent thermolytic cleavage of the base from the sugar[22]. When the phosphate group is attached to the 2'-position however, this cleavage does not occur. Since x-ray data and stereochemical analysis indicate that most of the pyrimidine nucleotides exist as the anti conformer in the solid state,[11] it is possible that the 2'-phosphate may interact strongly with the base in this preferred conformation, causing thermolytic cleavage of the C-N bond to be more difficult.

Conclusions

This study reveals some subtle and important effects of the phosphate group on intra and intermolecular interactions of nucleotides. Differential thermal analysis indicates that the position of the phosphate group markedly influences the polymerization process in the fused state, causing the thermogram for each isomer to be readily distinguishable. These enthalpic effects are apparently due to the influence of the phosphate group on the degree of alignment of each monomer, a factor which results in varying degrees of orientation that are conducive to self condensation at higher temperatures. The ease and extent of this process, therefore depends upon the intrinsic geometry of each structure and occurs with the greatest tendency for the more linear molecule, 5'-UMP, where the phosphate group is more available. Indeed the onset temperatures associated with the exothermic process of polymerization do reflect the ease of condensation:-

5'-UMP>3'-UMP>2'-UMP

Calorimetric studies also indicate a temperature limit for polymerization in all cases examined, except for 2'-UMP. This limit (340-350°C) appears to be associated with the thermal decomposition of uridine, 3'-UMP, and 5'-UMP, in which glycosidic cleavage results in the loss of uracil from the residual monomer units. Evidence for the presence of uracil is provided by the weak endotherm in the region of its melting point. The plausibility of such a thermolysis gains credence in the light of recent calorimetric studies, carried out in these laboratories, on deoxynucleosides[14].

The ability of the 2'-isomer to withstand thermolytic cleavage suggests additional intramolecular bonding between the sugar moiety and the base. It appears probable that when the phosphate group resides on the 2'-position it interacts with the adjacent base, stabilizing the latter against thermolysis. Indeed, even the unsubstituted 2'-OH group has been shown to possess characteristics which distinguish it from the 3'-OH. Warshaw and Cantor[23] have stressed the overwhelming importance of the 2'-OH in determining the structure of oligonucleotide stacked conformations. These researchers have attributed to the 2'-OH, the role of determining the preferred conformation of the sugar ring, which in turn can affect the orientation of the base about the glycosidic bond,[13] and permit the possibility of interaction between the base and the 2'-OH[24].

Our findings regarding the positional effects of the phosphate group on the relative ease of polymerization of the nucleotides and on the thermal stability of 2'-phosphate have important implications in the consideration of prebiotic synthesis of oligonucleotides. Thermal polymerization of nucleotides in solution, as carried out by Sulston et al.,[25] indicated a large portion of the 2'-5' and 3'-5' isomers. In view of the predominance of the latter in natural dinucleotides, these workers recognized that it was total specificity of the 3'-5' linkage which would be difficult to achieve under prebiotic conditions. To circumvent this problem, it was suggested that a preferential formation of the 3'-5' isomer might have resulted from a solid surface, of nonbiological origin, functioning as a contemporary enzyme. Alternatively, an intitial mixture formed indiscriminately of 2'5' and 3'-5' isomers might ultimately through an evolutionary process in the presence of a subsequently formed enzyme, result in the predominance of the 3'-5' isomer.

As a result of our studies we believe that it is possible to provide a more direct explanation for the predominance of the 3'-5' isomer within the limitations associated with prebiotic conditions. The thermal condensation of individual nucleotides in the fused state, show via the energetics associated with these reactions, that each process follows a different pathway. In particular the self-condensation of the 2'-nucleotide is much less favored than that of the 3' or 5' isomers in the lower temperature range. On the other hand, had pribiotic synthesis of oligonucleotides occurred in a cooling cycle from a temperature range above 340°C, the condensation of the 2' isomer would have been favored while the 3' and 5' isomers would have been degraded. Since the natural oligonucleotides favor the 3'-5' linkage, we would conclude that condensation had occurred during a heating cycle below 340", where polymerization of the 3' and 5' isomers is given preference.(21) Consequently, we may conclude that the controlling factor determining the predominant polymer lies in the unique property of 2' hydroxyl and the 2' phosphate group to participate in interaction with the base.

References

1. A. I. Oparin, "Proiskhozhdenie zhizni," Izd. Moskovskii Rabochii, Moscow, 1924; "The Origin of Life," The Macmillan Company, New York, New York 1938.

2. J. B. S. Haldane, "Rationalist Annual," 1929: Science and Human Life," Harper Bros., New York and London, pp 149 (1933).

3. J. Oro, N.Y. Acad. Sci. 108, 464 (1963).

4. C. Ponnamperuma and P. Kirk, Nature 203, 400 (1964).

5. C. Ponnampuruma and R. Mack, Science 148, 1221 (1965).

6. A. Schwartz and S. W. Fox, Biochim Biophys. Acta, 87, 696 (1964); 134, 9 (1967).

7. J. Moravek, Tetrahedron Lett., 18, 1707 (1967).

8. A. M. Bryan and P. G. Olafsson, J. Therm. Anal. 3, 421 (1971).

9. T. V. Waehneldt and S. W. Fox, Biochem. Biophys. Acta, 134, 1 (1967).

10. K. N. Fang, M. S. Kondo, P. S. Miller and P.O.P. Ts'o J. Am. Chem. Soc., 93, 6647 (1971).

11. M. Sundaralingam, Biopolymers 7, 821 (1961).

12. S. Arnott, D. W. L. Hukins, Nature 224, 886 (1969).

13. A. E. V. Haschemeyer and A. Rich, J. Mol. Biol., $\underline{27}$, 369 (1967).

14. P. G. Olafsson, A. M. Bryan, G. C. Davis and T. S. Anderson, manuscript submitted for publication.

15. O. Pongs and P.O.P. Ts'o, J. Am. Chem. Soc., $\underline{93}$, 5241 (1971).

16. E. Shefter and K. N. Trueblood, Acta Cryst., $\underline{18}$, 1067 (1965).

17. J. Moravek, Tetrahedron Lett., $\underline{35}$, 4167 (1966)

18. A. Beck, R. Lohrmann and L. E. Orgel, Science $\underline{157}$, 952 (1967).

19. M. W. Moon and H. G. Khorana, J. Am. Chem. Sco., 88, 1805 (1966).

20. J. Moravek and J. Skoda, Coll. Czech. Chim. Commun. $\underline{32}$, 206 (1967).

21. G. Schramm, H. G. Rotsh and W. Pollmann, Angew. Chemie Int. Ed. $\underline{1}$, 1 (1962).

22. S. Greer and S. Zamenhof, J. Mol. Biol. $\underline{4}$, 123 (1962).

23. M. M. Warshaw and C. R. Cantor, Biopolymers $\underline{9}$, 1079 (1970).

24. P.O.P. Ts'o, S. A. Rappaport and F. J. Bollum, Biochemistry, $\underline{5}$, 4133 (1966).

25. J. Sulston, R. Lohrmann, L. E. Orgel, H. Schneider-Bernhoehr, B. J. Weimann and H. T. Miles, J. Mol. Biol., $\underline{40}$, 227 (1969).

THERMAL DECOMPOSITION OF CEMENTITIOUS HYDRATES

J. N. Maycock, J. Skalny, and R. S. Kalyoncu

Martin Marietta Laboratories

Baltimore, Maryland 21227

INTRODUCTION

Portland cement is produced by intergrinding cement clinker with ca. 5% gypsum to a high surface area. The cement clinker is composed of various calcium minerals, e.g. calcium silicates, aluminates, ferrites, etc. with the principal minerals being tricalcium silicate, Ca_3SiO_5, dicalcium silicate, β-Ca_2SiO_4, tricalcium aluminate, $Ca_3Al_2O_6$, and a calcium alumino-ferrite solid solution, having a composition between $Ca_4Fe_4O_{10}$ and $Ca_4Fe Al_3O_{10}$.

When cement or clinker minerals are mixed with specific amounts of water, hydration products are formed whose chemical and physical properties vary with the conditions of hydration (time, temperature, pressure, etc.). Because of this sensitivity to the conditions of reaction with water, there are many hydrates of importance in cement chemistry. While the hydration charac-teristics of anhydrous minerals and the crystal habits of the hydrates have been studied for many years, little attention has been given to the energetics and kinetics of formation and decom-position of these hydrates.

Five hydrates have been selected for the present studies which are directed to determining their decomposition charac-teristics. These materials are: [1] _Tetracalcium aluminate carbonate 11-hydrate_ (commonly called carboaluminate), $3\,CaO.Al_2O_3.CaCO_3.11H_2O$, which may form if enough CO_3^{2-} ions are present in the cement/water system. The CO_3^{2-} may result from atmospheric CO_2 or the intentional presence of admixtures such as alkali carbonates or bicarbonates. Detailed studies of

697

the formation and properties of this hydrate were reported by several authors (1, 2, 3). Although its formation with calcium aluminates and high aluminous cements has been known for many years (4, 5) its formation with portland cement clinker, in the absence of sulfates, has been suggested only recently (6).

[2] Hexacalcium aluminate trisulfate - 32 hydrate (or ettringite), $3CaO. Al_2O_3. 3CaSO_4. 32H_2O$, is immediately formed upon contact of portland cement with water and may persist indefinitely or convert to tetracalcium aluminate monosulfate 12-hydrate (7, 8). Its corrected crystal structure has been reported recently (9).

[3] Tricalcium disilicate trihydrate (or afwillite), $3 CaO. 2 SiO_2. 3H_2O$, is a natural mineral (10) which can be obtained hydrothermally (11) or by ball-mill hydration of tricalcium silicates (12, 13). Other than ball-mill hydration, its formation at room temperature has not been reported. [4] An artificial and a [5] natural calcium silicate hydrate (C-S-H) were selected to show the similarities and differences between the two. Hydrated calcium silicates occur naturally or can be prepared synthetically with the ratios of CaO/SiO_2 and H_2O/SiO_2 and their crystal structures and crystallinities varying in wide ranges (14).

Most of the decomposition studies on these hydrates have been limited to observing changes in crystal structure as a function of temperature. Energetics of decomposition have been studied only for the simple products of hydration, such as calcium hydroxide and carbonate, which is formed in the presence of atmospheric CO_2 (15, 16, 17). A previous attempt to quantify the kinetics of decomposition of amorphous calcium silicate hydrates, formed during the reaction of tricalcium silicate with water, was made by the present authors (18).

The objective of this study is to gain a better understanding of the basic properties of calcium silicate and aluminate hydrates which may form under certain conditions in cementious systems. Detailed knowledge of the basic aspect of formation and degradation of these compounds eventually may lead to improvements in the engineering properties of silicious building materials.

EXPERIMENTAL TECHNIQUES

X-ray diffraction patterns were obtained by a Phillips XRD 3000, using Cu Kα radiation and graphite monochromator.

DTA/TGA studies were performed on a Mettler Thermoanalyzer 1, using 10-mg samples at a $10°C min^{-1}$ heating rate in flowing nitrogen (5 $l h^{-1}$) at sensitivities of 0.1 mg in^{-1} and 5 V in^{-1} (~$\frac{1}{2}°C$) for the TGA and DTA, respectively.

The kinetic studies were obtained on a Mettler DTA 2000 System, using 5 mg samples at heating rates 5, 10, 15 and 25°C min⁻¹. The sensitivities varied with the material studied. Indium was used to calibrate the instrument.

Morphology of the specimens was investigated with a JSM-U3 Scanning Electron Microscope (SEM), with energy dispersive X-ray analysis (EDAX) used for elemental analysis.

CHARACTERIZATION OF MATERIALS

$3\,CaO.\,Al_2O_3.\,CaCO_3.\,11\,H_2O$. A synthetic preparation was obtained by the courtesy of the Portland Cement Association, Skokie, Illinois. X-ray diffraction has shown all the d-spacings quoted in the literature for this compound (2, 7), with the basal spacing being 7.6 Å. No minor compounds were detected in the material. A typical DTA trace is shown in Fig. 1, curve A. TGA measurements have shown that 38% of the weight was lost between 20° and 1000°C. This is ca. 4% less than the theoretical loss, which is ca. 35% for H_2O and ca. 7.7% for CO_2. The weight loss associated with the third DTA peak of our sample CO_2 evolution is 7% and that for H_2O is 31%. This implies that some of the loosely bound water was lost before the thermal decomposition experiments were initiated. Of the 31% weight loss believed to be associated with dehydration of the sample, 16%, or approximately one-half, was decomposed between room temperature and 210°C (first DTA peak), and 15% was lost between 210°C and the beginning of the CO_2 evolution at 600°C.

$3\,CaO.\,Al_2O_3.\,3\,CaSO_4.\,32\,H_2O$. A synthetic preparation was also supplied by the Portland Cement Association. All d-spacings characteristic of this compound were located by X-ray diffraction (7). No minor compounds were present. A typical DTA trace is presented in Fig. 1, curve B. The TGA total loss was found to be 41% (compared with the theoretical 46%). Most of the water was lost before 200°C (ca. 85%); however, slight weight losses were found to correspond to DTA peaks at approximately 260°(1%) and 700°C(1%). The latter peak may correspond to decomposition of a small amount of $CaCO_3$; however, as mentioned earlier, its presence was not detected by X-ray diffraction. The former DTA peak had been shown previously, e.g. Midgley and Rosaman (8), and may be due to the presence of unidentified impurities or further degradation of the products of ettringite decomposition. The approximate chemical formula of ettringite, given by Moore and Taylor (9) as $Ca_6\,[Al(OH)_6]\,(SO_4)_3.\,26\,H_2O$, suggests such a stepwise decomposition.

Afwillite, $3\,CaO.\,2\,SiO_2.\,3\,H_2O$. A sample prepared hydrothermally at the Portland Cement Association. Free $Ca(OH)_2$ has been extracted. X-ray data were in complete agreement with

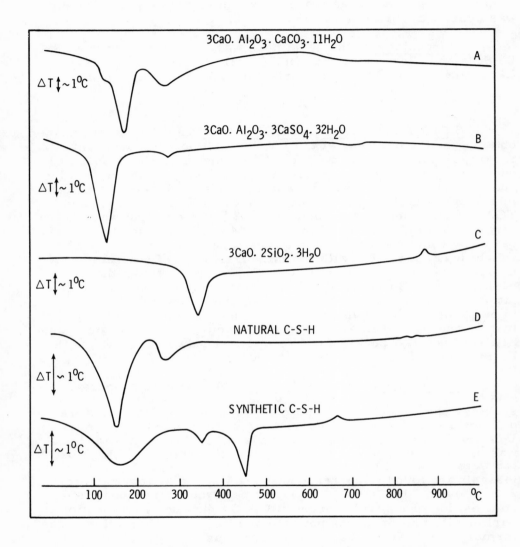

Fig. 1.   Differential thermal analysis traces for the five hydrates
investigated.   These data relate to a heating rate of
6°C min^{-1}, the sample in a platinum holder and a N$_2$
gas flow of 10 liters hour^{-1}.

those quoted in the literature (7, 13). No free $Ca(OH)_2$ or other compound was located by X-ray diffraction. The DTA pattern of the sample is shown in Fig. 1, curve C. The total TGA loss was 15.8% (15.7% theoretically). Most of the H_2O was lost between 300° and 360°C (75%), with additional gradual weight loss up to 700°C.

Naturally occuring calcium silicate hydrate composite. Found in Crestmore, California - was identified by powder X-ray diffraction as an intimate mixture of 14 Å tobermorite ($Ca_5Si_6O_{26}H_{18}$) and jennite, reported to have the approximate composition of $Na_2Ca_8Si_5O_{30}H_{22}$ (19). Scanning electron microscopy of the composite shows blade-shaped crystals, which, at lower magnifications, appear as fibrous aggregates (Fig. 2). At higher magnifications, however, the platelike character of the material is evident. EDAX did not reveal, even at very high sensitivities, the presence of Na; this absence is consistent with a report (14) that Na is not essential to the structure of jennite. The ratios of Ca to Si varied within a small range, and no significance could be attributed to the effects of this ratio on the morphology of the particles studied.

The material decomposes in two steps (Fig. 1, curve 4). The total loss was found to be 18% (20° to 1000°C), the majority of which was lost between ca. 50° and 200°C, with a peak at approximately 150°C. The second weight loss is broad and occurs between 150° and 400°C.

X-ray diffraction data for this material and information from the literature on jennite and 14 Å tobermorite are compared in Table I.

Synthetic calcium silicate hydrate. Prepared by Portland Cement Association from a mixture containing 2.65 moles of CaO per 1 mol of SiO_2 (C/S=2.65) hydrated in an excess of H_2O (w/s=9) at 25°C for 14 years. A DTA trace of the D-dried (21) sample is shown in Fig. 1, curve E, and shows several endotherms - a broad double peak between ca. 50° and 250°C, a smaller but sharper peak at approximately 340°C, and the $Ca(OH)_2$ decomposition peak at approximately 450°C. In addition, there is an exothermic peak at about 670°C.

X-ray diffraction data are given in Table II. As expected, $Ca(OH)_2$ and amorphous calcium silicate hydrate (CSH) are present. However, several peaks showing the presence of afwillite also were found. This is of special interest because, as mentioned earlier, afwillite previously had been synthetized only under hydrothermal or ball-milling conditions (7, 12, 13). The X-ray data are in agreement with the DTA/TGA studies showing an endotherm at ca. 340°C. As was also discussed earlier, pure afwillite

decomposes under the conditions used at about 335°C. The exothermic peak is visible at much lower temperatures than those
reported for afwillite (21) and CSH I (23, 24), suggesting that
neither kilchoanite (7) nor wallastonite (23, 24) are formed. X-ray
diffraction examination of the material heated to 1000°C revealed
primarily β -dicalcium silicate (larnite).

Table I

X-ray Powder Diffraction Data for the Natural C-S-H Composite

Observed d(Å)	I	Jennite (lit. 19) d (Å)	14 Å Tobermorite (lit. 20) d (Å)
14.0	10		14.0
10.5	10	10.5	
6.96	2		7.0
6.45	3	6.46	
5.52	8		5.53
5.20	2	5.20	
4.79	1	4.77	4.80
4.65	6		4.65
3.93 b	1	3.93	3.95
3.74	1	3.72	3.74
3.47	8	3.47	
3.29	4	3.29	
3.25	6		3.25
3.18	1	3.19	
3.07	3		3.07
3.05	3	3.04	
2.99	3		2.98
2.92	5	2.92	
2.82	6	2.83	
2.80	7		2.81
2.76	3	2.78	
2.72	5		2.72
2.66	7	2.66	
2.61	3	2.61	
2.52	1	2.52	2.52
2.42	1	2.43	2.41
2.31	3		2.32
2.27	1	2.27	2.27
2.16	1		2.15
2.08	1	2.08	2.08
2.05	1	2.04	

Fig. 2. Scanning Electron Micrographs of Natural C-S-H.

Table II

X-ray Powder Diffraction Data for the Synthetic C-S-H

Observed d(Å)	I	Ca(OH)$_2$ d(Å)	CSH I (lit. 7) d(Å)	Afwillite (lit. 20) d(Å)
12.5 b	4		12.5	
6.47	1			6.46
5.09	1			5.08
4.90	7	4.90		
3.75	1			3.75
3.18	3			3.19
3.12	4	3.12		
3.07	5		3.07	
2.85	2			2.84
2.80	2		2.80	
2.76	2			
2.74	3			2.74
2.62	10	2.62		
2.09	1		2.10	
1.945	2			1.949
1.927	5	1.927		
1.863	1		1.83	1.862
1.826	3			
1.796	4	1.796		

ENERGETICS OF DECOMPOSITION

It is general practice to obtain kinetic data of solid state decompositions by evaluating a series of experiments performed under isothermal conditions. To save time, it would be of considerable value if differential thermal analysis/differential scanning calorimetry could be used to provide these kinetic data; but, kinetics is a quantitative subject which requires very precise techniques. Until very recently, DTA experiments have employed fairly large samples , ca. 200 to 500 mg, in order to observe the required thermal event. This requirement immediately casts suspicion on resultant kinetic data, because it assumes that the heat capacity of the sample is constant over a fairly wide temperature range and that the temperature within the sample is constant.

The heat capacity of interest is that of the sample holder plus the sample. This parameter can change considerably during the course of a reaction because of varying amounts of transformed and untransformed material. Overcoming this potential problem requires either a massive sample holder (which limits detection sensitivity) or the use of very small samples (which require extremely sensitive detection techniques). The variability of temperature within a large sample can be overcome by using a very small sample and the appropriate electronic detection sensitivity. A system capable of acheiving the desired sensitivity with a small sample is the Mettler DTA 2000 system, which, for this study, employed 5-mg samples and, when needed, an optimum ΔT sensitivity of $115 \mu V / °C$.

Any kinetic analysis of DTA/DSC data requires a knowledge of the sample temperature. Therefore, in the data presented, all of the temperatures were corrected for instrument lag by the relationship

$$T_s = T_p + \frac{\Delta U}{S} - 0.4 \frac{d T_p}{dt}$$

where T_p is the correct sample temperature; T_p is the temperature readout on the DTA trace; ΔU the height of the endotherm in μV at T_p ; $S = \Delta U / \Delta T$ calculated from a calibration with pure indium and is basically a sensitivity factor; and $d T_p / dt$ is the heating rate.

The DTA/DSC data for the decomposition of the minerals previously discussed were analyzed by the Rogers and Morris method (25) where,

$$- E = R \frac{\ln d_1 - \ln d_2}{1/T_1 - 1/T_2} = \frac{4.58 \log (d_1 / d_2)}{1/T_1 - 1/T_2}$$

and where d_1 are any two distances from the baseline at the associated absolute temperatures T_1 and T_2. Since the distances are used as a ratio, the proportionality constants (i. e. , the constants that relate distance to the rate of heat absorbed) are cancelled. Therefore, a plot of log d against reciprocal temperature will yield an Ahrrenius plot. Figure 3 illustrates a typical DTA/DSC trace for the decomposition of ettringite from which

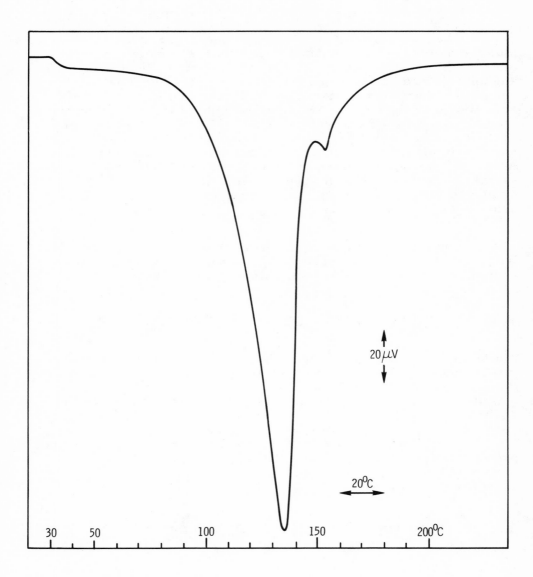

Fig. 3.   Differential thermal analysis trace for ettringite
            obtained in the Mettler TA 2000 using Al crucibles, a
            static air atmosphere and a heating rate of   5°C min^{-1}.

the Arrhenius plot of Figure 4 was derived. The data of Figure 4 show a good consistancy of analysis for the three heating rates used to investigate the decomposition of ettringite. Analysis of all the other minerals gave data of the quality shown in Figure 3.

As a further qualitative check on the validity of the energies of decomposition calculated by the Rogers and Morris technique, the data also were analyzed by the Kissinger method (26). This technique employs the variation in peak temperature with heating rate to determine the activation energy; it is based

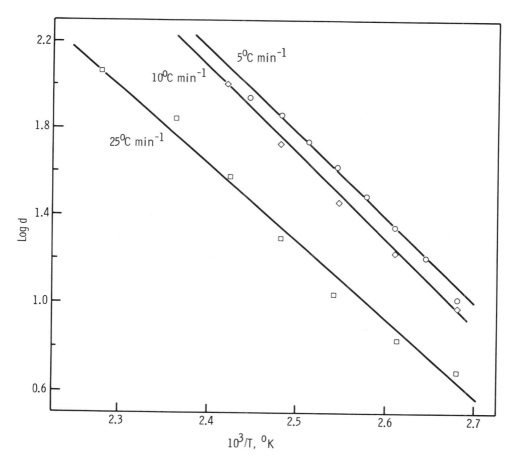

Fig. 4. Arrhenius plot of ettringite decomposition using the Rogers and Morris analytical approach.

upon the premise that the maximum reaction rate occurs at the
apex of the peak. However, the maximum rate of reaction actually occurs between the low temperature inflection point and the
peak temperature. In fact, the peak temperature is simply the
point at which the rate of heat evolution or absorption by the sample matches exactly the rate of heat transfer to the sensing element. However, with the experimental arrangement of the instrument used--i. e. small heat capacity aluminum cups, small sample
sizes (5 mg) and high thermal sensitivity--the maximum rate of
reaction should be very close to the peak temperature of the
thermal event. Figure 5 is an Arrhenius plot using the Kissinger
technique for the decomposition of ettringite.

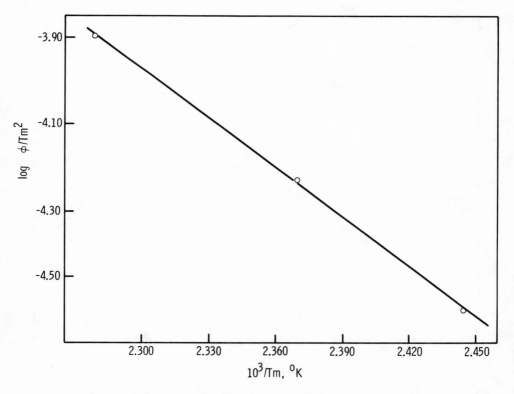

Fig. 5.   Arrhenius plot of ettringite decomposition using the
           Kissinger analytical method.

Table III

Activation Energies and Enthalpies of Dehydration

Material	Corresponding DTA Peak Temperature from Fig. 1 °C	Activation Energies of Dehydration kcal/mole					Enthalpies of Dehydration cal/gm
		Kissinger Method	5°C/min	10°C/min	15°C/min	25°C/min	
Carbo-aluminate	~170 ~260	13.5 16	17 15	16 14	– –	17 –	98 –
Ettringite	~130	17.5	17.3	17.5	–	16	148
Afwillite	~340	40	–	42	39	40	45
Natural C-S-H	~150 ~270	9 34	– 24	– –	10 –	14.5 28	55 –
Synthetic C-S-H	~160 ~340	4.5 42	5.7 –	– –	5 –	– –	33 –

RESULTS

A compilation and comparison of activation energies for the thermal decomposition of the hydrates studied is shown in Table III. Inspection of this table shows a general consistency of data between the Rogers and Morris technique used for different heating rates and the energy values found by the Kissinger method. This does not necessarily prove that the Kissinger approach is quantitatively correct, but it does substantiate our assumption that, for the equipment used, the maximum rate of reaction is close to the peak temperature. Also included in Table III are the enthalpies of dehydration for the minerals studied.

It is also interesting to find that the energies related to the afwillite mineral agree very well with the trace afwillite suspected in the synthetic calcium silicate hydrate. This finding, determined by both X-ray and the thermal analyses, is the first documented evidence for the existence of afwillite in this system under these preparation conditions.

ACKNOWLEDGEMENTS

The authors would like to acknowledge the extremely helpful cooperation of the Mettler Instrument Corporation.

REFERENCES

1.   R. Turriziani and G. Schippa, Ricerca Sci. 26 (1956), 2792.

2.   E. T. Carlson and H. A. Berman, J. Research NBS 64A (1960), 333.

3.   W. Dosch, H. Keller and H. zur Strassen, Chemistry of Cement, Vol. II, 72 Proc. Vth Int. Symp. , Tokyo, 1968.

4.   G. Schippa, Ricerca Sci. 28 (1957), 2120.

5.   J. Farran, Rev. Materiaux Construct. et Trav. Publ. , C, 490, 155; 492, 191 (1956).

6.   S. Brunauer, J. Skalny, I. Odler add M. Yudenfreund, Cement Concrete Res. 3 (1973), 279.

7.   H. F. W. Taylor (Editor), The Chemistry of Cements, Academic Press, London & New York, 1964.

8.   H. G. Midgley and D. Rosaman, Chemistry of Cement, Vol. 1, 205, Proc. IVth Int. Symp. , Washington , 1960.

9. A. E. Moore and H. F. W. Taylor, Acta Cryst. B26 (1970), 386.

10. G. Snitzer and E. H. Bailey , Amer.   38 (1953), 629.

11. L. Heller and H. F. W. Taylor, J. Chem. Soc. (1952), 1018, 2535.

12. S. Brunauer, L. E. Copeland and R. H. Bragg, J. Phys. Chem. 60 (1956), 112.

13. D. L. Kantro, S. Bruanuer, C. H. Weise, J. Colloid Science 14 (1959), 363.

14. H. F. W. Taylor, Chemistry of Cement, Vol. II, 1, Proc. Vth Int. Symp. , Tokyo, 1968.

15. N. G. Dave and S. K. Chopra, J. Amer. Ceram. Soc. 49 (1966), 575.

16. R. Sh. Mikhail, S. Bruanuer and L. E. Copeland, J. Colloid Interface Sci. 21 (1966), 394.

17. H. T. S. Britton, S. J. Gregg and G. W. Winsor, Trans. Faraday Soc. 48 (1952), 63.

18. J. N. Maycock and J. Skalny, Thermochimica Acta 5 (1973).

19. A. B. Carpenter, R. A. Chalmers, J. A. Gard, K. Speakman and H. F. W. Taylor, Amer. Min. 51, (1966), 56.

20. L. Heller, H. F. W. Taylor, Crystallographic Data for the Calcium Silicates, H. M. S. O. , London, 1956.

21. L. E. Copeland and J. C. Hayes, ASTM Bull. 149 (1953), 70.

22. L. Heller, Chemistry of Cements, Proc. III rd Int. Symp. , 77, 237, London, (1952).

23. D. Lawrence, Nature 201 (1964), 607.

24. D. Lawrence, Proc. 7th Conf. Silicate Ind. , 259, Budapest (1963).

25. R. N. Rogers and E. D. Morris, Anal. Chem. , 38, (1966), 412.

26. H. E. Kissinger, Anal. Chem. 29, (1957), 1702.

THERMAL DECOMPOSITION STUDIES OF SODIUM AND POTASSIUM TARTRATES

Alfred C. Glatz[*] and A. Pinella[**]

[*]Voland Corporation, New Rochelle, N.Y.

[**]Western Electric, No. Andover, Mass.

I. INTRODUCTION

In a previous publication on the thermal decomposition of hydrated and deuterated Rochelle Salts (1), a mechanism was proposed to explain the observed thermal data that involved assumptions of the thermal decomposition of both sodium and potassium tartrates. This study showed that endothermic transitions were obtained in the temperature ranges of 108-125°C, and 158-180°C, which were attributed to the dehydrations of sodium and potassium tartrates, respectively. Also, the weight losses obtained at 250°C were attributed to the decomposition of the tartrates to the oxalates with the deposition of carbon. A search of the available literature showed that Mattu and Pirisi (2) published the only previous paper on the alkali tartrates. They presented DTA thermograms of sodium and potassium tartrates, but the data was not sufficient to permit the analysis of the thermal decomposition of these compounds. It is, therefore, the purpose of this paper to study the thermal decompositions of sodium and potassium tartrates by modern thermoanalytical techniques (DTA and TGA), and to relate these studies to the previous investigation of the Rochelle Salts.

II. EXPERIMENTAL METHODS

A. <u>Materials</u> - Sodium and potassium tartrates, neutral reagent grade crystals, (Matheson Coleman and Bell) were used directly without further purification.

B. <u>Thermal Analysis Equipment</u> - A Chyo Simultaneous

DTA/TGA System (Model TRA3L) was used for simultaneous scans in
air and inert gas atmospheres. A heating rate of 10°C/min. was
employed.

Also, a Voland Model 1100 Thermalanalyzer with the Model
1100-11 TGA Attachment was used for more accurate TGA scans
of the decompositions. The heating rate employed was 5°C/min.
These thermograms were used for calculating the kinetic parameters of the decompositions.

C. X-Ray Studies - The x-ray diffraction studies were
performed with a Norelco Diffractometer and the observed d-
spacings were compared with the ASTM X-Ray Diffraction Card File.

III. EXPERIMENTAL RESULTS
The thermal decomposition of $K_2C_4H_4O_6 \cdot \frac{1}{2} H_2O$ (KTar) by
simultaneous DTA/TGA in air is shown in Figure 1. It can be
observed that the endotherm at 171°C has a weight loss associated with it, which is due to the dehydration of KTar. Then,
there is an exotherm beginning at approximately 250°C, followed
by a sharp endotherm at 262°C, and another exotherm at approximately 300°C. This series of enthalpic behavior is due to the
conversion of the tartrate to the oxalate (exotherm at 250°C),
followed immediately by the vaporization of water (endotherm
at 262°C. and subsequent weight loss), and the conversion of
the oxalate to the carbonate (exotherm at 300°C).

The thermal decomposition of $Na_2C_4H_4O_6 \cdot 2 H_2O$ (NaTar) by
simultaneous DTA/TGA in air is shown in Figure 2. It can be
observed that the general enthalpic behavior of this thermogram
is nearly the same as that observed for KTar. The only difference being the endotherm and corresponding weight loss observed
at 70°C. This is probably due to the loss of some of the water
of crystallization by efflorescence, as is observed for the
Rochelle Salts (3). The endotherm at 119°C. corresponds to the
completed dehydration of NaTar. The exotherm beginning at approximately 240°C is then due to the conversion of the tartrate to
the oxalate. This is immediately followed by the endotherm at
247°C, which is due to the vaporization of water with an associated weight loss. The immediate exotherm at about 275°C is
due to the conversion of the oxalate to the carbonate. Powder
x-ray diffraction studies at both 280° and 500°C for NaTar
verified that the decomposition product was Na_2CO_3. Also
visual inspection at 300°C of the decomposition products for
both NaTar and KTar indicated a black carbon deposit.

A more accurate display of the respective **weight losses for**

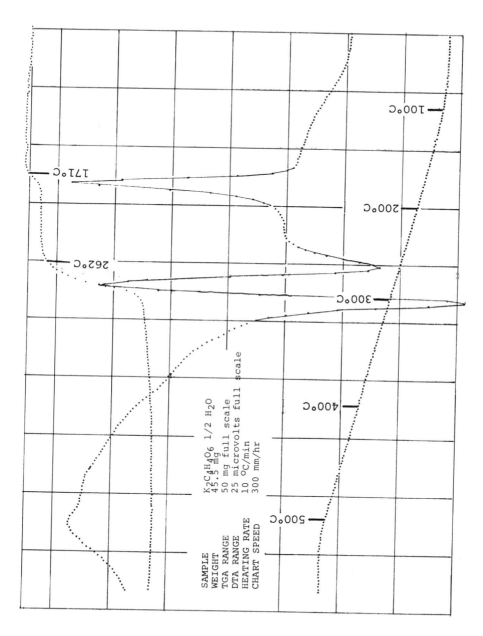

FIGURE 1 - SIMULTANEOUS DTA/TGA SCAN OF $K_2C_4H_4O_6 \cdot \frac{1}{2}H_2O$

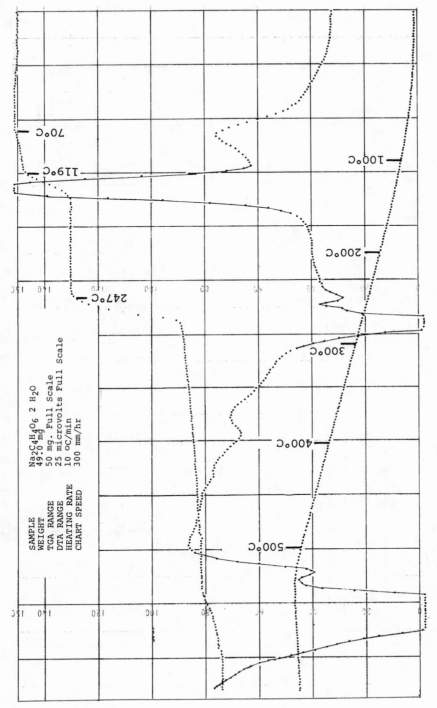

FIGURE 2 — SIMULTANEOUS DTA/TGA SCAN OF $Na_2C_4H_4O_6 \cdot 2H_2O$

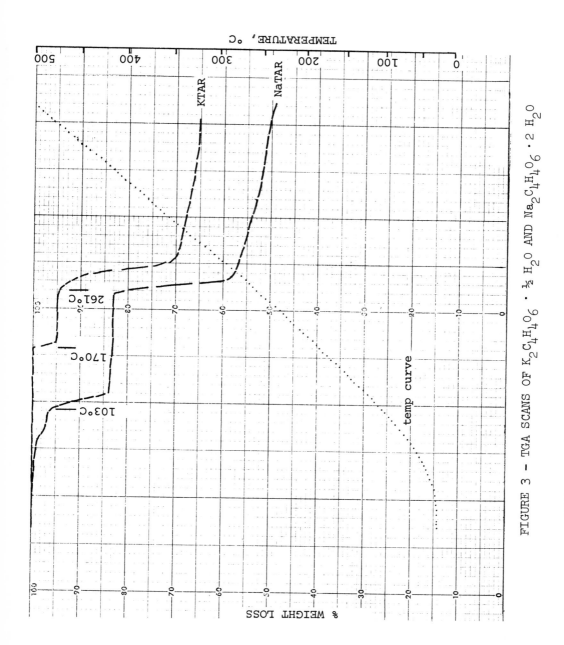

FIGURE 3 — TGA SCANS OF $K_2C_4H_4O_6 \cdot \frac{1}{2} H_2O$ AND $Na_2C_4H_4O_6 \cdot 2 H_2O$

KTar and NaTar was obtained with the Voland Model 1100-11 TGA
System. These scans are shown on Figure 3. The corresponding
Percent Weight Losses for dehydration, decomposition to the
oxalate, and decomposition to the carbonate were obtained
directly from these TGA scans and are given in Table I below.

TABLE I

PERCENT WEIGHT LOSSES FOR KTar AND NaTar

		Total % Wt. Loss	
		Measured	Calculated
1.	Dehydration		
	KTar	5.0	3.82
	NaTar	16.1	15.64
2.	Decomposition to the Oxalate		
	KTar	30.0	19.2
	NaTar	41.9	31.3
3.	Decomposition to the Carbonate		
	KTar	35.0	31.0
	NaTar	51.0	56.5

The calculated values given in Table I were determined from the
following stoichiometric equations,

$$ATar \cdot nH_2O \longrightarrow ATar + nH_2O$$

$$ATar \longrightarrow AC_2O_4 + 2C + 2H_2O$$

$$AC_2O_4 \longrightarrow ACO_3 + CO$$

where, $A = K_2$ or Na_2

It can be observed that there is good agreement for the percent
weight losses for both the dehydration and conversion to the
carbonate reactions. But for the conversion to the oxalate
reaction, the agreement is poor. This is consistent with the
conversion of the tartrate to the oxalate, and its immediate
conversion to the carbonate, i.e. the tartrate is converted
directly to the carbonate. This result is consistent with the
DTA/TGA scans shown in Figures 1 and 2. It can be observed that
as soon as the tartrate is converted to the oxalate (endothermic)
there is observed an immediate exothermic reaction for the conversion to the carbonate. This result is also supported by the
x-ray studies.

The TGA scan in Figure 3 can be used to calculate the water of crystallization lost by efflorescence for NaTar. This calculation shows that,

$$Na_2C_4H_4O_6 \cdot 2H_2O \xrightarrow{70^\circ} Na_2C_4H_4O_6 \cdot (2-x)H_2O + xH_2O$$

where x = 0.396

IV. DISCUSSION

These studies have shown that the thermal decomposition in air of $K_2C_4H_4O_6 \cdot \frac{1}{2}H_2O$ (KTar) and $Na_2C_4H_4O_6 \cdot 2H_2O$ (NaTar) occur via equations (1) and (2), respectively.

For KTar

a) $K_2C_4H_4O_6 \cdot \frac{1}{2}H_2O \xrightarrow{171^\circ} K_2C_4H_4O_6 + \frac{1}{2}H_2O$

b) $K_2C_4H_4O_6 \xrightarrow{262^\circ} K_2C_2O_4 + 2C + 2H_2O$ (1)

$\quad\quad\quad\quad\quad\quad \longrightarrow K_2CO_3 + CO$

For NaTar

a) $Na_2C_4H_4O_6 2H_2O \xrightarrow{70^\circ} Na_2C_4H_4O_6 \cdot 1.6H_2O + 0.4H_2O$

b) $Na_2C_4H_4O_6 \cdot 1.6H_2O \xrightarrow{119^\circ} Na_2C_4H_4O_6 + 1.6H_2O$ (2)

c) $Na_2C_4H_4O_6 \xrightarrow{247^\circ} Na_2C_2O_4 + 2C + 2H_2O$

$\quad\quad\quad\quad\quad\quad \longrightarrow Na_2CO_3 + CO$

The kinetics of the dehydrations and decompositions of KTar and NaTar were determined from the TGA scans given in Figure 3. The data were plotted according to the theory of Horowitz and Metzger (4). The plots for NaTar are given in Figure 4. It can be observed that reasonable straight lines are obtained which indicates that the reactions are characterized by first order kinetics. The activation energies for the various reactions were obtained from the slopes of the linear plots and are given in Table II on the following page. It is interesting to observe that these reactions are characterized by relatively high activation energies. The high activation energies for the decomposition reactions would be expected, since they represent the conversion of the tartrate directly to the carbonate. But for the dehydration reactions it is an unexpected result.

IV. CONCLUSIONS

This study of the thermal decomposition of KTar and NaTar essentially verified the previous mechanism for the decomposi-

TABLE II

ACTIVATION ENERGIES FROM KINETIC STUDIES

		E* (Kcal/mole)
1.	KTar	
	a) Dehydration	93.8
	b) Decomposition	72.2
2.	NaTar	
	a) Dehydration	64.2
	b) Decomposition	126.2

tion of the Rochelle Salts (1), by explaining the endotherms in the temperature ranges of 108° - 125° and 158° - 180°C. These endotherms were shown to be due to the dehydrations of NaTar and KTar, respectively. In addition, it has been shown that the decomposition of the alkali tartrates occurs over a narrow temperature range in a non-flowing air environment, and probably consists of consecutive reactions. TGA studies of the dehydrations

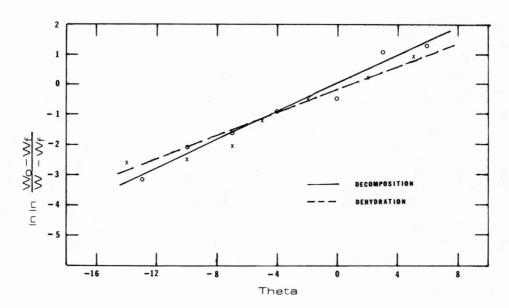

FIGURE 4 - TGA KINETIC DATA PLOT FOR $Na_2C_4H_4O_6 \cdot 2 H_2O$

and decompositions of KTar and NaTar indicate that these reactions are characterized by relatively high activation energies and first order kinetics.

In addition, this study showed that some caution should be exercised in using NaTar as a standard for Karl Fisher determinations, because of the efflorescence of some of the water of hydration.

V. REFERENCES

1. A. Glatz, I. Litant, and B. Rubin in "Analytical Calorimetry", Ed. by R. Porter and J. Johnson, Vol. 2, Plenum Press, New York, 1970, pgs. 255-268.

2. F. Mattu and R. Pirisi, Rend. Seminar fac. Sci. Univ. Cagliari, 25, 96-117, (1955)

3. W. P. Mason, "Piezoelectric Crystals and their Applications to Ultrasonics", D. Van Nostrand Co., New York, 1950, pgs. 114-136.

4. H. Horowitz and G. Metzger, Anal Chem., 35, 1464, (1963)

THE ENTHALPY AND HEAT OF TRANSITION OF Cs_2MoO_4 BY DROP CALORIMETRY

D. R. Fredrickson and M. G. Chasanov

Argonne National Laboratory
9700 S. Cass Avenue
Argonne, Illinois 60439

INTRODUCTION

The enthalpy measurements on Cs_2MoO_4 reported in this paper were carried out as part of a calorimetric program to study compounds of interest in reactor technology. Cs_2MoO_4 may be produced in the urania-plutonia fuel elements of a liquid-metal fast-breeder reactor (LMFBR) since both cesium and molybdenum have been observed in the oxide-clad interface region of the fuel pin. Therefore, thermodynamic data are needed to predict the conditions of formation of Cs_2MoO_4 and its stability.

The high-precision drop-calorimetric system has been in operation for some time involving measurements on more than a dozen different compounds. Recently the furnace has been disassembled and rebuilt with various improvements being made.

APPARATUS

The calorimetric system, Figure 1, consists of (1) a furnace with a molybdenum core heated by tantalum wire and operated under a high vacuum, (2) a copper-block calorimeter which uses a quartz-crystal thermometer for temperature measurement, (3) a gate to isolate the calorimeter from the furnace except during a drop, and (4) a drop mechanism for transfer of the sample capsule from the furnace to the calorimeter.

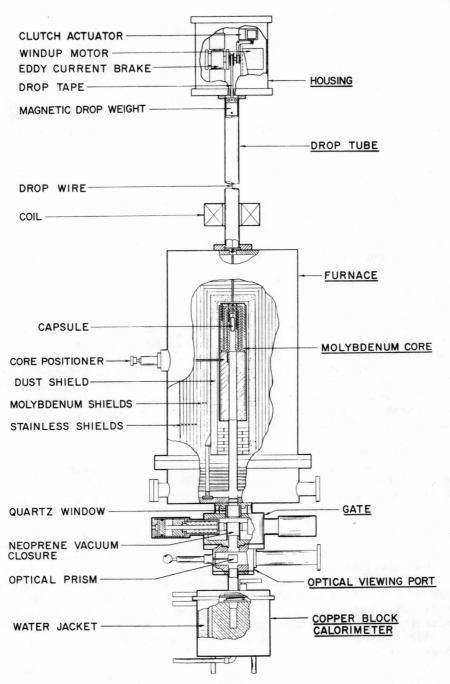

CLUTCH ACTUATOR
WINDUP MOTOR
EDDY CURRENT BRAKE
DROP TAPE
HOUSING

MAGNETIC DROP WEIGHT

DROP TUBE

DROP WIRE

COIL

FURNACE

CAPSULE

MOLYBDENUM CORE

CORE POSITIONER
DUST SHIELD
MOLYBDENUM SHIELDS
STAINLESS SHIELDS

QUARTZ WINDOW
GATE

NEOPRENE VACUUM
CLOSURE

OPTICAL PRISM
OPTICAL VIEWING PORT

WATER JACKET
COPPER BLOCK
CALORIMETER

Fig. 1.   Resistance-Heated Drop-Calorimetric System

Furnace

One of the most crucial requirements that must be met in high precision drop calorimetry is that of a uniform and accurately known sample temperature. The attainment of a uniform sample temperature is facilitated by the minimization of thermal gradients in the furnace heating zone through low power input and proper radiation shielding. Special consideration was given to these factors in the design of the furnace. The furnace core, Figure 2, is divided into two separately heated sections. The upper section is 6 in. long and consists of a heated shield, a main shield, and an inner shield; the lower section is 8 in. long and consists of a single heated shield. The three component upper section is designed to smooth out temperature gradients around the capsule, while the lower heated shield is designed to allow the sample to accelerate to approximately 245 cm/sec before entering the relatively cold space between the furnace and the calorimeter. Tantalum wire (20 and 25 mil diameter, respectively) insulated with alumina tubing is used for the upper and lower heaters. Surrounding the whole core is a "dust shield" consisting of a tantalum shell with an inner space filled with alumina grain; this dust shield is equivalent to many very closely spaced radiation shields. A series of three molybdenum and six stainless steel radiation shields of low thermal capacity surround the dust shield.

The molybdenum furnace core, Figure 2, contains six Pt, Pt-10% Rh thermocouples for temperature measurement and control purposes. The temperatures of the upper and lower sections of the core are controlled automatically.

During an experiment, the difference between the control thermocouple voltage and the temperature-setting bucking voltage is amplified and recorded. A signal proportional to the deviation of the recorder from a set point is sent to a current adjusting type control unit, (L&N CAT-60) which produces an output current proportional (in the range 0-5 mA) to the input signal. This current is fed to a transistorized power controller (Barber-Coleman SCR) which controls the dc power to the heater. Two identical control units are used, one for the upper section of the core and the other for the lower section of the core; this arrangement maintains a furnace temperature to within 0.1° of the control setting.

The temperature of the inner shield Fig. 2, is measured by three N.B.S. calibrated (International Practical Temperature Scale of 1968) Pt, Pt-10% Rh thermocouples which are situated at different locations so that a temperature profile around the capsule may be obtained. Upon proper equilibration, readings obtained from these thermocouples show that a temperature gradient of less than one

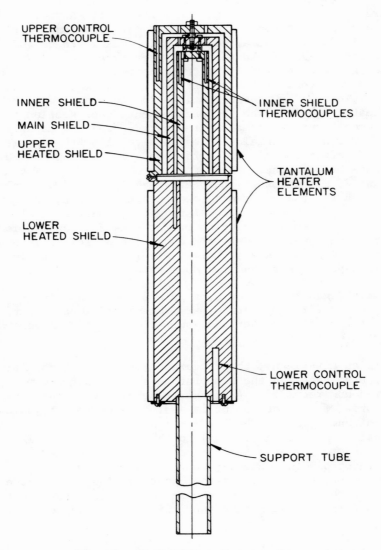

Fig. 2.   Molybdenum Furnace Core

degree exists in the inner shield. The temperature of the suspended capsule is, therefore, taken to be the same as the inner shield. A temperature differential of less than one degree exists between the inner shield and the heated shield under control conditions.

Calorimeter

The calorimeter consists of a 10-kg cylindrical block of copper with a chromium-plated exterior. This cylinder contains a well with a replaceable metal lining for receiving the sample. This well is part of the furnace vacuum system when the gate (Fig. 1) is open, and when the gate is closed the calorimeter well has its own vacuum system. Radiation from within the well is blocked (except during the sample drop) by a remotely operable pair of shutters. Machined into the bottom of the copper block is a hole about 1 cm from the side for a quartz crystal thermometer, and six holes, equally spaced about 1 cm from the receiver well, for an electrical calibration heater. A 1-cm air space separates the copper block from a chromium -plated brass jacket. Constant temperature ($\pm$0.002°) water is circulated through the jacket.

Gate

Between the furnace and calorimeter there is a vacuum tight gate. (Figure 1, neoprene vacuum closure). The seal is accomplished by pneumatic compression of a 2 in. length of neoprene tubing. To change a sample capsule without disturbing the furnace temperature and vacuum, (1) the capsule is lowered into the calorimeter, (2) the gate is closed around the drop wire, (3) the calorimeter is brought to atmospheric pressure, (4) the calorimeter is lowered and (5) the capsule changed.

Drop Mechanism

The drop mechanism is contained in the vacuum housing and drop tube shown in Fig. 1. The capsule hangs in the furnace on a 20 mil tantalum wire that is suspended from the magnetic weight positioned in the drop tube. The weight in turn is suspended by a metal tape wound on the drum of the eddy current brake. When the weight is released, it falls the length of the drop tube and allows the capsule to fall into the calorimeter. As the capsule approaches the calorimeter, the magnetic weight passes a reed switch on the outside of the drop tube causing an electric pulse to actuate the eddy current brake. The capsule is decelerated and comes to rest in the calorimeter.

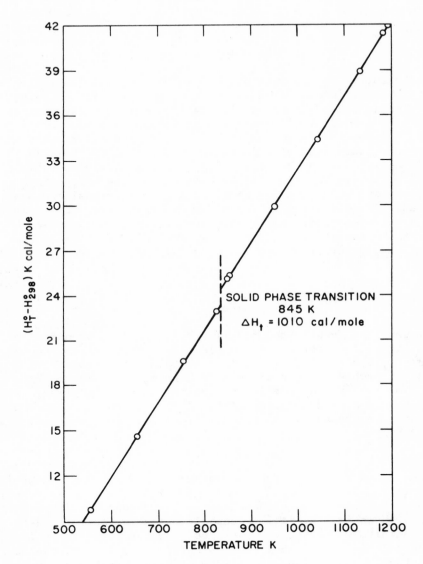

Fig. 3.   Enthalpy of Cs_2MoO_4

EXPERIMENTAL

Sample and Capsule

The sample of crystalline Cs_2MoO_4 used in these experiments was a 99.9% pure salt which was further purified[1] by recrystallization from concentrated aqueous solution. A cylindrical nickel capsule with a 0.032 in. wall thickness and 0.5 in. diameter was used to contain the powdered sample. The internal volume of the capsule was 5 cc; the capsule length, 2 in. The capsule had a filling tube of dimensions 3/16" O.D., 0.010 in. wall thickness, and 1.25 in. length.

Procedure

The nickel capsule was filled with 11.09770 g of Cs_2MoO_4 ($M = 425.7484$ g mol^{-1}); the filling tube was then crimped and welded closed under vacuum. All of the operations were carried out in a glovebox filled with flowing high-purity helium.

Approximately 40% of the measured heat was due to the Cs_2MoO_4; the rest was due to the nickel capsule. The enthalpy of an empty capsule with similar dimensions and surface appearance was measured in a separate series of experiments and an empirical expression was determined from the data to represent the enthalpy of the nickel capsule. This expression was used to correct for the enthalpy of the capsule at each experimental temperature.

Results

The experimentally measured enthalpies for Cs_2MoO_4 above the reference temperature 298.15K are listed in column 2 of table 1, and shown graphically in figure 3. It is evident that a transition occurs in the system, as shown by the discontinuity of the curve. The transition from α Cs_2MoO_4 to β Cs_2MoO_4 has been reported to occur at 845K.[2] Four experimental points were obtained below the transition temperature and a straight line was used to represent the data. The equation of this line, obtained by the method of least squares is:

$$(H^{\circ}_T - H^{\circ}_{298.15}) = 49.4407 \ T - 1.77672 \times 10^4 \ \text{cal/mole} \ (550 - 845 \ K) \tag{1}$$

The standard deviation for equation (1) is 214 cal/mole or 1.2%. This deviation is larger than normally encountered for this calorimetric system; however it appears to be mainly due to the amplification of relative experimental errors when dealing with the relatively small heats measured in this temperature range.

TABLE 1

ENTHALPY OF Cs_2MoO_4

Temperature	$(H^\circ_T - H^\circ_{298.15})$	$(H^\circ_T - H^\circ_{298.15})$	% Dev
K	observed	calc. eq. (1)	
556.1	9764.72	9726.79	−0.4
658.0	14604.30	14764.79	+1.1
753.1	19696.31	19466.60	−1.2
825.8	22953.78	23060.93	+0.5
		calc. eq. (2)	
848.4	25184.63	25168.01	−0.07
848.4	25120.17	25168.01	+0.19
950.8	29920.24	29850.34	−0.23
1044.4	34367.36	34371.13	+0.01
1136.1	38928.81	39023.23	+0.24
1181.1	41349.56	41386.92	+0.09
1190.9	42005.58	41908.72	−0.23

The following equation with the constants determined by the method of least squares represents the enthalpy of Cs_2MoO_4 over the temperature range 845 to 1191K

$$(H^\circ_T - H^\circ_{298.15K}) = 22.1054\ T + 1.31283 \times 10^{-2}\ T^2 - 3035.74\ cal/mol \tag{2}$$

The standard deviation for equation (2) is 82 cal/mol or 0.2%.

The heat associated with the solid-phase transition at 845K can be determined by extrapolating eq. (1) up to 845K and extrapolating eq. (2) down to 845K. This yields a value of about 1010 cal/mol for the heat of the $\alpha \to \beta$ transition.

We had planned to make measurements through the melting point of Cs_2MoO_4 but the capsule developed a leak above 1200K and further measurements have been suspended.

REFERENCES

1. O'Hare, P. A. G. and Hoekstra, H. R., _J. Chem. Thermo._, 5, 851-6 (1973).
2. Hoekstra, H. R., _Inorg. Nucl. Chem. Letters_, 9, 1291-1301 (1973).

THE DISSOCIATION ENERGY OF NiO AND VAPORIZATION AND

SUBLIMATION ENTHALPIES OF Ni

Milton Farber and R. D. Srivastava

Space Sciences, Inc.

135 W. Maple Ave., Monrovia, California 91016

The vaporization thermodynamics of nickel oxide have been controversial for a number of years. Johnston and Marshall (1) determined the vaporization of NiO by heating a nickel ring coated with the oxide in a high vacuum. They assumed that the only species which transported oxygen was NiO(g). From these weight loss data in the temperature range 1438 to 1566 K, Johnston and Marshall calculated a ΔH_O of 117.05 ± 1 kcal/mole for the heat of sublimation of NiO(s). Subsequently, Brewer and Mastick (2) re-examined the data of Johnston and Marshall and concluded that the O_2 pressures calculated would agree with a treatment of their data of the vaporization of NiO(g) to O_2 instead of NiO(g). Thus Brewer and Mastick determined that the dissociation mechanism is the only hypothesis which fits the data of Johnston and Marshall. Brewer and Mastick also performed three effusion experiments of NiO in a beryllium crucible at temperatures of 1816 K and 1782 K. From an analysis of the effused material they found that vaporization by dissociation is the chief method by which NiO vaporizes. They therefore concluded that the vapor pressure data of Johnston and Marshall would be an upper limit; they calculated a D_O of NiO(g) of $\leqslant 99$ kcal/mole.

Huldt and Lagerquist (3) studied the dissociation of NiO spectroscopically in flames. However, since they found a very large amount of NiO dissociation and were unable to determine a definitive value for the D_O of NiO(g), they reported a limiting value of $\leqslant 97$ kcal/mole.

A mass spectrometric study for the bond energy of NiO has been reported by Grimley, et al (4) resulting from the vaporization of NiO(s) in the temperature range 1575 to 1709 K. They found that the

731

conclusions of Brewer and Mastick (2) were valid and that NiO(s),
upon heating, dissociated primarily to the elements since solid Ni
was present on the lid of their effusion cell. The partial pressures
calculated from the mass spectrometric intensities for O_2 were at
least 50 times greater than those for NiO(g). From their data they
reported a third law D_0 of 86.5 ± 5 kcal/mole for NiO(g), with a
second law value of 103 ± 10 kcal/mole. A similar discrepancy of at
least 10 kcal/mole existed in their reported vaporization values for
nickel.

Numerous thermodynamic data have been reported for the sublimation and vaporization of Ni. Heats of sublimation values at 298 K
include 104.0 ± 0.2 kcal/mole by Gulbransen and Andrew (5), 102.72
± .02 by Morris, et al (6), 101.3 ± 0.3 by Johnston and Marshall (1),
100.8 ± 1.3 by Bryce (7), and 98.0 ± 1.4 by Jones and Langmuir (8).
Grimley, et al (4) reported a third law value of 101.3 kcal/mole and
a second law value of 112.4 kcal/mole. Rutner and Haury (9) recently
reported third and second law values of 102.6 ± 2 and 115.8 ± 5.5
kcal/mole, respectively. Hultgren, et al (10) recommend an average
value of 102.67 ± 1.4 kcal/mole for ΔH_{s298}, while Stull and Sinke (11)
recommend 101.26 kcal/mole. Oriani and Jones (12) reported a melting temperature of 1725 ± 4 K and a ΔH of fusion of 4.21 kcal/mole,
Wust, et al (13) reported a value of 3.3 kcal/mole, and Margrave (14)
reported 3.6 kcal/mole. Hultgren, et al (10) and Stull and Sinke (11)
recommend the single value of Oriani and Jones (12).

In order to resolve some of these discrepancies an effusion-
mass spectrometric study was undertaken to investigate the vaporization of NiO(s) and the reaction thermodynamics of nickel metal with
oxygen vapor.

EXPERIMENTAL PROCEDURES AND METHODS OF CALCULATION

The dual vacuum chamber quadrupole mass spectrometer apparatus
employed in these experiments has been described previously (15).
An alumina effusion cell was used since elemental Ni was found to
remain in its original condition at the conclusion of the experiments.
The Al_2O_3 container had been employed previously by Grimley, et al
(4) with no evidence of crucible reaction except for a blue coloration
on the walls of the cell. The alumina effusion cell was 25.4 mm long
and was attached to an alumina rod. The cell had an inside diameter
of 6.8 mm and employed an elongated orifice 0.986 mm diam. by
6.6 mm long for beam collimation. An alumina annulus was connected
from the cell to a low-pressure oxygen tank through which O_2 was
metered into the effusion cell at a constant flow rate. This method
was used for the experiments involving the reaction of O_2 with
solid Ni.

The reactions involved in the determination of the dissociation

energy of nickel oxide were entirely gas phase and were not concerned with the state of the condensed elements and compounds in
the cell. It was recognized that solid solutions might be occurring to
some extent between the NiO and Al_2O_3 as well as the NiO and Ni.
However, examination of the compounds at the conclusion of the
experiments showed that a considerable amount of the solid NiO and
elemental Ni remained unreacted. Therefore, it was concluded that
the Ni vapor effusing from the cell was in equilibrium with elemental
nickel in either the solid or liquid state.

Experiments were performed to confirm this and to establish that
either the solid or liquid nickel existed at unit activity throughout
the intensity measurements. The effusion-mass spectrometer method
has been employed previously at this laboratory to study phase
changes of aluminum and boron (16) and the vanadium oxides (17).
Continuous intensity-time studies were made during the experiments
for periods as long as one hour with no measurable changes in the
ion intensity, indicating a continuous source of supply of the vapor
species involved.

The ion intensities were identified by their masses, isotopic
distribution and appearance potentials. The shutterable, or chopped,
portion only of the ion intensities was directly recorded. In order
to minimize contributions due to fragmentation, ionizing electron
energies of 1 to 3 eV above the appearance potentials were used to
insure that the species were the parent ions. Appearance potentials
found for the species were 9.5 ± 1 eV for NiO, 7.5 ± 1 eV for Ni,
and 12 ± 1 eV for O_2, all in good agreement with previously reported
values (4, 18).

Cross-section and multiplier efficiency corrections were made
for the neutral species. Few experimental cross-sections for formation of positive ions are reported in the literature. However,
theoretical estimates have been made for the various elements and
molecules (19-22). The relative maximum ionization cross-sections
for single ionization of the elements involved in this study were
taken from Lin and Stafford (20). The cross-sections for the molecular
species were calculated by multiplying the sum of atomic cross-
sections by an empirical factor of 0.7 (23). Pottie (21) stated that
various experimental studies have shown that the additive rules of
Otvos and Stevenson (22) for the cross-section sum should be
lowered by a factor somewhat less than 1. The relative ionization
cross-sections thus obtained were Ni = 3.99, NiO = 3.71 and O_2 =
2.4. These are probably good to ± 30 percent. The electron multiplier efficiency correction is a function of the square root of the
molecular weight of the ions impinging upon the electron multiplier.

Second law equilibrium data were obtained by multiplying the
intensity, I, by the absolute temperature, T, for each species. The
heat of reaction was obtained from a Van't Hoff ($\log K_1$, $1/T$) plot.

The partial pressures for Ni(g) were determined from the second law plots of the Ni sublimation data. The measured ion currents for NiO and O_2 were correlated with the corresponding partial pressures, using the relation

$$p_i = \frac{I_i \, (\sigma\gamma)_{Ni}}{I_{Ni} \, (\sigma\gamma)_i} \, p_{Ni}$$

where σ is the relative molecular cross-section (19-21,24) and γ is the electron multiplier gain (the multiplier gain correction is based on the square root of the molecular weight), and p_{Ni} is the partial pressure of nickel. Various methods have been presented involving more elaborate corrections in converting ion intensities of absolute pressures (19,22,25-27). However, the practice here was the measurement of the intensities of a species at 1 to 3 eV above its appearance potential. This method eliminates discrepancies which occur when a fixed ionization potential is employed (26) for all the species involved. As shown previously (28), the shapes of the curves which relate ion intensities as functions of the ionizing electron energies vary significantly among different species.

RESULTS AND DISCUSSION

Sublimation and Vaporization Enthalpies of Nickel

Mass spectrometer intensity data were obtained for the vaporization and sublimation of Ni as

$$Ni(s) = Ni(g) \tag{1}$$

and
$$Ni(l) = Ni(g) \tag{2}$$

employing both the dissociation of NiO and the reaction of elemental Ni with O_2. The log IT, where I is the Ni^+ intensity, was plotted against $1/T$ in both the liquid range of 1740 to 1986 K and the solid range of 1583 to 1723 K (see figure 1). The Van't Hoff plots yielded a heat of sublimation of 99.4 kcal/mole at an average temperature of 1653 K. Employing the thermal data of Stull and Sinke (11) this reduced to 102.2 ± 2 kcal/mole at 298 K, which is in agreement with the recommended values of Stull and Sinke (11) and Hultgren, et al (10) of 101.26 and 102.67 ± 1.4 kcal/mole, respectively. It is also in agreement with the third law ΔH_o of 101.3 kcal/mole reported by Grimley, et al (4) in the only other mass spectrometer study. However, the second law plot of the Grimley, et al (4) data as shown in figure 1 yields a ΔH of sublimation of 112 kcal/mole. As can be seen, the Ni vapor pressures in both the current data and those of Grimley, et al, are close, although the second law slopes differ by more than 10 kcal.

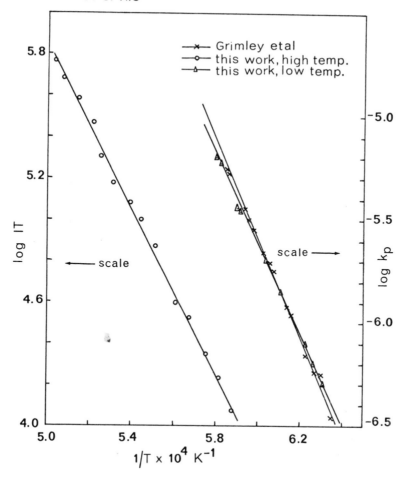

Fig. 1. Heats of Vaporization and Sublimation of Ni
(ΔH_v, temperature range 1740 - 2000 K;
ΔH_s, temperature range 1583 - 1723 K)

Figure 1 also presents the log IT plot of the Ni$^+$ intensities versus 1/T resulting from the dissociation of NiO(g) in the temperature range 1740 to 1986 K. A ΔH of vaporization at an average temperature of 1863 K of 92.2 kcal/mole was obtained. The difference in the heat of vaporization and heat of sublimation at the melting point yielded a heat of fusion of 4.6 kcal/mole.

Dissociation Energy of NiO

Two separate studies were conducted to obtain the bond energy of NiO: the dissociation and vaporization of solid NiO in the temperature range 1703 to 2000 K and the reaction of O_2 with solid nickel in the temperature range 1583 to 1723 K.

Ion intensities were corrected into partial pressures as described. Partial pressures calculated for the O_2 + Ni(s) reaction are presented in Table I. Thermodynamic data for the reaction

$$NiO(g) = Ni(g) + 1/2O_2(g) \qquad\qquad (3)$$

are also presented in Table I. The free energy functions employed for the elements O_2 and Ni(g) were taken from Stull and Sinke (11). The fef for NiO(g) was taken from Brewer and Chandrasekharaiah (29) and was also the value used by Grimley, et al (4). A least squares second law heat of reaction of 24.6 kcal/mole was calculated at an average temperature of 1654 K. This value reduced to 298 K was 25.9 + 3 kcal/mole. The corresponding third law ΔH_{298} was 24.9 ∓ 1.5 kcal/mole. A resulting ΔH_{f298} of NiO(g) was calculated to be 76.3 + 1.5 kcal/mole, with an energy of dissociation of 83.5 + 1.5 kcal/mole.

Calculations were also made from the intensity data obtained in the heating of solid NiO. Table II lists the data in the temperature range 1916 to 2000 K where the nickel is derived from the dissociation of NiO in the liquid state.

The data in Table II yielded a third law ΔH_{298} for reaction (3) of 22.6 + 2 kcal/mole in the temperature range 1916 to 2000 K, resulting in a ΔH_{f298} value of 78.6 + 2 kcal/mole for NiO(g). (A summary of the thermodynamic data for the D_o of NiO(g) is presented in Table III.)

In a previous mass spectrometric study Grimley, et al (4) reported second and third law values for D_o of 103 + 10 and 86.5 + 5 kcal/mole, respectively. Their third law value is within 2 kcal of that obtained in this study. The partial pressures of Grimley, et al (4) are within a factor of 2 of those obtained in this study. However, some error in their data may be attributed to their choice of relative ionization cross-sections chosen for Ni (24.4), NiO (27.7), O_2 (6.58), and O (3.29), as taken from Otvos and Stevenson (22). This Ni/O_2 ratio of 3.7 is twice the ratio of 1.7 employed in these current studies. This discrepancy could account for a difference of 1 to 2 kcal in the ΔH value. The relative cross-sections of Ni and O reported by Mann (19) are 3.99 and 1.31, respectively. A measurement of the ionization cross-sections of silver reported by Crawford

Table 1

THERMODYNAMIC DATA FOR THE REACTION

$$NiO(g) = Ni(g) + 1/2 O_2(g)$$

T (°K)	P (atm)			$\log K_p$	$-\dfrac{\Delta(G-H_0)}{T}$	$-\dfrac{\Delta(G-H_{298})}{T}$	ΔH_0	ΔH_{298}
	Ni	NiO	O_2					
1583	5.00×10^{-7}	3.85×10^{-8}	7.20×10^{-5}	-0.960	11.03	11.28	24.4	24.8
1593	6.30×10^{-7}	4.64×10^{-8}	7.20×10^{-5}	-0.940	11.05	11.27	24.4	24.8
1603	7.90×10^{-7}	5.38×10^{-8}	7.14×10^{-5}	-0.910	11.05	11.32	24.4	24.8
1635	1.40×10^{-6}	8.26×10^{-8}	7.02×10^{-5}	-0.850	11.07	11.28	24.4	24.8
1655	1.95×10^{-6}	1.04×10^{-7}	7.00×10^{-5}	-0.815	11.09	11.32	24.5	24.9
1688	3.40×10^{-6}	1.58×10^{-7}	7.02×10^{-5}	-0.745	11.10	11.34	24.5	24.9
1695	3.98×10^{-6}	1.77×10^{-7}	6.89×10^{-5}	-0.730	11.11	11.35	24.5	24.9
1718	5.88×10^{-6}	2.43×10^{-7}	6.82×10^{-5}	-0.700	11.12	11.35	24.6	25.0
1723	6.30×10^{-6}	2.47×10^{-7}	6.72×10^{-5}	-0.679	11.12	11.35	24.5	24.9

Table II

VAPORIZATION OF NiO(g) = Ni(g) = $1/2O_2$(g)

| T (°K) | P (atm) | | | $\log K_p$ | $-\dfrac{\Delta(G-H_0)}{T}$ | $-\dfrac{\Delta(G-H_{298})}{T}$ | ΔH_0 | ΔH_{298} |
	Ni	NiO	O_2					
1916	6.69×10^{-5}	7.92×10^{-7}	1.13×10^{-4}	-.046	11.40	11.42	21.88	22.28
1943	9.20×10^{-5}	1.31×10^{-6}	1.59×10^{-4}	-.053	11.37	11.57	22.56	22.96
1970	1.35×10^{-4}	1.98×10^{-6}	2.20×10^{-4}	+.004	11.36	11.56	22.34	22.74
1986	1.60×10^{-4}	2.01×10^{-6}	2.30×10^{-4}	+.083	11.37	11.57	21.83	22.23
2000	1.93×10^{-4}	2.55×10^{-6}	2.31×10^{-4}	+.061	11.35	11.55	22.14	22.54

and Wang (30) yielded $(4.65 \pm 1) \times 10^{-16}$ cm^2, which is considerably smaller than the 34.8 reported by Otvos and Stevenson (22); Pottie (21) obtained an experimental value of 2.4 for O_2.

The large difference in the second law values of the two studies is similar to the difference in the second law sublimation heats of Ni. The second and third law agreement in the current studies supports the lower value for the bond energy of NiO and establishes a D_0 of 83.5 $\pm$ 1.5 kcal/mole.

Table III

THERMODYNAMIC DATA FOR THE DISSOCIATION OF NiO

$$NiO(g) = Ni(g) + 1/2 O_2(g)$$

ΔH_0 (III)	24.5 $\pm$ 1.5 kcal/mole
ΔH_{298} (III)	24.9 $\pm$ 1.5 kcal/mole
ΔH_{1654} (II)	24.6 $\pm$ 3.0 kcal/mole
ΔH_{298} (II)	25.9 $\pm$ 3.0 kcal/mole
ΔH_{f298} NiO(g)	76.3 $\pm$ 1.5 kcal/mole
D^0_{298} (NiO)	84.4 $\pm$ 1.5 kcal/mole
D^0_0 (NiO)	83.5 $\pm$ 1.5 kcal/mole

THERMODYNAMIC DATA FOR THE REACTION OF Ni(s) WITH $O_2(g)$

$$NiO(g) = Ni(g) + 1/2 O_2(g)$$

ΔH_0 (III)	22.2 $\pm$ 2.0 kcal/mole
ΔH_{298} (III)	22.6 $\pm$ 2.0 kcal/mole
ΔH_{f298} NiO(g)	78.6 $\pm$ 2.0 kcal/mole

REFERENCES

1. H. L. Johnston and A. L. Marshall, J.Am.Chem.Soc. <u>62</u>, 1383 (1940).
2. L. Brewer and D. F. Mastick, J.Chem.Phys. <u>19</u>, 834 (1951).
3. L. Huldt and A. Lagerquist, Z.Naturforsch <u>9a</u>, 358 (1954).
4. R. T. Grimley, R. P. Burns and M. G. Inghram, J.Chem.Phys. <u>35</u>, 551 (1961).
5. E. A. Gulbransen and K. F. Andrew, J. Metals <u>11</u>, 71 (1959).
6. J. P. Morris, G. R. Zellars, S. L. Payne and R. L. Kipp, U.S. Bur. of Mines Report 5364 (1957).
7. G. Bryce, J. Chem. Soc. <u>2</u>, 1517 (1936).
8. H. A. Jones, I. Langmuir and G. M. J. Mackay, Phys. Rev. <u>30</u>, 201 (1927).
9. E. Rutner and G. L. Haury, Tech. Report AFML-TR-72-217 (1973).
10. R. Hultgren, R. L. Orr, P. D. Anderson and K. K. Kelley, Selected Values of Thermodynamic Properties of Metals and Alloys, Wiley & Sons (1963).
11. D. R. Stull and G. C. Sinke, Thermodynamic Properties of the Elements, Adv. Chem. Series No. 18 (ACS, Washington D. C. 1956).
12. R. A. Oriani and T. S. Jones, Rev.Sci.Inst. <u>25</u>, 248 (1954).
13. F. Wust, A. Meuther and R. Durrer, Forsch.Gebiete Ingenieurw. VDI-Forschungsh. <u>204</u> (1918).
14. J. L. Margrave, Faraday Discussions of the Chemical Society, Symposium No. 8, to be published.
15. M. Farber, M. A. Frisch and H. C. Ko, Trans. Faraday Soc. <u>65</u>, 3202 (1969).
16. M. Farber and M. A. Frisch, First Int'l. Conf. Calorimetry and Thermodynamics, Warsaw, Poland, Aug. 1969, Proceedings, pp 443-456.
17. M. Farber, O. M. Uy and R. D. Srivastava, J.Chem. Phys. <u>56</u>, 5312 (1972).
18. U.S. Dept. of Commerce, Nat. Bur. of Standards Publ. NSRDS-NBS 26, Ionization Potentials, Appearance Potentials, and Heats of Formation of Gaseous Positive Ions, (1969).
19. J. B. Mann, J. Chem. Phys. <u>46</u>, 1646 (1967).
20. S. Lin and F. E. Stafford, J. Chem. Phys. <u>47</u>, 4667 (1967).
21. R. F. Pottie, J.Chem. Phys. <u>44</u>, 916 (1966).
22. J. W. Otvos and D. P. Stevenson, J.Am.Chem.Soc. <u>78</u>, 546 (1956).
23. M. Farber and R. D. Srivastava, J.C.S. Faraday I, <u>69</u>, 390 (1973).
24. A. C. H. Smith, E. Caplinger, R. H. Neynaber, E. W. Rothe and S. M. Trujills, Phys. Rev. <u>127</u>, 1674 (1962).
25. R. E. Honig, J. Chem. Phys. <u>22</u>, 126 (1968).
26. J. Drowart and P. Goldfinger, Angnew. Chem. <u>6</u>, 581 (1967).
27. M. G. Inghram, R. J. Hayden, and D. C. Hess, Nat. Bur. of Standards Circ. 522, 257 (1953).
28. M. Farber, R. D. Srivastava and O. M. Uy, J.C.S. Faraday I, <u>68</u>, 249 (1972).

29. L. Brewer and M. S. Chandrasekharaiah, UCRL-8713, (Rev.) (1960).

30. C. K. Crawford and K. L. Wang, J.Chem.Phys. <u>47</u>, 4667 (1967).

THERMODYNAMIC PROPERTIES OF REX₃,AuCu₃-TYPE,INTERMETALLIC COMPOUNDS

THERMODYNAMIC PROPERTIES OF REX_3, $AuCu_3$-TYPE, INTERMETALLIC COMPOUNDS

A.Palenzona and S.Cirafici

Institute of Physical Chemistry,University of

Genoa,Italy

THEORY

The crystallographic data of many rare earths (RE) intermetallic compounds are well established but in most cases their thermodynamic properties are not known.Accurate thermodynamic data are valuable not only as a quantitative measure of the relative stability of alloys but as basic informations for testing theories of the metallic state.We have therefore developed a method suitable for the direct determination of the heats of formation and heats of fusion based on using a conventional differential thermal analysis apparatus.If the heat of formation of a compound can be measured by a reaction involving direct combination of the elements,the results are likely to be more accurate than if an indirect method is employed,such as vapour pressure or electromotive force measurements or solution calorimetry.Another advantage of the direct method compared with other traditional techniques is the speed with which the measurements can be carried out:a single determination taking no more than one or two hours.

The method of dynamic differential calorimetry (DDC),first described in 1958,has received a satisfactory theoretical treatment by Faktor and Hanks (1) who used this method to evaluate the heats of formation of rare earth arsenides (REAs)(2).Briefly,the sample (S) and the reference material,a one gram ingot of Mo (R), are placed in two molybdenum containers with covers, which rest on two differential thermocouples,supported

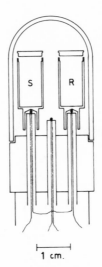

Fig.1—Crucible assembly for dynamic differential calori-
      metry. S=sample,R=reference.

by alumina rods.The whole is enclosed by an alumina cap
and a third thermocouple between the crucibles records
the temperature of the enclosure.(Fig.1).Following the
theory developed by Faktor and Hanks,the sensitivity
(S) of DDC,i.e.the peak area produced by unit enthalpy
change in the sample,can be written as follows:

$$\frac{1}{S} = \frac{\Delta H}{\int_0^t \Delta T dt} = A + BT_m^3$$

where:

ΔH = enthalpy change in the sample,

$\int_0^t \Delta T dt$ = peak area,

A,B = constants containing the transmission coef-
        ficients for conduction and radiation,and

T_m = temperature of the central thermocouple at
       which the peak area can be divided into
       equal parts (see Fig.4).

    We have tested such a theory in the temperature
range 200°—1100°C and have applied the DDC method to
REX$_3$ compounds,where X=Sn,Pb,Tl and In.These phases
have been choosen first of all because these metals
react easily at low temperature with RE,the reaction is
highly exothermic and the heat evolved is sufficient to
melt the sample;secondly,because the heats of formation
of some of these compounds have been measured by other

Table 1—Reference materials.(T_f=°C ; ΔH_f=Kcal/mole)

Subst.	T_f	ΔH_f	Refer.
Sn	232	1.645 ± 0.016	5
Bi	271	2.755 ± 0.018	5
Cd	321	1.530 ± 0.04	3
Zn	420	1.755 ± 0.020	5
Sb	631	4.690 ± 0.1	4
Mg	650	1.970 ± 0.016	5
NaCl	801	6.7 ± 0.1	3
Ag	961	2.650 ± 0.1	3
Au	1063	3.05 ± 0.1	3
Cu	1083	3.1 ± 0.1	3

authors using different methods,and their result supplies a test for the reliability of the method itself.
Moreover the data obtained for these compounds,which
crystallize all with the AuCu₃-type of structure,could
provide a measure of the relative stability of these
phases.

EXPERIMENTAL

The apparatus used was a conventional DTA equipment obtained from Netzsch (West Germany) of the same
type used by Faktor and Hanks in their investigation.
Crucibles and covers were of molybdenum which is inert
in respect to partner elements and rare earths.A two-
point recorder gave simultaneous plots ot the differential temperature and the temperature of the enclosure.
The operating atmosphere was ultra-pure argon.The furnace could be programmed for linear heating rates from
0.1 to 100 deg/min.up to 1500°C.Thermocouples were
Pt/Pt-10%Rh.

Calibration curves were obtained using elements
and compounds of known heats of fusion in the temperature range 200°-1100°C and their values are reported
in Table 1.For each substance,samples corresponding to
thermal effects of 25-30 cal. were employed and only
fusion peaks were considered.The furnace was heated at
10 or 20 deg/min. and each substance was tested four
times.The peak areas were determined by a planimeter
so that the reciprocal sensitivity was espressed in
cal/cm²;Fig.2 reports a plot of these values against

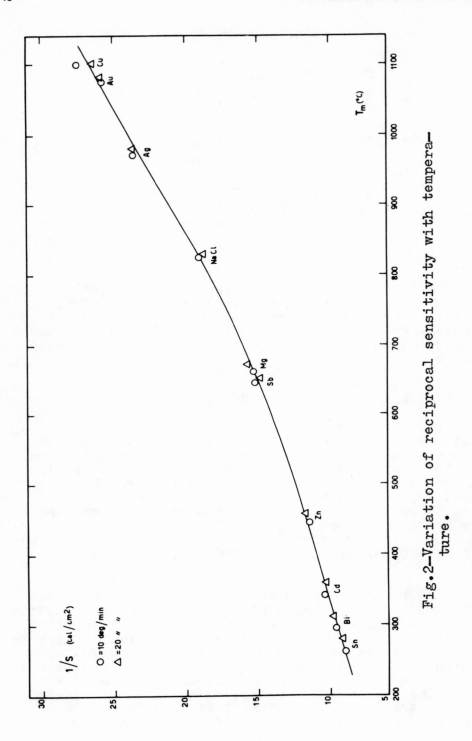

Fig.2—Variation of reciprocal sensitivity with tempera—
ture.

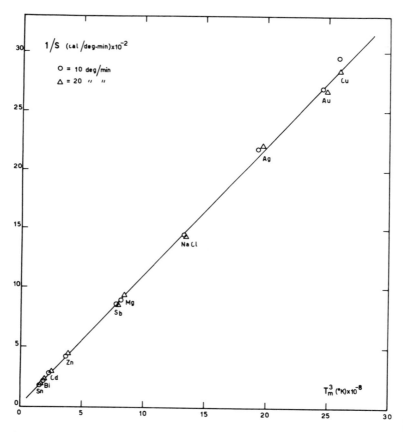

Fig. 3—Reciprocal sensitivity against the cube of the absolute temperature.

temperature. The reciprocal sensitivities in cal/cm^2 were converted into cal/deg.min., as required by the theoretical analysis, by using the three conversion factors: recorder sensitivity ($cm/\mu V$), thermoelectric power of the thermocouple at the temperature in question ($\mu V/deg$) and chart speed (cm/min). Thus a plot of 1/S against T_m^3 (Fig.3) appears linear as the theory requires. In the evaluation of the heats of formation and heats of fusion, single curves for 10 and 20 deg/min were employed.

The major source of errors in the calibration curves lies in the existing discrepancies in the values of the values of the heat of fusion for the reference materials. From the examination of the literature data (3-10) it appears evident that for a certain element the known values of the heat of fusion differ from one another by $\pm$ 3-5% at least, each one being known with an accuracy of $\pm$ 1-3%. Therefore the maximum error introduced from calibration curves and from standard points was estimated to be $\pm$ 4%. The values reported in Table 1 were selected from a critical survey of the known data and correspond in most cases to the most recent determination.

REX$_3$ compounds, AuCu$_3$-type, are well known structurally, their crystallographyc properties can be found in current literature and some of them have been recently investigated determining their heats of formation (Table 2).

Table 2—Heats of formation of some REX$_3$ compounds from literature. (ΔH=Kcal/mole)

Comp.	ΔH	Refer.	Comp.	ΔH	Refer.
LaSn$_3$	−60.60	11	SmSn$_3$	−53.80	14
CePb$_3$	−41.60[*]	12	EuSn$_3$	−50.48	15
PrIn$_3$	−55.30	13	GdSn$_3$	−49.65	16
PrPb$_3$	−34.80[*]	12	YbSn$_3$	−42.10	17
NdPb$_3$	−39.20[*]	12			

[*]these values are affected by a 20% error (see the text).

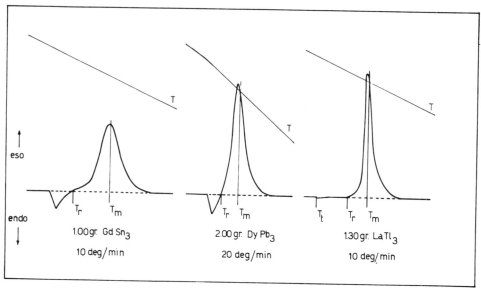

Fig.4—Examples of thermograms obtained for the reaction
between rare earths and X elements.

RE metals and partner elements used in this research were obtained from Koch—Light Laboratories,England,
and had a purity of 99.9 and 99.999% respectively.
Samples ranging from 0.75 to 2.50 gramme each were prepared from stoichiometric amounts of powders of the two
metals (50 mesh) well mixed and pressed directly in the
Mo containers (1.5 Kbar).Manipulation of oxidizable substances such as light RE,Pb and Tl was carried out in
a dry box filled with ultra—pure argon.The reaction
between RE and other metals starts at low temperature,
is very fast especially for the lighter RE from La to
Sm,and the thermal effect due to the melting of the X
element must be considered in the evaluation of the heat
of formation of the corresponding compound.The other RE
react at higher temperature and the two thermal effects,
i.e. melting of the partner element and heat of reaction,
are widely indipendent.Typical examples are reported in
Fig.4.To ensure reproducibility in the results,several
samples for different heating rates were prepared and
at least two "good" samples were considered in the evaluation of the heats of formation.Table 3 reports the
values obtained.

After the reaction has ended the samples were examined micrographically:only alloys which contained a
single phase were considered and subsequent x—ray exa-

Table 3—Heats of formation for REX$_3$ compounds at 298°K
     (Kcal/mole)

RE	RESn$_3$	REIn$_3$	REPb$_3$	RETl$_3$
La	−60.25	−49.94	−49.18	−43.27
Ce	−49.08	−47.88	−46.89	−42.97
Pr	−55.94	−53.37	−50.31	−43.77
Nd	−65.96	−	−49.81	−39.82
Sm	−53.55	−	−47.28	−37.91
Eu	−	−	−41.94	−
Gd	−48.78	−	−42.88	−35.22
Tb	−	−	−41.90	−34.08
Dy	−	−	−40.88	−32.60
Ho	−	−	−39.14	−
Er	−	−	−35.73	−
Tm	−	−	−33.85	−
Yb	−45.86	−	−35.37	−26.21

mination showed the presence of the REX$_3$ phase with no
extra lines due to other phases or to unreacted metals.
The values of the lattice constants are in good agreement with literature values .

     As the heats of formation are obtained at different
temperature,they should be reported,by means of the
Kirchoff's relation,at 298°K corresponding to the reaction:
$$RE(s) + 3X(s) = REX_3(s) + \Delta H_f$$
The calculation would imply the knowledge of the heat
capacities for RE,liquid and solid X element and for
the compound but,if nothing whatever is known about the
heat capacity of the compound,assuming the validity of
the Neumann and Kopp's rule,ΔC_p may be considered to be
zero for condensed reactions without affecting the results too much,especially in this case where the ΔT is
small(3).The heats of formation were therefore corrected only for the heat of fusion of X elements and the
values so obtained are reported in Table 3.Assuming a
total error of $\pm$ 5–6% for our values we observe good
agreement with those obtained by other authors who used
different methods(Table 2).For CePb$_3$,PrPb$_3$ and NdPb$_3$
the values were obtained by vapour pressure measurements of lead between 993°–1073°K,but the error in their
determination,as quoted by the same authors,was $\pm$ 20%.

     Heats and entropies of fusion of REX$_3$ compounds
were determined by preparing a new series of one gram

Table 4—Melting temperature (°C), heats (Kcal/mole) and entropies (cal/deg.mole) of fusion for REX₃ compounds.

	RESn₃			REPb₃			RETl₃		
	T_m	ΔH_m	ΔS_m	T_m	ΔH_m	ΔS_m	T_m	ΔH_m	ΔS_m
La	1138	11.50	8.15	1145	20.26	14.28	1050	20.28	15.33
Ce	1160	16.45	11.45	1130	14.68	10.46	1070	18.01	13.41
Pr	1155	18.01	12.61	1120	20.05	14.34	1060	20.92	15.69
Nd	1138	17.67	12.52	1105	20.81	15.10	1050	19.40	14.66
Sm	1072	15.95	11.86	970	19.78	15.91	850	17.44	15.53
Eu	784	10.14	9.59	790	12.25	11.54	—	—	—
Gd	—	—	—	—	—	—	965	15.20	12.28
Tb	—	—	—	—	—	—	950	12.54	10.25
Dy	—	—	—	—	—	—	925	14.98	12.50
Yb	805	15.30	14.19	740	13.08	12.91	585	9.71	11.32

samples and sealing them by arc welding under argon atmosphere in Mo crucibles; this procedure is necessary to avoid losses of metals (Sm,Eu,Tm,Yb) at high temperature via sublimation. At the melting point of the alloys, which are in good agreement with the known values, the heats of fusion were determined by means of the calibration curves and hence the entropies of fusion were calculated; these values are reported in Table 4.

DISCUSSION

Generally RE form with Sn,Pb,Tl and In series of isomorphous compounds, all crystallizing with the AuCu₃-type of structure; the only exception being: EuIn₃, LuPb₃, (Tb,Dy,Ho,Er,Tm,Lu)Sn₃. Compounds of Eu and Yb, from magnetic measurements and lattice constant values were proved to contain Eu^{2+} and Yb^{2+} and the values of the corresponding heats of formation agree with this assumption. In effect, if these rare earths were present in these phases in the trivalent state, their heats of formation, according to Gschneidner [18], should be greater (more positive) of about 23 and 9 Kcal/g.at. respectively of the neighbouring trivalent RE compounds; the values of 23 and 9 Kcal/g.at. being the energies required for the promotion of the extra 4f electron to a valence or outer electron level. This condition is never verifyied for the examined compounds; only for YbSn₃ some indications exist of a possible valency state for Yb slightly different from two [19].

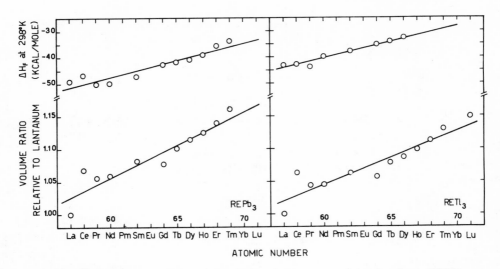

Fig. 5—Heats of formation and volume ratios for REPb₃
and RETl₃ compounds vs. atomic number.

    The examination of the obtained data and the compa-
rison with other"series",such as RECd (20) and REMg (21),
lead to the conclusion that these REX₃ are even more
stable phases with increasing stability in the order Tl,
Pb,In and Sn.

    Recently Robinson and Bever (22) have examined all
existent thermodynamic data for intermetallic compounds
and have shown that some indications concerning the type
of bonding could be obtained simply by considering the
magnitude of these data.The results obtained in this research suggest for these compounds a chemical bond pre-
dominantly of the metallic type with ionic or covalent
contributions very small and increasing in the series
Tl,Pb,In and Sn.

    The general trend in the heats of formation for
REX₃ compounds is a decrease from La to Lu and this behaviour,in agreement with the trend in the melting points of these phases (Table 4),can be correlated with
the lanthanide contraction in the compounds and compared to the contraction observed in the pure metals,as
given by Gschneidner (23).The unit cell volumes of the
REX₃ phases were divided by the atomic volume of the pure metal (RE) and these volume ratios were then divided
by the volume ratio relative to the lanthanum compound
giving a scale relative to lanthanum (1.000). The resultant values (Fig. 5) increase with RE atomic number,

Table 5—Comparison of the experimental ΔH values with those calculated from the equation of Miedema. (ΔH=Kcal/g.at.)

	$\Delta H_{calc.}$	$\Delta H_{exp.}$
LaSn₃	−12.7	−15.1
LaIn₃	−11.0	−12.5
LaPb₃	−12.0	−12.3
LaTl₃	−12.7	−10.8

i.e.,the lanthanide contraction in the compounds is less severe than in the pure metals and a decrease in the heats of formation can be expected in going from La to Lu.This result is well evident for REPb₃ (24),RETl₃ and at a less extent for RESn₃ (25) ,due to the non existence of some of these phases;REIn₃ are not yet completely examined but a similar behaviour can be expected.This same consideration is valid if we consider Ba,Eu and Yb as members of a "baride" family of divalent elements. Again,at least for isomorphous compounds, the volume ratio relative to barium and heats of formation vs. atomic number should show the same trend as for trivalent RE (24).

According to a recent work of A.R.Miedema (26),the heats of formation of binary alloys can be evaluated using a simple relation,which in the case of REX₃ phases can be written as follows:

$$\Delta H = f(c)\left[-Pe(\Delta \phi^*)^2 + Q(\Delta n_{ws})^2 - R\right]$$

where $f(c)$ = symmetrical function of the concentration,

$\Delta \phi^*$ = difference in the electronegativity parameter,

Δn_{ws} = difference in the electron density at the boundary of the Wigner—Seitz cells and

P,Q,R = experimental constants.

Although the relation don't take into account the particular structure adopted by a compound,it is possible to found a good agreement between observed and calculated values (Table 5).In these calculations the average values of ϕ^* and n_{ws} for RE were used but undoubtedly a better knowledge of these two functions,both for RE and partner elements,could lead to a higher degree of agreement with the experimental values.

We have tried to correlate the heats of formation of these REX₃ compounds with a parameter which could take into account the dimensions of the involved elements,such as metallic,covalent or ionic radii.The best

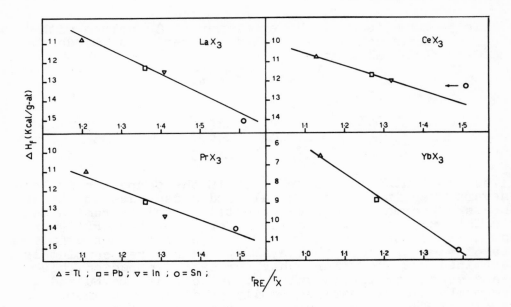

Fig. 6—Heats of formation vs. ionic radii ratio.

result is obtained on plotting the heats of formation of the REX_3 phases for a given rare earth, vs. the radii ratio $r(RE)/r(X)$ where $r(RE)$ is the ionic radius of the divalent or trivalent rare earth considered and $r(X)$ is the ionic radius of the X element corresponding to its higher valency state. The values of the ionic radii are taken from Ahrens (27). In all cases (Fig. 6) a straight line is obtained, both for trivalent and divalent RE and also for those not reported in the figure. At a lesser extent a similar correlation seems to be valid using the covalent radii but not with the metallic radii. The scatter in the value of $CeSn_3$ is not surprising as Ce in this compound has a valency higher than three (28), which causes a decrease in the ionic radius with a better agreement to the straight line. Even if is not well clear the meaning of this correlation, it can be used to evaluate the heats of formation of the remaining iso-morphous REX_3 phases which, owing to experimental diffi-culties, could not be determined. All the experimental values were collected together and fitted to a straight line whose equation is:

$$\Delta H_f = -10.52\ r(RE)/r(X) + 1.53\ (\pm 0.84\ \text{Kcal/g.at.})$$

for trivalent RE and,

$$\Delta H_f = -11.31\ r(RE)/r(X) + 4.77\ (\pm 0.39\ \text{Kcal/g.at.})$$

for divalent RE and the missing data could be evaluated.

Such a correlation not only occours for these REX_3 phases but examining the literature data for the heats of formation of binary alloys we could find other "series" of compounds for which it holds too,such as: Na_3(As,Sb,Bi); Sm(Zn,Cd,Hg);(Li,Na)(Sn,Pb,Tl);(Mg,Ca,Sr,Ba)(S,Se,Te); Ga_2(S,Se,Te)$_3$;Mg_2(Ge,Sn,Pb) and others.

As concluding remarques we can say that the DDC method can be employed with reasonable confidence in the determination of the heats of formation and heats of fusion of intermetallic compounds.The method appears to prove itself competitive with other traditional techniques both for the total error ($\pm$ 5–6%) which affects the results and for the speed and ease of operation,but only highly exothermic and fast reactions,which lead to the complete formation of the desired compound can be examined.The heats of formation of particular "series" of compounds can be correlated with the ionic radii ratios of the involved elements and such correlation can be used in the prediction of some of these values whose experimental determination is actually impossible.

ACKNOWLEDGEMENTS

The authors wish to thank Prof.A.Iandelli for his help and useful suggestions given during the development of the present research.

REFERENCES

1.M.M.Faktor and R.Hanks,Trans.Faraday Soc.63(1967)1122.
2.M.M.Faktor and R.Hanks,Trans.Faraday Soc.63(1967)1130.
3.O.Kubaschewski,E.L.Evans and C.B.Alcock,Metallurgical Thermochemistry,Pergamon Press,London 1967.
4.R.Hultgren,R.L.Orr,P.D.Anderson and K.K.Kelly,Selected values of thermodynamic properties of metals and alloys,Wiley,New York,1963.
5.P.Chiotti,G.J.Gartner,E.R.Stevens and Y.Saito,J.Chem. Eng.Data,11(1966)571.
6.L.Malaspina,R.Gigli and V.Piacente,Gazz.Chim.Ital., 101(1971)197.
7.P.D.Garn,Thermoanalytical Methods of Investigation, Academic Press,New York,1965.
8.C.J.Smithells,Metal Reference Book,Butterworths,London,1967.
9.K.A.Gschneidner,Solid State Physics,Vol.16,Academic Press,New York,1964.

10.Metals Handbook,8th ed.,Vol.1,A.S.M.,1964.
11.J.R.Guadagno,M.J.Pool,S.S.Shen andP.J.Spencer,Trans.
   Met.Soc.AIME,242(1968)2013.
12.P.P.Otopkov,I.Gerasimov and A.M.Evseev,Dokl.Akad.
   Nauk.SSSR,139(1961)616.
13.V.A.Degtyar' ,A.P.Bayanov,L.A.Vnuchkova and V.V.Se-
   rebrennikov,Russ.J.Phys.Chem.,45(1971)1032.
14.A.Percheron,J.C.Mathieu and F.Trombe,C.R.Acad.Sci.
   266 C (1968)848.
15.A.Bacha,C.Chatillon—Colinet,A.Percheron,J.C.Mathieu
   and J.C.Achard,C.R.Acad.Sci.276 C (1973)995.
16.A.Bacha,C.Chatillon—Colinet,A.Percheron and J.C.Ma-
   thieu,C.R.Acad.Sci.274 C (1972)680.
17.C.Chatillon—Colinet,A.Percheron,J.C.Mathieu and J.
   C.Achard,C.R.Acad.Sci.270 C (1970)473.
18.K.A.Gschneidner,J.Less—Common Metals,17(1969)13.
19.A.Iandelli,private communication.
20.E.Veleckis,LJohnson and H.Feder,USAEC Report n°
   ANL—7175,(1966)154.
21.J.R.Ogren,N.J.Magnani and J.F.Smith,Trans. TMS—AIME,
   239(1967)766.
22.P.M.Robinson and M.B.Bever,in J.H.Westbrook(ed),
   Intermetallic Compounds,Wiley,New York,1967,p.38.
23.K.A.Gschneidner,J.Less—Common Metals,17(1969)1.
24.A.Palenzona and S.Cirafici,Thermochim.Acta,6(1973)455.
25.A.Palenzona,Thermochim.Acta,5(1973)473.
26.A.R.Miedema,J.Less—Common Metals,32(1973)117.
27.L.H.Ahrens,Geochim.et Cosmochim.Acta,2(1952)155.
28.J.R.Cooper,C.Rizzuto and G.L.Olcese,J.Phys.(Paris),
   Colloq.,C1,suppl.n°2—3,Tome 32,Fevrier—Mars 1971,
   C 1—1136.

THERMAL AND MICROSCOPICAL STUDY OF THE CONDENSED PHASE

BEHAVIOR OF NITROCELLULOSE AND DOUBLE BASE MATERIALS

Scott I. Morrow

Propellant Division
Feltman Research Laboratory
Picatinny Arsenal
Dover, N.J. 07801

ABSTRACT

Methods were devised for using a polarizing microscope to study the condensed phase behavior of propellants. Some of the phenomena associated with this condition for nitrocellulose, catalyzed nitrocellulose, and double base propellant both with and without catalysts was elucidated. At the same time techniques of differential scanning calorimetry (DSC), thermogravimetric analysis (TGA), and thin layer chromatography (TLC) were brought to bear on the problem. Optically and thermally measurable differences in catalyzed and uncatalyzed materials tended to be small and were probably not as great, for instance, as is to be found with certain cases of composite propellants. Nevertheless, clues were obtained as to ways the catalysts were affecting condensed phase behavior. Unraveling these problems fully remains a formidable task.

INTRODUCTION

Perhaps the greatest barrier to elucidating fully the mode of action of catalysts in combustion of solid propellants is the problem of isolating events therein for detailed study. The catalyst is buried in a complex matrix of binder, fuel, and oxidizer, comprising a system that can exhibit a number of changes incident to the process of ignition and combustion. In addition the catalyst itself, can undergo physical and chemical transformations which are obscured by violent processes occuring in the surrounding propellant. It is not surprising, therefore, that investigators have tended to favor essentially indirect techniques

for studying catalysis in solid propellants. For example, Kirby
(1) used differential scanning calorimetry (DSC) in studying the
condensed-phase heat of reaction of double base propellants.
Waesche (2) also reported results on decomposition of composite
solid propellants involving catalysis using DSC and differential
thermal analysis (DTA). Results in thermoanalytical and electrical conductivity investigations in catalytic studies related to
solid propellants have been published by Freeman and Rudloff (3).
Apparently, however, no serious attempt has been made to use
techniques of polarizing microscopy in studying solid propellant
combustion.

This paper involves research with a newly developed, DTA-
light photometer-equipped polarizing microscope on the condensed
phase behavior of solid propellant materials. It also describes
conjoint use of thermoanalytical techniques such as DSC and TGA
for studying catalytic phenomena.

EXPERIMENTAL

Observations with the Polarizing Microscope

Thin Sections from Propellant Strands. A series of four
double base type propellants were sectioned with a microtome. Even
in thin sections, 0.01 to 0.02 inches thick, these samples were
quite opaque. The first, number 6103 did not have a catalyst,
whereas 6104, 6105, and 6106 contained catalysts as shown in Table
1. A thin section was mounted on a microscope slide with two DTA
sensor thermocouples, one for the sample, the other for reference,
beneath a cover slip held in place by silicone cement. This assembly was then mounted in the Mettler FP-52 microfurnace and heated
at a rate of 10°C/min. These procedures have been discussed in a
previous paper by this author (4).

Figure 1 shows DTA thermograms of 6103 and 6104 propellants
and Figure 2 those for all four propellants. The sequence of
events involved in these experiments is represented by the photomicrographs shown in Figures 3 and 4.

Tables 2 and 3 summarize pertinent data obtained in experiments with the polarizing microscope and microfurnace where DTA
was not used. Omission of DTA facilitated weighing as well as
measurement of thickness of the samples on the slides with a micrometer before and after heating.

Thin Sections from Double Base Sheet. A translucent sheet
of double base propellant, fairly thin at the edge, was used for
these experiments. Sections were cut from the thin portion and

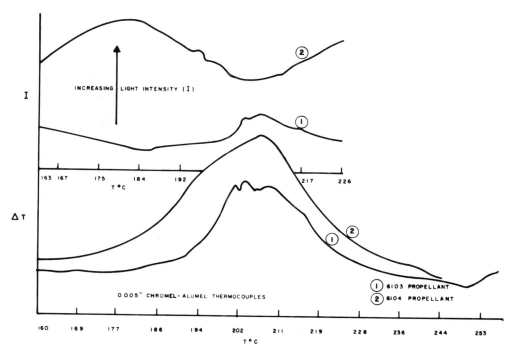

Fig.1.   DTA-Transmitted light study of double base type propellants
with a polarizing microscope and Mettler hot stage and DTA
amplifier.

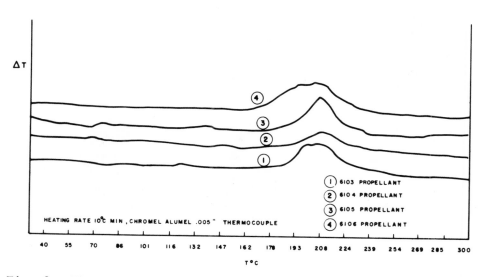

Fig. 2.   Thermograms of double base type propellants, uncatalyzed
and catalyzed by DTA equipped polarizing microscope.

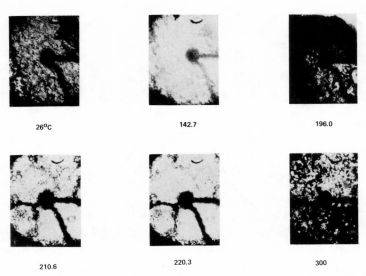

Fig. 3.   Condensed phase behavior of 6103 propellant during
          microscopical DTA pyrolysis.

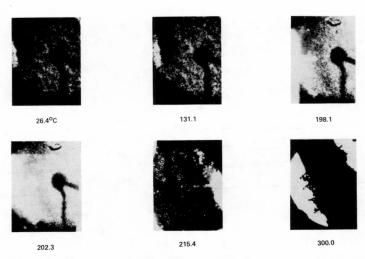

Fig. 4.   Condensed phase behavior of 6104 propellant during
          microscopical DTA pyrolysis.

TABLE 1

COMPOSITION OF DOUBLE BASE PROPELLANT STRANDS

Propellant Number	6103	6104	6105	6106
Nitrocellulose,(12.6%N)%	57.9	56.7	56.7	56.7
Nitroglycerine,%	28.1	27.5	27.5	27.5
Triacetin,%	12.0	11.8	11.8	11.8
2-Nitrodiphenylamine,%	2.0	2.0	2.0	2.0
Lead Salicylate,%	----	2.0	----	----
Lead Salicylaldehyde,%	----	----	2.0	----
Lead Stannate,%	----	----	----	2.0

TABLE 2

CONDENSED PHASE BEHAVIOR OF DOUBLE BASE TYPE PROPELLANTS ON THE METTLER HOT STAGE, HEATING RATE 10°C/MIN. TO 225°C

Propellant	6103	6104	6105	6106
Loss in Weight,%	79.4(5) Note 1 78.8(3) Note 2	78.3(5) 76.6(3)	80.8(4) 79.9(3)	79.7(4) 75.9(2)
Degree in Thickness,%	93.1(2)	88.8(2)	81.7(2)	88.9(2)
Formation of Liquid Prior to Melting	No	Yes	No	No
Onset of Melting, T°C	167.9(3)	166.1(5)	167.3(5)	169.9(4)
End of Bubbling, T°C	218.2(2)	215.8(2)	216.3	215.7

NOTE 1: Number in parenthesis indicates number of experiments used to determine the mean value shown for the parameter in question.

NOTE 2: Larger samples consisting of several slices of propellant were used to give greater weight to improve weighing accuracy.

TABLE 3

CONDENSED PHASE BEHAVIOR OF DOUBLE BASE TYPE PROPELLANTS ON
THE METTLER HOT STAGE, HEATING RATE 10°C/MIN. TO 225°C(Note 1)

Propellant	6103	6104	6105	6106
Weight of Sample,mg.	4.73	4.73	3.85	4.07
Thickness of Sample, Inches	0.01240	0.01428	0.01308	0.01008
Loss in Weight,%	77.4	80.7	83.5	81.1
Thickness of Pyrolyzed Residue,Inches	0.00068	0.00136	0.00368	0.00132
Decrease in Film Thickness %	94.5	90.5	71.9	86.9

NOTE 1. Samples were square pieces microtomed from strands placed
under cover slip on slide.

TABLE 4

CONDENSED PHASE BEHAVIOR OF DOUBLE BASE PROPELLANT ON THE
METTLER HOT STAGE,HEATING RATE 10°C/MIN TO 225°C IN TUBES AT
AMBIENT PRESSURE

Experiment No.	Onset of Rosetting °C	Onset of Active Bubbling, °C	End of Bubbling °C	Loss in Weight,%
1(Note 1)	181.5	198.2	219.3	-----
2	182.7	197.0	-----	36.1
3	181.3	196.9	216.5	-----
4	182.6	198.2	214.3	-----
5	180.6	197.8	213.2	-----
Average	181.7	197.6	215.8	-----

NOTE 1. Atmosphere in tube was air.

TABLE 5

CONDENSED PHASE BEHAVIOR OF DOUBLE BASE PROPELLANT ON THE METTLER HOT STAGE, HEATING RATE 10°C/MIN. TO 225°C IN TUBES PRESSURED WITH HELIUM

Experiment No.	Onset of Rosetting, °C	Onset of Active Bubbling, °C	End of Bubbling, °C	Loss in Weight, %	Pressure atm.
1	180.7	196.4	---	41.4	41.4
2	181.2	195.2	---	39.3	39.3
3	181.4	193.5	221.5	36.3	36.3
4	191.4	---	222.7	34.1	34.1
5	180.1	193.6	---	63.3	37.1
6	188.8	199.6	225.2	30.4	30.4

Average (Note 1)

NOTE 1: Due to variations in pressure, an average value cannot be obtained for the different parameters.

TABLE 6

CONDENSED PHASE BEHAVIOR OF 12.6% N NITROCELLULOSE FILM
UPON HEATING IN OPEN GLASS CAPILLARY TUBES AT 10°C/MIN. IN METTLER
FP 52 MICROFURNACE TO 250°

Experiment No.	Sample Weight, mg.	Onset of Rosetting, °C	Onset of Total Bubbling, °C	End of Bubbling, °C	Loss in Weight, %
1	1.12	183.0	199.7	221.4	----
2	1.06	182.1	199.0	219.8	----
3	1.85	182.7	199.0	219.4	55.6
4	1.31	184.8	-----	219.2	51.9
5	1.07	185.2	199.2	218.2	48.8
6	1.64	184.6	198.3	219.8	53.6
7	1.22	184.8	199.2	218.8	47.8
8	0.96	184.0	199.0	219.6	51.0
Average	1.27	183.9	199.05	219.5	51.4

TABLE 7

CONDENSED PHASE BEHAVIOR OF 12.6% N NITROCELLULOSE FILM
AT 10°C/MIN. IN METTLER HOT STAGE IN HELIUM AT 34 ATMOSPHERES IN GLASS
CAPILLARIES

Experiment No.	Sample Weight, mg.	Onset of Rosetting, °C	Onset of Total Bubbling, °C	End of Bubbling, °C	Loss in Weight,%
1	1.70	182.6	193.5	214.1	74.4
2	1.36	187.5	198.5	214.0	----
3	1.32	182.8	194.7	213.1	50.8
4	1.81	184.5	197.1	216.2	57.6
5	1.49	183.7	194.6	212.5	----
6	1.43	184.3	195.7	214.5	57.7
7	1.70	183.6	194.7	212.6	55.8
8	1.54	181.5	196.3	211.6	67.2
Average	1.54	183.8	195.6	213.6	60.6

TABLE 8

CONDENSED PHASE BEHAVIOR OF NEW 12.6% N NITROCELLULOSE FILM WITH
2.0% LEAD SALICYLATE CATALYST UPON HEATING IN OPEN GLASS CAPILLARY TUBES
AT 10°C/MIN. IN METTLER FP 52 MICROFURNACE TO 250°C

Experiment No.	Sample Weight, mg.	Onset of Rosetting, °C	Onset of Total Bubbling, °C	End of Bubbling, °C	Loss in Weight,%
1	0.86	192.3	200.9	216.8	48.5
2	2.03	181.3	198.0	217.0	51.0
3	0.64	193.1	201.1	217.4	46.2
4	0.69	178.9	198.4	217.5	----
5	2.02	180.7	198.3	218	55.8
6	1.58	182.8	199.6	217.6	46.0
7	1.71	182.5	199.2	217.3	52.4
8	1.71	180.3	199.1	217.2	50.0
Average	1.40	184.0	199.3	217.4	50.0

TABLE 9

CONDENSED PHASE BEHAVIOR OF 12.6% N NITROCELLULOSE FILM WITH
2.0% ADDED LEAD SALICYLATE CATALYST UPON HEATING AT 10°C/MIN
IN METTLER HOT STAGE IN HELIUM AT 34 ATMOSPHERES IN GLASS CAPILLARY TUBES

Experiment No.	Sample Weight, mg.	Onset of Rosetting, °C	Onset of Total Bubbling, °C	End of Bubbling, °C	Loss in Weight, %
1	0.990	186.5	197.9	216.3	48.0
2	1.05	182.2	196.2	212.7	47.8
3	0.990	180.8	196.4	216.3	52.3
4	0.493	192.4	199.1	221.8	52.7
5	0.886	185.6	197.8	216	45.0
6	1.98	179.7	194.6	213.7	54.2
7	1.70	182.0	196.7	214.2	----
8	0.981	184.4	198.0	215.8	----
Average	1.13	184.2	197.1	215.8	50.0

TABLE 10

COMPARISON OF THE CONDENSED PHASE BEHAVIOR OF NEW PURE
12.6 N NITROCELLULOSE FILMS AND ONES WITH 2.0% LEAD SALICYLATE
CATALYST ADDED AT NORMAL AND SUPERATMOSPHERIC PRESSURES

Table Data Taken From, Material	Pressure Atm.	Gas Phase	Onset of Resetting 0, Note 1	Onset of Bubbling, $^{\circ}C$	End of Bubbling $^{\circ}C$	Loss in Weight, %
6, N.C. Film	1	air	183.9_8	199.0_8	219.5_7	51.4_6
8, Catalyzed Film	1	air	184.0_8	199.3_8	217.4_8	50.0_7
7, N.C. Film	34	He	183.8_8	195.6_8	213.6_8	60.6_6
9, Catalyzed Film	34	He	184.2_8	197.1_8	215.8_8	50.0_6

NOTE 1: All values in the table are average ones derived from a series of experiments, the number of which are denoted by the subscript to the lower right of the figure.

TABLE 11

APPEARANCE OF TLC RINGS FROM
THIN FILM PYROLYTIC RESIDUES

Sample	Nitrocellulose	Nitrocellulose with 2% Lead Salicylaldehyde
Start of Experiment		
Appearance of Spot in Artificial Light	*Negligible Yellow Color	*Definite Yellow Color
Visibility in Ultraviolet Light 2537 Å 3660 Å	Strongly visible *Weak	Strongly visible *Medium Strong,Yellowish
End of Experiment after Separation of Components		
Appearance in artificial light	*1. Colorless Center Spot 2. Slightly yellow ring at edge of center spot 3. Outer white ring	*1. Center spot less yellow than at start 2. Yellow ring outside center spot more faint 3. Outer white ring
Visibility in Ultraviolet light		
Inner spot 2537 Å 3660 Å	Dark Faint	Dark Faint
Outer ring 2537 Å 3660 Å	Dark *Faint	Dark *Strong

Iodine Test   *(The two samples differed)
*Denotes where differences occur between the two samples.

mounted under a cover slip on a glass slide. Ostensibly the
previous double base strand material, 6103, and this sample should
have been quite similar. Yet they differed substantially as indicated by their gross appearance. This sample was composed of:
90% nitrocellulose, 10% nitroglycerine, and 2% nitrodiphenylamine.
Experiments were carried out at both ambient and superatmospheric
pressure in capillary tubes as described by Morrow (5) previously.
Tables 4 and 5 summarize the results.

Nitrocellulose Thin Films with and without Additives. Films
of 12.6% nitrocellulose were cast from acetone for study at ambient
vs. higher pressures according to methods described before by
Morrow (4). Results of an investigation in capillary glass tubes
of unmodified nitrocellulose and modified film are summarized in
Tables 6-10. Figure 8 shows the appearance of residues of four
different films pyrolyzed under cover slips on slides in the microfurnace.

THERMOANALYTICAL OBSERVATIONS

Nitrocellulose Thin Films with and without Additives

A DuPont 900 Differential Thermal Analysis unit equipped with
differential scanning calorimeter (DSC) and the 950 thermogravimetric analysis (TGA) accessory was used for these experiments. In
the DSC mode small slices of films weighing 2-5 milligrams were
heated under ambient conditions in open aluminum micro pans. Various heating rates and sensitivity settings were used to find
optimum operational conditions. Typical DSC thermograms of a pure
nitrocellulose film, and that with alumina, lead salicylate, and
lead salicylaldehyde respectively are shown in Figure 6. Results
with TGA are shown in Figure 7. A platinum basket-type holder was
used in the TGA experiments.

Thin Layer Chromatography

Thin Film Pyrolysis. Pyrolyzed residues of a pure and lead
salicylaldehyde modified nitrocellulose film were examined by
TLC techniques. Samples were heated to 170 at 10°C/min., to 175
at 3°C/min., and held at 175 isothermally for 30 minutes on the
Mettler hot stage. After cooling the residues were dissolved in
tetrahydrofuran. The tests were carried out with a variety of
solvents. The data in Table 11 was obtained with a tetrahydro-
furan-petroleum ether mixture pipetted onto a square sheet of Eastman Chromagram paper No. 6060 silica gel with fluorescent indicator. This gave concentric rings. Ultraviolet light and iodine
development were used to make the rings visible.

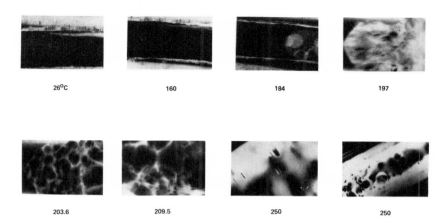

Fig. 5.   Condensed phase behavior of double base propellant at
ambient pressure in capillary glass tube.

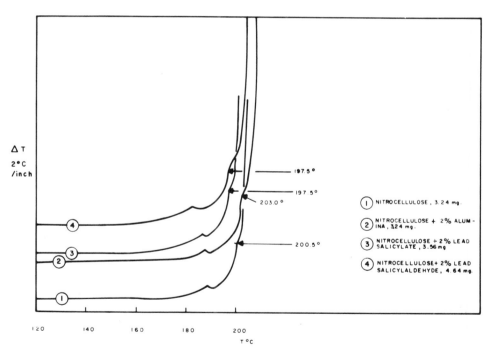

Fig. 6.   Differential scanning calorimetry of catalyzed vs. pure
12.6 % N Nitrocellulose films.

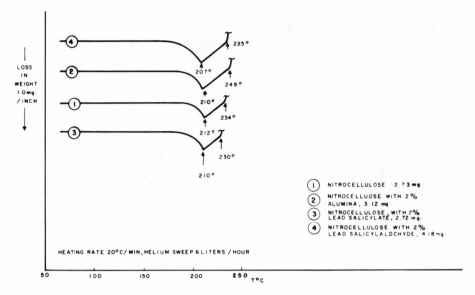

Fig. 7. Thermogravimetric analysis of catalyzed vs. pure
12.6 % N Nitrocellulose films.

PURE NITROCELLOUSE FILM   NITROCELLULOSE WITH ALUMINA ADDITIVE

NITROCELLULOSE WITH LEAD SALICYLATE CATALYST   NITROCELLULOSE WITH LEAD SALICYLALDEHYDE CATALYST

Fig. 8. Residues from pyrolysis of films to 225°C.

DISCUSSION

In order to carry out this research it was necessary to develop new ways to use a polarizing microscope for our purposes. This author's previous papers (4,5) highlight these developments. Thermoanalytical work, on the other hand, was carried out in an essentially conventional manner.

As can be seen from Figure 1 the DTA-light photometer polarizing microscope system is an effective means for characterizing propellant materials. In particular, the light photometer revealed a substantial difference in the light transmission behavior of unmodified, 6103, vs. lead salicylate promoted, 6104, propellant. Slight differences in the DTA thermograms of 6103-6 propellants are evident in Figure 2, but these were not explored at length. Due to the opaque, inhomogeneous nature of these thin sections of strands it was felt that they were unsuitable for extended study with a microscope. Our experience with them in fact caused us to seek ways to study the action of the catalysts on the principal energetic material in the strands, namely nitrocellulose itself. From photographic studies, such as those seen in Figures 3 and 4 we came to believe that the catalysts probably affected the viscosity and surface tension of the propellant as it became liquid at higher temperatures. Also there seemed to be possible differences, which were difficult to characterize without using high speed photomicrography, such as in the bubble structure of the various materials as they decomposed. We first observed the phenomena, which we named "rosetting", with the translucent double base sample referred to in Tables 4-5 and illustrated in Figure 5. Localized semi-liquification seems to develop during the process of the "rosetting" transformation, which takes a variable length of time to come to completion from sample to sample. When we first saw this happening, we wondered if it was due to the nitrocellulose component. We later proved that this was the case. The experiments with the 6103-6 series of propellants were carried out with an older Mettler microfurnace than the new FP-52 unit used in all the other work. In spite of the great accuracy and reliability of the temperature measuring systems in these instruments, very slight differences (i.e. a few tenths of a degree) might have existed between them. These experiments indicated that it is more difficult to demonstrate catalytic effects in double base type than in certain cases of composite propellant reported by Waesche (2).

It was much easier to study the transparent or highly translucent nitrocellulose films that we prepared than strand sections. Aluminum oxide, of course, appeared as a granular material dispersed in the nitrocellulose. Lead salicylate particles could not be detected in the dry films. Lead salicylaldehyde on the other hand, seemed to dissolve in part and also to cause formation of

small brownish clumps, which could be seen in the dried film.
According to Lenchitz (6) the aluminum oxide used here (corundum)
is noncatalytic. He has theorized that it has a physical effect.
Our results showed that it did behave differently than the lead
catalysts. Both lead salicylate and salicylaldehyde containing
films developed a brown color prior to the "rosetting" transformation. Aluminum oxide containing films did not. One might advance
the argument that the brown color is due to release of nitrogen
dioxide. If this were the case, a free radical organic species
could also be expected to be present in the film.

     Since only small differences were detected in the behavior
of pure and modified nitrocellulose film at ambient and elevated
pressure with microscopical techniques, Tables 6-10, the DSC-TGA
thermoanalytical investigation was carried out. Figures 6 and 7
are typical of the results obtained in air with DSC and in helium
with TGA. We think that the DSC traces indicate the "explosion
temperature" of the films. This is a somewhat unusual result inasmuch as we have not experienced such a seemingly clearcut measure
of this phenomenon with other types of materials, such as inorganic perchlorates and nitrates. It is possible that use of an inert
gas atmosphere will modify this behavior. Bizarre phenomena which
we cannot explain fully yet was encountered in the TGA investigations. The appearance of the burned residues of films, Figure 8,
of lead salicylate and salicylaldehyde-containing films was different from those left from non-additive-containing and alumina-con-
taining films. These films were mounted under cover slips on a
slide and heated at $10^{\circ}C$/min., in the Mettler microfurnace. In
TGA experiments, fragile, hollow, black, bubble-like boules were
formed in case of lead catalyzed films whereas flat residues were
obtained from pure and alumina containing nitrocellulose. Kubota
and co-workers (7) have also observed these spheres in combustion
experiments. The principal conclusions to be drawn from TLC sepa-
rations of residues from this film pyrolyses are that those from
pure nitrocellulose and from lead salicylaldehyde promoted film
were different. Exact definition of these chemical differences is
a matter for future research.

SUMMARY

     Insight into phenomena in the condensed phase combustion be-
havior of nitrocellulose, promoted nitrocellulose, double base,
and promoted double base materials has been obtained by use of the
polarizing microscope in connection with thermoanalytical procedures. The advantage of the microscope has been shown to be that
it is capable of detecting and affording measurements related to
transformations, phase changes, and decomposition behavior in this
stage of combustion. Some further insight into the condensed
phase behavior of propellants has been gained. There were indica-

tions that physical effects of the catalysts may be important. Relative importance of physical vs. chemical effects (catalytic) remain a matter of conjuncture at this time.

CONCLUSIONS

It was apparent that new polarizing microscope techniques worked out here are a valuable adjunct, if not "raison d'etre", for propellant combustion studies. Evidence of both physical and chemical effects of the catalysts was obtained. Such work obviously requires further pursuit and elaboration by bringing additional analytical techniques to bear upon this most complex of problems in catalyst technology.

REFERENCES

1. Kirby, C.E., "Investigation of Flameless Combustion Mechanism of M-2 Double Base Propellant", April 1971, NASA TND-6105, NASA, Washington, D.C.
2. Waesche, R.H., "Research Investigation of the Decomposition of Composite Solid Propellants", FR H910476-36, July 1969, United Aircraft Research Laboratories, East Hartford, Connecticut.
3. Freeman, E.S., and Rudloff, W.K., "Mechanisms of Solid State Catalytic Processes Related to Combustion", FR C6107-15, Contract No. DAAA 21-67-C-0798, July 1968, IIT Research Institute, Chicago, Illinois.
4. Morrow, S.I., "A DTA-Light Photometer Polarizing Microscope System for Hot Stage Microscopy", The Microscope, January-April 1973, Vol. 21 No. 1, pg. 29-38.
5. Morrow, S.I., "Microscopical Combustion Studies of Nitrocellulose Thin Films in Pressurized Capillary Tubes", presented at Inter/Micro '73, Cambridge University, Cambridge, England, 16-20 July 1973; in print in The Microscope.
6. Lenchitz, C., private communication, this laboratory 1973.
7. Kubota, N., Ohlemiller, T.J., Caveny, L.H., and Summerfield, M., "The Mechanism of Super-Rate Burning of Catalyzed Double Base Propellants", Aerospace and Mechanical Sciences Report No. 1087, Requisition NR 092-516/8-8-72 (473) March 1973.

THERMAL ANALYSIS OF DICYCLOPENTADIENYL ZIRCONIUM DIBOROHYDRIDE AND BIS(TRIPHENYLPHOSPHINE)BOROHYDRIDO COPPER(I)

B.D. James[*], B. Annuar, J.O. Hill and R.J. Magee[**]

*Department of Chemistry, University of Queensland, Brisbane, Queensland, **Department of Inorganic and Analytical Chemistry, La Trobe University, Bundoora, Victoria, Australia

INTRODUCTION

Considerable interest has centred on transition metal borohydride complexes since their first discovery in 1949 by Hoekstra and Katz(1). Several compounds containing the borohydride anionic species bonded directly to a transition metal have been successfully prepared and isolated with varying degrees of stability(2). Such complexes are non-dissociative in solution and decompose at various temperatures both in solution and in the solid state to form the free metal and numerous other products. Considerable doubt focusses on the mode of attachment of the central metal ion to the borohydride ligand; systems involving up to three hydrogen bridges between metal and boron are plausible (Fig. 1). Other ligands bonded directly to the metal considerably influence the degree of stability of the metal-borohydride linkages. It has been suggested(3) that cyclopentadienyl ligands bonded to the metal result in a large thermal stabilization of the borohydride groups, as in dicyclopentadienyl titanium diborohydride. The dicyclopentadienyl metal diborohydrides are readily synthesised by reacting

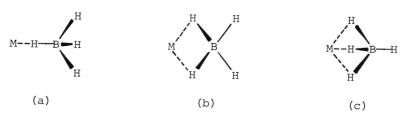

(a)   (b)   (c)

Fig. 1

the corresponding dicyclopentadienyl metal dichlorides with excess lithium borohydride in ether under nitrogen. Since pyrolytic studies(4) form the only indication of the thermal stability of these complexes it was considered that a complete thermal analysis of the zirconium and copper complexes would provide quantitative thermodynamic stability data for such complexes and also indicate the mechanism of thermal decomposition. This paper reports the thermal analysis (TGA, DTA) of dicyclopentadienyl zirconium diborohydride, $(C_5H_5)_2Zr(BH_4)_2$, and bis(triphenylphosphine)borohydrido copper(I). Thermal decomposition mechanisms are proposed for both complexes.

EXPERIMENTAL

Preparation of $(C_5H_5)_2Zr(BH_4)_2$ and $(P\emptyset_3)_2CuBH_4$

Dicyclopentadienyl zirconium diborohydride, $(C_5H_5)_2Zr(BH_4)_2$, was prepared(5) by smoothly stirring dicyclopentadienyl zirconium dichloride with lithium borohydride in ether under dry nitrogen according to the reaction

$$(C_5H_5)_2ZrCl_2 + 2LiBH_4 \longrightarrow (C_5H_5)_2Zr(BH_4)_2 + 2LiCl$$

After filtering off the resulting lithium chloride, $(C_5H_5)_2Zr(BH_4)_2$ was recovered from the filtrate as a white solid which when purified by sublimation in vacuo at 110-115°C yield a very pale yellow solid of m.p. 155°C. The compound is diamagnetic and not susceptible to oxidation but slowly hydrolysed by moisture.

$(P\emptyset_3)_2CuBH_4$, was prepared(4) by adding sodium borohydride in small portions to the colourless solution of copper sulphate in ethanol containing triphenylphosphine until effervescence and the precipitation of $(P\emptyset_3)_2CuBH_4$ ceases. The colourless compound is soluble in benzene, chloroform or tetrahydrofuran and the solutions are stable up to 43-50°C but deposit copper on further heating. The complex crystallizes from $CHCl_3$-ethanol mixture (m.p. 177°) (decomposed) or from benzene-ethanol mixture with one molecule benzene of crystallization (m.p. 187°C (decomposed)).

Thermal Analyses

Three different commercial thermal analysers were used.

(1) A Stanton TR-01 Thermobalance with a nickel crucible. A heat-resistant glass cylinder through which the effluent gas would pass was fitted to the furnace, so as to enclose the sample holder. Heating was carried out in an argon atmosphere at a flow rate of 100 ml/min. and a heating rate of 4°C/min.

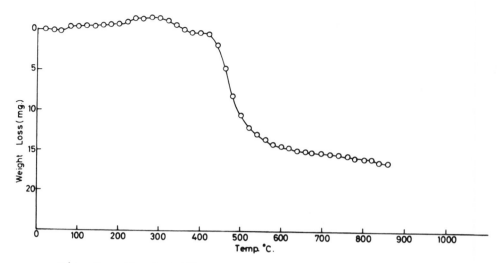

Fig. 2. Stanton TG curve of $(C_5H_5)_2Zr(BH_4)_2$ in argon.

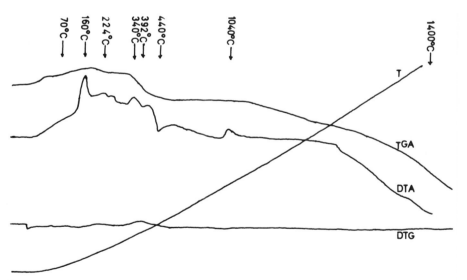

Fig. 3. DTA-TG-DTG curves of $(C_5H_5)_2Zr(BH_4)_2$ in nitrogen on the Rigaku-Denki Thermobalance.

For $(C_5H_5)_2Zr(BH_4)_2$ strict precautions were taken to minimise contact with air when loading and transferring the loaded crucible to the thermobalance.

(2) A Rigaku-Denki Thermobalance (Type, Thermoflex). Heating (20°C/min.) was performed in a nitrogen atmosphere (flow rate 150 ml/min.) and alumina was used as reference sample (DTA).

(3) A Mettler Thermobalance (Type, Thermoanalyser 1). Argon atmosphere.

RESULTS AND DISCUSSION

Structure of $(C_5H_5)_2Zr(BH_4)_2$ and $(P\emptyset_3)_2CuBH_4$

The crystal structure of $(C_5H_5)_2Zr(BH_4)_2$ is not known. Infrared studies in the present work indicate that the borohydride groups are bonded to the zirconium atom by way of two-hydrogen bridges(6) of the type shown in Fig. l(b). Solution nmr spectra give evidence of a rapid exchange process which renders bridge and terminal hydrogens indistinguishable. It appears that the two-point attachment of the borohydride group to the zirconium atom is preferred to three-point attachment, (Fig. l(c)), which is exhibited in $Zr(BH_4)_4$, since the infra-red spectrum of $(C_5H_5)_2Zr(BH_4)_2$ shows a band at 1123 cm^{-1} which is attributed to the deformation of terminal BH groups. In contrast, $Zr(BH_4)_4$, which is crystallographically known to have a three-point attachment of the borohydride groups to the metal(3), has only BH terminal groups and shows no band at $\sim$1100 cm^{-1} in the infra-red spectrum(3). Other bands associated with the BH terminal and bridging bonds of $(C_5H_5)_2Zr$-$(BH_4)_2$ in the infra-red occur at 2386 cm^{-1}, 2149 cm^{-1} and 1295 cm^{-1}.

In the case of bis(triphenylphosphine)borohydrido copper(I) the X-ray crystal structure has been determined(2). It consists of loosely packed mono-molecular units and there is a two-point attachment of the borohydride group to the copper atom. The coordination geometry may be described as quasi-tetrahedral with significant departures from this idealized configuration particularly in that the angles about the copper atom deviate markedly from the tetrahedral value of 109.47°. The deviations result from various steric interactions involving the phenyl groups and hydrogen bridge atoms.

Thermal Analysis of $(C_5H_5)_2Zr(BH_4)_2$

All three TGA profiles (Figs. 2, 3 and 4) are essentially similar. During the initial stages of heating, the sample undergoes a gradual increase in weight followed by a more rapid decrease in weight which results in a hump in the profile. Such a broad

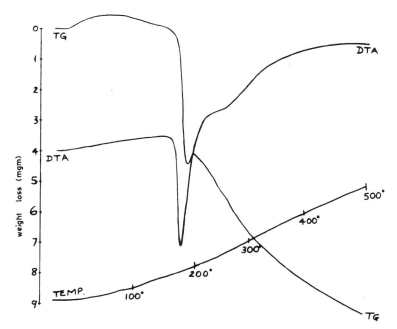

Fig. 4. Mettler TG-DTA curves of $(C_5H_5)_2Zr(BH_4)_2$ in argon.

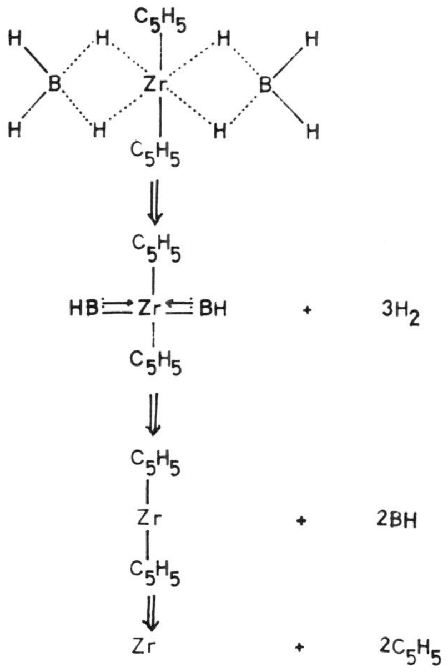

Fig. 5. Thermal decomposition mechanism of dicyclopentadienyl zirconium dibrohydride.

hump occurs in the region 70 to 380° in the Stanton thermogram
(Fig. 2); 70 to 340° in the Rigaku-Denki thermogram (Fig. 3) and
20 to 120° in the Mettler thermogram (Fig. 4). The endothermic
DTA peak at 160° in the Rigaku-Denki thermogram (Fig. 3) indicates
that the initial increase in weight by the sample is due to slow
absorption of the carrier gas on the surface of the sample. The
dissimilarities in the shapes and extents of the hump in the three
TGA curves are due to additional bouyancy factors and different
flow rates of carrier gas employed in conjunction with the various
thermal analysers. Desorption of carrier gas by the sample results
in a lowering of the hump as sample temperature increases. The
temperature at which the primary decomposition of dicyclopentadienyl
zirconium diborohydride is registered differs according to the
thermobalance used: (430° Stanton, 350° Rigaku-Denki and 170°
Mettler thermograms respectively). Such a noncoincidence of decomposition temperatures for the three thermograms is the result of
different heating rates employed - the most rapid heating rate
corresponds to the lowest decomposition temperature (Mettler thermobalance). Also the different sample sizes used in conjunction
with each thermal analyser lead to dissimilarities in depth and
detail in the three TGA profiles. The Mettler thermogram (Fig. 4)
clearly indicates that the primary decomposition of dicyclopentadienyl zirconium diborohydride is a two-stage process. The overall
weight loss corresponds to the removal of two BH_4 units and the
initial weight loss indicates the removal of six hydrogen atoms
from the molecule.

The Rigaku-Denki TGA/DTA profiles constitute the best indicator
of the overall thermal decomposition of the complex. On the basis
of all the accumulated thermoanalytical data, the following thermal
decomposition mechanism is proposed for dicyclopentadienyl zirconium
diborohydride (Fig. 5).

The initial increase in weight shown on the TGA profile at
approximately 160° and the corresponding endothermic peak in the
DTA profile is indicative of adsorption of carrier gas by the
sample. Between 350° and 500° two BH_4 units are lost in a two stage
decomposition process involving the initial loss of six hydrogen
atoms followed by the loss of two BH entities. Such a two stage
decomposition mechanism is confirmed by the two exothermic peaks
appearing in the DTA profile at 392° and 440°. The TGA profile
shows a nearly flat plateau between temperature limits of 500 and
1040° which represents the region of thermal stability of the
metallocene. The gradual slope in the exothermic direction of the
DTA curve in this temperature region confirms that dicyclopentadienyl zirconium cracks slowly. The DTA endothermic peak at 1040°
denotes rupture of the main skeleton of the ring system. Thence
decomposition is more rapid and zirconium metal is the non-degradeable residue at approximately 1400°.

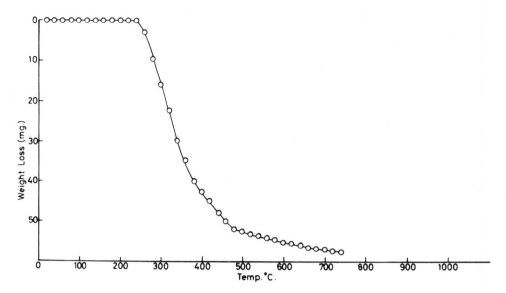

Fig. 6.   Stanton TG curve of $(P\phi_3)_2CuBH_4$ in argon.

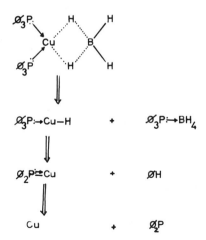

Fig. 7.   Thermal decomposition mechanism of bis(triphenylphospine) borohydrido Copper I

In support of the proposed mechanism, a present pyrolytic study of dicyclopentadienyl zirconium diborohydride revealed that only hydrogen was evolved when the compound was heated in a sealed tube for four hours and repeated heating indicated that the cyclopentadienyl rings were retained long after elimination of the hydridic hydrogen atoms. Also, the infrared spectrum of the residue obtained from the pyrolysis tube indicated no change in the vibration frequencies of the ring system whereas the BH terminal stretching frequency was reduced.

Thermal Analysis of Bis(triphenylphosphine)borohydrido Copper(I)

The TGA profile of $(P\emptyset_3)_2CuBH_4$ (Fig. 6) indicates three stages of thermal decomposition. The notable absence of a hump in the profile during the initial stages of heating suggests no adsorption of carrier gas by the sample. Rapid decomposition commences at 240° and slackens at 380°. At 480°, the rate of decomposition decreases further and the ensuing degradation is negligible up to the final metallic product. The following thermal decomposition mechanism is proposed (Fig. 7). The initial weight loss corresponding to the primary decomposition of $(P\emptyset_3)_2CuBH_4$ represents the abstraction of triphenylphosphine borine $\emptyset_3PBH_3$. The weight loss corresponding to the second thermal transition in the temperature range 380 to 480° represents the evolution of precisely one molecule of benzene. The first transition complex, $\emptyset_3PCuH$, presumably transposes to a second transition complex $\emptyset_2PCu$ which then degrades slowly in the argon atmosphere to yield metallic copper as the final non-degradeable residue. In support of the proposed mechanism, Davidson(4) has reported that the pyrolysis products of bis(triphenylphosphine)borohydrido copper(I) are triphenylphosphine borine, triphenylphosphine, hydrogen and metallic copper. The intermediate $\emptyset_3PCuH$ is not detected. The abrupt nature of heating in the pyrolysis process probably weakens the coordinating property of the lone pair on phosphorus such that fracture of $\emptyset_3PCuH$ yielding triphenylphosphine copper and hydrogen is immediately complete.

REFERENCES

(1) H.R. Hoekstra and J.J. Katz, J. Am. Chem. Soc., 71, 2488 (1949).
(2) S.J. Lippard and K.L. Melmed, Inorg. Chem., 6, 2223 (1967).
(3) B.D. James and M.G.H. Wallbridge, Progress in Inorg. Chem., 11, 99 (1970).
(4) J.M. Davidson, Chem. and Ind. (London), 2021 (1964).
(5) R.K. Nanda and M.G.H. Wallbridge, Inorg. Chem., 3, 1798 (1964).
(6) B.D. James, R.K. Nanda and M.G.H. Wallbridge, J. Chem. Soc. (A), 182, 1966.

ACKNOWLEDGEMENTS

We gratefully acknowledge the Chemistry Department, Georgia Technical College, Atlanta, Georgia, U.S.A. and Dr. Han, Department of Chemical Engineering, Monash University, Clayton, Victoria, Australia for supplying the dicyclopentadienyl zirconium diborohydride, Mettler and Rigaku-Denki thermograms respectively.

ABSTRACT

Thermoanalytical data (TGA/DTA) have been obtained for dicyclopentadienyl zirconium diborohydride and bis(triphenylphosphine)-borohydrido copper(I). A thermal decomposition mechanism is proposed for each complex.

SLOW REVERSION OF POTASSIUM NITRATE

Paul D. Garn

Department of Chemistry
The University of Akron
Akron, Ohio 44325

Potassium nitrate exists in several crystalline forms depending upon both the temperature and the pressure (1,2). The three easily observed forms are the orthorhombic KNO_3II, stable under standard conditions, a trigonal form, KNO_3I, stable above 127.7°C at atmospheric pressure (3), and a second trigonal form, KNO_3III, which is typically formed when KNO_3I is cooled. The persistence of this form varies with, among other things, its degree of confinement; that is, the expansion of the crystallites transforming from I→III may be unrestrained in a shallow pan, but in a sample cup or well the points of contact are under pressure, so a confined sample transforms more quickly to the orthorhombic phase (4).

Recently, Deshpande et al. (5) observed that samples which had been cooled from phase I and reheated soon thereafter showed evidence of a double peak as it transformed again to phase I. They concluded that the transition occurred in two steps, II→III→I. This would imply a thermodynamic stability for KNO_3III which had not been observed by other workers. The work reported by Garn, Diamondstone and Menis (4) was undertaken to test the conclusion of Deshpande et al. or, if necessary, explain the observation. While the explanation for the variations in behavior was found, the transitory existence of KNO_3III on heating was neither confirmed nor refuted experimentally, although Garn et al. concluded from other evidence that the heating transformation was a single step. By a different approach, this work does show that the metastable III does not form from the orthorhombic II during heating.

It is well established that the metastable trigonal III transforms to I at a slightly lower temperature than does the orthorhombic II (1, 6, 7). The III→I enthalpy change is about half that of the II→I change. Kracek (1) found 1059 and 557 cal/mole for the II-I and III-I transitions. Gray (7) found lower values. It follows, as Gray pointed out, that the degree of conversion of III→I can be measured by determining the area of the heating peak.

More important for the immediate purpose, if the III→I and II→I peaks are distinguishable, it can be ascertained unequivocally whether or not the II→III transition occurs. If the II→III transition does occur, the first peak would be larger if more II were present, that is, if the total area became greater. But if the II→III transition does not occur, the first peak should become smaller as the total area increases.

EXPERIMENTAL MATERIALS

The potassium nitrates tested were the NBS-ICTA Standard Reference Material, Mallinckrodt Analytical Reagent (used in the First International Test Program (8), Matheson-Coleman-Bell ACS Reagent, two batches of sized particles from the NBS material along with the parent material, Merck Reagent, and sections of single crystals from a melt and from solution crystallization.

EXPERIMENTAL PROCEDURE

Samples usually weighing about 10 mg. were run in a Perkin-Elmer DSC-1B under a series of conditions of heating, cooling and soaking. In general, the heating run was terminated ca. 410°K by switching the scanning drive to the reverse position quickly, stopping the chart drive, switching to the selected cooling rate (for example, 32°/min.), and lifting the recorder pen. The scan drive was stopped when the selected temperature was reached. The time interval was measured from that point provided the I→III exotherm had been completed. At the end of the time interval, the sample was brought rapidly (for example, 32°/min.) to about 390°K, at which point the rate switch was set to the selected scan rate. The recorder pen was dropped and the chart drive begun $(394-396^{\circ})$. The selected temperatures or soak times were in a systematic progression but an early run would be repeated occasionally to guard against any systematic variation in the sample, such as decreasing crystallite size.

In some runs, the sample was taken to ca. 440°K. In others, the sample was "conditioned" ca. 410°K before cooling. These variations are identified.

RESULTS

The curves obtained with the several materials varied in the
detailed shapes of the peaks. Selected sets of curves shown here
are descriptive of the general behavior. For processes of this
type, sensitivity to even low levels of impurities can be expected.
For a given batch of material, good reproducibility was obtained.

Gray has already shown that slower cooling enables a greater·
amount of material to transform to KNO$_3$II. The transformation is
now shown to be <u>time</u> dependent rather than <u>cooling rate</u> dependent.

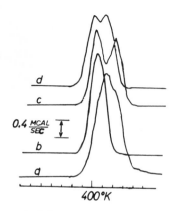

Figure 1.   DSC curves at 2°/min. for Mallinckrodt KNO$_3$. <u>a</u>, First
heating; <u>b</u>, Cooled to 380°K and rerun; <u>c</u>, Cooled to 360°K and reheated; and <u>d</u>, Cooled through transition, held 15 min. at 390°K
and rerun.

Figure 1 shows the variation in heating curves from various cooling
treatments. The area of Curve <u>1b</u> is 59% that of <u>1a</u> (53% corresponds to zero III→II transformation), indicating that the III→II
change proceeds fairly rapidly for this material. Cooling to 360°K
yields 74% of the original area and a very noticeable diminution
of the first (III→I) peak. The low temperature favors nucleation.
Yet even at 390°K, the reversion to the orthorhombic form proceeds;
Curve <u>1d</u> has noticeable peak separation and 70% of the original
area.

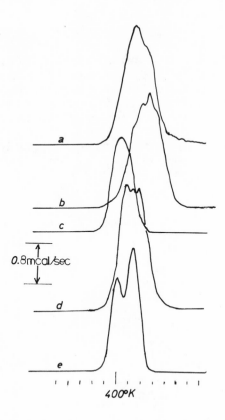

Figure 2. DSC curves at 2°/min. for Mallinckrodt KNO_3. a, First heating; b, Held at 360°K 20 hrs.; c, Cooled to 380°K and rerun; d, Held at 360°K 60 min.; and e, Held at 380°K 60 min.

The III→II transition will go to completion at 360°K. Curve b of Figure 2 has >99% of the area of Curve a. The latter is also Curve a of Figure 1. Cooling to 380°K and rerunning yields 55% of the first-run area, close to the expected 53%.

Now note that Curve 2d shows evidence of a third peak. This is a persistent phenomenon. One might postulate transitory existence of KNO_3IV. Formation of metastable states is far from a rarity, as the ambient-pressure occurrence of KNO_3III attests. However, one may also postulate that the KNO_3II is not in a single well-defined energy state. That is, formation of II may not yield the lowest energy state or some further change might follow the initial III→II transition. The evidence clearly favors the latter.

Note that Curves 2a and b are both broader than d or e. These
runs were made after protracted periods at ambient temperature and
at 360°K, as compared to the one hour at 360°K for d or 380°K for
e. The area of 2d was 102% of 2a, a phenomenon which recurred.
The higher temperature preceding Curve 2e may well allow more
facile movement, but certainly a lesser difference in free energy
between III and II and a lesser frequency of nucleation. The
lesser rate of III→II is hence not surprising. The area is 71%
of 2a.

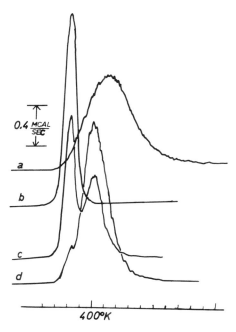

Figure 3. DSC curves at 2°/min. for KNO₃, -200+325 mesh. a,
First heating; b, Cooled to 380°K and rerun; c, Held 10 min. at
360°K; and d, Held 30 min. at 360°K.

        Figures 3 and 4 show data on the smaller (-200+325 mesh) NBS
test sample. The general behavior is similar to that already shown
but some different observations were made. Curve 3a is, again, the
first heating. The material had been kept sealed for ca. 2 years
after aging at 383°K for 16 hours. The overall peak is embellished
with smaller peaks due to individual particles. Curve 3b (Area =

50% of 3a) shows no such small peaks. Materials held long enough
for substantial III→II transformation characteristically show small
peaks. This author (9) had attributed them simply to individual
particles in contact with the thermocouple. From this work, it is
apparent that the particles have measurable different transition
temperatures. This is shown quite generally by the increasing
breadth of the peak with greater conversion to the orthorhombic
(II) state.

Curve 3c shows the partial conversion to II at 360°K (Area =
67% of 3a). The III→I peak is still well defined, since only one-
third has transformed. In 30 min. at 360°K, the III→I peak is
hardly discernible. The area of 3d is 95% that of 3a.

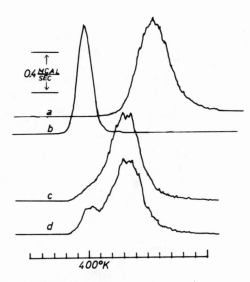

Figure 4. DSC curves at 2°/min. for a sample of -200+325 mesh KNO₃,
aged at 385°K overnight. a, First heating; b, Cooled to 380°K and
rerun; c, Held 30 min. at 360°K; and d, Held 10 min at 360°K.

The variation in transition temperatures of particles was
tested with this same material. A specimen was held overnight at
385°K in a stream of dried nitrogen. The stability of the orthorhombic phase was clearly increased. The peak of Figure 4a is
about 6° higher than that of 4b, as compared to about 2° in other
cases. It is clearly an effect of recent history. Thirty minutes
at 360°K (Fig. 4c) yielded an area 101% that of 4a. A shorter

time, 4d, shows a small amount of III→I transition and 97% of the
area of 4a.

   The single crystal shows distinctly different behavior. The
material transforms on cooling directly to KNO₃II -- at first. The
stresses due to the anisotropic lattice expansion and contraction
appear to prevent formation of the metastable Phase III. This is
in accord with the earlier finding (4), that physical restriction
caused strain which induced reversion to II.

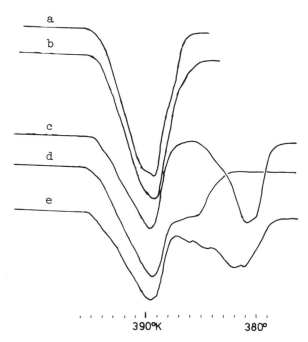

Figure 5. DSC cooling curves at 4°/min. for solution-grown single
crystal of KNO₃. a, First cooling; b, Fourth cooling; c, Heated
to 440°K, cooled from 404°K; d, Cooled immediately from 413°K;
and e, Annealed 100 min. at 413°K.

   However, if a single piece is cycled through the transition
a number of times, it is no longer a single crystal, but many
closely fitting crystallites. The second peak on cooling shows
only a suggestion of separation from the first in the first several
heatings (Figure 5) and the area remains nearly the same (ca. 5%
less) as in the heating peak.

   If, however, the sample is heated to ca. 440°K, the cooling

peaks are clearly separated (Fig. 5c). The measured area, using
a single base line across the two peaks, is slightly higher (4%)
than that of the heating peak. The effect is not permanent, because subsequent heating at 413°K and cooling at the same rate immediately brings the second peak close to the first (Fig. 5d).
The basic cause appears to be an annealing of the high temperature
Phase I so that the trigonal III formed from the trigonal I (by a
unidirectional compression of the lattice with expansion perpendicular to that axis) is also well annealed. This can be inferred
from the similarity of the curve from annealing 100 min. at 413°K
(Fig. 5c) to that from the 440°K heating.

In all the subsequent reheatings, the areas under the peaks
were within a few per cent, indicating that the III→II transformation goes essentially to completion.

The single crystal subjected to a number of transformations
is no longer transparent and it cracks easily. Kracek (1) had
noted spontaneous cracking with larger specimens.

It was easily perceivable that there were small differences
in base line with differing degrees of completion of the III→II
transition. To verify this, a sample of the -200+325 mesh material
was heated to 410°K, cooled to 360°K and heating begun (at 32° per
min.) as soon as the recorder pen ceased rapid movement (at 2 mcal
per sec.). In other runs, the start of the heating was delayed by
various times. The heat capacity was higher for these runs when
the sample was immediately rerun because the Phase III has a
higher heat capacity (10). Delays before reheating decreased the
heat capacity. The supply of heat from the transformation is not
large enough to compensate for the difference in heat capacity.

The rate of transformation could, in principle, be measured
by "isothermal DSC" just as rates of polymer curing (11) and other
processes have been measured. The change is unsteady, showing
peaks rather than a steady change, so any numerical values would
not be firmly based. That part of the program was not continued.

CONCLUSIONS

When KNO_3I has transformed to KNO_3III, the further transformation to KNO_3II may occur rapidly if the crystallites are under
physical restraint, as in the single crystal. This is in accord
with the earlier report (4).

If the III→II transformation is not rapid, the heating transformations III→I and II→I can be clearly separated. Increased
III→II transformation results in increased total area but a

decreased III→I peak. This demonstrates the absence of any II→III
transformation on heating.

Particles in the orthorhombic phase develop inhomogeneities
in energy, presumably due to relief of strain, which causes a
broad initial peak. Recently transformed material, even mixtures
of II and III, show narrower temperature ranges of transition than
well aged KNO_3II.

Differences in behavior between the materials tested
are principally in their responses to the heating and cooling
regimes. The temperatures of the heating transitions obtained
immediately after cooling are highly reproducible for each sample
tested. The differences between the samples warrant the reminder
that the NBS-ICTA certification of a Standard Reference Material
was for only that batch of potassium nitrate.

ACKNOWLEDGMENTS

The author is indebted to the Alexander von Humboldt-Stiftung
and to the National Science Foundation for support of this work
and to Prof. S. Haussühl, Institut fur Kristallographie der
Universität zu Köln for the single crystal potassium nitrate.

The experimental measurements were performed at the
Mineralogisch-Petrographisches Institute der Universität zu Köln,
Federal Republic of Germany.

LITERATURE CITED

1. F. C. Kracek, J. Phys. Chem. 34, 225 (1930).
2. E. Rapoport and G. C. Kennedy, J. Phys. Chem. Solids 26, 1995
    (1965) cited in Group III, Vol. 3, "Ferro- and antiferroelectric
    Substances", LANDOLT-BORNSTEIN NUMERICAL DATA AND FUNCTIONAL
    RELATIONSHIPS IN SCIENCE AND TECHNOLOGY (edited by K. -H- and
    A. M. Hellwege), Springer-Verlag, Berlin, Heidelberg,
    New York, 1969.
3. U. S. National Bureau of Standards Circular 500 (1952).
4. P. D. Garn, B. I. Diamondstone, and O. Menis, J. Thermal
    Analysis 6 (1974) in press.
5. V. V. Deshpande, M. D. Karkhanavala, and U. R. K. Rao, J.
    Thermal Analysis 6 (1974) in press.
6. O. Menis and J. T. Sterling, Proc. Symposium on the Current
    Status of Thermal Analysis, Gaithersburg, Md., 1970, NBS Spec.
    Pub. 338, pg. 61. Edited by Oscar Menis.
7. A. P. Gray, "A brief study of the phase behavior of KNO_3 in a
    differential scanning calorimeter", Perkin-Elmer Corp.,
    Aug. 1972.

8. H. G. McAdie, in THERMAL ANALYSIS (Robert F. Schwenker, Jr.
    and Paul D. Garn, eds.), Vol. 1, p. 693, and Vol. 2, p. 1499,
    Academic Press, New York, 1969.
9. P. D. Garn, Anal. Chem. 41, 447 (1969).
10. H. Miekk-Oja, Ann. Acad. Sci. Fennicae, Ser. A1, Math. Phys.
    No. 7, 1941, cited in LANDOLT-BORNSTEIN. See Ref. 2.
11. H. Kambe, I. Mita, and K. Horie, in THERMAL ANALYSIS (Robert
    F. Schwenker, Jr. and Paul D. Garn, eds.), Vol. 2, p. 1071,
    Academic Press, New York, 1969.

THERMOANALYTIC MEASUREMENTS OF FIRE-RETARDANT ABS RESINS

H. E. Bair

Bell Telephone Laboratories, Incorporated

600 Mountain Avenue, Murray Hill, New Jersey   07974

INTRODUCTION

Previously we have shown that it is possible to detect the type and amount of each component within a polyblend by either measuring the component's glass temperature, T_g, and the magnitude of the increase in specific heat, ΔC_p, occurring at the transition or determining the component's melting point, Tm_o, and the corresponding heat of fusion, ΔH_f.[1,2] In our laboratory this technique has become a reliable and rapid quantitative tool for the analysis of impact modified polyblends.[2,3,4] However, these previous measurements have relied upon the immiscibility of the individual polyblend components to produce separate phases which can be identified by calorimetry. It is the purpose of this paper to describe the quantitative thermal analysis of several commercial ABS resins which have organic fire-retardant (FR) additives which are soluble in either the glass and rubber phases or in only the matrix of the resin.

FR additives, which are soluble in the rubbery polybutadiene (BD) phase, or the glassy matrix of poly (styrene-co-acrylonitrile) (SAN) of an acrylonitrile-butadiene-styrene (ABS) resin, will lower or raise the BD or SAN glass temperature, depending on the respective position and relative difference in the glass temperatures between the additive and the polymeric component. Thus the FR additives were separated from the FR ABS resins and identified.[5] Subsequently, the effect of the FR on the SAN and BD T_g as a function of concentration was determined.

797

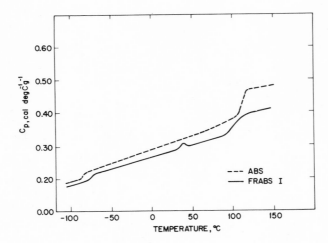

Figure 1 - Specific heat behavior of ABS and FR ABS I.

EXPERIMENTAL

Several ABS resins have been studied, and they include the following:

1. ABS;

2. FR ABS I - a purple FR ABS containing an organic halide, FR I, and Sb_2O_3;

3. FR ABS II - a white ABS resin blended with an organic halide, FR II, and Sb_2O_3; and

4. FR ABS III - a gray ABS with the same FR additives as the previous FR ABS II.

In addition, a concentration series was prepared by milling known amounts of FR I in an ABS. Lastly, FR I was mixed into polybutadiene (BD) and into a commercial styrene-acrylonitrile copolymer (SAN).

Glass transition temperatures and the concentrations of SAN and BD in the ABS samples were measured by differential scanning calorimetry (DSC) using a Perkin Elmer DSC-2 instrument.[2] The

onset of the discontinuity of C_p at T_g was recorded as the glass
temperature and listed in the tables as T_{gi}. The termination of
the linearly increasing C_p at T_g is denoted T_{gf}. Thus the width of
T_g, ΔT, is the difference between T_{gf} and T_{gi}. The heating rate in
all cases was $40°C$ per minute.

Volatilization studies of the organic FR additives were performed on a Perkin Elmer TGS-1. Flexure tests were made on the ABS
and FR ABS specimens in a Perkin Elmer TMS-1.

RESULTS AND DISCUSSION

Comparative Thermal Behavior of ABS and FR ABS I

In Figure 1 the heat capacity, C_p, of ABS without any FR additives is compared with that of FR ABS I over a temperature range of
$-110°$ to $+150°C$. The two expected glass transitions for the BD and
SAN phases of the ABS resin were observed at $-88°$ and $104°C$,
respectively.[2] From the magnitude of the increase in C_p, ΔC_p, the
ABS resin was estimated to contain 13 weight percent BD and 78
weight percent SAN. In contrast to this behavior the FR ABS I
resin has not only a lower C_p over the entire temperature range,
but also the BD transition was shifted $12°$ higher to $-76°$ and the
SAN T_g was lowered $11°$ to $93°$ (Table I).

ΔC_p at T_g for the BD phase of FR ABS I equals 0.012 cal$°C^{-1}g^{-1}$,
which is slightly higher than the value of 0.011 cal$°C^{-1}g^{-1}$, which
was measured for the BD in the original ABS. However, ΔC_p for the
SAN phase of the FR ABS I was found to be 0.054 cal$°C^{-1}g^{-1}$, which
is significantly lower than the value of 0.074 cal$°C^{-1}g^{-1}$ for SAN
in the non-flame-retardant ABS. In addition, the FR resin has a

TABLE I

GLASS TRANSITION BEHAVIOR OF FR ABS RESINS

	BD			SAN		
	T_{gi} °C	T_{gf} °C	ΔC_p cal$°C^{-1}g^{-1}$	T_{gi} °C	T_{gf} °C	ΔC_p cal$°C^{-1}g^{-1}$
ABS	-88	-78	0.011	104	115	0.074
FR ABS I	-76	-62	0.012	93	104	0.054
FR ABS II	-88	-78	0.015	86	99	0.059
FR ABS III	-88	-77	0.013	84	97	0.047

small melting transition near 40° with a heat of fusion of 0.13 cal/g. The latter transition may be due to several percent of a chlorinated hydrocarbon.

The shifts in T_g indicate that the organic halide, FR I, is miscible in both phases of the ABS resin. We have examined the effect of FR I on the T_g's of BD and SAN after milling known amounts of FR I into an ABS resin. In addition, the C_p behavior of blends of FR I and SAN and FR I and BD was studied. The T_g of FR I is 35°C. The glass transition data are listed in Table II.

In Figure 2 the relationship between T_g of BD and SAN and the weight fraction, w_1, of FR I is depicted. The lowering of the SAN T_g is about the same for equal amounts of FR I blended into either ABS or SAN. However, the increase of the BD T_g with increasing concentration of FR I is greater for the rubber in the blended FR I/ABS compositions than in the FR I/BD blend. This apparent anomaly is probably caused by a greater solubility of FR I in BD than SAN.

TABLE II

T_g IN BLENDS OF FR I AND ABS

	BD			SAN		
w_1 of FR I	T_{g_i}, °C	T_{g_f}, °C	ΔT, °C	T_{g_i}, °C	T_{g_f}, °C	ΔT, °C
0.00	−88	−80	8	104	118	14
0.05	−86	−78	8	102	117	15
0.10	−81	−70	11	96	112	16
0.15	−78	−64	14	94	111	17
0.20	−70	−56	14	92	108	16

T_g OF FR I/BD BLENDS AND FR I/SAN BLENDS

	BD				SAN			
w_1 of FR I	T_{g_i} °C	T_{g_f} °C	ΔC_{p} cal°C^{-1}g^{-1}	ΔT °C	T_{g_i} °C	T_{g_f} °C	ΔC_{p} cal°C^{-1}g^{-1}	ΔT °C
0.00	−79	−73	0.090	6	101	108	0.099	7
0.09	−74	−64	0.077	10	91	101	0.087	10
0.17	−71	−63	0.077	8	90	101	0.085	11
0.28	−67	−58	0.067	9	82	93	0.073	11
0.50	−55	−41	0.055	14	69	84	0.050	15

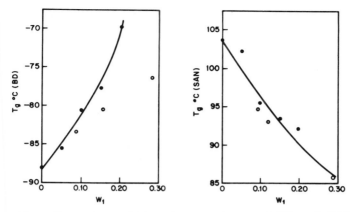

Figure 2 - The glass transition temperature of BD and SAN plotted
against weight fraction, w_1, of FR I.

A) ● FR I/ABS Blends; o FR I/BD Blends

B) ● FR I/ABS Blends; o FR I/SAN Blends

The decrease in ΔC_p of BD and SAN in the FR I/SAN and FR I/BD mixtures is roughly proportional to the increasing concentration of
flame-retardant (Table II); also the width of the glass transition
was found to increase with increasing amounts of FR I.

Examination of the plot of T_g's versus w_1 (Figure 2) for the
amount of FR I needed to produce the observed shifts in T_g for FR
ABS I indicates about 14.5 weight percent of that FR had been
added to this particular FR ABS. ΔC_p measurements were corrected
for the attenuation caused by the FR. After these corrections it
was estimated that FR ABS I contains 62 weight percent SAN and 16
weight percent BD. Thus the ratio of BD to SAN in FR ABS I has
been increased by approximately 40 percent over that found in ABS.
Apparently, the increased concentration of rubber in FR ABS I is
the chief reason for the impact strength being improved to 3.5 ft-
lbs/in over the value of 1.0 registered by an early FR ABS which
contained 10 to 11 percent BD.

The FR concentrates used in producing self-extinguishing ABS
resins depend upon an organic halide compound which vaporizes
before the degradation of the ABS structure. The vaporization
behavior of the FR additive used in FR ABS I is no exception to
this rule. In Figure 3 the rate of volatilization of FR I has been
plotted against reciprocal temperature. Note that at processing
temperatures (∿230°C) and atmospheric pressure, FR I volatilizes at
10 weight percent per minute.

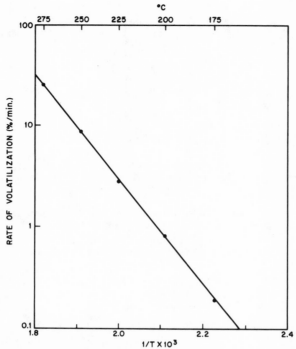

Figure 3 - Logarithm of rate of volatilization of FR I plotted
against reciprocal absolute temperature.

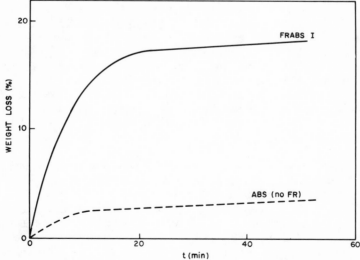

Figure 4 - The percentage weight loss of FR ABS I and ABS at 275°C
in nitrogen are plotted against time.

This appreciable rate of volatilization can be used to gain an independent measure of the amount of FR I in FR ABS I. In Figure 4 the difference in amount of weight loss which occurs at 275°C after 30 minutes between FR ABS I and ABS is 14 weight percent and is attributed to the loss of FR I. This value agrees well with our earlier estimates based upon shifts in T_g.

The amount of Sb_2O_3 in FR ABS I was determined thermogravimetrically by pyrolyzing the resin at 450°C. The remaining residue equaled 6.8 weight percent of original sample. Nondispersive X-ray (NDX) spectrometry of the residue indicated mainly the presence of Sb atoms. Since O atoms are not detectable by this instrument, it is presumed that the Sb is present as an Sb_2O_3. The lowering of the T_g of the SAN matrix of FR ABS I and the relatively large increase in the amount of BD should produce a softer material with lower flexural strength. These inferences are borne out in a flexure test. Small (20-mil-thick and 100-mil-wide) specimens of FR ABS I and ABS were supported on knife edge supports of a flexure platform at a maximum fiber stress of 264 psi. The deflection of the specimen was monitored continuously at a heating rate of 5° per minute. At 60° FR ABS I began to deflect more than ABS. This difference increased with increasing temperature (Figure 5).

Thermal Analysis of FR ABS II

A comparison of the thermal behavior of FR ABS II and ABS revealed no shift in the T_g of BD but an 18° lowering of the SAN T_g

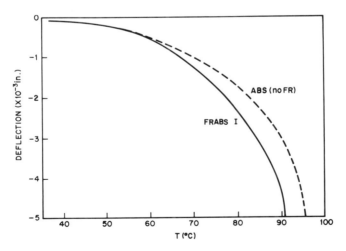

Figure 5 - The deflection of 20-mil-thick films of ABS and FR ABS I under a maximum fiber stress of 264 psi plotted against temperature. Heating rate was 5° per minute.

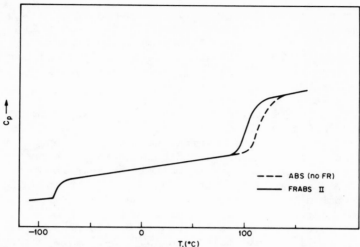

Figure 6 - Comparative thermal behavior of ABS and FR ABS II.

to 86°C (Figure 6).   From ΔC_p measurements (Table I) FR ABS II is estimated to contain 17 weight percent rubber and 65 weight percent SAN.

Thermal analysis indicates the organic flame-retardant is soluble in the SAN and not in the BD phase of FR ABS II.   Since we

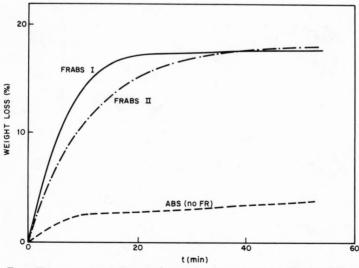

Figure 7 - The percentage weight loss of ABS, FR ABS I, and FR ABS II at 275°C in nitrogen plotted against time.

TABLE III

	ABS		FR Additives			Impact Strength
	SAN Weight Percent	BD Weight Percent	FR I Weight Percent	FR II Weight Percent	Sb$_2$O$_3$ Weight Percent	ft-lbs/in at 24°C
ABS	78	13	--	--	--	4-5
FR ABS I	62	16	14.3	--	6.8	3-4
FR ABS II	55	15	--	16.7	12.4*	2.5
FR ABS III	65	17	--	14.4	5.2	6

* The value is assumed to include several weight percent of white colorant.

have not investigated the effect of this organic FR on T_g, we cannot estimate the amount of FR II in FR ABS II from the position of its glass temperature. Nevertheless, we can compare the relative weight loss of FR ABS II to ABS at 275° and assume the difference equals the amount of FR II. From the data in Figure 7 we estimate FR ABS II has about 14 weight percent of an organic halide. The complete analysis of FR ABS II and other FR ABS samples is listed in Table III.

CONCLUSION

Several commercial fire-retardant (FR) ABS resins have been found by thermal analysis to contain about 14 weight percent of an organic halide and 6 weight percent of antimony oxide. In one FR ABS the FR additive is soluble in both the BD and SAN; in another FR ABS the FR is miscible in only the SAN phase of the ABS. Consequently, the T_g of the SAN is lowered by as much as 18°. In order that the impact strength of the FR ABS resin be comparable to ABS, the relative amount of BD to SAN in the FR ABS has been increased by 30 to 40 percent above that incorporated in ABS. The increased rubber content and depressed T_g of the self-extinguishing resins yield a softer composite material with lower flexural strength.

ACKNOWLEDGMENTS

The author thanks D. J. Boyle, J. T. Ryan, and E. W. Shieh for impact strength values; P. G. Kelleher for the commercial resins and T. F. Reed for all other FR blends used in this study; and L. L. Blyler, S. Matsuoka, and T. F. Reed for stimulating discussions.

REFERENCES

1. H. E. Bair, Polymer Eng. and Sci. 10, 247 (1970).

2. H. E. Bair in "Analytical Calorimetry," Volume 2, edited by R. S. Porter and J. F. Johnson, Plenum Press, New York, Page 51 (1970).

3. T. F. Reed, H. E. Bair, and R. G. Vadimsky, Polymer Preprints 14, 1074 (1973); in "Recent Advances in Polymer Blends, Grafts and Blocks," edited by L. H. Sperling, Plenum Press (1974).

4. H. E. Bair, Polymer Eng. and Sci., in press.

5. P. G. Kelleher, private communication.

B. Annuar, Department of Inorganic and Analytical Chemistry, La Trobe University, Bundoora, Victoria, Australia

T. Arikawa, Tokyo University of Agriculture and Technology, Koganei, Tokyo, Japan

H.E. Bair, Bell Telephone Laboratories, Inc., Murray Hill, New Jersey 07974

P.A. Barnes, Department of Chemistry, Leeds Polytechnic, Yorkshire, England

E.M. Barrall II, IBM, Research Laboratory, San Jose, California 95193

W.D. Bascom, Naval Research Laboratory, Washington, D.C. 20375

R.L. Berger, National Heart and Lung Institute, Bethesda, Maryland 20014

E. Berlin, Diary Products Laboratory, Agricultural Research Service, U.S. Department of Agriculture, Philadelphia, Pennsylvania 19118

G.L. Bertrand, Department of Chemistry, University of Missouri-Rolla, Rolla, Missouri 65401

S.R. Betso, Department of Chemistry, University of Georgia, Athens, Georgia 30602

R.L. Blaine, DuPont, Wilmington, Delaware 19898

G. Berthon, Laboratoire de Thermodynamique Chimique et Electro-chimie, 86022 Poitiers, France

L.D. Bowers, Department of Chemistry, University of Georgia, Athens, Georgia 30602

W.P. Brennan, Perkin-Elmer Corp., Norwalk, Connecticut 06856

A.M. Bryan, Department of Chemistry, SUNY, Albany, New York 12222

W. Bryant, Air Force Institute of Technology, Wright-Patterson Air Force Base, Ohio 45433

T.E. Burchfield, Department of Chemistry, University of Missouri-Rolla, Rolla, Missouri 65401

R.H. Callicott, Department of Chemistry, University of Georgia, Athens, Georgia 30602

P.W. Carr, Department of Chemistry, University of Georgia, Athens, Georgia 30602

B. Cassel, Perkin-Elmer Corp. Norwalk, Connecticut 06856

E. Catalano, Lawrence Livermore Laboratory, University of California, Livermore, California 94550

A.C. Censullo, Department of Chemistry, The Pennsylvania State University, University Park, Pennsylvania 16802

M.G. Chasanov, Argonne National Laboratory, Argonne, Illinois 60439

C.C. Chen, Department of Food Science, Rutgers University, New Brunswick, New Jersey 08903

J.J. Christensen, Center for Thermochemical Studies, Brigham Young University, Provo, Utah 84602

S. Cirafici, Institute of Physical Chemistry, University of Genoa, Italy

R.G. Craig, University of Michigan School of Dentistry, Ann Arbor, Michigan 48104

G.E. Cranton, Imperial Oil Enterprises Ltd, Sarnia, Ontario, Canada

J.A. Currie, Villanova University, Villanova, Pennsylvania 19085

B.L. Dawson, IBM, San Jose, California 95193

E.A. Dorko, Air Force Institute of Technology, Wright-Patterson
     Air Force Base, Ohio 45433

A. A. Duswalt, Hercules Incorporated, Research Center, Wilmington,
     Delaware 19899

D. J. Eatough, Center for Thermochemical Studies, Brigham Young
     University, Provo, Utah 84602

J. C. English, Lawrence Livermore Laboratory, University of
     California, Livermore, California 94550

M. Farber, Space Sciences, Inc. Monrovia, California 91016

H. J. Ferrari, Lederle Laboratories Division, American Cyanamid
     Co., Pearl River, New York 10965

J. Feuer, 328-3 University Village South, Gainesville, Florida
     33603

W. H. Flank, Union Carbide Corporation, Tarrytown, New York 10591

J. H. Flynn, Institute for Materials Research, Washington,
     D.C. 20234

L. Forlani, Centro di Biologia Molecolare, Universita di Roma,
     Rome, Italy 00185

D.R. Fredickson, Argonne National Laboratory, Argonne, Illinois
     60439

R. Fuchs, Department of Chemistry, University of Houston,
     Houston, Texas 77004

G.E. Gajnos, Polymer Science and Engineering, University of
     Massachusetts, Amherst, Massachusetts 01002

C. Giavarini, Institute of Applied and Industrial Chemistry,
     University of Rome, v.Eudossiana 18, 00184 Italy

P. S. Gill, DuPont Company, Wilmington, Delaware

A.C. Glatz, Voland Corporation, New Rochelle, New York

S. Go, Department of Chemistry and Institute of Molecular Bio-
     physics, Florida State University, Tallahassee, Florida
     32306

A. H. Guenther, Air Force Weapons Laboratory, Kirtland Air Force
    Base, New Mexico

L. D. Hansen, Center for Thermochemical Studies, Brigham Young
    University, Provo, Utah 84602

I. R. Harrington, The Pennsylvania State University, University
    Park, Pennsylvania 16802

W. C. Herndon, Department of Chemistry, The University of Texas
    at El Paso, El Paso, Texas 79968

J. O. Hill, Department of Inorganic and Analytical Chemistry,
    La Trobe University, Bundoora, Victoria, Australia

B. Howard, Chemistry Department, Long Island University, Brooklyn,
    New York 11201

H. W. Hoyer, Department of Chemistry, Hunter College, New York,
    New York 10021

R. M. Izatt, Center for Thermochemical Studies, Brigham Young
    University, Provo, Utah 84602

B. D. James, Department of Chemistry, University of Queensland,
    Brisbane, Queensland, Australia

T. E. Jensen, Center for Thermochemical Studies, Brigham Young
    University, Provo, Utah 84602

D. L. Jernigan, Mary Hardin Baylor College, Belton, Texas 76513

D. E. Johnson, IBM, Research Laboratory, San Jose, California
    95193

J. Jordan, Department of Chemistry, The Pennsylvania State
    University, University Park, Pennsylvania 16802

R. S. Kalyoncu, Martin Marietta Laboratories, Baltimore, Mary-
    land 21227

H. Kambe, University of Tokyo, Komaba, Meguro-ku, Tokyo, Japan

F. E. Karasz, Polymer Science and Egnineering, University of
    Massachusetts, Amherst, Massachusetts 01002

E. Karmas, Department of Food Science, Rutgers University,
    New Brunswick, New Jersey 08903

E. W. Kifer, Koppers Company, Inc., Research Department,
    Monroeville, Pennsylvania 15146

E. Kirton, Department of Chemistry, Leeds Polytechnic, Leeds
    LS1 3HE, Yorkshire, England

P. G. Kliman, Nutrition Institute, Agricultural Research Center,
    Beltsville, Maryland

F. Kloos, Department of Chemistry and Institute of Molecular
    Biophysics, Florida State University, Tallahassee,
    Florida 32306

L. H. Leiner, Research Department, Koppers Company, Inc.,
    Monroeville, Pennsylvania 15146

P. F. Levy, Dupont, Wilmington, Delaware

J. A. Lynch, Department of Chemistry, The Pennsylvania State
    University, University Park, Pennsylvania 16802

R. J. Magee, Department of Inorganic and Analytical Chemistry,
    La Trobe University, Bundoora, Victoria, Australia

L. Mandelkern, Department of Chemistry and Institute of Molecular
    Biophysics, Florida State University, Tallahassee,
    Florida 32306

M. A. Marini, Department of Biochemistry, Northwestern University
    Medical and Dental Schools, Chicago, Illinois 60611

E. E. Marti, Central Research Services Department, Ciba-Geigy
    Limited, Basle, Switzerland

C. J. Martin, Department of Biochemistry, The Chicago Medical
    School, University of Health Sciences, Chicago, Illinois
    60612

J. N. Maycock, Martin Marietta Laboratories, Baltimore, Maryland
    21227

J. L. McAtee, Jr., Baylor University, Waco, Texas 76703

K. Miasa, Tokyo University of Agriculture and Technology,
    Koganei, Tokyo, Japan

R. E. Mitchell, Departments of Chemistry, Texas Tech University,
    Lubbock, Texas 79409

S. Miyata, Tokyo University of Agriculture and Technology,
    Koganei, Tokyo, Japan

S. I. Morrow, Propellants Division, Feltman Research Laboratory,
        Picatinny Arsenal, Dover, New Jersey   07801

S. Nevin, Department of Chemistry, Hunter College, New York,
        New York   10021

F. Noel, Research Department, Imperial Oil Enterprises Ltd.,
        Sarnia, Ontario, Canada N7T 7M1

T. Ohnishi, Department of Anesthesiology, Hahnemann Medical
        College and Hospital, Philadelphia, Pennsylvania 19102

P. G. Olafsson, Department of Chemistry, State University of
        New York at Albany, Albany, New York   12222

A. Palenzona, Institute of Physical Chemistry, University of
        Genoa, Italy

N. J. Passarello, Lederle Laboratories Division, American
        Cyanamid Co., Pearl River, New York   19065

N. Pathmanand, Villanova University, Villanova, Pennsylvania
        19085

E. M. Petruska, Villanova University, Villanova, Pennsylvania
        19085

P. Peyser, Naval Research Laboratory, Washington, D.C.   20375

A. Pinella, Western Electric, North Andover, Massachusetts

J. M. Powers, School of Dentistry, The University of Michigan,
        Ann Arbor, Michigan   48104

T. L. Regulinski, Air Force Institute of Technology, Wright-
        Patterson Air Force Base, Ohio   45433

D. W. Rogers, Chemistry Department, Long Island University,
        Brooklyn, New York   11201

H. M. Rootare, L. D. Caulk Company, Milford, Delaware   19963

K. Sakaoku, Tokyo University of Agriculture and Technology,
        Koganei, Tokyo, Japan

J. W Sherbon, Department of Food Science, Cornell University,
        Ithaca, New York   14850

J. Simon, Institute for General and Analytical Chemistry,
   Technical University, Budapest, Hungary

J. Skalny, Martin Marietta Laboratories, Baltimore, Maryland
   21227

E. B. Smith, Department of Chemistry, University of Georgia,
   Athens, Georgia 30602

D. M. Speros, Lighting Research Laboratory, General Electric Co.,
   Cleveland, Ohio 44112

B. R. Sreenathan, Department of Biochemistry, University of
   Health Sciences, Chicago, Illinois 60612

R. D. Srivastava, Space Sciences, Inc., Monrovia, California
   91016

G. L. Stutzman, The Pennsylvania State University, University
   Park, Pennsylvania 16802

Y. Takashima, Tokyo University of Agriculture and Technology,
   Koganei, Tokyo, Japan

A. E. Tonelli, Bell Laboratories, Murray Hill, New Jersey 07974

R. W. Tung, Villanova University, Villanova, Pennsylvania 19085

S. R. Urzendowski, Air Force Weapons Laboratory, Kirtland Air
   Force Base, New Mexico

R. Viswanathan, Department of Applied Physics and Information
   Science, University of California, San Diego, La Jolla,
   California

D. H. Waugh, Department of Chemistry, The Pennsylvania State
   University, University Park, Pennsylvania 16802

L. A. Williams, Chemistry Department, Long Island University,
   Brooklyn, New York 11201

E. M. Woolley, Department of Chemistry, Brigham Young University,
   Provo, Utah 84602

R. Yokota, Institute of Space and Aeronautical Science, University
   of Tokyo, Komaba, Meguro-ku, Tokyo, Japan

N. S. Zaugg, Department of Chemistry, Brigham Young University,
   Provo, Utah 84602

INDEX

A

Abiogenic synthesis of biopoly-
    mers 685
Acetylated-δ-chymotrypsin 425
Activation energies 697
Adiabatic calorimetry 81
Afwillite 697
Alkali tartrates 713
α-approximation 283
ASTM tests for polymer charac-
    terization 293
Aminobenzoic acid compound 321
Amorphous polyethylene 89
Anhydride-epoxy system 537
Antioxidants 305
    stability 305
Apatites 397

B

Beer's law 249
Benesi-Hildebrand equation 249
Benzazole derivatives 621
Blood analysis 147
Boltzmann factor 89
Bond formation 119
Bound water 489
Bovine albumin 457
Burning process 199

C

Calorimeters, adiabatic 45
    isoperibol titration 7
    isothermal titration 7
    precision solution 381
Calorimetry 1, 7, 45, 81, 127,
    217, 237, 249, 381, 479,
    723, 743
Carbon dioxide 57
Catalysis in solid propellants
    757
Cations 649
Cementitious hydrates 697
Chloride determination 217
Cholesteryl myristate 103
Cholesterol palmitate 103
Chymotrypsin ionization
    reactions 425
Chymotrypsinogen-A 425
Complexes in solution 283
Computer automated labora-
    tory 69
Condensed phase behavior 757
Crystallization of poly-
    ethylene
    self-seeding nucleation 579
    under high pressure 603
Cumene cracking 649
Cyclopentane 103
Cytotoxic agents 237

D

Dehydration 497
Dental waxes 349
Dicalcium phosphates 397
Differential freezing techni-
    que 57
Differential scanning calori-
    metry 103, 127, 147, 293,
    305, 465, 479, 497, 537,
    555, 569, 611, 621, 661,
    685
  theory 17
Differential thermal analysis
    57, 69, 579
Dimethyl phthalate 185
Dipalmitol lecithin 103
Dissociation curves, proteins
    407
Dissociation energy 731
DNA gels 465
Dotriacontane 103
Double base materials 757
Drop calorimetry 723
Dynamic differential calori-
    metry 743

E

Effluent gas analysis techni-
    que 57
Electron microscopy, hot stage
    363
Energy of rehydration 381
Enthalpimetric analysis 207,
    217
Enthalpy 17, 119, 283
  of fusion 593
  of solution 473
  of sublimation 731
  of transfer 473
  of transition 443
Entropy 119
  of fusion 89
  titration 249
Equation of state studies 165

F

Fertilizers 671
Fire retardants 199
  additives 797
Floating method 1

Flory theory for melting point
    depression 569
Free energy 119
Freeman-Carroll method 537
Fusion 89

G

Gas generating reactions 137
Gel phase chromatography 611
Gem-diol 443
Glass temperature 797
Glass transition temperature
    17, 537
Gruneisen parameter 165

H

Heat:
  capacity 81, 17, 147
  of dilution 489
  of dissolution 127
  of formation 743
  of fusion 127, 569, 621, 661,
    743
  of immersion 381
  of ionization 425
  of reaction 537
  of solution 1, 397
Helix-coil transition 443, 465
Hexa-aminecobalt (III)-penta-
    chlorocuprate(II) 103
High dilution method 283
Hot-stage electron microscopy
    363
Human enamel 397
Hydrogen bonding solutes 479
Hydroxyapatites 381

I

Imidazo compound 321
Imidazole rings 621
Imidazolidinone compound 321
Impurities detection 321
Insect repellents 185
Insulin 237
Interfacial dislocation net-
    works 593
Intermetallic compounds 743
Intermolecular complexes 249
Ionic constants, Proteins 407

Iron hematoporphyrin complexes
  217
Isomeric uridine phosphates
  685
Isoperibol calorimeters   237
Isotactic polypropylene   89,
  569
Isothermal calorimeters   237
Isothermal dehydration curve
  489

              J
Johnson noise   45

              K
Kaolinite   363
Katharometer   57
Kinetics:
  of dehydration   713
  homogeneous   511
  oxidation   305
  polymer decomposition   629
  polymer phase transforma-
    tions   537
  reaction   505
Kirchoff's relation   743
Kissinger method   697

              L
Laser beams   81
Lysine   425

              M
Mass Spectrometer   731
Melting behavior:
  of polymers   293
  of oligomers   621
Metal borohydride complexes 777
Methylamino compound   321
Microcalorimeter   1
Milk fat systems   661
Minocycline   321
Mixed crystals characterization
  555
Molecular sieve zeolites   649
Morphology, polyethylene   579

              N
Newton's Law of cooling   17
Nickel-Iron alloy   103
Nickel nitrate hexahydrate   103
Nickel oxide   731
Nitrocellulose   757
Nitrofurantoin   321
Nyquist criteria   45
Nystatin   321

              O
Olefins   207
Oxazepine compound   321
Oxidation kinetics   305

              P
Petroleum oil   305
Phase rule   103
Phase transitions   103
Polarizing microscope   757
Polybenzimidazole   621
Polyblends   797
Poly-γ-benzyl-L-glutamate   443
Polyethylene   199, 579, 593,
  603
Polylactones   611
Polymer conformations   89
Polymer crystallinity   293
Polymer model   611
Polymerization reactions   537
Polymethylmethacrylate   629
Polypropylene   569
Polystyrene   1, 199
Polytetrafluoroethylene   103,
  629
Propellants   757
Protein   407, 425, 457
Protein denaturation 147, 237
Protic acids   425
Proton ionization   119
Pure base method   283
P-xylene puriss   127
Pyridine - chloroform complex
  283

              Q
Quench experiment   579

R

Rare earths 743
Reaction calorimetry 249
Rogers and Morris technique 697
Rubber elasticity 1

S

Sickle cell trait 147
Silver oxide 57
Solution calorimetry 127
Solvent-solute interactions 473
Specific heat of cheese whey 497
Steady state calorimetry 81
Steroid compound 321
Supercooling 17
Superphosphates 671

T

Thermal analysis 321, 649
  computer automated 69
Thermal:
  expansion 165
  decomposition 511, 697, 713,
    777
  denaturation of calf thymus
    465
  evolution analysis 185, 199
  history 293
  properties of milk fat 661
  titrimetry 407, 425, 457
Thermistor sensitivity 45
Thermogravimetric analysis 137,
    611, 629
Thermomechanical analysis 349
Thiadiazole compound 321
Thin layer chromatography 757
Titration calorimetry 7, 45, 217
    237, 479
Transitions, first order 17
  second order 17
Triazole compound 321
12-Phosphotungstic acid 457
2,4-dinitrophenylhydrazine 611

U

Ultrasonic sound velocity measure-
    ments 165

V

Van't Hoff's law 17
Vapor pressure 185
Vaporization enthalpies,
    nickel 731
Volatilization studies 797
Volume dilatometry 569

W

Weibull function 505
Wheatstone bridge circuit 207

Z

Zeolite, absorption by 237
Zimm-Bragg co-operativity
    parameter 445